W9-CSP-619

ION EXCHANGE DEVELOPMENTS AND APPLICATIONS

Proceedings of IEX '96

Papers presented at the SCI Conference IEX '96 - Ion Exchange Developments and Applications organised by the Separation Science and Technology Group of the Society of Chemical Industry, held on 14 - 19 July 1996, at Churchill College, UK.

Organising Committee

Dr J. A. Greig (Co-Chairman)	The NutraSweet Kelco Company
Mr M. A. Sadler (Co-Chairman)	Consultant
Professor T. Stephenson (Co-Chairman)	University of Cranfield
Dr T. V. Arden	Consultant
Mr C. Bainbridge	Dow Chemical Company
Mr K. Blaxall	Purolite International Limited
Dr R. R. Harries	Powergen plc
Dr K. Huddersman	De Montfort University
Dr M. T. Hughes	Bayer plc
Mr K. R. Schmidt	Ecolochem International Inc
Mr R. A. F. Scott	Rohm and Haas (UK) Limited
Professor M. Streat	Loughborough University of Technology
Ms H. Ward	Ecolochem International Inc
Mr M. S. Verrall	SmithKline Beecham Pharmaceuticals

Refereeing Committee

Dr T. V. Arden	Dr T. deV. Naylor
Mr C. Bainbridge	Mr M. A. Sadler
Mr J. Bayley	Mr K. R. Schmidt
Mr K. Blaxall	Mr R. A. F. Scott
Professor M. Cox	Dr M. J Slater
Dr H. Eccles	Professor T. Stephenson
Dr R. R. Harries	Professor M. Streat
Dr K. Huddersman	Mr M. S. Verrall
Mr D. Naden	

Ion Exchange Developments and Applications

Proceedings of IEX '96

Edited by

J. A. Greig
The NutraSweet Kelco Company, UK

THE ROYAL
SOCIETY OF
CHEMISTRY
Information
Services

Sep/Ae
chem

Special Publication No. 182

ISBN 0-85404-726-3

A catalogue record for this book is available from the British Library

© SCI 1996

All rights Reserved
No part of this book may be reproduced or transmitted in any form or
by any means—graphic, electronic, including photocopying, recording,
taping, or information storage and retrieval systems—without written
permission from the Society of Chemical Industry, 14/15 Belgrave Square,
London SW1X 8PS, UK.

Published by The Royal Society of Chemistry,
Thomas Graham House, Science Park, Cambridge
CB4 4WF

Printed in Great Britain by Hartnolls Ltd, Bodmin, UK

TD 757
.5
I 66
1996
CHEM

Preface

This volume contains the papers presented at the Seventh International Ion Exchange Conference organised by the Separation Science and Technology Group of the Society of Chemical Industry and held at Churchill College, Cambridge, UK, in July 1996.

This conference follows highly successful events spanning over forty years:-

Ion Exchange and its Applications	London University	1954
Ion Exchange in the Process Industries	Imperial College	1969
The Theory and Practice of Ion Exchange	Cambridge University	1976
Ion Exchange Technology	Cambridge University	1984
Ion Exchange for Industry	Cambridge University	1988
Ion Exchange Advances	Cambridge University	1992

All papers - oral and poster - presented at this conference are published in these Proceedings. It contains many new and interesting contributions to the science, engineering and implementation of ion exchange technology, and it is hoped that it will be of equal interest to researchers and to practising industrials.

Over the last forty years of Ion Exchange Conferences, we have seen the emergence of an international community of research and industrial co-workers. From the early papers on water treatment and metallurgical industries, successive conferences have broadened the scope to include topics such as bioprocesses, membranes and catalysis. Approximately the same format has been adopted as in 1992 with the subject matter and therefore the contents of the book subdivided into eight separate topic areas: water treatment; environmental and pollution control; nuclear; resin developments; fundamentals; inorganic materials; hydrometallurgy and separations. Judging by the response to this conference, while polymer based resin development is still at the forefront of many applications, the synthesis and use of inorganic materials, particularly in relation to the environment and areas of pollution control, is experiencing an upsurge in interest.

Since 1984, the Organising Committee has honoured an authority in ion exchange with the presentation of the IEX Award. Previous recipients have been:-

1984	T. R. E. Kressman
1988	F. G. Helfferich
	R. Kunin
1992	T. V. Arden
	F. Martinola
	D. Weiss

At this conference, Professor Michael Streat of Loughborough University of Technology will be recognised for his contribution to the academic and industrial research and development of ion exchange and adsorption processes.

The Chairmen acknowledge the willing help of the Conference Organising Committee; whose assistance, supported by their companies and universities, makes this event happen. We express our gratitude to the members of the Committee; Tom Arden, Colin Bainbridge, Kevin Blaxall, Katherine Huddersman, Richard Harries, Michael Hughes, Ken Schmidt, Bob Scott, Michael Streat, Michael Verrall, Heather Ward and the many other colleagues who also helped with refereeing. In addition, we are indebted to the staff at SCI, in particular, Anne Potter, Monique Heald, Kathryn Potter and Anne Borcherds for putting the conference and this book together.

J. A. GREIG
The NutraSweet Kelco Company

M. A. SADLER
Consultant

T. STEPHENSON
Cranfield University

Co-Chairmen, IEX '96

Acknowledgements

The SCI and the Conference Committee are most grateful for the support of the following companies in the organisation of IEX '96 and the sponsoring of events held during the meeting:

Bayer plc
Dow Separation Systems, Dow Chemical Company Limited
Ecolochem International Inc.
Purolite International Limited
Rohm and Haas (UK) Limited

Contents

Environmental and Pollution Control

Nuclear

Resin Developments

Fundamentals

Inorganic Materials

Hydrometallurgy

Separations

REGENERATION OF CONDENSATE POLISHING PLANT RESINS BY THE USE OF THE "RESIN ON RESIN" TECHNIQUE AT AGHADA GENERATING STATION

D. J. O'Sullivan and P. Powell
Aghada Generating Station ESB, Co Cork, Ireland

H. R. Bolton and E. K. Bullas
Thompson Kennicott, Rolls-Royce Industrial Power Group, Wolverhampton, WV4 6JY, UK

M. A. Sadler
Consultant, Portishead, BS21 8JT, UK

1 INTRODUCTION

Ultrapure water is used in many industries with the major users being the power and semi-conductor manufacturing industries. Their current and future demands in terms of water quality have been discussed in previous papers and it is clear that there will be a requirement for water containing even lower levels of impurities than the "ultrapure" water currently being produced.[1] The needs of different industries obviously vary with the power industry, and particularly the nuclear sector, being primarily concerned about the levels of potentially corrosive inorganic impurities such as chloride and sulphate. The importance of removing impurities such as particulate and colloidal matter is recognised but the techniques required are different from the ones employed for removal of ionic impurities. The aim of the current studies is, however, to further develop the understanding of ion exchange processes at these very low impurity levels and to identify the mechanisms responsible for the leakage of ionic impurities from beds.

2 SOURCES OF IMPURITIES IN WATER FROM CONDENSATE POLISHING PLANTS

Nuclear Power Stations such as those with Pressurised Water Reactors or Boiling Water Reactors endeavour to operate with the highest possible quality of water in their steam generators. Some of these stations claim to be using feedwaters with sodium and chloride levels of less than 5 ng/kg. This quality is achieved by the use of the well established technique of condensate polishing which is, essentially, the purification of all the returning condensate by ion exchange, usually deep beds (~ 1 m) of mixed anion and cation resins. The condensate polishing plant recently installed by Thompson Kennicott on a modern PWR handles approximately 1.1 m³/s so it will be appreciated that large volumes of resins are employed.

The traces of impurities that remain in condensate after it has been polished by ion exchange are believed to be contributed by several mechanisms including flow distribution problems in the service vessels, to kinetic leakage of ions through the ion exchange bed and to equilibrium (or elution) leakage. These mechanisms have been discussed previously. The problem of mechanical design of ion exchange beds, and particularly the distribution of the influent and collection of the treated water has received considerable attention and service vessel designs have been improved. Kinetic leakage, whereby influent ionic impurities pass through the bed, to contaminate the treated water, is largely a function of flow rate and organic fouling of the resins. It is still a very common problem with condensate polishing plants, which operate at linear velocities of 100 to 120 m/h, and can only be overcome by maintaining resins in good "kinetic condition" - and this usually means timely resin replacements.

Equilibrium leakage is a well known effect. It consists, essentially, of the release of traces of ionic impurities held by the resins into the water being treated. This movement, the reverse of what is normally expected, obviously depends on the levels of impurities present in the water and on the resins. It takes place to satisfy the equilibrium conditions governed by the well known equation and is controlled by the selectivity coefficient for the ions being exchanged, i.e.

$$K_H^{Na} = \frac{[R - Na][H^+]}{[R - H][Na^+]}$$

Thus, at equilibrium the concentration of sodium in the water phase is proportional to that in the resin phase. Equilibrium leakage, therefore, should exert a controlling influence over the ionic impurity levels in the final effluent. In order to further improve this quality it is obviously necessary to reduce the ionic impurities held, by ion exchange, on the resins themselves, i.e. to further improve regeneration. It is accepted that this is a very simplistic view and that other mechanisms may become dominant at the low ng/kg impurity levels.

When considering these ultra-trace leakage mechanisms the contribution from resins themselves must not be overlooked. It is known that cation resins release traces of sulphonated organic compounds ranging from simple molecules such phenolsulphonic acid to sulphonated macro-molecules (sulphonated polystyrene) of molecular weights of over 15,000 and in some cases up to 100,000 g/mole.[2] Fisher claims that the functional sulphonate groups on cation resins are also slowly lost resulting in the appearance of sulphate in the water phase.[3] Anion resins are known to release traces of impurities such as amines but these are of less concern to the power industry. It must be stressed that all of these resin impurity releases are believed to take place at extremely low levels and normally only become of consequence when attempting to reduce the impurities to low ng/kg concentrations.

3 REGENERATION OF RESINS FOR USE IN CONDENSATE POLISHING

Assuming that problems due to poor flow distribution and to effects such as resin fouling are overcome by good design and operation the obstacle to achieving even higher water qualities remains that of resin regeneration. A modern power station employing volatile alkali conditioning of its steam/water circuit, such as a 1100 MWe Pressurised Water Reactor, can use in its polishing plant about 100 m³ of ion exchange resins with the cation resins requiring regeneration every 3 to 7 days depending on design. The option of using "throwaway" resin cannot therefore be considered. Calculations show that to reduce equilibrium leakage of sodium and chloride, from a polishing plant operated with resins in the H-OH form, to less 0.1 ng/kg it is necessary to reduce the chloride on anion resin to less than 0.05% of the functional sites in the chloride form and the sodium on cation resin to less than 0.007%. These simple calculations use the accepted selectivities of 18 and 1.7

although there are indications that at the very low resin phase impurity levels being considered the selectivities will be higher. The calculated values nevertheless form a useful guide for these studies.

The sequence of operations in the regeneration of mixed bed resins is, of course, separation, regeneration proper (i.e. elution of impurities by the use of acid or alkali) and rinsing. The separation of mixed resins has received considerable attention as processes devised for simple deionisation plants are inadequate for ultrapure water plants. Improved separation procedures include the use of inert resins as buffer zones, "interface isolation" separation in a high density ammonium sulphate solution (Amsep) and processes such as Purosep.[4,5] A commonly used high efficiency separation/regeneration process is the proprietary "Conesep" procedure which can achieve separation giving about 0.05% cation in anion resin and about 0.2 - 0.3% anion in cation resin.[6] Now, cross contamination of this level will, on a 1:1 mixed bed, result in at least 0.05% of the cation resin being contaminated with the sodium hydroxide regenerant used for the anion resins and conversely 0.2 - 0.3% of the anion resin being contaminated by the acid used for cation resin regeneration. Thus, separation problems alone will prevent the target resin purities being achieve unless the effects of cross contamination are corrected.

To achieve sodium levels as low as 0.007% using conventional regeneration techniques requires the use of a very great quantity of acid. The total actually needed will obviously depend on the levels of sodium on the resin prior to regeneration. However, Dr A Miller in work carried out for the Electric Power Research Institute showed that diaminoethane very effectively displaced sodium from cation resins. This finding was developed in work at Thompson Kennicott into a two stage procedure for removing sodium from cation resins with analysis showing that less than 0.003% sodium remaining on the resin. A low percentage of the amine remained after regeneration, about 7%, but for power production purposes this is not considered to be a problem. It is even more difficult to remove traces of chloride from strongly basic anion resin by conventional regeneration processes. However, a two stage process involving its displacement by carbonate followed by elution of the carbonate by sodium hydroxide solution is known to very effectively reduce chloride levels to less than 0.1% and probably much lower. Thus, the indications are that even with contaminated resins it should be possible to elute sodium and chloride to the extent required for the preparation of water containing sub ng/kg levels of impurities. In practice polishing resins only become badly contaminated in the event of the ingress of impurities following, say, a condenser leak. It seems likely that once they have been purified by a two stage regeneration procedure that most subsequent regenerations can be adequately carried out using a high level conventional regeneration procedure. The two stage regeneration process should then only be required following in-leakage events.

However, the problem of separation remains together with the difficulty of completely rinsing resins free of all traces of regenerants. This problem becomes acute with some anion resins and is thought to be due to the fact that they become fouled with organic matter which incorporate weakly acidic exchange sites that act to retain sodium. Cation resins show retention of the acids used in their regeneration so both forms of resins have the ability to carry traces of impurities in the form of retained regenerants into the mixed bed.

4 "RESIN ON RESIN"

The use of a technique termed "Resin on Resin" has been proposed to overcome the problem of regenerant residuals and also to address the problem of imperfect separation. As described in an earlier paper this was first used in South Africa by J B Conlin[7] and later on an experimental basis at Aghada Generating Station, Ireland. It was subsequently adopted by Moneypoint Power Station, Ireland and is also being used by several PWRs in the USA. It involves the separation of the mixed bed resins in the normal way and then regeneration and rinsing of the anion resin component. The regenerated anion resin and the untreated cation resin are then remixed and allowed to stand for about three hours with air mixing for 5 minutes every 30 minutes. This allows any residual sodium regenerant still retained by the anion resin to be removed by the cation resin. In the subsequent separation most of the cross contaminating cation resin, which had been converted to the sodium form during the anion resin regeneration, should be removed with the bulk of the cation resin so that the process also addresses the question of imperfect separation. A proposal for extending the process so that anion resin is used to purify cation resins from the effects of retained regenerant and cross contamination has been made by Thompson Kennicott.[8] Workers in Japan have successful used, on an experimental basis, anion resins to cleanse cation resins of anionic organic impurities and cation resins to cleanse anion resins of cationic organic impurities.[9,10,11] Thus, if impurity leakages into ultrapure water are dominated by equilibrium leakage effects, it should be possible to devise an appropriate regeneration procedure to allow regeneration of mixed bed resins to the extent required for the preparation of water with impurity levels of less than 0.1 ng/kg. In actual plant use, there will always be other impurity sources such as the motive water used to transfer and rinse resins, the air used for resin cleaning and the regenerants employed.

5 PLANT TRIALS

5.1 Aghada Generating Station

The Electricity Supply Board of Ireland (ESB) has always taken an interest in the preparation and treatment of high purity water. Although it does not employ nuclear power stations, it recognised that the resin purities being projected could have an application on conventional power stations by facilitating the operation of their condensate polishing plants in the economical ammonium form. A collaborative ESB - Thompson Kennicott trial to explore the applicability of the extended "Resin on Resin" technique for purifying both cation and anion resins was arranged and took place at Aghada Generating Station, Ireland.

Aghada Generating Station has a 270 MWe gas fired once through boiler unit operating at reheat and stop valve conditions of 538/538°C and 16/3.8 MPa. The unit is sea water cooled and employs a titanium tubed condenser with single aluminium bronze tube plates. The condensate polishing plant, which was designed and installed by Thompson Kennicott, consists of two 100% service units. The external two vessel regeneration plant was equipped with the first Conesep separation/regeneration to be commercially installed and which has performed well, without major maintenance, for 16 years. Early tests on the Conesep at Aghada showed that cation in anion resin cross contamination after separation was 0.2 - 0.3% before secondary separation which reduced it to less than 0.07%. The steam/water circuit at Aghada G.S. is conditioned by the use of ammonia dosed to give a pH of 9.25 @ 25°C. The polishing plant is operated with the resins in the H$^+$ and OH$^-$ forms respectively and is normally taken out of service when the polished water shows the first indication of ammonia break through, i.e. when the outlet conductivity shows an increase to 0.07 μS/cm @ 25°C over its steady state value of 0.055 to 0.056 μS/cm @ 25°C.

5.2 Analytical

Chemical monitoring of the steam/water circuit is performed by the use of high purity water conductivity meters, on line sodium analysers and a laboratory based ion chromatograph. Plant waters are fed directly to this instrument, an up-dated two channel Dionex 16, which loads automatically from the flow. For the trial, it was equipped with AS2 separation columns with AG2 concentrator columns for anion analysis and with a CS1/CG1 combination for cation analysis. Eluents were 8 millimolar sodium carbonate and 10 millimolar hydrochloric acid with suppression by use of micro-membranes. These separation and elution conditions were selected after careful examination of alternative ion chromatographic procedures. Sample volumes of 600 to 1000 mL were used for both the actual determinations and for the regular (daily) check calibrations the standards for which were prepared by a sequential dilution method using syringe pumps and streams of standards flowing to waste.

Ion chromatography was the only technique available at Aghada that offered the sensitivity required. It is widely used for this type of application but even so the bias and precision of data so acquired are difficult to characterise. This is a much wider issue and beyond the scope of the present paper. During the course of the work approximations were made, to act as guides, of the standard deviations associated with single determinations of sodium, (1.7 ng/kg based on 44 determinations), chloride (2.7 ng/kg based on 46 determinations) and sulphate (5.5 ng/kg based on 46 determinations). These values suggest that the confidence limits associated with the means should be sufficiently narrow to allow meaningful comparisons to be made.

5.3 The Trials

It was decided to conduct the trial using a charge of resins which had been installed in vessel A two months previously. This consisted of 5.0 m^3 of a macroporous cation resin (12% DVB) and 2.5 m^3 of macroporous anion resin. The anion resin had been supplied in the chloride form and prepared for service by multiple regenerations. This reduced the chloride to a satisfactory level for normal station use but for the "Resin on Resin" trial efforts were made to reduce to further reduce the ionic chloride content of the resin. The separated anion resin was treated with 5% sodium bicarbonate solution, using 500 g NaHCO$_3$/litre resin, at ambient temperature. This was followed by a thorough rinse and regeneration with 4% sodium hydroxide solution at 384 g/litre resin. This process reduced the chloride content of the resin to 0.2%. The value, before treatment was about 8%. The cation resin was regenerated using 5% sulphuric acid at 384 g/litre of resin which was three times the quantity normally used by the station. This reduced the sodium content of the cation resin to 0.04%. Following the bicarbonate treatment, the regenerated and mixed anion and cation resins were remixed and put into service with the purity of the polished water being carefully monitored. These results obtained on this "Preliminary

Trial" were used for comparison purposes with those later obtained on the "Resin on Resin" trial.

Preparations for the "Resin on Resin" trial itself commenced as soon as the preliminary service run of about 8 days had been completed. The exhausted resins were separated in the Conesep vessel and the cation resin transferred and both resins cleaned by air scouring and back washing. The cation resin was then regenerated using 384 g H_2SO_4/litre resin and rinsed to less than 1 µS/cm at 25°C. It was then recombined with the anion resin, which had not been regenerated, in the Conesep vessel and allowed to remain in contact for 3 hours with air mixing for 5 minutes every 30 minutes. The mixed resins were again separated and the anion resin regenerated using 384 g NaOH/litre resin. 1 m^3 of H^+ form cation resin was then drawn from a specially provided holding tank and mixed with the freshly regenerated anion resin in order to cleanse it of any remaining sodium hydroxide regenerant. After a 3 hour standing period, with air mixing every 30 minutes, the resins were separated and the 1 m^3 of cation resin returned to its holding tank via the interface tank. It is to be noted that the main charge of cation resin was not involved in these operations. The regenerated and "purified" anion and cation resins were finally remixed and put into service.

Resin and water samples were taken at critical stages during the regeneration programme and the on-stream instrumentation monitored water purity during the subsequent 11 day service run of the mixed bed. At the water purity levels existing in the Aghada steam/water circuit, instruments such as conductivity and sodium monitors were simply recording their baseline values so that only available technique capable of making measurements at the low ng/kg level was ion chromatography (IC). It was realised that the precision and bias of ion chromatography measurements at the concentrations being considered needed to be quantified and an approximation of precision was subsequently made although the bias was not identified.

6 RESULTS

6.1 Sodium

The average sodium in polished water was shown to improve from 13 ng/kg to 2 ng/kg by the use of the "Resin on Resin" procedure. It must be appreciated that even the 13 ng/kg figure is lower than normally attained at Aghada owing, no doubt, to the effect of the high regeneration level used in preparing the resin. Sodium levels routinely attained at Aghada are about 25 ng/kg. The sodium levels over the first and second days in service average 2.4 ng/kg so that a post regeneration "wash-out" effect was not obvious. Resin analysis indicated that 0.08 to 0.1% of the exchange sites were in the sodium form and calculations suggest that a leakage of 0.7 ng/kg should have been seen. The correlation between a predicted 0.7 and a measured 2 ng/kg is satisfactorily close bearing in mind the analytical difficulties.

6.2 Chloride

As Aghada employs sulphuric acid as the cation resin regenerant, it was not expected that chloride leakage from the condensate polishing plant would decrease by the use of "Resin on Resin". An improvement from a leakage of 10 ng/kg, obtained from the highly regenerated bicarbonate treated resins, to an average 5 ng/kg from the same bed after the use of "Resin on Resin" in the course of the regeneration was interesting, particularly, as the respective resin analysis did not show a difference. These indicated that the chloride leakage on both tests should, from equilibrium leakage considerations, have been about 0.5 ng/kg. The possibility that another source of chloride exists must therefore be considered. A route involving the trace release of organic chlorides from the resins themselves, with a proportion being oxidised so releasing Cl⁻, deserves investigation. The finding of noticeable leakages of organic chloride impurities from the occasional batch of resin is reported from time to time. Is it possible that, at the ultra-trace level most resins leak some organic chloride - at least during the early stages of their life. Chloride levels in polished water routinely found at Aghada are about 10 ng/kg.

6.3 Sulphate

The sulphate results are more difficult to explain in that resins treated by the "Resin on Resin" procedure showed more than twice the leakage than those obtained on the prior service run, i.e. without "Resin on Resin"! Thus, on the test run obtained from the highly regenerated bicarbonate treated resins, average sulphate leakage was measured at 26 ng/kg and on the test run after "Resin on Resin" treatment, it increased to 62 ng/kg. Clearly the technique was failing to reduce sulphate leakage.

The critical question obviously concerns the source of the measured sulphate. This is not known but it is possible to speculate. Laboratory tests on hydrochloric acid regenerated cation resin gave encouraging results suggesting that the "Resin on Resin" approach could be usefully applied to the post regeneration purification of cation resins. Attempts to repeat these tests with sulphuric acid regenerated cation resins were prevented by analytical difficulties, mainly the problem of determining traces of sulphuric acid remaining on the cation resin.

It is well known that cation resins release traces of sulphonate organics with the complexity of the actual compounds ranging from simple compounds such as benzene sulphonic acid to macro molecules of sulphonated polystyrene with molecular weights of over 100,000 g/mole. It is also known that some of these macromolecules, those of mid molecular weight it is believed, are taken up by anion resins but fail to be removed by regeneration. In fact they act to foul the anion resins. It is possible that a fraction of the sulphonated organic impurities fouling the resins, or indeed still held by the cation resin, become oxidised during the regeneration process and subsequently release sulphate into the treated water. The treatment of anion resins with bicarbonate possibly acted to cleanse the resins of some at least of these sulphonated impurities so lessening the subsequent in-service release of sulphate. During the following service run, the anion resins may have again picked up sulphonates shed by the cation resins which the "Resin on Resin" treatment failed to dislodge. The treatment may, in fact, have made things worse in that *new* cation resin was used to "cleanse" the anion resin which, although well rinsed, could have contaminated the anion resin.

7 DISCUSSION

The use of the "Resin on Resin" technique to strip remaining traces of sodium hydroxide regenerant from anion resins and from cation resin entrained in the anion resin was shown to be very successful. Given the known difficulties in assigning confidence limits and bias to ion chromatographic measurements at the levels in question, the agreement between measured values and those predicted from resin analysis is very encouraging. The average sodium concentration during the test run was 2 ng/kg with calculations based on resin analysis indicating 0.7 ng/kg. These calculations are also based on the assumption that equilibrium leakage is the only source of sodium and, at these ultra-trace levels, this is questionable. By further improving conditions of regeneration by using nitrogen for the various procedures now employing air, by using very high purity acids and by employing water with very low sodium content for acid dilution, resin movements, cleaning and rinsing it may be possible to reduce sodium levels in treated water to pg/kg levels.

The success in reducing sodium leakage was not followed by an equal success in reducing sulphate leakage, indeed the procedure seems to have increased it! It is possible that the sulphate found in the polished condensate at Aghada did not originate as regenerant sulphate but as sulphonates shed by either the main charge of cation resin or the new cation resin used to "cleanse" the anion resin. This is obviously important for, if true, could represents a obstacle to the achievement of sub ng/kg water qualities. The Aghada trials employed cation resins in the main charge that had only been in service for a short period of time, less than three months. The cation resin provided for cleansing the anion resin had not been previously used but had been thoroughly rinsed although it was then stored for a short period awaiting use. Thus, any sulphonate/sulphate release from cation resins may have been exaggerated by the experimental and plant conditions. It must be remembered that these trials were carried out on an operating power station and had to be fitted in with operational demands. It will be interesting to see if these relatively high leakages still occur, even to a lesser extent if older resins, well rinsed of sulphonate impurities, are used or if the "Resin on Resin" procedure is reversed. The possibility of partial separation of the mixed bed giving a cation rich bottom layer must also be taken into account when considering sulphate/sulphonate leakage mechanisms.

A significant improvement in sulphate leakage is sought so that levels of the impurity at least match those of sodium and chloride. Different techniques for reducing the effect are worth considering. The technique successfully adopted at Susquehana Boiling Water Reactor (BWR), in which they used a layer of anion resin below the mixed bed to retain any sulphate being released, is interesting in this connection. The work of Foutch and his co-workers in computer modelling of the ion exchange process in very high purity water is also important.[11] On their assumed model they say that, at ultra-trace levels, sulphate generation from cation resins "accounts for most of the equilibrium leakage observed from a more homogeneous mixed bed". The model incorporates a rate equation governing the sulphate generation, based on work of Fisher and Burke which reflects the effect of temperature on the degradation.[12] The results of their computer simulation has been compared with plant studies carried out at Susquehanna BWR and is claimed to be consistent. As the search for even higher water qualities continues, attention should therefore be directed towards understanding the stability of cation resins and the plant conditions which influence this stability and possibly to innovative methods of containing any sulphate leakage. The work is continuing.

References

1. M. A. Sadler, H. R. Bolton and E. K. Bullas, Paper IWC-92-3, *International Water Conference*, Pittsburgh, October 1992.
2. J. R. Stahlbush, R. M. Strom, J. B. Henry and N. E. Skelly, "Ion Exchange for Industry", M. Streat, ed., Ellis Horwood, Chichester, 1988.
3. S. Fisher, Electric Power Research Institute Condensate Polishing Workshop, New Orleans, USA, September, 1993.
4. D. C. Auerswald and F. M. Cutler, Electric Power Research Institute Condensate Polishing and Water Purification Workshop, Scottsdale, Arizona, USA, June 18-20, 1991.
5. J. E. Earls, *Ultrapure Water*, 1988, October, 15.
6. UK Patent No GB2027 610 B.
7. J. B. Conlin, Eskom Technology Centre, South Africa, Private Communication, 1992.
8. UK Patent Application 9221947.6.
9. W. Agui, M. Takeuchi, M. Abe and K. Oginao, *J. Jpn Oil Chem. Soc.*, 1988, **37**, 1114.
10. W. Agui, M. Takeuchi, M. Abe and K. Oginao, *J. Jpn Oil Chem. Soc.*, 1990, **39**, 307.
11. G. L. Foutch, S. Pondugula and D. J. Morgan, *Ultrapure Water*, 1994, September, 55.
12. S. Fisher and E. Burke, Electric Power Research Institute Plant Chemists Meeting Palo Alto, California, Feb 25-26, 1993.

ION-EXCHANGE REGENERATION MODELING FOR ULTRAPURE WATER APPLICATIONS

V. Chowdiah and G. L. Foutch

School of Chemical Engineering
Oklahoma State University
Stillwater, OK 74078
USA

1 INTRODUCTION

Mixed-bed ion-exchange units are used for production of ultrapure water economically. Effluent water quality from the mixed bed is determined by service cycle and regeneration cycle performance. To obtain the desired level of water purification, attention needs to be paid to both modes of operation. The objective of this paper is to present a model for mixed-bed regeneration applicable for ultrapure water processes.

Resin regeneration efficiency is the primary parameter controlling the lower limit of impurities in service-cycle water. Resin cross contamination, insufficient regenerant-resin contact time, and resin fouling contribute to increased impurity leakage. Cation and anion exchange resin separation, prior to regeneration, is critical to minimizing cross contamination of the resin beads and subsequent regeneration by the wrong regenerant. Sufficient time for regeneration and rinse-down is also necessary to obtain satisfactory service-cycle performance.

Any of the regeneration-cycle problems will translate into increased equilibrium leakage during the service cycle because of the higher residual impurities on the ion-exchange resin. There are three mechanisms for residual impurity loading during regeneration: 1) Cross-contaminated resin beads will be exhausted (undesireable) after regeneration. For example, cation-exchange resin sites will be converted to the sodium form if the anion-exchange resin regenerant, NaOH, contacts some cation-exchange resin beads. Similarly, the anion-exchange resin will pick-up chloride if contacted with HCl (regenerant for cation-exchange resin), or sulphate if sulphuric acid is used. 2) Incomplete regeneration leaves some of the exchange sites unconverted to the appropriate hydrogen or hydroxide form. 3) Insufficient rinse-down time introduces regenerant chemicals to the counter resin when the resins are remixed prior to service. Hence regeneration is a key to obtaining good service-cycle water quality.

Regeneration contact time is usually determined, in practice, from nomographs or equations supplied by the ion-exchange resin manufacturer. By these methods the amount of chemical indicated may be excessive in order to assure sufficient regeneration. Use of excess regenerant chemicals is a problem for condensate polishing units[1], and leads to increased rinse water requirements and additional disposal costs[2]. In order to minimize chemical usage the amount of regenerant and the regenerant-resin contact time are frequently modified by operators based on their plant experience.

This paper presents a model for strong base anion-exchange resin regeneration. Specifically, regeneration of chloride form resin by sodium hydroxide solution. The ion-exchange regeneration process has been described using a kinetic model. Since the concentration of the regenerant is high, the system is particle diffusion limited. Resin phase diffusion is assumed to be the slowest step and hence controls the rate of exchange. Regeneration is operating under unfavorable equilibrium. Stoichiometry, equilibrium, exchange rate, and process configuration govern the performance of an ion-exchange resin regeneration system. Cation-exchange resin regeneration, from previous work[3], is compared with anion-exchange resin regeneration. Some of the process parameters affecting regeneration have been studied.

The model is not limited to applications for the production of ultrapure water. The model developed here may be applied to other ion-exchange resin regeneration applications, using system specific transport properties. For example, anion-exchange resin is frequently used to remove nitrate from drinking water. The anion-exchange resin is regenerated with sodium chloride or potassium chloride solution[4,5,6]. This process is attractive if it minimizes cost and reduces environmental burden. This regeneration model can be used to study the process economics and optimize operation.

2 COLUMN MODEL DEVELOPMENT AND SOLUTION

Weak base anion-exchange resin regeneration has been modelled using reaction-diffusion coupled models[7,8]. For the case of weakly basic anion-exchange resins, there is a chemical reaction between the fixed group on the resin and the counterions. The exchange is accompanied by the water neutralization reaction. This makes the regeneration process pH dependent. There is a sharp boundary, inside the resin particle, between the reacted or regenerated shell and the unreacted core; the sharp loading profiles inside the bead have been photographed[9]. Strongly basic anion-exchange resin regeneration can be modelled using only equilibrium and mass-transfer limited mechanisms.

The rate-controlling mechanism in regeneration is intraparticle diffusion, due to the high concentration of regenerant solutions[10]. Diffusivity of the ion in the resin particle is an important parameter required for modelling this system. Mathematical description of ionic diffusion is based on either Fick's law or the Nernst-Planck equation. The Nernst-Planck equation accounts for the electrostatic effects present in a solution of electrolytes. Applicability of Nernst-Planck equation for ion exchange has been demonstrated elsewhere[10]. The complexity of the model increases if the Nernst-Planck equation is used to describe diffusivity because of the presence of the electrostatic term. However, the concentration dependence of diffusivity is modelled well by the Nernst-Planck equation.

An effective diffusion coefficient can also be used to model the system using Fick's law. In this case the effective diffusion coefficient is independent of the concentration of the ions and must be defined for each ion. It varies due to interactions between the ions, the solvent, and the resin structure. There is no method available for quantifying all the interactions within the exchanger.

The effective diffusion coefficient is a best-fit coefficient from experimental data. There are no correlations linking liquid-phase diffusivity to resin-phase diffusivity. Such a correlation can only account for the structural effects of the resin particle on diffusion coefficient and other ion-solvent interactions; however, the resulting correlation will be very system specific and have little predictive value. An effective diffusion coefficient can be obtained by matching the average resin phase concentration profile calculated using the Nernst-Planck equation and Fick's model[3].

An effective diffusion coefficient is used to model the column regeneration system using the following assumptions: 1) isothermal and isobaric operation, 2) plug flow, 3) constant liquid-film resistance, 4) uniform bed porosity, 5) uniform spherical particles, 6) uniform initial saturation of the resin particle in the bed, and 7) negligible axial dispersion.

The material balance and the rate expression are the governing equations of the column model. The differential material balance of an ionic species, in the column can be written as:

$$\frac{u}{\varepsilon}\frac{\partial C_{Cl}}{\partial z} + \frac{\partial C_{Cl}}{\partial t} + \frac{(1-\varepsilon)}{\varepsilon}\frac{\partial q_{Cl}}{\partial t} = 0 \tag{1}$$

The resin (solid) phase material balance on the ionic species is:

$$\frac{\partial q_{Cl}}{\partial t} = \frac{3D_e}{R^3}\int_0^R \frac{1}{r^2}\frac{\partial}{\partial r}\left(r^2\frac{\partial \bar{C}_{Cl}}{\partial r}\right)r^2\,\partial r \tag{2}$$

Initial condition (t = 0) are:

$$C_{Cl} = 0 \qquad\qquad\qquad 0 \le z \le L$$

$$q_{Cl} = \bar{C}_0 \qquad\qquad\qquad 0 \le r \le R \tag{3}$$

Boundary conditions for the bed are:

$$C_{Cl} = 0 \qquad\qquad\qquad z = 0$$

$$C_{Cl} = C_{Cl,\,effluent} \qquad\qquad z = L \tag{4}$$

Boundary conditions for the resin are:

$$\frac{\partial \bar{C}_{Cl}}{\partial r} = 0 \qquad\qquad\qquad r = 0 \tag{5}$$

$$\frac{\partial \bar{C}_{Cl}}{\partial r} = \frac{K_F}{D_e}\left(C_{Cl}^* - C_{Cl}\right) \qquad r = R$$

Integrating equation 2 with boundary conditions (equation 5) we obtain

$$\frac{\partial q_{Cl}}{\partial t} = \frac{3K_F}{R}\left(C_{Cl}^* - C_{Cl}\right) \tag{6}$$

Equations 1 and 6 are nondimensionalized using the following variables

$$X = \frac{C_{Cl}}{C_{Cl,Feed}} \qquad\qquad Y = \frac{q_{Cl}}{Q} \qquad\qquad X^* = \frac{C_{Cl}^*}{C_{Cl,Feed}}$$

$$\tau = \frac{C_{Cl,Feed}}{Q}\frac{D_e}{R^2}\left(t - \frac{\varepsilon z}{u}\right) \qquad\qquad \xi = \frac{D_e(1-\varepsilon)z}{uR^2} \tag{7}$$

The nondimensionalized equations are

$$\frac{\partial X}{\partial \xi} + \frac{\partial Y}{\partial \tau} = 0 \tag{8}$$

$$\frac{\partial Y}{\partial \tau} = \frac{3 K_F R}{D_e} \left(X^* - X \right)$$ (9)

The system is also subject to the following constraints:

$$\sum X_i = 1.0$$ (10)

$$\sum Y_i = 1.0$$ (11)

where $i = OH^-$ and Cl^-.

2.1 Numerical Solution of the Column Model

Equations 8 and 9, with the constraints of Equations 10 and 11, form the set of equations to be solved in order to describe the regeneration process. Equation 1, a partial differential equation, has been reduced to a system of ordinary differential equations (Equation 8). Equations 8 and 9 are solved by the method of characteristics. This method involves solution along one variable while holding the other constant. The solution is extended by incrementing the variables. In this case, the equations are first solved using a constant τ (time-distance) and incrementing ξ (distance). The procedure is repeated by updating τ with a small increment and solving over ξ. A Runge-Kutta method is used to start the solution, then an Adams-Bashforth fourth order method is used for the complete solution.

3 DISCUSSION

As noted earlier, the regeneration model has assumed that the transport of ions can be modelled using an effective diffusion coefficient assumed to be constant through the exchange process. Diffusion data used in this study is presented in Table 1. The effective diffusivity used in the model corresponds to the chloride value for intraparticle diffusivity given in Table 1. The film mass-transfer coefficient is calculated using available literature mass-transfer correlations[12].

Table 1 *Diffusion Data*

Ionic Species	Liquid-phase diffusivity (from conductance data) x 10^{-9} m²/s	Intraparticle diffusivity[11] (anionic resin: SA10A) x 10^{-10} m²/s
Hydroxide	5.32	6.44
Chloride	2.04	2.12

Diaion SA10A: Strong base anion-exchange resin (gel type) with 8% crosslinkage and quaternary ammonium groups

Strongly basic anion-exchange resins of type I are hard to regenerate completely[13]. High selectivity of the anion-exchange resin for chloride makes regeneration difficult. This leads to operation of regeneration under unfavorable equilibrium, and requires a concentrated regenerant solution. Also, high chloride selectivity of the anion-exchange resin necessitates the use of high-purity regenerants (Rayon grade caustic solution).

Complete regeneration will be dictated by the equilibrium characteristics and penetration of regenerant into the resin (stereoscopic effects). The equilibrium relation used here is given by:

$$X = \frac{\alpha Y}{1 + Y(\alpha - 1)} \tag{12}$$

In Equation 12, α is the binary selectivity coefficient which is appropriate for describing the preference of the ion-exchange resin at low influent concentrations, like those encountered in the service cycle. Using the binary selectivity coefficient may not be the best choice for modelling regeneration, but in this case, lack of equilibrium data has forced the use of this expression. Equation 12 should be replaced in the model when sufficient and accurate equilibrium data (for regeneration) is available.

Equation 12 is presented graphically in Figure 1. An anion-exchange resin isotherm for chloride-hydroxide exchange ($\alpha = 21.0$) is shown in Figure 1. From Figure 1, the anion-exchange resin achieves complete equilibrium, but is practically unachievable in a real-plant situation. For comparison, a cation-exchange resin equilibrium isotherm—sodium-hydrogen exchange ($\alpha = 1.5$)—is also presented in Figure 1.

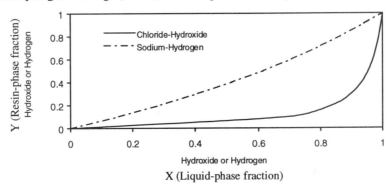

Figure 1 *Ion-exchange equilibrium isotherms*

Table 2 *Input Values Used for the Simulations*

Parameter	Value Used	Suggested Value[14]
Temperature (K)	298	
Column Diameter (m)	1.5	
Resin Bed Depth (m)	1.0	
Resin Particle Diameter (m)	0.0006	
Resin Capacity (keq/m^3)	1.2	
Flow Rate ((m^3)/(s)(m^3 of resin))	1.15	0.57 to 1.15
Regenerant Concentration (keq/m^3)	1.5	1.0 to 2.0

Table 2 presents the input parameters used for simulations, but the model is not restricted to these values. The suggested values for flow rate and regenerant concentration are also given in Table 2.

The effect of regeneration process parameters on resin-phase loading of hydroxide is presented in Figures 2 through 4. Figure 2 shows the progress of hydroxide loading on the anion-exchange resin with bed depth as a function of time. Since regeneration is occurring under unfavorable equilibrium, the concentration profile has a dispersive character. As the front progresses through the column it spreads continuously. At 25 mins, only 10% of the bed had been fully regenerated. Figures 3 and 4 present the effect of regenerant concentration and flow rate on the resin-phase loading profile. From Figure 4, doubling the flow rate resulted in a three-fold increase in the fraction of the bed that was completely regenerated.

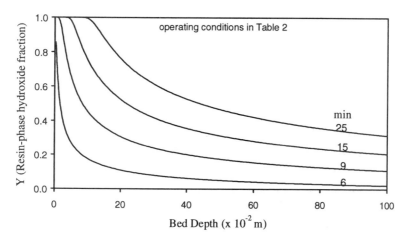

Figure 2 *Progress of hydroxide loading on the anion-exchange resin*

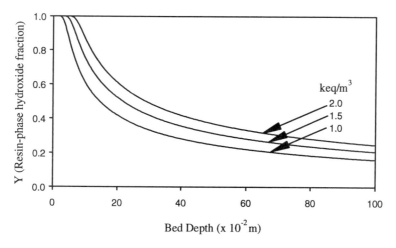

Figure 3 *Effect of regenerant concentration on the loading profile (at 15 mins)*

3.1 Comparison with Cation Resin Regeneration

Cation-exchange resin regeneration has been modelled in an earlier paper[3] where experimental data was available to generate the equilibrium isotherm. In this case, anion-exchange resin regeneration modelling, equilibrium data are lacking. In both cases, as resin-phase concentration increases equilibrium becomes the limiting factor.

A direct comparison of the two regeneration processes is not possible. The cation-exchange resin and regenerant properties are different. For purposes of illustration, the model is solved assuming that Table 2 represents the properties of both regeneration processes; only the equilibrium isotherms are assumed different and are presented in Figure 1 (the cation-exchange resin is being converted from sodium form to the hydrogen form). At 15 mins of column operation, 5% of the anion-exchange resin bed was fully regenerated, while 70% of the cation-exchange resin bed was completely regenerated. A comparison of the selectivity coefficients for cation and anion exchange accounts for the differences in the loading profile—the chloride-ion is more difficult to remove from the anion-exchange resin than a sodium-ion from the cation-exchange resin. Hence a longer time is taken to regenerate the anion-exchange resin.

Cation-exchange resin regeneration equilibrium, from Figure 1, is more favorable compared to the anion-exchange resin regeneration equilibrium. This results in the cation-exchange resin being regenerated better and easier than the anion-exchange resin. Also, equilibrium results in different resin-phase loading profiles; hydrogen loading on the cation-exchange resin is sharp, and the resin loading fronts are less dispersive than compared to the hydroxide loading fronts.

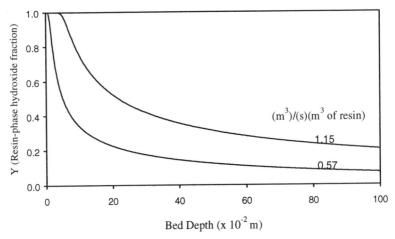

Figure 4 *Effect of flow rate on resin-phase loading profile (at 15 mins)*

4 CONCLUSIONS

Anion-exchange resin regeneration, chloride being replaced by hydroxide, has been modelled. A numerical solution for the column model has been developed. Experimental data is required to validate the model predictions. The model needs system specific equilibrium data to more accurately describe the equilibrium isotherm. With ion-exchange equilibrium information, the model can be used to simulate a particular regeneration operation and study the effect of process parameters on anion-exchange resin regeneration efficiency. Instead of trial-and-error experimentation, simulation studies can be performed to yield information on parameter sensitivity and aid in the design of experiments.

Nomenclature

C_{Cl}	liquid phase concentration of chloride	keq/m^3
C_{Cl}^*	interfacial concentration of chloride	keq/m^3
$\bar{C}_{Cl}$	resin phase concentration of chloride	keq/m^3
$\bar{C}_o$	initial resin phase concentration of chloride	keq/m^3
D_e	effective diffusivity	m^2/s
K_F	film-mass transfer coefficient	m/s
i	hydroxide, chloride (subscript)	
Q	capacity of the resin	keq/m^3
q_{Cl}	average resin phase concentration of chloride	keq/m^3
r	radial distance within the resin bead	m
R	radius of the resin bead	m
T	temperature	K
u	superficial velocity	m/s
X	liquid-phase fraction	
Y	resin-phase fraction	
z	bed depth	m
α	binary selectivity coefficient	
ε	bed porosity	
τ	dimensionless time	
ξ	dimensionless distance	

References

1. G. J. Crits, "Ion Exchange Advances-Proceedings of IEX'92", M. J. Slater, ed., Elsevier Applied Science, London, 1992, p. 81.
2. M. G. Kuriychuk and R. T. Gallupe, *Pulp and Paper Canada*, 1990, **91(12)**, 153.
3. V. Chowdiah and G. L. Foutch, *Ind. Eng. Chem. Res.*, 1995, **34**, 4040.
4. J. P. Van der Hoek and A. Klapwijk, *Waste Management*, 1989, **9**, 203.
5. M. Dore, Ph. Simon, A. Deguin and J. Victot, *Wat. Res.*, 1986, **20(2)**, 221.
6. F. Dossier, J. P. Croue, and M. Dore, *Environmental Technology*, 1993, **14**, 567.
7. M. G. Rao and A. K. Gupta, *A. I. Ch. E. Symposium Series*, 1982, **219(78)**, 96.
8. W. Höll and H. Sontheimer, *Chem. Eng. Sci.*, 1977, **32**, 755.
9. W. Höll and R. Kirch, *Desalination*, 1978, **26**, 153.
10. F. Helfferich, "Ion Exchange", McGraw Hill, New York, 1962.
11. T. Kataoka and H. Yoshida, *A. I. Ch. E. J.*, 1988, **34(6)**, 1020.
12. P. N. Dwivedi and S. N. Upadhyay, *Ind. Eng. Chem. Process Des. Dev.*, 1977, **16(2)**, 157.
13. M. A. Sadler, "Ion Exchange Processes: Advances and Applications", A. Dyer, M. J. Hudson, and P. A. Williams, eds., The Royal Society of Chemistry, Cambridge, 1993, p. 15.
14. V. R. Davies, *Chemical Engineering Progress*, January 1994, 63.

ION EXCHANGE RESINS IN NUCLEAR POWER STATIONS

J. C. Bates

Nuclear Electric Ltd.
Barnwood
Gloucester GL4 7RS

1 INTRODUCTION

Nuclear power stations routinely use ion exchange resins to control the water chemistry of several systems in the plant. These vary from purifying the raw water supplies to controlling the reactor power. This paper reviews some applications of ion exchange resins at nuclear stations and highlights the importance of understanding practical problems as well as theoretical concepts, in order to optimise their use. The role of ion exchange in achieving the plant requirements are presented for each system.

2 RAW WATER TREATMENT

Every steam driven electric generating plant requires an adequate supply of high purity water to replace losses from the steam/water circuit during normal operation and for supplying several ancillary systems. Nuclear plants require particularly good quality water. This make up water is usually produced by a combination of appropriate water treatment processes designed to remove essentially all of the contaminants from the raw water supply.

2.1 Make Up Plant Designs

The design and operation of the twelve make up plants (MUP) in Nuclear Electric have recently been reviewed. The results of this survey are summarised in Table 1. The installed MUP at each site was designed for optimum operation with the original raw water supply. However, the source of raw water has recently been changed by the suppliers at some sites, on both a short and long term basis. Raw water conductivities have increased from 20 to 200 μS/cm at one site and by 350 to 1100 μS/cm at another.

This has had a significant effect on the operation of the MUP; the output of the plant has been reduced to less than required in some cases. The survey revealed that ion exchange was the dominant process used to remove soluble impurities. Some stations that had reverse osmosis (RO) equipment installed initially have since decommissioned it. At one site unadvised dosing of the raw water by the suppliers destroyed the RO membranes and as the conventional MUP had adequate capacity the membranes were not replaced. Various stations are reviewing the performance of their current equipment with a view to updating to cope with the changes in water supplied. Mobile water treatment units have been used at some sites to complement the existing MUP, or to replace it under certain conditions.

2.2 Removal of Organics

Organic material in the raw water supply to power stations is the origin of a wide range of operating problems, both in make up water production and in maintaining the steam/water circuit water quality by condensate polishing.

The removal of organic matter has been investigated at all stages of make up water production by the CEGB and Nuclear Electric: pretreatment, ion exchange and final clean up. Knowles et al[1] carried out laboratory investigations into the various forms of pretreatment i.e. alum dosing, scavenger resins, activated carbon, pH, polyelectrolytes and contact time on the removal of organic matter passing through a sand filter. Ozone treatment of raw water and its subsequent effects on water treatment have also been studied[2]. The organic removal achieved by each process was not consistent and was particularly governed by the raw water supply.

The water emerging from MUP ion exchange beds can still contain high levels of organics even after passing through all the above treatment stages. Combined ultra-violet irradiation and hydrogen peroxide treatment of this demineralised water reduced the Total Organic Carbon (TOC) concentration in pilot plant tests[3]. A full scale version of this equipment was fitted to one plant, it did produce some improvements but it is not currently used due to the high operating costs.

Membrane filtration of make up water has been shown to be an effective, albeit expensive, means of reducing TOC levels at some foreign stations[4]. Reverse osmosis and electrodialysis reversal systems are also effective at removing certain organics; these techniques are not currently used on Nuclear Electric stations.

3 CONDENSATE POLISHING PLANT

The term condensate purification (or polishing) is used to describe the treatment of condensed steam from turbines operating in the power industry. The condensate polishing plant (CPP) is installed at this point in the steam water circuit to remove any impurities before the condensate is fed back into the boilers. Any ingress of soluble and insoluble contaminants into the feedwater will increase the risk of corrosion within the boiler and turbine sections.

All CPP use ion exchange resins to remove the impurities, either as beads or in powder form, as the membrane systems cannot treat the very high flow rates involved. Sizewell B, AGRs and the two Magnox stations with once through boilers operated by Nuclear Electric, all incorporate CPP to achieve the stringent feedwater qualities demanded of these stations.

There are several possible sources of soluble impurities in the steam/water circuit including cooling water ingress in the condenser, poor quality make up water and breakdown of organic material in the boilers. Insoluble impurities are usually corrosion products from the materials in the system, with copper and iron oxides being the predominant species.

As the CPP determines the final feedwater quality great effort has been expended within Nuclear Electric, and previously the CEGB, to investigate the principles and practical considerations governing the performance of this plant[5]. The wide range of CPP designs used at Nuclear Electric is now presented together with some of the more important findings of these investigations.

3.1 Nuclear Electric CPP Installations

The condensate polishing plant is normally situated immediately after the condenser hotwell in the steam/water circuit, this is the coolest part of the circuit where the resins will have the longest life and work most efficiently. It is also in the best position to remove impurities arising from condenser leaks. At all Nuclear Electric stations the CPP treats 100% of the condensate flow, however, various percentages of feedwater flow bypass the condenser via steam separator drains and bled steam lines, 45% at Sizewell 'B' and up to 10% at Wylfa.

Sadler collated the details of the CPP installed at CEGB stations in 1978[6]. A range of CPP designs are in service at different stations, varying in the sequence of the ion exchange beds, the type and quantities of resins used, pre-treatment and regeneration equipment. A summary of the plants used at Nuclear Electric Magnox, AGR and PWR stations is given in Table 2.

Three basic CPP designs are used, mixed bed, cation/mixed bed and mixed bed/pre-coat filter. Each of these could incorporate pre-filters to reduce the iron oxide loading onto the subsequent resin beds. All three designs can produce water of the required quality. However, as the main loading on the plant is normally ammonia, there is a requirement for a substantial quantity of cation resin in all these designs. In those systems not having a lead cation bed this can restrict the volume of anion resin used in the mixed bed. This increases the anion resin specific flow rate, which varies between 160 and 350 anion bed volumes per hour at the different plants. This can lead to problems of chloride slip when treating condenser leaks, as discussed in Section 3.4.

The more modern naked mixed bed CPP systems have resin bed depths of 1.2 and 1.6 m, compared to 0.9-1.0 m at the older plants. This allows extra cation resin to be used without unduly limiting the volume and height of anion resin required for chloride removal.

3.2 Flow Distribution Problems

Despite all the research into the theoretical performance of ion exchange resins in CPP it has been found that poor flow distribution in the service units is the major limiting factor on the operating capacity of CPP resins. Pressure drop limitations restrict the depth of resin used to 1.0 to 1.6 m, resulting in the development of larger diameter beds, up to 3.6 m. The designs of the water inlet distributor and the bottom collector systems are critical. Investigations at Heysham 1 showed that the dished bottoms of the vessels were affecting the flow distribution, reducing the operating capacity and limiting the ammonia concentration that can be dosed into the feedwater[7]. At Hinkley Pt B minor changes to the inlet distributors improved the operating capacity of CPP beds from 0.5 to 1.0 eq/L.

3.3 Mixed Bed Regeneration Problems

At Nuclear Electric plants regeneration of exhausted CPP mixed bed resins is always carried out externally to the service vessels, to minimise the risk of contaminating the feed system with regenerant chemicals. Some of these plants use a single regeneration vessel where regenerant cross contamination can occur due to (a) poor centre collector design, (b) poor resin separation and (c) incorrect resin level. Others transfer the anion resin to a separate regeneration vessel which avoids bulk contamination with sulphuric acid, due to incorrect valve operation or poor centre lateral design. However, not all the anion resin is usually transferred, normally a 0.02 to 0.05 m layer of anion resin is left behind in the cation regeneration vessel and is sulphated during the injection of sulphuric acid. To avoid these problems the Conesep regeneration system is now employed on the more recent plants and has been back fitted to one other station.

The Conesep process consists of a separation/anion regeneration vessel, a cation regeneration vessel and a resin interface vessel. The exhausted mixed bed charge is returned from the service unit to the Conesep vessel where it is initially conventionally backwashed into two layers. The cation resin is then transferred hydraulically, from the bottom of the vessel, to the cation regeneration unit. The cation/anion resin interface is detected automatically by conductivity, so that changes in resin charge volume do not affect the efficiency of the process[8].

Tests on Conesep systems installed at Aghada G.S.[9] and Oldbury P.S.[10] showed that cation-in-anion resin contamination can be reduced to less than 0.06% and anion in cation to less than 0.5%. Conesep systems are also installed at Heysham 2 and Sizewell B CPP. Back fitting of this system is viable as the service vessels can be retained, only the regeneration plant has to be replaced.

3.4 Kinetic Leakage Through Mixed Beds

The high flowrate operation of CPP mixed beds means that ion exchange kinetics can limit the process. A particular problem has been the effects of progressive deterioration in anion exchange kinetics in many CPPs[11]. This has led to prolonged bed rinse down after regeneration and unacceptable anion leakages into the treated water, especially during condenser leaks. The deterioration of anion resin kinetics in CPP can occur undetected as, under normal conditions, the only loading is ammonia on the cation resin and the mixed bed can therefore produce excellent quality water. However, when an unexpected condenser leak occurs there is an immediate chloride and sulphate loading on the anion resin. If this resin is in poor condition these anionic impurities can pass through the mixed bed, contaminating the feed.

A great deal of research was carried out in NE and the CEGB to investigate anion resin kinetics. Harries summarises the results of this work and details a standard kinetic test method[12]. Sulphate exchange occurs slower than chloride exchange, this effect is accentuated at high pH and with fouled resins[12]. After determining the anion resin mass transfer coefficient, by the standard test method, predictions can be made on the actual performance of that resin under plant conditions. As a result it has been proposed that in new CPP designs the anion resin specific flow rate should be limited to 200 anion bed volumes per hour (ABV/hr), to reduce the susceptibility to anion resin fouling[5]. Of the existing plants Station B, which operates at 350 ABV/hr, is particularly prone to organic fouling due to the high flow rate and the raw water supply.

Resin kinetic performance is now recognised as a key parameter and the ASTM and UNIPEDE are preparing reference methods for its determination.

4 CHEMICAL MONITORING

One of the key developments in ion exchange over the last twenty years has been the analytical technique of ion chromatography. Hamish Small's paper at the 1976 Cambridge conference[13] showed how ion exchange chromatography had been refined using special resins so that trace ions could be separated and quantified. This ability to be able to measure trace concentrations of impurities, less than 1 µg/kg, in water has led to a greater understanding of the water treatment processes used in power generation. As the current management phrase says: *"If you can't measure it, you can't manage it"*.

This technique has been investigated and adopted by Nuclear Electric for a wide range of analyses. Standard methods have been devised and verified for cations and anions in boiler feedwater[14] and for transition metals in PWR primary coolant[15]. It is mainly this technique that has enabled trace water treatment processes to be monitored routinely at the sub µg/kg level allowing tighter specifications to be written for various systems.

5 SPENT FUEL ELEMENT COOLING PONDS

After removal from the reactor, spent fuel elements are stored in water filled ponds for a period in order to allow radioactive decay prior to transport off site for processing. Ion exchange resins are used to control the chemical conditions of the pond water to minimise corrosion and to remove radioactive species from the pond water.

In Magnox ponds the water is maintained at pH 11.6 with 200 mg/kg sodium hydroxide to control corrosion of the fuel cladding. The pond water treatment plant includes a cation bed to remove the dosed sodium, a degasser to extract carbon dioxide and a mixed bed to remove chloride and sulphate. A caesium removal unit is available, as required, to treat the water entering the plant.

The same principles apply to AGR ponds but the operational chemistry is completely different. In this case the pond is dosed with boric acid to give 1250 mg/kg of boron which is required to control criticality. The pH of the system is maintained at pH 7.5 to minimise corrosion. Two non regenerable mixed beds are used to remove impurities arising in the pond. As the anion resins operate through into the borate form high purity resins are required and the removal capacity of the beds for chloride is limited.

6 PWR PRIMARY CIRCUIT CONTROL

Ion exchange resins are essential for controlling the chemistry of the active primary circuits of Pressurised Water Reactors. This control is achieved by passing a small side stream of water through the chemical and volume control system (CVCS). The CVCS design varies between plants but a typical design would be two parallel mixed beds followed by a single cation bed.

The CVCS has to perform several functions under varying solution conditions. During early cycle operation the boron concentration falls from 1200 to 900 mg/kg and is then gradually reduced by 3 mg/kg per day throughout the period of power operation. Lithium is continuously generated in the circuit and has to be removed to maintain a steady concentration of 2.2 mg/kg throughout most of the cycle reducing to 0.71 mg/kg at the end. A mixed bed is then used to remove all the active corrosion products released when the reactor comes off load at the end of the operating cycle. The operating performance of CVCS resins was described by Kashiwai[16].

Hoffman has described the variation in the borate ion equilibria during these changes and the importance of resin selectivity in determining the performance of borate form anion resin[17].

Nuclear Electric have reviewed the CVCS operating regimes at several PWR's to identify the optimum option for Sizewell B. Reduction in active resin discharges will give significant cost savings. The original operating procedure would result in 2.55 m^3 of active resin being discharged per year.

This review indicated several options for reducing resin discharges. Various stations have shown that a single mixed bed has sufficient capacity, for active corrosion products, to operate during one cycle of normal operation followed by off load clean up. Removal of lithium by a hydrogen/borate form mixed bed was also proved to be a feasible option. This would allow the cation bed to be kept in reserve and mainly used during refuelling. A cation bed used in this manner could last for several cycles. Using both these options would at least halve the volume of resin discharged from the CVCS per year giving significant cost savings.

7 GENERATOR STATOR COOLING

All Nuclear Electric stations have water cooled stators in their alternators. Recent investigations have shown that improvements in the chemical control of these cooling systems would give greater corrosion protection[18]. There is now a requirement to maintain the pH of these systems greater than 7 while keeping the conductivity below 1.0 μS/cm. This can be readily achieved by using a mixed bed to remove impurities and dosing the effluent with alkali[19]. However this option was not favoured as the dosing system added extra complications and the possibility of high conductivities if it failed.

A control system first devised and implemented by the South African utility, ESCOM, based on separate cation and anion beds[20] has recently been back fitted to several Nuclear Electric plants. The system consists of a cation bed followed by an anion bed. During normal operation the cation bed is bypassed with all the flow passing through the anion bed. This removes all the anionic impurities, mainly carbon dioxide with traces of chloride and sulphate, and allows the cations, mainly sodium, to pass through. This results in an elevated pH with no dosing requirement. The conductivity at the outlet is after anion column conductivity. As the conductivity and pH of the system would increase continuously in this mode the cation bed is put into service occasionally to control the system within the required chemical parameters. A further refinement in control is applied at some stations where the bypass of the cation bed is set so that a steady pH and conductivity can be maintained for long periods without regular attention.

An integrated stator coolant control and monitoring package, for dissolved oxygen, pH and conductivity has been developed and is available for other utilities.

8 SUMMARY

This paper has highlighted some of the systems in nuclear power plants which require ion exchange resins for chemistry control. Other important systems, not mentioned, are also dependent on ion exchangers. Although membrane processes are now becoming more competitive, particularly for treating raw water supplies, ion exchange will continue to have a key role in nuclear power generation. Without ion exchange resins these systems could not be optimised, in many cases they could not be operated at all.

References

1.   G. Knowles and K. Tittle, *Effluent and Water Treatment Journal*, 1980, July, 317.
2.   K. Tittle et al., in "Ion Exchange Technology", D. Naden and M. Streat, eds., Ellis Horwood, Chichester, 1984.
3.   J. D. Tyldesley and G. Knowles, *Water Chemistry of Nuclear Reactor Systems*, 5, BNES, London, 1989.
4.   J. D. Willerson, *Proc. 55th International Water Conference*, Pittsburgh, 1994.
5.   M. A. Sadler and M. R. Darvill, EPRI Report NP-4550, 1986.
6.   M. A. Sadler, EPRI CPP Workshop, Pheonix,1978.
7.   K. Tittle et al, *Water Chemistry of Nuclear Reactor Systems*, 2, BNES, London, 1980.
8.   J. R. Emmett and P. M. Grainger, *Water Chemistry of Nuclear Reactor Systems*, 2, BNES, London, 1980.
9.   M. A. Sadler et al, *Proc. Amer. Power Conf.*, 1983, 45.
10.   M. A. Sadler et al, in "Ion Exchange for Industry", M. Streat, ed., Ellis Horwood, Chichester, 1988.
11.   M. Ball   and R. R. Harries, in "Ion Exchange for Industry", M. Streat, ed., Ellis Horwood, Chichester, 1988.
12.   R. R. Harries, in "Ion Exchange for Industry", M. Streat, ed., Ellis Horwood, Chichester, 1988.
13.   H. Small and J. Solc, in "The Theory and Practice of Ion Exchange", SCI, 1976.
14.   D. B. Smith, in "Recent Developments in Ion Exchange", P. A. Williams and M. J. Hudson, eds., Elsevier, 1987.
15.   M. D. H. Amey and G. R. Brown, in "Recent Developments in Ion Exchange", P. A. Williams and M. J. Hudson, eds., Elsevier, 1987.
16.   T. Kashiwai et al, *Water Chemistry of Nuclear Reactor Systems*, 2, BNES, London, 1980.
17.   B. J. Hoffman and M J Gavaghan, in "Ion Exchange for Industry", M. Streat, ed., Ellis Horwood, Chichester, 1988.
18.   M. Moliere et al, *Corrosion Science*, 1990, **30**, 189.
19.   K. Schleithoff et al, *VGB Kraftwerkstechnik*, 1990, **70**, No. 9.
20.   ESCOM, Chemistry Standards for Water Cooled Generator Windings, NWS 1061, 1992.

Table 1 *Make up Plant Designs and Raw Water Supplies*

STATION	RAW WATER		M.U.P.	RESIN LIFE	PROBLEMS / COMMENTS
	SOURCE	CONDUCTIVITY	DESIGN	Years	
		µS/cm			
A	Surface	750	C/DG/SB/MB	4 - 8	Single stream limits flexibility, raw water variations. New plant I/S 1995
B	Well	350-1100	C/WB/DG/MB	C 11,A 5	
C	Upland Surface	150	C/A/MB	C 10,A 5	Stratified anion bed containing WB/SB resins
D	River Severn	600	C/WB/DG/SB/MB	~6	Changes in water supply, organic fouling
E	Bore hole	1100	WA/C/WB/DG/MB	8+	
F	Lowland Surface	200-350	SF/C/DG*/SB/UV*/MB		Organic fouling, organo-chloride slip
G	Well	350-1100	C/DG/MBA/MB	5 - 7	High ΔP on cation beds
H	Bore hole	950-1100	C/DG/SB/MB	4 - 8	Cation stratabed intermixing, sticking valves
I	Upland Surface+River	80-200	C/SB/MB		Changes in raw water quality
J	Upland Surface+River	150-200	SF/C/SB/MB	6	Acid dilution tank leaks, organics
K	Upland Surface	150	C/DG*/SB/MB	4 - 8	Changes in raw water supply
L	Well	650-1000	C/DG/SB/MB/VDG		Cation stratabed, pre cartridge filter

C = Cation Unit, DG = Degasser, WB = Weak Base Unit, SB = Strong Base Unit, MB = Mixed Bed Unit, SF = Sand Filter, OS = Organic Scavenger
MBA = Mixed base acrylic, UV = Ultra Violet Unit, VDG = Vacuum degasser, WA = Weak Acid Cation Unit, * = Not normally used

Table 2   *Condensate Polishing Plant Designs*

STATION	DESIGN	CATION BED VOL (resin type) m³	MIXED BED VOL m³	DIA m	DEPTH m	MAX FLOW m³/h	C:A RATIO (resin type)	ANION RESIN SPECIFIC FLOW ABV/hr	COMMENTS
A	MB		1.7	1.52	0.94	220	1(M):1(M)	258	Morpholine form, Conesep Regen
B	MB		5.66	2.59	1.07	380/594	2.33(G):1(G)	350	Monosphere resins, 5AP dosing, organic fouling
C	C/MB	6.68(G)	6.68	3.04	0.92	720	1(M):1(M)	216	Manual regen control, high ΔP with 1 stream O/S
D	(P)/MB		8.1	3.04	1.12	720	2(M):1(M)	267	Precoat units bypassed
E	MB/P		6.75	3.04	0.93	720	2(M):1(M)	320	Effluent discharge problems
F	MB		11.8	3.06	1.6	866/954	2(M):1(M)	320	Conesep base been replaced
G	C/MB	6.68(G)	6.68	3.04	0.92	720	1(M):2(M)	161	Difficult to see separation in MB regen vessel
H	MB		12	3.6	1.24	995	1.4(G):1(M)	200	Conesep Regen, 4 x 50% beds per turbine

C = Cation Bed, MB = Mixed Bed, P = Precoat Unit, (G) = Gel Resin, (M) = Macroporous Resin, 5AP = 5 amino pentanol

OPERATION OF WORLD'S LARGEST TRIPOL® CONDENSATE POLISHING VESSELS AT STANWELL POWER STATION (QUEENSLAND) AUSTRALIA

R. N. R. Robinson
AUSTA Electric, Stanwell Power Station, Queensland, Australia

L. A. Chapple
Thames Water Asia/Pacific Pty. Ltd., Sydney, Australia

1 INTRODUCTION

AUSTA Electric was incorporated in January 1995 from the Generating Division of Queensland Electricity Commission (QEC) and is responsible for the Generating Plant owned by the Queensland State Government. Total Generating capacity operated by AUSTA Electric is 5,200MW in a total of 12 Power Stations including Coal Fired, Hydro Electric and Gas Turbine generating facilities.

In the late 1970's population and industrial growth predictions resulted in AUSTA Electric planning for 3 new power stations to be constructed over a 10-15 year period to give a significant increase in available power capacity. These stations were Tarong, Callide B and Stanwell, details of which are shown in Table 1.

	GEN. CAPACITY	COMMISSIONED
Tarong	4 x 350 MW	1983 - 1986
Callide B	2 x 350 MW	1987 - 1989
Stanwell	4 x 350 MW	1992 - 1996

Table 1 *Generating Capacity of AUSTA Electric 3 newest Coal Fired Power Stations*

The boilers are all drum type, rated at 18MPa, pulverised coal fired with single stage reheat. The Unit Cycle chemistry is all volatile treatment based with a normal operating pH of 9.32 - 9.35.

In the design of each station it was decided that Condensate Polishing would be incorporated as an integral part of the condensing and feed heating plant. This incorporation of condensate polishing was not unusual in power stations in Australia at the time with all States incorporating Polishing Plants into Power Station designs from the late 1960's. However one major departure from the Australian precedent was to specify a condensate polishing plant which could operate in the Ammonia Cycle as well as the Hydrogen Cycle.

The decision to operate the Condensate Polishing Plant in the Ammonia Cycle was based on projected cost savings in the NPV calculations.

2 PLANT DESIGN

The Power Stations at Tarong and Callide B were commissioned before the specification for Stanwell was issued. The Operational experience at these power stations confirmed the decision to operate Ammonia Cycle condensate polishing plants and resulted in Stanwell Condensate Polishing Plant specification as detailed in Table 2.

Number of Vessels	1 x 100% per unit (4 total)
Number of Regeneration Systems	1
Total No. of Resin Charges	5
Regeneration Levels	
Sulphuric Acid	$160kg/m^3$ Low (used in H cycle operation - no leaks only) $320kg/m^3$ High
Caustic Soda	$80kg/m^3$ Low (used in H cycle operation - no leaks only) $160kg/m^3$ High
Resin Cross Contamination	0.5% anion in cation in Cation Regenerator 0.1% cation in anion in Anion Regenerator
Resin Removal from Service Vessel	99.99% of total charge to be transferred out to Regeneration area
Design Flow	300L/s
Design Temperature - normal	35-41°C
Operational Mode	Booster Loop with unvalved bypass
Ammonia capacity to ammonia break	Not less than 160kg (as $CaCO_3$) total per resin charge.

Table 2   *Condensate Polishing Plant Specifications for Stanwell Power Station*

The product water quality required out of the Condensate Polisher under leak and normal operation were not requested in the specification. Due to their experience at Tarong and Callide B Power Stations, the AUSTA Electric design engineers were certain that a Condensate Polishing Plant meeting the design conditions set out in Table 2 would produce the required water quality under all operating conditions.

3 PLANT SELECTION

During evaluation of the condensate polishing plant tenders, AUSTA Electric determined that Thames Water Asia/Pacific's patented Tripol® Condensate Polishing system gave the lowest capitalised costing.

This type of plant uses separated cation and anion beds and was at variance with Tarong and Callide B plants where mixed bed polishers were installed.

As the only Tripol® system in commercial usage at the time was at Muja Power Station in Western Australia. AUSTA Electric taking into consideration the performance history of this installation plus the results from the large scale pilot plant operation at several power stations, namely:

a)   Gladstone by AUSTA Electric,
b)   Drakelow by C.E.G.B.,
c)   Vales Point by Pacific Power,
d)   San Onofore by Southern California Edison.

decided that these plants demonstrated the required product water quality (as shown in Table 3) could be achieved and the performance of the Tripol System would be as good if not better than a mixed bed system.   Since Stanwell commenced operation two other installations at Loy Yang B and Seraya II Power Station have also been commissioned.

Ammonia capacity - total kg NH₃ as CaCO₃		250
Ammonia capacity - total kg NH_3 as $CaCO_3$		250
Acid Conductivity	μS/m	less than 8
Chloride	μg/L as ion	less than 0.5
Sulphate	μg/L as ion	less than 0.5
Sodium	μg/L as ion	less than 0.5

Table 3   *Guaranteed Product Water Quality*

The information available concerning condenser leakage during ammonia cycle operation shown by above pilot units demonstrated that the following sodium leak performance (as shown in Table 4) would be achieved during Ammonia Cycle operation.

Polisher Influent Sodium	*Sodium Capacity*
10μg/L	2.0kg as Na
25μg/L	3.0kg as Na
50μg/L	4.6kg as Na
100μg/L	7.7kg as Na
500μg/L	9.0kg as Na (Note 4)
1000μg/L	9.0kg as Na (Note 4)

1.   Ratio of other cations to sodium in polisher influent as per "Normal Water" to Demin Plant.
2.   Capacity based on outlet pH of service unit being maintained at 9.3.
3.   Chloride and Sulphate not to exceed 1.0μg/L.
4.   Limited by total anion capacity.

Table 4   *Sodium Leak Performance during Ammonia Cycle Operation*

Further to this leakage performance, crud removal levels of 90% with influent above 50ppb and down to 5ppb with influent levels of less than 50ppb were guaranteed.

Figure 1 *Simplified resin transfer schematic*

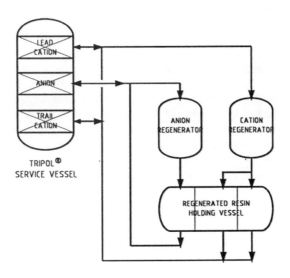

Figure 2 *Tripol service vessel diagrammatic internal arrangement*

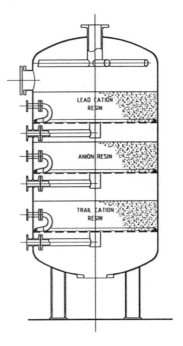

4 PLANT INSTALLATION

The theory of Tripol® operation has previously been described.

The system is based on maintaining separate cation and anion beds in both service, regeneration and resin holding vessels. Separation of the cation and anion resins ensures that cross contamination of resins does not occur and conversion of some cation resin to sodium from which sodium will leak off during Ammonia Cycle will be averted.

A simplified set up for service, regeneration and resin holding system is shown in Figure 1 and the service diagrammatic vessel internal arrangement in Figure 2.

The installation at Stanwell is significant in the fact that these are the largest diameter vessels of this type of polisher system installed anywhere in the world. Muja Power Station service vessels are 1700mm in diameter whilst the vessels at Stanwell have a diameter of 2800mm to handle the flow of 300L/s.

Details of the plant supplied are outlined in Table 5.

In common with normal Australian practice, the CPP regeneration plant is located adjacent to the make-up demineralising plant.

SERVICE VESSELS	
Dimensions	2800mm dia. x 3500 height o.s.
Vessel Area	6.14m^2
Service Velocity	176m^3/h/m^2
Resin Volumes/Depths	
Lead Cation	2.05m^3/334mm (H$^+$ Form)
Anion	2.75m^3/449mm
Trail Cation	2.05m^3/334mm (H$^+$ Form)
Resin Support	Laser cut stainless steel screens at 250kPa design Differential Pressure
Inlet Distributor	5 armed radial distributor with slotted holes (before modification) 5 "loud speaker" type distributors (final)
REGENERATION VESSELS	
Cation Regenerator	1378mm dia. x 4900mm height o.s.
Anion Regenerator	1498mm dia. x 4900mm height o.s.
Resin Holding Vessel (with 3 compartments)	1800mm dia. x 4700mm length o.s.

Table 5 *Installed Plant Design Information*

This results in being able to utilise common resources such as sulphuric acid and caustic soda storage systems, hotwater tank etc., and the water treatment plant operator is able to work in one area rather than several separate locations. Into and out of service control of polisher service vessels is controlled by the plant technicians in the respective Monitoring Centres for the units.

The locating of the resin regeneration equipment adjacent to the make-up demineralising plant results in long resin transfer pipes from the various service areas to the regeneration plant. At Stanwell Power Station, the longest resin transfer distance is 600 metres. Transfer of resins from the service area in the Turbine Hall to the regeneration area and back again is achieved hydraulically.

Resin removal from the Service Vessel is by application of top and bottom transfer water and a ringmain situated on the outside of the laser cut screen at each resin level.

5 PLANT PERFORMANCE

During commissioning of the condensate polishing plant on Unit 1, significant disturbance of the lead cation resin bed was observed. The resultant capacity to ammonia break was found to be significantly lower than the design requirement due to this bed disturbance. However the product water quality in both hydrogen and ammonia cycle operation was better than the guaranteed levels.

Modification to the radial arms were undertaken, with additional holes being drilled to lower exit velocity. This did not give sufficient lowering of disturbance of lead cation bed and a new type of inlet distribution system was installed with 2 sets of perforated plate in a loudspeaker type design to lower exit velocities even further and more evenly distribute the flow. This design was modelled prior to installation, but actual performance once installed was not as good as expected. Some minor modifications were made to this design to give more flow to the centre of the unit and capacity to design level and above was achieved.

The effect of the disturbed lead cation bed was clearly seen in measurements of capacity to ammonia break as well as in the transition time from ammonia break to full ammoniation of the cation resin was reduced from seven to two days at full unit load.

Total cation capacity to ammonia break with various distributor systems is shown below in Table 6.

Original Design	170 - 180
Mod. 1	200 - 210
Mod. 2	220 - 230
Final	250 - 310

Table 6   *Total Cation Capacity to Ammonia Break (kg as CaCO$_3$)*

The resin transfer was found to be incomplete with some resin left on the middle half of the service vessel screens. An inner ringmain was added, fed from the outer ringmain, this enabled complete resin removal to be achieved with additional incorporation of switching between ringmain and bottom transfer water.

Due to the unknown recirculation flow in the bypass line around the polisher service vessel and no automatic monitoring of the actual feed conductivity into the service vessel, calculation of the polisher capacity to ammonia break could not be undertaken using any of the service vessels' instrumentation.

A direct conductivity monitor is located upstream of the CPP Booster Loop. This measures conductivity of the condensate going to the polishing plant and is before the junction where condensate is mixed with recycled polished condensate and fed to the service vessel.

This conductivity was therefore used to calculate the ammonia load as per standard practice with polisher systems. During calculations, it was found that Thames and AUSTA Electric figures did not correlate with variances of up to 10% noted in some calculated values, the AUSTA Electric figures being lower. AUSTA Electric data was derived from equations given in reference 5, whilst Thames information was generated 20 years previously in their laboratories in London.

Due to its derivation from the latest known information, the equation AUSTA Electric used to calculate ammonia concentration is:

$$NH_3 = 0.0012967DC^2 + 0.629885DC - 1.00897 \tag{1}$$

with NH_3 in $\mu g/L$ and DC in $\mu S/m$.

As part of testing, condensate samples were sent to external laboratories for analysis of ammonia to compare with values calculated from conductivities. The analytical result scatter did not prove or disprove the use of equation (1) in calculating ammonia content from direct conductivity reading.

There is no flow monitoring close to this conductivity monitor, the only flow device in the circuit being located downstream of the polisher service vessel on the inlet to the deaerator. This device was used to infer the untreated condensate flow into the polisher as there was negligible water take off between the condensate polisher and the flow monitoring point.

In plant operation, in both hydrogen and ammonia cycle on all 4 units, product water quality as shown in Table 7 was typically observed. This confirms values measured on the San Onofone Pilot Plant during Hydrogen Cycle operation of below 0.1µg/L achievable for sodium, chloride and sulphate.

		Lab		Analysers	
		H CYCLE	**NH$_3$ CYCLE**	**H CYCLE**	**NH$_3$ CYCLE**
Sodium	(µg/L)	<0.1	0.2 - 0.3	<0.1	0.2 - 0.4
Chloride	(µg/L)	<0.1	0.2	-	-
Sulphate	(µg/L)	<0.1	0.2	-	-
Silica	(µg/L)	<2	3	<2	3
Conductivity	(µS/m)				
Direct	[	6.0 - 6.5	-	6.0 - 6.5	-
Cation	[	5.5 - 6.0	6.0 - 6.5	5.5 - 6.0	6.0 - 6.5

Table 7 *Typical Product Water Quality Measured*

The crud removal capacity of the polisher was measured over several service runs in Hydrogen cycle. A sampling test rig was set up to collect particulate matter, on the inlet and outlet lines of a polisher service vessel, on a 0.1 micron filter paper. Sampling is operated until at least 10mg of crud is collected on the filter on the inlet side of the service vessels. Both samples were weighed to the nearest 0.1mg.

During start-up, levels of 80-90µg/L of crud were measured in the polishing plant feed whilst in normal operation this falls to 10-20µg/L. Polished condensate had measured crud level of less than 5µg/L under both feed conditions. This equates to greater than 90% crud removal at the higher of the 2 feed levels (i.e. start-up).

Further crud removal trials undertaken during Ammonia Cycle operation and simulated condenser leak tests will be conducted in the near future and these results will be available in the first half of 1996.

6 CONCLUSIONS

The Stanwell condensate polishing system, apart from the initial inlet distribution problem, has given good plant performance in both Hydrogen and Ammonia Cycle. Reliability and availability has been good.

Leakage of sodium is very low in Hydrogen and Ammonia Cycle operation and no 'sodium blip' is seen when going from Hydrogen through to Ammonia Cycle. The simplified regeneration systems of separate bed technology is also an advantage.

This type of service vessel can be built up to 2800mm diameter and rated at 176m/h and give good quality condensate in both Hydrogen and Ammonia cycle provided the distributor system is adequate.

References

1   P. W. Renouf, Development of Condensate Polishing in Australia.
2   G. Crits, Technical Discussions on Condensate Polishing and Ammonex in Europe, 1974.
3.   T. A. Peploe, J. H. Smith, The Tripol Process - A New Approach to Ammonia Cycle Condensate Polishing.
4   M. Ball, R. Burrows, The Tripol Condensate Purification Process.
5   S. Kerr, D. Ryan, Ammonia - pH - Conductivity Relationships.
6   D. C. Auerswald, F. M. Cutler, S. S. Simmons, Parts per Trillion without Mixed Beds.
7   D. J. P. Swainsbury, Tripol Condensate Polishing Plant - Operational Experience.

ULTRAPURE WATER FOR THE MANUFACTURE OF SEMICONDUCTOR DEVICES

J. Keary and A. D. Mortimer

Elga Ltd.
Lane End
High Wycombe
Bucks HP14 3JH

1 INTRODUCTION

The Microelectronics industry has always been one of the most demanding in terms of the quality of the raw materials it uses in its manufacturing processes. One of these raw materials, ultrapure water, has been the subject of tightening quality standards which have perhaps been more related to developments in analytical technique detection limits than demonstrable improvements in finished product yield. Basically the 'best' available is required and the best available is having impurity levels at less than the limits of detection of the most sensitive analytical methods.

It is not so long ago that simply '18 megohm' resistivity was the single quality standard required from purified water used in the semiconductor manufacturing process, but in recent years extra parameters, including Total Organic Carbon (TOC), particles, bacteria, specific dissolved ions and gases, have been brought into consideration with some having limits at the parts per trillion level.

As one would expect, the size and complexity of the water purification systems required to achieve these stringent purity levels have progressively developed at the same rate as the limits have tightened.

This Paper reviews the history of these developments, based on the experiences of one manufacturing facility, and describes in detail the most recent plant commissioned at this Facility.

2 THE BACKGROUND

One of the more recent large treatment systems installed by Elga is at a facility which manufactures state of the art semiconductor devices using submicron technology. Elga's first involvement with this facility came in the late 70's when a system was installed with a process treatment train as outlined in Figure 1. This was designed to produce rinsing water for 3 inch silicon wafer fabrication processes. When the focus switched to 4 inch wafers in the mid 80's a new, higher specification system was required, satisfying a demand of 8m³/hr (Figure 2). This process configuration was developed from the experiences gained from the operation of the first plant together with extensive 'front end' pilot plant trials to arrive at an optimum pretreatment arrangement. This system was implemented and after 5 years successful operation, the original plant was upgraded to bring it in line with the later system. In 1994 a significantly increased demand for water, resulting from expansion plans, meant that a new water purification system was needed.

Figure 1 *Late 70's system (3 inch silicon wafer fabrication)*

Pretreatment	(Depth Filtration
	(Organic Scavenging Resin
	(Base Exchange Softening
Make Up	(Reverse Osmosis (Hollow Fibre Membranes)
	(Two Bed Deionisation (Co-current)
Polishing	(Mixed Bed Deionisation (Regenerable)
Recirculation	(Cartridge Filtration
	(UV Disinfection
	(ABS Distribution Loop

Figure 2 *Mid 80's system (4 inch wafer IC fabrication) (8m³/h usage)*

	(Polyelectrolyte Dosing *
	(Multimedia Filtration *
Pretreatment	(Organic Scavenging Resin
	(Base Exchange Softening
	(Water Heating *
Make Up	(Reverse Osmosis (thin film composite, spirally wound Membranes *)
	(Mixed Bed Deionisation (Regenerable) *
	(Mixed Bed Deionisation (Regenerable)(2 Series, 1 standby)
Polishing	(Cartridge Filtration
Recirculation	(UV Disinfection
	(Cartridge Filtration
	(ABS Distribution Loop

*   Items subsequently retrofitted to the Late 70's system

3 THE PROJECT SPECIFICATIONS

Enquiry documents were issued in early 1994 and after an extensive consultation period, a comprehensive process train was devised to meet the client's requirements (see Figure 3). The incoming mains water had always proved difficult to treat due to the high and variable levels of organic and colloidal material present throughout the year (see Table 1). In particular the silt density index (SDI), which is an empirical measure of membrane-fouling colloidal material, is almost always immeasurably high in the raw water.

The required quantity of water at point of use was specified at 27 m³/hr which, given the reject/backwash requirements through the system equated to a feed flowrate of 50 m³/hr. One further requirement was that the plant should be designed for, and be easily upgradable to, 125% of the initial value - a not inconsiderable challenge given the designers' objective of avoiding 'dead legs' that may need to be included for later plant additions and the desire to keep flow velocities high.

The treated water specification (see Table 2) represents what was then one of the most stringent ever requested in a semiconductor manufacturing environment. Under the circumstances, the issue of process guarantees was approached at the tender stage with a mixture of experience, educated guesswork and 'gut feeling' since, in some cases, the difference between meeting and failing the water quality specification could be as little as a few parts per trillion! Certain determinands represented 'breaking new ground' so close client consultation was needed to agree the differentiation between target and guarantee values.

Figure 3 *Mid 90's system (6 inch silicon wafer fabrication)(27 m³/hr usage)*

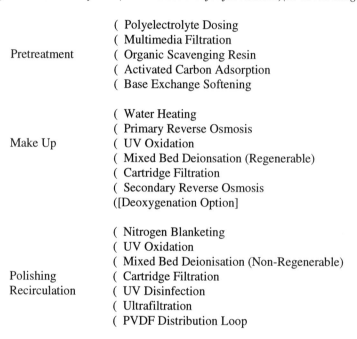

Pretreatment	(Polyelectrolyte Dosing (Multimedia Filtration (Organic Scavenging Resin (Activated Carbon Adsorption (Base Exchange Softening
Make Up	(Water Heating (Primary Reverse Osmosis (UV Oxidation (Mixed Bed Deionsation (Regenerable) (Cartridge Filtration (Secondary Reverse Osmosis ([Deoxygenation Option]
Polishing Recirculation	(Nitrogen Blanketing (UV Oxidation (Mixed Bed Deionisation (Non-Regenerable) (Cartridge Filtration (UV Disinfection (Ultrafiltration (PVDF Distribution Loop

4 THE SYSTEM FEATURES AND BENEFITS

Perhaps the most important feature of the system was its flexibility in terms of upgradability. As mentioned earlier a capacity of 125% of initial flow capability had been build into the design by provision for extra process units such as RO modules. Indeed, on commissioning the pretreatment system was operating at 133% of its initial capacity and at the same time exceeding design calculations in terms of SDI reduction. In addition, since the original project completion, the client has now decided that a dissolved oxygen removal process is required and this has recently been installed into the treatment train.

One of the more interesting features of this plant is the remote monitoring system which, via a modem link, enables operating performance data to be transmitted back to the Head Office in High Wycombe. There, the data is received and used to assess trends

Table 1   *Feedwater Analysis (Maximum Values)*

Parameter	Value	Parameter	Value
Colour	15 Hazen Units	Turbidity	1.0 NTU
Suspended Solids	17 mg/L	Silt Density Index	80-150
pH	7.2-7.8	Conductivity	615 μS/cm
Total Dissolved Solids	430 mg/L	Total Organic Carbon	5 mg/L
Bacteria	60-100CFU/mL	Silica	10 mg/L
Sodium	12 mg/L	Potassium	3 mg/L
Chloride	32 mg/L	Nitrate	1 mg/L
Sulphate	200 mg/L	Aluminium	1 mg/L
Copper	0.08 mg/L	Iron	0.7 mg/L
Manganese	0.05 mg/L	Zinc	1 mg/L
Phosphate	0.1 mg/L	Lead	0.02 mg/L
Calcium	170 mg/L	Magnesium	80 mg/L

Table 2   *Initial Treated Water Specification*

Parameter	Value	Parameter	Value
Resistivity	18.2 MΩ	Total Organic Carbon	<5 μg/L
Trihalomethanes	<3 μg/L	Ptcls/ ltr 0.1-0.2μm	<1000
Ptcls/ ltr 0.2-0.3μm	< 500	Ptcls/ ltr 0.3-0.5μm	<10
Ptcls/ ltr over 0.5μm	<1	Bacteria per 100 mL	Nil(<1)
Silica	1 μg/L	Ammonium	0.3 μg/L
Aluminium/Phosphate	0.2 μg/L	Potassium/Bromide/Nitrate	0.05 μg/L
Sulphate/Manganese/Nitrate	0.05μg/L	Calcium/Magnesium/Lithium	0.05 μg/L
Residue	0.05μg/L	Sodium/Chloride	0.025μg/L
Chromium/Iron/Zinc	0.02μg/L	Copper/Lead	0.002μg/L
Dissolved Oxygen	10 μg/L		

in water quality variations in relation to routine maintenance procedures. Performance parameters routinely monitored through the system include volume, pressure, flow and resistivity.

This has great potential for the future, where water treatment plants can be monitored by the manufacturers, whose process engineers can then provide a 'Babysitting Service' and advice to end users on optimisation of performance and give early warning of the need for Clean in Place (CIP) operations, cartridge filter/resin changes or UV lamp replacement before water quality/quantity is compromised.   Furthermore this could ultimately extend to the full operation of the plant, with the 'experts' being responsible for not only design, manufacture, installation and commissioning but also the continuous remote control of this plant.

Given that this was the third water treatment system installed at the site by Elga, then all the lessons learned from the experiences of operating the first two were incorporated into the project. In particular, the multistreaming aspects of the plant meant that routine maintenance could be carried out with minimum impact on flow and quality of water to point of use.

The use of special high purity non-regenerable ion exchange polishing cylinders in the recirculation system has lead to consistently high resistivity water being produced without the need for routine regeneration. Indeed the original resin charges used in commissioning are still performing well after almost a year in service, thus providing a major running cost benefit. Final purification to eliminate particles is achieved in a hollow fibre ultra-filtration system. Special double skinned polymeric membranes not only ensure the lowest possible particle counts but are also intrinsically clean devices which will not contribute to the TOC or inorganic impurity levels.

5 THE PLANT PERFORMANCE

5.1 Overview of Final Water Quality

When commissioned the whole water treatment plant was subjected to an intensive 5 day validation period to prove system operation and that purity levels achieved specification.

Approximately 90 days after commissioning, a further 2 day water quality validation test was carried out to demonstrate that the treatment plant was maintaining the required degree of purification. The results obtained are presented in Table 3. The consistent achievement of levels within the required acceptance standards for all parameters was not only pleasing but total justification for the comprehensive "no compromise" approach to the system design which was not the least expensive submitted to the client.

5.2 Total Organic Carbon (TOC) Removal

The TOC removal pattern through the system highlighted an interesting trend (see Table 4). The single most effective pretreatment unit is the organic scavenging resin which removes 71% of the incoming TOC. The combination of activated carbon adsorption and softener resin achieved a further 16% reduction (this equating to a further 64% removal across just the two units) with the Reverse Osmosis units taking out 81% of the remaining feedwater TOC (equating to a 95% removal over the RO membrane). The reduction of trace TOC from 18.5 ppb to 3 ppb is largely achieved by the polishing RO and the photooxidising UV/Ion Exchange Plant.

5.3 Trihalomethanes (THM) Removal

The THM removal pattern was rather different (see Table 5) in that the pretreatment system had very little effect on the non-brominated compounds (principally chloroform) but the primary RO did achieve a 33% reduction. Further significant removals were achieved by the secondary RO and photooxidising UV units with the final level of 4 ppb $CHCl_3$ contributing about 15% of the residual TOC, although these values may well have been falsely high due to the exaggerated response of the Anatel TOC analyser to ions such as chloride.

Table 3   *Final Water Acceptance Test Results (May 1995)*

Parameter	Acceptance Level (μg/L unless stated)	Measured Value (μg/L unless stated)
(1). Cl	<0.1	<0.05
(1). Br	<0.1	<0.06
(1). NO_2	<0.1	<0.06
(1). NO_3	<0.1	<0.05
(1). PO_4	<0.1	<0.1
(1). SO_4	<0.1	<0.05
(1). K	<0.1	<0.05
(1). NH_4	<0.1	<0.1
(1). Ca	<0.1	<0.06
(1). Na	<0.05	<0.01
(1). Mg	<0.1	<0.05
(2). Li	<0.03	<0.01
(2). B	<2	<0.1
(2). Al	<0.05	<0.05
(2). Cr	<0.03	<0.03
(2). Cu	<0.05	<0.04
(2). Fe	<0.05	<0.05
(2). Mn	<0.02	<0.002
(2). Zn	<0.06	<0.03
(2). Pb	<0.01	<0.005
(3). SiO_2	<1	<0.05
(4). Particles/Litre 0.1-0.2μm	<300	64
(4). Particles/Litre >0.2μm	<74	23
(5). Bacteria CFU/100mL	<6	1

(All Measurements Off Line except Particle counts)

(1). Ion Chromatography   (4). PMS Laser Counter
(2). AAS/ICP-AES/ICP-MS   (5). Total Viable Count @ 27°C
(3). AAS Carbon Furnace

Table 4 *TOC Reduction Through the Treatment System (July 1995)*

Treatment Process	TOC ($\mu g/L$)
Mains Water	4140
Ex Multimedia Filtration	3930
Ex Organic Scavenging Resin	997
Ex Adsorption & Softening	355
Ex Primary Reverse Osmosis	18.5
Ex Mixed Bed Deionisation	19.8
Ex Secondary Reverse Osmosis	7.3
Ex Polishing System	~ 3

(Analysis Off-Line by Modified Dohrman DC80 TOC Analyser)

Table 5 *THM Reduction Through the Treatment System (July 1995)*

Treatment Process	THM ($\mu g/L$)			
	$CHCl_3$	$CHCl_2Br$	$CHCl Br_2$	$CH Br_3$
Mains Water	40	13	3	<1
Ex Multimedia Filtration	39	13	3	<1
Ex Organic Scavenger Resin	36	13	3	<1
Ex Adsorption & Softening	44	9	<1	<1
Ex Primary RO	28	6	<1	<1
Ex Mixed Bed DI (Regenerable)	19	1	<1	<1
Ex Secondary RO	10	<1	<1	<1
Ex Polishing System	4	<1	<1	<1

(Analysis Off-Line by Headspace Gas Chromatography)

5.4 Other Parameters

With regard to other performance parameters, it was found that the incoming water SDI of 90 was being reduced to 4.4 by the multimedia filter with upstream dosing of a polyelectrolyte material to promote coagulation of colloids. A further reduction to 2.8 was achieved after the organic scavenging resin.

The primary RO unit, operating at 75% recovery and 20°C, achieved an overall salt rejection of 98% (317 μS/cm feed; 6.5 μS/cm permeate). Given the high purity level of the water being fed to the secondary RO it was not possible to assess salt rejection 18.2 MΩ feed and permeate!). Although as stated earlier its primary function was to remove TOC.

6 THE FUTURE

With regard to the particular site described the short term future will see the final commissioning of a new membrane deoxygenation system. This is designed to reduce dissolved oxygen from almost full saturation level to less than 10 ppb and utilises a relatively new process where the oxygen is drawn through a hydrophobic gas-permeable membrane by a combination of nitrogen and vacuum. Inital Start up trials look very promising, indicating levels of 3 to 7 ppb being achieved in the treated water.

In a more general context, future developments in water treatment for semiconductor manufacturing applications are likely to include some or all of the following :

- Continuing development of high quality mixed bed polishing ion exchange resins which not only achieve complete removal of inorganic ions but also do not add any particles or TOC causing compounds into the treated water.

- Wider application of membrane based degassing systems, for both oxygen and nitrogen removal.

- New Reverse Osmosis membranes capable of high rejection and high fluxes at lower operating pressures. Although these membranes have higher capital costs than the conventional ones, their use does produce savings in pump size and running costs.

- More extensive use of crossflow microfiltration (CMF) and ultrafiltration (UF) in pretreatment applications in place of the more conventional media based processes (MM, OS, AC, CF etc). Although more capital intensive, MF and UF produce a more consistent output which is far less susceptible to variations resulting from raw water quality changes.

- More efficient TOC reduction to sub-ppb levels

- Elevated temperature distribution for control of organisms and greater solvating power of the Ultrapure Process water.

A PRELIMINARY COMPARISON OF TYPE-II AND TYPE-III STRONG BASE ANION EXCHANGE RESINS AT THE NEW PLYMOUTH POWER STATION - NEW ZEALAND

W. S. Williamson
Contact Energy Ltd., New Plymouth Power Station
Breakwater Road, New Plymouth, New Zealand

J. Irving
Purolite International Ltd., Cowbridge Road
Pontyclun, Mid Glamorgan CF72 8YL, Wales, UK

1 INTRODUCTION

Laboratory test work on two new strong base anion resin types based on work by Bauman and McKellar [4] and Soldano and Boyd[5] has previously been reported[1,2,3]. These reports showed one of these had functional groups which offered important improvements over existing Type-I and Type-II strong base resin products. Although Type-I and Type-II resins offer a reasonably good combination of regenerability and low residual silica, the Type-I resins require a higher consumption of regenerant chemical and the Type-II are less thermally stable. Both acrylic strong base and macroporous styrenic Type-II resins have been used at New Plymouth Power Station (NPPS). Acrylic strong base resins are generally preferred because of their high resistance to organic fouling, good silica removal and good regenerability. However thermal degradation gives rise to carboxylic groups which can produce extended rinses[6]. In practice at NPPS the performance characteristics of acrylic resins have been similar to the Type-II resins with a useful service life of 3 - 4 years treating the high silica New Plymouth water. Use of acrylic resins at moderate to high silica loadings carries the risk of silica precipitation within the resin structure.

The so called Type-III Purolite A555 macroporous strong base anion exchange resin was found to have thermal stability significantly better than Type-II and acrylic resins when laboratory tested[1]. Regenerability was also close to that of Type-II or acrylic resins while silica removal approached that of Type-I resins[1]. It is believed that Purolite have developed a resin with superior characteristics to those traditionally used at NPPS.

NPPS was interested to test this new product in an effort to improve the economics of running its demineralisation plant. Use of a resin type which provided a 50% or more increase in operating life over Type-II strong base resins while also providing comparable operating capacity would be a significant advance over current practice.

The demineralisation plant at NPPS has two parallel trains each consisting of a cation unit, anion unit and mixed bed polishing unit. The two trains share a common degasser and common regeneration equipment. Raw water is piped from the local municipal supply. The bicarbonate removal efficiency of the degasser is better than 85%. The main anion vessels have a working diameter of 1.37 m giving an operating bed depth of 0.95 m.

The new anion exchange resins were installed in the NPPS main anion units on 27 October 1994. The Purolite A510 Type-II macroporous resins (1.4 m³) were installed in

Train 1 and the Purolite A555 macroporous resins (1.4 m³) were installed in Train 2. A comparison of the performance of these two resins has therefore been possible.

2 OPERATING PARAMETERS

Service cycle end points at NPPS are controlled using on-line direct conductivity (K_o) and colourimetric silica analysers. The cycle end point for the cation unit is set at and reflected by a direct conductivity of 40 μS.cm^{-1} measured at anion unit outlet. The anion cycle end point is set to a silica breakthrough point of 4.5 mg.L^{-1}. The normal service flow rate is 16.5 L.sec^{-1} for both the cation and the anion vessels.

Note that for the purposes of more reliable, consistent comparison of resin performance the average silica leakage values referred to in this paper are calculated for that part of the service cycle when silica leakage is less than or equal to 0.5 mg SiO_2.L^{-1}.

Mixed bed polishing unit end points are K_o = 0.3 μS.cm^{-1} and SiO_2 = 0.020 mg.L^{-1}. The mixed bed end points are very rarely reached since only ~ one third of the available mixed bed operating capacity is consumed during each service cycle. Higher silica leakage from the anion units has always been acceptable because of this under utilisation of available mixed bed capacity.

Ongoing plant product water quality and ion exchange resin performance assessment is being carried out with the use of on-line plant data system (PI) and by manual grab sampling to provide additional detail.

The main cation unit uses a single step, single concentration acid injection carried out at ambient temperature. The mixed bed uses a single step, single strength acid injection carried out at ambient temperature.

The mixed bed anion resin and main anion unit are regenerated with caustic soda in thoroughfare mode. Caustic is heated to 90°C in a batch dilution tank and is injected onto the mixed bed at ~55°C after 50% dilution. The regenerant caustic then passes onto the main anion unit resin with another 50% dilution bringing the main anion unit peak regenerant temperature to ~30°C. The regeneration level for the main anion unit anion exchange resins is ~100 g NaOH/L resin.

All vessels are regenerated and operated in the co-flow mode.

3 ANION RESIN OPERATING CONDITIONS & PERFORMANCE

The 'new' Type-II and Type-III strong base anion resins have been in service for almost 12 months at the time of writing this paper.

There have been circumstances during the trial so far that have influenced how and how much information has been collected at various stages in the first 12 months of resin operation.

Despite periods of low plant operation and cation unit limitations due to calcium sulphate precipitation valid data has been collected. Representative data is supplied here to indicate the performance of the Type-II and Type-III resins.

3.1 Raw Water Analysis

A key characteristic of the water supply at NPPS is that it is relatively 'thin' and it contains a high proportion of dissolved silica. Raw water analyses representative of the trial period to date are given in Table 3.

Table 1 *Raw Water Analysis*

Date	27/10/94	15/11/94	25/11/94	28/11/94	20/04/95	2/05/95	30/05/95	8/06/95	15/08/95
pH	7.9	8.1	8.0	8.0	7.9	8.4	8.1	8.0	8.1
Ko [Lab] mS/cm	140	145	129	144	176	149	152	142	135
KH+ [Lab] mS/cm	299	265	241	253	294	267	279	258	253
Na [mg/L]	8.6	7.6	6.7	7.3	8.5	7.6	7.5	7.8	7.7
K [mg/L]	2.6	2.2	2.2	2.4	2.5	3.1	3.0	2.7	3.0
Mg [mg/L]	5.8	4.3	3.6	4.3	5.5	4.5	5.0	4.3	3.6
Ca [mg/L]	15.7	12.9	10.3	12.0	11.9	13.7	14.2	12.8	12.9
Fluoride [mg/L]	0.7	0.5	0.5	0.8	0.9	1.0	0.9	0.8	1.2
Chloride [mg/L]	10.5	9.7	9.8	10.7	11.3	7.5	10.8	11.1	11.8
Nitrate [mg/L]	0.4	0.6	-	-	-	0.1	-	-	1.1
Phosphate [mg/L]	-	-	-	-	-	-	-	-	-
Sulphate [mg/L]	18.7	14.7	15.9	17.8	20.3	14.1	20.0	17.5	12.3
Bicarbonate [mg/L as CaCO$_3$]	44.6	34.0	27.7	34.6	45.8	34.3	36.0	34.5	31.6
Silica [mg/L]	24.7	21.1	25.7	21.3	27.1	21.7	21.9	21.0	19.1

3.2 Anion Resin Operating Throughput

The initial Type-III resin operating capacity expectation was that it would have slightly less capacity than the Type-II resin when new. It was then expected that, as the Type-II resin lost its strong base capacity more quickly than the Type-III resin, we would see a drop off in Type-II resin operating capacity while the Type-III resin maintained its capacity for up to 50% longer.

The operating throughputs for both anion units are charted in Figure 1 below. The trend indicates a higher productivity for the Type-III resin.

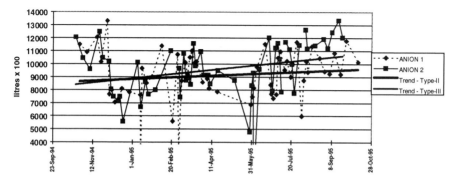

Figure 1 *Anion cycle throughputs throughout the trial period to date*

Cation capacity limitations over a very dry spell in the summer months (December-end '94 - April '95) prevented us from extending the anion cycle to operational exhaustion, preventing any extensive comparison of Type-III resin operating capacity vs Type-II. These problems originated from an increase in raw water cations and from a calcium sulphate precipitation problem which developed during this period. The calcium sulphate problem was overcome by reducing the main cation unit regenerant acid strength. This significantly improved cation operational throughput. However, cation limitation remained a problem until month 5 (April 1995) when extensive rain broke the local dry spell and reduced raw water total cation levels.

Design calculations from Purolite show that the plant is cation limited. Initially 1150 m³ were being processed by both resins in the service cycle. This is considerably more than the Purolite design programme predicts for the Type-II resin. After the initial longer runs lower throughputs followed due to cation limitation. Results for September and October 1995 would indicate that the Type-III resin is still capable of delivering the same performance. Over the last three months the Type-II resin has shown some 20% loss in throughput in comparison to Type-III. This data alone does not indicate if this is a function of resin degradation.

Figure 2 offers a reason why the performance of the Type-II has changed. The loss in strong base capacity (SBC) is more marked than for the Type-III resin. Comparison with the reported laboratory stability data at 80°C would have indicated a SBC loss approximately twice that of the Type-III. Thus a fall to 93.5% SBC for the Type-III should have produced a value of 87% SBC for the Type-II in the same time frame. The same comparison for data at 50°C over 6 months showed 96% SBC for the new Type-III and 89% SBC for the Type-II, a factor of 2.5% higher. Thus, from a loss of 6.5% SBC for Type-III, a loss of 6.5 x 2.75 would be predicted resulting in a predicted SBC value for Type-II of 82.1% compared with the 83.5% actually obtained. The 50°C laboratory model more closely fits the available trial data than the 80°C model. Also, 9 months at 30°C is equivalent to 7-9 months at 50°C in a static model, but only 1.5 - 3 weeks at 80°C. It is possible that the mechanism of degradation changes as the temperature increases.

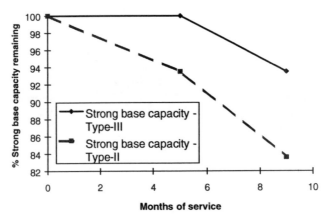

Figure 2 *Strong base capacity performance of Type-II vs Type-III resins*

3.3 Anion Resin Operational Silica Leakage

The Type-III strong base anion resin was expected to provide a significantly lower level of silica leakage than the Type-II strong base resin under the same conditions.

The comparison of the first service cycles for each anion unit indicated that the silica leakage from the Type-III strong base resin was much the same as that from the Type-II strong base resins.

Despite the indications in the first service cycles and the erratic regeneration conditions on Anion 2 the Type-III resins generally performed to provide an operational silica leakage 40% - 65% lower than the Type-II resins.

Table 2 shows the average silica leakage performance for each anion unit based on available plant data over the first 12 months of the trial. It is clear from Table 2 that the silica leakage is more or less stable over the trial period reported for the Type-III resin. Values are very good considering that the operating end point is continued to 4.5 mg.L^{-1}. The operational silica leakage for the Type-II resin is close to acceptable limits in many plant at 0.45 mg.L^{-1}. The effect of the anion resins on conductivity is currently less clear although analysis to date indicates that the anion outlet conductivity from both anion resins does seem to correspond with alkali generated by cations leaked from the main cation units.

Cation leakage from both main cation units into the anion influent normally comprises sodium and potassium ion. Total cation leakage onto the anion unit is typically 1.0 - 1.75 mg.L^{-1} as $CaCO_3$. This is generally reflected by anion conductivity values.

Table 2   *Average Silica Leakage Performance on a Monthly Basis*

Anion 1, Type-II	SiO2 (mg/L)		Anion 2, Type-III	SiO2 (mg/L)	
	Average	Min		Average	Min
October '94	0.23	0.14	October '94	0.23	0.16
November ' 94	0.27	0.16	November ' 94	0.09	0.02
December '94	0.35	0.29	December '94	0.17	0.12
January '95 *	0.31	0.21	January '95 *	-	-
February '95 *	-	-	February '95 *	-	-
March '95 *	0.35	0.20	March '95 *	0.25	0.10
April '95 *	0.30	0.20	April '95 *	0.13	0.10
May '95 *	0.23	0.20	May '95 *	-	-
June '95	0.34	0.24	June '95	0.26	0.09
July '95	0.28	0.21	July '95	0.19	0.13
August '95	0.34	0.28	August '95	0.13	0.10
September '95	0.44	0.29	September '95	0.13	0.09
October '95	0.45	0.17	October '95	0.12	0.06

*   Minimal or no information collected due to lack of plant operation or due to cation limited service cycles

Figures 3 & 4 show typical silica leakage and conductivity profiles throughout the service cycle for each of the anion. Figure 5 in particular shows silica excursions linked in turn to conductivity excursions during brief periods with higher cation concentrations in the anion influent.

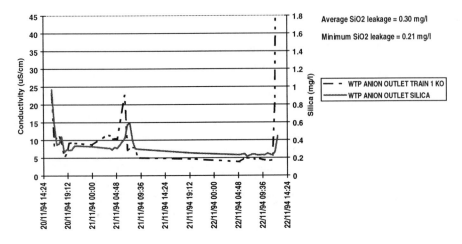

Figure 3 *Anion 1 - Type-II, November 1994*

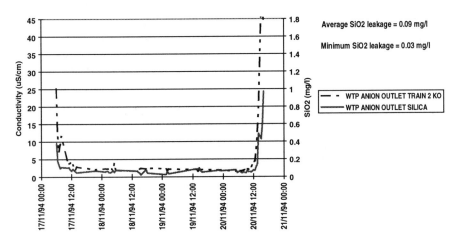

Figure 4 *Anion 2 - Type-III, November 1994*

3.5 Anion Resin Rinse Down Characteristics

Comparison of the regeneration final rinse requirements for the Type-II and Type-III resins indicate that little difference between the Type-II and Type-III strong base resins.

3.6 Anion Resin Expansion

Resin bed depths were measured at the time of initial installation, at operational exhaustion, and after regeneration to determine the expansion characteristics of the resins in service. Type-III resins appear to expand significantly going from the exhausted to the regenerated state.

The Type-II strong base resin showed a permanent expansion at operational exhaustion of 1.6%. Expansion from operational exhaustion to operational regeneration is 3%. Total overall expansion from complete exhaustion in the chloride form [as supplied new] for Type-II resin is 5%.

The Type-III resin showed an expansion from new to operational exhaustion of 13%. Expansion from operational exhaustion to operational regeneration is 9%. Total overall expansion from complete exhaustion in the chloride form [as supplied new] into the regenerated state for Type-III resin is 22%.

The much higher resin expansion figures for the Type-III resin are to be expected. The resin behaves more like a Type-I resin in its silica removal properties because the hydroxide is more highly hydrated and more highly ionised. The expansion is even greater than for a Type-I resin because the Type-III resin is more highly converted to the hydroxide form[2,3]. The Type-II resin on the other hand has the hydroxyl group from the dimethylethanolamine structure which partially satisfies the hydration needs of the hydroxide ion, so reducing the swelling and increasing the regenerability of the Type-II.

3.7 Pressure Drop in Service

Pressure drop across the Type-II resin appears to be normally 0.5 bar higher than that across the Type-III resin under the same flow conditions as indicated in Table 3.

Table 3 *Pressure Drop (bar) across Anion Resins in Service at 12 L/s*

Resin Bed	Inlet Pressure	Outlet Pressure	Apparent Pressure Drop
Anion 1 - Type-II	4.1	3.4	0.7
Anion 2 - Type-III	4.1	3.9	0.2

3.8 Organic Fouling

The local water supplies in the New Plymouth area are notorious for high loadings of dissolved organic materials. In theory, the organic fouling potential can be very high. No specific resin testing has been carried out so far to determine the extent to which the Type-III resins may have become organically fouled. However, ongoing good operational capacity over the first 12 months of service suggests that the Type-III resins handle organic loading at least as well as Type-II macroporous resins. The intensity of colour released into caustic regenerant by organics released from the Type-III resins appears the same as the colour released from the Type-II under equivalent regeneration conditions.

4 CONCLUSIONS

In these first operational trials the first 12 months has produced promising performance results from the Purolite A555 Type-III compared with Purolite A510 Type-II Strong Base anion exchange resin.

The Type-III resins appear to be demonstrating the expected strong base capacity retention advantage over Type-II resins after 12 months. This compares favourably with reported laboratory test results. The Type-III resins are also demonstrating significantly improved silica leakage performance over Type-II resins. This also reflects reported laboratory test results.

Ongoing scrutiny of the operating characteristics of the Type-III resins are planned for the next 3 - 5 years. Specific analysis will be carried out to determine the extent to which Type-III resins become organically fouled.

The 20% advantage in the throughput experienced over months 11 and 12 of operation, if continued, together with a 50% reduction in silica leakage and a 1-2 year extended life span would make this product commercially attractive. The advantages sort by NPPS resulting from potential reduction in size of the plant, reduced costs arising from saving in resin repurchases, reduced chemical consumption and from reduced load on following mixed beds would be realised.

Acknowledgment

Staff of Production Chemical Services, Operations and Maintenance sections, New Plymouth Power Station, Contact Energy Ltd, for their co-operation and contribution towards producing the trial information.
Contact Energy Limited for their permission to use the trial information.
Dr. J.A. Dale and co-workers of Purolite International Ltd who developed the novel Type-III anion exchange resin.

References

1.   J. Irving & J. A. Dale, *IWC Pittsburgh*, IWC-92-34, 1992.
2.   J. Irving & J. A. Dale, "Ion Exchange Advances" (Proceedings of IEX'92), M. J. Slater, ed., SCI Elsevier, 1992. p. 33.
3.   J. Irving & J. A. Dale, *IWC Pittsburgh*, IWC-93-41, 1993.
4.   W. C. Bauman & R. McKellar, U.S. Pat. 2614099, 1952.
5.   B. A. Soldano & G. E. Boyd, *J. Amer. Chem. Soc.*, 1953, **75**, 6099.
6.   R. M. Wheaton & W. C. Bauman, *Ind. & Eng. Chem.*, 1951, **43**, 1088.
7.   A. C. Cope & E. R. Trumball, *Org. Reactions*, 1960, XI, 317.

ASSESSMENT OF CATION EXCHANGE RESINS WITH RESPECT TO THEIR RELEASE OF LEACHABLES

K. Daucik

ELSAM
Denmark

1 INTRODUCTION

Power plants showed the first serious interest in leachables from ion exchange resins in connection with the marketing of a new type of resins - tough gel resins. The use of these resins gave rise to serious problems because of an extremely high rate of leachable release. Particularly cation resin leaching aromatic sulphon derivatives originating from low polymerized or depolymerized sulphonated polystyrene[1,2]. Sulphate is yielded from these derivatives at the conditions of the boiler.

The interest in leachables continued even after withdrawal of the tough gels from the market. It has been established that several problems with the purity of the water/steam cycle can be traced to leachables[3,4,5]. Sulphate has been found to be a dominating cycle contaminant of units where the traditional contaminants originating from cooling water inleakages are eliminated by well-performing full flow Condensate Polishing Plant (CPP).

However, some people might still be misled about the origin of sulphate if they use sulphuric acid for regeneration of cation exchange resin. Our calculations indicate that with a reasonable rinse of the resin after regeneration ($\kappa \leq 0.1$ µS/cm) the sulphate contamination originating from the regenerant will at least be of a magnitude lower than sulphate originating from cation resin leachables.

2 EXPERIMENTAL

2.1 Analytical Method

A number of analytical methods has been developed to estimate the amount of leachables from resins. A comparison of the different analytical procedures is shown in Table 1. It is apparent that all the procedures involve an extraction of the resin sample in water at a defined temperature. The temperature as well as the time of extraction vary in the different methods.

In the early stage of our involvement, the method of Southern California Edison (SCE) was chosen for evaluation of the tendency of resins to release leachables. After a short period of time a modification appeared to be necessary. The results of the SCE method showed an extreme variation from one product to another and from one charge to another when analysing new resins, which was due to the quality of the rinsing carried out

Table 1 *Comparison of Different Leachable Analysis Procedures*

Procedure	1	2	3
Pretreatment	Rinse 10 BV	Rinse 10 BV	Rinse <0.3µS/cm
Extraction Time (h)	16	16	24
Extraction Temperature (°C)	50	60	60
Evaluation TOC (mg/kg DR)	0.51	1.52	3.28

DR - Dry Resin

by the producer and not the quality of the resin. Thus, all samples were subjected to a standard pretreatment, thereby eliminating the leachables which would be rinsed anyway in case of commissioning of new resins following our standard procedure. The pretreatment is shown in Table 2. After this modification, the results were lower and seemed to give a reasonable measure of the resin behaviour in the first period of operation.

Using the same method for the evaluation of the properties of used resins brought about some surprises. The pretreatment of samples resulted in an increase of the results of the following analysis. This apparent anomaly was interpreted as a consequence of a change in the ionic form of the resin during pretreatment. The sample of used resin from CPP - irrespectively of it being taken after or before regeneration - will be in hydrogen or ammonium form. The resin is converted into sodium form during pretreatment. It has been shown that sodium form resin leaches more than hydrogen and even ammonium form resin[6,7]. Our findings confirmed this relationship.

Table 2 *Procedure for Pretreatment*

Resin	New	Used
5% HCl	2 BV	
Rinse	2.5 BV	
2% NaOH	5.5 BV	
Rinse	5 BV	
10% HCl	15 BV	15 BV
Rinse	2.5 BV	2.5 BV
2% NaOH	20 BV	20 BV
Rinse	<0.3 µS/cm	<0.3 µS/cm

2.2 Problems with the Aanalysis

However, after using this procedure for a couple of years, a new surprise appeared when a dramatic rise of the leachables released from a 10-year-old resin was observed within 1 year. An investigation showed that the rise was not real, but due to the sampling of the resin. When the resin was sampled after regeneration, the result of the leachables analysis was considerably higher than the result from resin sampled just before regeneration. This phenomenon is not due to the resin form as both samples had been subjected to the standard pretreatment.

Table 3 shows an example of the cation resin from the CPP analysed before and after regeneration, subjected to as well as not subjected to pretreatment. The Table also comprises a result of an analysis made after repeated pretreatment.

Table 3 *Leachables from Cation Resin from CPP (mg TOC/kg DR) after Different Pretreatments*

Sampling	Before Regeneration	After Regeneration
Rinse with Water	7	14
Standard Pretreatment	46	212
Twice Standard Pretreatment		55

3 RESULTS AND DISCUSSION

3.1 Mechanism of the Leachable Extraction

These results indicate that the effect of a single pretreatment is neither additive nor multiplicative. A kind of memory effect seems to be attached to the result of the leachables analysis. The effect of the first and second pretreatment seems to be contradictory. However, it can be explained by the following hyphothesis:

• Leachables from new resins are mainly residues from the production including monomers and oligomers, side products, production aids, solvents, etc., altogether initial leachables.

• Leachables from used resins are mainly decomposition products from the resin backbone, which are continuously being produced within the whole volume of any bead. Let us call them permanent leachables.

- Figure 1 shows a UV-spectrum of leachable extracts from typically new and used (CPP) cation resins. Used cation resins from CPP show a typical absorbance at 225-230 nm. This absorbance has been recognized as typical for decomposition products of cation exchange resins[2]. In leachable extracts from new resins, this absorbance is camouflaged by background absorbance of initial leachables. In leachable extracts from resins used in demin. plants, the camouflage is caused by absorbed organic compounds from raw water.

- The characteristic absorbance of permanent leachables can be used for their quantification. A good correlation has been established between the absorbance at 230 nm (A_{230}) and TOC in the leachable extracts from cation resins used in the CPP:

$$TOC = b \times A_{230}$$

$$where\ b = 2.1\text{-}2.4\ (mg/L)$$

- During normal operation of a CPP, the produced permanent leachables are continuously being extracted from the surface of beads by means of condensate. The distribution of the produced leachables within a bead will then look almost like Curve 1 on Figure 2.

- Due to the changes of resin volumes during regeneration or pretreatment, the leachables inside the beads will redistribute. Curve 2 characterizes the distribution after this exercise. The concentration on the bead surface is now higher than before treatment. However, the average concentration in the beads is lower than before treatment because part of the leachables would be removed by the sponge like an effect of the treatment. A repetition of the pretreatment would again remove some of the leachables from the beads but there would not be much to redistribute. Thus, the result of a leachable analysis after the second pretreatment is lower than before. Curve 3 on Figure 2 illustrates the new situation.

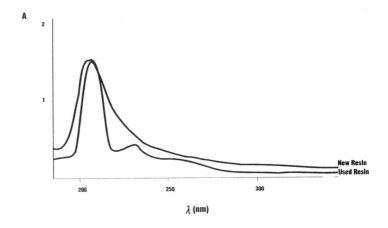

Figure 1 *Typical spectra of cation resin leachable extracts*

3.2 Kinetics of the Leachable Analysis

The memory effect on the result of the leachable analysis prevents any sensible evaluation of CPP resins, unless completely uniform sampling conditions and pretreatment are assured. This limitation devaluates the previous experiences as the sampling conditions have not been properly defined in the past.

Furthermore, the storage time between sampling and analysis would, according to the hypothesis, also has an impact on the result of the analysis. Table 4 illustrates some examples. Further improvement of the analytical method for leachables was desirable.

A kinetic study was made on a used cation exchange resin from the CPP. The resin was sampled before and after regeneration. The study was made on the two samples as received, and also after standard pretreatment. Figure 3 shows how the concentration of leachables in the extract (measured as A_{230}) increased with time. It is remarkable that the major differencies between the curves occur in the first 5 hours of the extraction. After 24 hours of extraction, the curves seem to approach the same slope, indicating that the rate of leachable release is almost equal. This would suggest that the repeated extraction from the same sample of resin would be a better measure of permanent leachables and even independent of the sampling conditions. Figure 4 shows an example of repeated analyses of used and new resins.

Table 4 *Effect of Storage on Release of Leachables*

Sample	A	B	C	D	D
Type	Gel	Gel	Macro	Macro	Macro
Age	New	New	New	Used	Used
Regeneration				After	Before
Leachables as received TOC (mg/kg DR)	150	136	28	12.0	0.5
Storage type Time (days)	Dry 29	Dry 166	Dry 36	Wet 180	Wet 180
Leachables after storage TOC (mg/kg DR)	220	629	41	12.8	2.6
Increase TOC (mg/kg day) (%/day)	2.4 1.6	3.0 2.2	0.4 1.3	0.004 0.04	0.01 2.3

It is obvious that two 24-hour extractions do not bring the resin in equilibrium independently of the sampling and pretreatment. However, the leaching seems to stabilize at reasonably uniform levels after 3 to 5 days of extraction. Sampling after regeneration still produces slightly higher results.

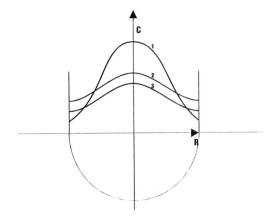

Figure 2 *Distribution of leachables within a resin bead*

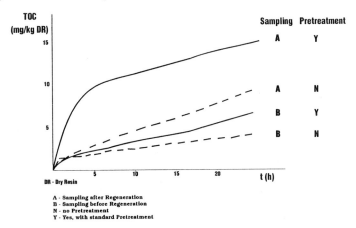

A - Sampling after Regeneration
B - Sampling before Regeneration
N - no Pretreatment
Y - Yes, with standard Pretreatment

Figure 3 *Kinetics of leachable extraction*

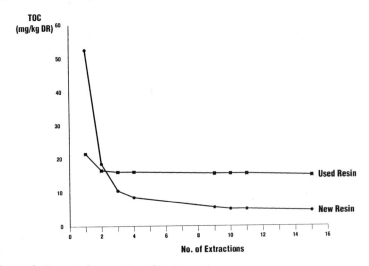

Figure 4 *Repeated extraction of cation resins*

3.3 Advanced Pretreatment

The assessment of resins on the basis of several repeated extractions is rather time consuming. An alternative way of bringing the resins into comparable conditions was considered.

The effect of an extended pretreatment was investigated. The resin samples were cycled between hydrogen and sodium form by short (10 min.) treatment with 10% HCl and 10% NaOH respectively. The effect of cycling was disappointing. Several possible reasons could contribute to the unexpected effect of cycling.

- The effect of cycling at ambient temperature might be negligible comparing to the effect of repeated extractions at 60°C.

- The contact time (10 min.) might be too short to develop the sponge effect.

- The sponge effect might be dependent on the contact time with high concentration reactant.

4 CONCLUSION

A method for assesment of ion exchange resins with respect to leachables is needed. Several methods have been developed to measure initial leachables. Measurement of permanent leachables is rendered complex by accumulation of leachables within beads and their postponed release. A method predicting the release of permanent leachables from new as well as used resins must involve pretreatment of a sample where a kind of equilibrium state is established.

A mathematical model of the extraction process is needed to ensure registration of the right properties and to make an extrapolation of laboratory data to the conditions of service.

References

1. D. C. Auerswald, "Ion Exchange for Industry", M. Streat, ed., Ellis Horwood Limited, Chichester, England, 1988, p. 11
2. J. R. Stahlbush and Co., *Proc. IWC-87*, Pittsburgh, 1987.
3. K. Daucik, *Proc. EPRI 4th Int. Conf. on Fossil Plant Cycle Chem.,* Atlanta, Sept., 1994.
4. C. C. Scheerer, R. B. Mitzel, *Proc. EPRI Int. Conf. on Fossil Plant Cycle Chem.,* Baltimore, 1991.
5. K. Wieck-Hansen, "Ion Exchange Advances" (Proceedings of IEX'92), M. J. Slater, ed., Elsevier Applied Science, New York, 1992, p. 128.
6. S. Fisher, G. Otten, *Proc. IWC-86-20*, Pittsburgh, 1986.
7. F. M. Cutler, *Ultrapure Water*, Sept., 1988.

AN ECONOMIC COMPARISON OF REVERSE OSMOSIS AND ION EXCHANGE IN EUROPE

P. A. Newell, S. P. Wrigley, P. Sehn and S. S. Whipple

Liquid Separations Technical Service and Development
The Dow Chemical Company
D-77834 Rheinmuenster
Germany

1 INTRODUCTION

The comparison between ion exchange and reverse osmosis membrane technology as options for water treatment applications has been the subject of a number of studies and continues to generate a high level of interest in the literature[1]. Most of these studies have been carried out in the USA with little reported in other regions. As technical developments in both product areas continue to be made, such as counter-flow regeneration packed bed systems, narrow particle sized ion exchange resins and high rejection, lower energy membranes, the need for monitoring economic performance remains. In addition, external factors such as water costs and disposal, power and chemical costs continue to change and are different around the world, further affecting the economics.

The purpose of this paper is to make a comprehensive economic comparison reflecting recent technical developments and using European costs and practices.

2 DESIGN BASIS

Four feed water salinities between 1.6 to 9.4 eq/m^3 (80 to 470 ppm as CaCO$_3$) were used, with a water analysis based on an averaged composition from a number of European sites. These waters cover the ranges found from the Nordic region (low TDS), through those in Central Europe (medium) to Southern Europe (high). The ratio of sodium to total cations was taken at 30% (the range was found to vary between 10 and 50%) and % alkalinity at 50% (range between 20 and 80%) for all salinities. Silica was set at 10 ppm throughout.

The chemical and utility costs in Table 1 were used to obtain a base case cost of producing water. The chemical and power costs were then changed in a sensitivity study to determine the effect on the overall economics of the different systems.

Two different types of water treatment systems were used : ion exchange only (IX) and reverse osmosis followed by ion exchange polishing mixed beds (RO-IX). The systems were sized to continuously produce mixed bed quality water (<1 μS/cm and 10 ppb SiO$_2$) at flow rates of 50 and 200 m^3/hour net. Operating costs include chemicals, power, labour and maintenance, together with water and waste water, which are an increasing consideration in water treatment economics[2,3]. Surface water was used for both size plants with pretreatment consisting of flocculation, clarification and sand filtration.

Acid and antiscalant were dosed prior to RO and 5 micron filters were used. The assumptions for the cost analysis are given in Table 1.

Water Analysis, eq/m^3				
Total	1.6 eq/m^3	3.2 eq/m^3	6.4 eq/m^3	9.6 eq/m^3
Ca	0.96	1.92	3.84	5.76
Mg	0.16	0.32	0.64	0.96
Na	0.48	0.96	1.92	2.88
HCO$_3$	0.80	1.60	3.20	4.80
SO$_4$	0.48	0.96	1.92	2.88
Cl	0.32	0.64	1.28	1.92
SiO$_2$ (ppm)	10	10	10	10
Temp (C)	15	15	15	15
pH	7	7	7	7

Costs :	Energy	6.5 p/kWh	Sulphuric acid	52 £/te
	Steam	9.5 £/te	Scale inhibitor	3080 £/te
	Caustic soda	195£/te 100%	Lime	65 £/te
	Feed water	4 p/m^3		

Deprecation of Capital	15 years
System Sizes	50 m^3/h and 200 m^3/h
System Operating Rate	360 days/year
Product Water Quality ex Mixed Bed	<10 ppb Na, <10 ppb SiO$_2$ and <0.1 µS/cm

Table 1 *Basis and Assumptions for Cost Analysis*

Identical water storage facilities were assumed for all systems, with 12 hour treated water storage for the 50 m^3/h case and 5 hours for the 200 m^3/h case. The cost of neutralising the waste water was included together with the disposal costs of the waste effluent.

Labour costs were not assumed to vary across the options studied and are based on 1.5 manyears at £18,000/manyear. Annual maintenance costs were estimated at 3% of the capital cost of the plant.

The capital number used includes the cost of an installed plant to an end-user but does not take into account the cost of land, buildings or taxes. The capital was depreciated linearly over 15 years.

Three bed ion exchange system: The basis for the IX plant design is given in Table 2.

The IX plant is a packed bed counter-flow regenerated design consisting of 2 x 100% streams with cation-degasser-layered bed anion-mixed bed polishers containing uniform particle sized resins. Details of layout, vessel sizing, resin volumes and operating conditions are given in Figure 1.

Operational Sequence		Specification
Pretreatment	Flocculation/clarifier	
	Sand filtration	
Demineralised Water Train	Cation resin unit	Uniform size (UPS) gel strong acid cation
	Degasifier	
	Layered bed anion unit	UPS weak MP/strong base gel anion
	Mixed bed	UPS gel strong acid cation/strong base anion
Demineralised Water Storage		
Waste Neutralisation		Neutralise to pH 7
Regeneration	Cation regeneration	Counterflow H_2SO_4 or HCl
	Anion regeneration	Counterflow NaOH
	Time between regens	24 hours - 1.6 eq/m³ case
		12 hours - 3.2 & 6.4 eq/m³ cases
		8 hours - 9.6 eq/m³ case
	Regeneration time	2 hours
Resin Life	Cation resin	8 years
	Anion resin	4 years

Table 2   *Bases and Assumptions for Cost Analysis for IX System*

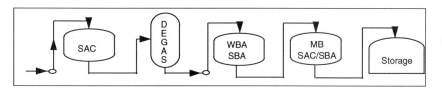

50 m³/h	**Cation**	**Anion Layered Bed**				**Mixed Bed**	
Ionic Load	SAC Volume liters	WBA Volume liters	SBA Volume liters	Cycle Time hours	SAC Volume liters	SBA Volume liters	Cycle Time hours
1.6 eq/m³	2725	1150	1175	24	575	850	450
3.2 eq/m³	2775	1075	750	12	575	850	450
6.4 eq/m³	5575	1750	1275	12	575	850	450
9.6 eq/m³	5775	1725	1250	8	575	850	450
200 m³/h							
Ionic Load	SAC Volume liters	WBA Volume liters	SBA Volume liters	Cycle Time hours	SAC Volume liters	SBA Volume liters	Cycle Time hours
1.6 eq/m³	10725	4550	4650	24	2100	3175	450
3.2 eq/m³	11025	4200	3075	12	2100	3175	450
6.4 eq/m³	22125	6800	5275	12	2100	3175	450
9.6 eq/m³	22925	6850	5000	8	2100	3175	450

Figure 1 *Layout for IX system*

Service run-lengths of 24, 12, 12 and 8 hours were taken for the four different salinities and reflect typical practices in the respective European regions. For the low salinity feeds, silica loading is the limiting factor for the anion sizing and a maximum of 10g SiO_2/L resin end-of-cycle was set. Warm caustic was also used for the 1.6 and 3.2 eq/m³ feeds due to the higher silica loads. Mixed bed sizing was set according to specific flow conditions and a silica limit of 1.5g/L resin and is constant for a given flow rate.

Regeneration with both 60g/L H_2SO_4 and 50g/L HCl were considered, but it was found that the cost per unit of treated water were similar for both acid regenerants, as the increased chemical efficiency of HCl and lower resin inventories are off-set by the higher cost of the chemical. Only the economics using H_2SO_4 are therefore reported.

Reverse osmosis/mixed bed system: The basis for the RO/IX plant design is given in Table 3.

Operational Sequence		Specification
System Pretreatment	Flocculation/clarifier	
	Sand filters	
RO Train Pretreatment		Acid & antiscalant addition
		5-Micron cartridge filter
RO Membranes	Type	Thin film composite, spiral wound
	Life	3 years
	Recovery	80% in two stages
	Feed pressure	14 Bar (200 psi)
Degasifier		
Ion Exchange Polishing	Mixed-bed	UPS Strong acid cation/Strong base anion - gel
Waste Neutralisation		Neutralise to pH 7

Table 3 *Bases and Assumptions of Cost Analysis for RO-IX System*

This consists of 1 x 100% line with RO-degasser-mixed bed polisher for 50 m³/hour and 2 x 50% lines for 200 m³/hour. A system recovery of 80% was used in every case. Details of layout, numbers of elements and operating conditions are given in Figure 2. The mixed bed design is the same as for the IX system: only the service run-length is shorter due to the higher expected average silica leakage from the RO compared to the layered bed anion.

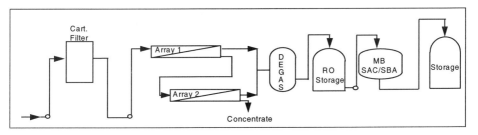

| **Thin Film Composite RO System** | | | | | **UPS Resin Mixed Beds** | | |
| **50 m³/h** | Array 1 | | Array 2 | | | | | |
Ionic Load	PV	Total Elements	PV	Total Elements	Feed Pressure Bar	SAC Volume liters	SBA Volume liters	Cycle Time hours
1.6 eq/m³	5	30	2	12	18	575	850	125
3.2 eq/m³	5	30	2	12	18.1	575	850	125
6.4 eq/m³	5	30	2	12	18.7	575	850	125
9.6 eq/m³	5	30	2	12	19	575	850	125
200 m³/h								
1.6 eq/m³	18	108	8	48	19.5	2100	3175	125
3.2 eq/m³	18	108	8	48	19.7	2100	3175	125
6.4 eq/m³	18	108	8	48	20.4	2100	3175	125
9.6 eq/m³	18	108	8	48	20.7	2100	3175	125

Figure 2 *Layout for RO-mixed bed ion plants*

Capital estimates for both systems at each salinity and flow rate are summarised in Table 4.

		M £ Sterling			
Feed Water (eq/m³)		**1.6**	**3.2**	**6.4**	**9.6**
IX Plant	50 m³/h	475	465	500	500
	200 m³/h	760	755	860	865
RO-IX Plant	50 m³/h	560	560	560	560
	200 m³/h	935	935	935	935

Table 4 *Purchased Equipment (Installed & Commissioned) Capital Estimates[6]*

These estimates are derived from the individual equipment component costs used to make up the plant. The capital cost of the IX plant increases with water salinity, due mainly to larger vessel sizes. The RO capital cost is higher than IX, due in part to pretreatment requirements, but is almost independent of water salinity.

3 RESULTS AND DISCUSSION

Base cases: The costs to produce water at each salinity and flow rate are given in Table 5 for the ion exchange base case and in Table 6 for RO/IX.

Cost of Treated Water p/m³	50 m³/h				200 m³/h			
Feed eq/m³	1.6	3.2	6.4	9.6	1.6	3.2	6.4	9.6
Operating Costs								
Chemicals								
Sulphuric Acid	0.7	1.4	2.9	4.5	0.7	1.4	2.8	4.4
Caustic Soda	1.2	2.1	3.9	5.8	1.2	2.1	3.8	5.7
Lime	0.3	0.6	1.1	1.7	0.3	0.6	1.1	1.7
Energy	1.0	1.0	1.1	1.2	1.9	1.9	2.0	2.2
Resin replacement	1.6	1.4	2.1	2.1	1.6	1.4	2.1	2.1
Raw water/Effluent	4.5	4.5	4.7	4.9	4.4	4.5	4.7	4.9
Labour	6.2	6.2	6.2	6.2	1.5	1.5	1.5	1.5
Maintenance	3.3	3.2	3.5	3.5	1.3	1.3	1.5	1.5
Depreciation (15 years)	11.1	10.9	11.7	11.7	4.5	4.4	5.0	5.1
Total Cost	29.9	31.4	37.1	41.6	17.4	19.2	24.7	29.1

Table 5 *Cost to Produce Water for IX System*

The results of the base case calculation given in Table 5 indicate that the cost to produce water using only ion exchange increases with feed TDS as expected, principally due to regenerant chemical costs. The effect of increasing plant size is to lower the cost to produce water as the regenerant costs increase proportionately but capital, raw water, labour and maintenance costs are relatively lower for the larger plant.

The costs for IX vary between 30-42 p/m³ at 50 m³/hour and 17-29 p/m³ at 200 m³/hour. At 50 m³/hour, operating costs account for ~70% of the total cost with regenerants, raw water, labour and maintenance making the most significant contributions. At 200 m³/hour, operating costs increase to ~80%.

Cost of Treated Water p/m³	50 m³/h				200 m³/h			
Feed eq/m³	1.6	3.2	6.4	9.6	1.6	3.2	6.4	9.6
Operating Costs								
Chemicals								
Sulphuric Acid	0.05	0.05	0.05	0.05	0.04	0.0	0.0	0.0
Caustic Soda	0.25	0.25	0.25	0.25	0.24	0.2	0.2	0.2
Lime	0.01	0.01	0.02	0.03	0.01	0.0	0.0	0.0
Antiscalant	0.7	1.1	1.9	1.9	0.7	1.1	1.9	1.9
Energy	5.1	5.1	5.2	5.3	5.1	5.1	5.2	5.3
Resin replacement	0.4	0.4	0.4	0.4	0.3	0.3	0.3	0.3
Membrane replacement	2.8	2.8	2.8	2.8	2.8	2.8	2.8	2.8
Raw water/Effluent	5.4	5.4	5.4	5.4	5.4	5.4	5.4	5.4
Labour	6.2	6.2	6.2	6.2	1.5	1.5	1.5	1.5
Maintenance	3.9	3.9	3.9	3.9	1.6	1.6	1.6	1.6
Depreciation (15 years)	13.1	13.1	13.1	13.1	5.5	5.5	5.5	5.5
Total Cost	37.9	38.3	39.1	39.2	23.3	23.7	24.6	24.6

Table 6 *Cost to Produce Water for RO-IX System*

Table 6 shows that the cost of producing water using RO/IX is also dependent on feed TDS, but much less so than for the IX system. This is due to the fact that the main cost contributors (power, water, labour, maintenance and capital) are relatively constant over the water salinity range considered.

RO/IX costs are around 38-39 p/m³ at 50 m³/hour and 23-25 p/m³ at 200 m³/hour. Operating costs are 72-80% of the total cost for the two plant sizes.

The total costs to produce water are plotted against feed salinity in Figure 3 for 50 m³/hour and in Figure 4 for the 200 m³/hour base cases.

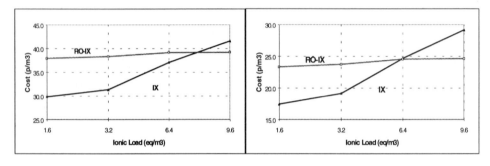

Figure 3   *Cost of treated water*   **Figure 4**   *Cost of treated water*
                50 m³/h plant - base case   *200 m³/h plant - base case*

The break-even point for the two technologies is at 7.9 eq/m³ TDS for the 50 m³/h case and 6.3 eq/m³ for 200 m³/h. It should be emphasised that these break-even points are based on the assumptions specified in the base case given in Table 1. A sensitivity study was therefore ran to assess the effect of the changes in the cost of water, power and chemicals on the economics.

Sensitivity study: The sensitivity of power costs is shown in Figure 5 for 50 m³/hour and Figure 6 for the 200 m³/hour case.

Power costs were varied over the range 3.3-9.8p/kWh (base case = 6.5p/kWh). For the RO/IX system, this resulted in a change in the cost to produce water of +/-2.5p/m³ compared to the base case, thereby affecting the break-even point with IX at 50 m³/h to 7.9+/-1.5 eq/m³ and at 200 m³/h to 6.3+/-1.2 eq/m³.

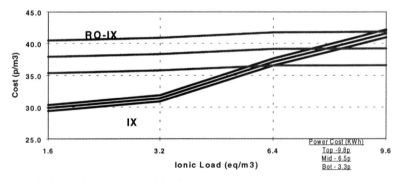

Figure 5   *Cost of treated water 50 m³/h - power sensitivity*

The sensitivity of caustic regenerant price on IX economics is shown in Figure 7 for the 50 m³/hour plant.

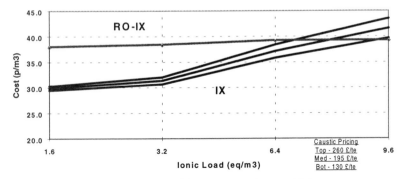

Figure 6 *Cost of treated water 50 m³/h - caustic price sensitivity*

The effect on RO/IX economics is minimal. Over the range 130-260 £/ton 100% NaOH, the break-even point for a 50 m³/h plant ranges from 6.9-9.3 eq/m³.

Finally the effect of raw water/effluent cost is shown in Figure 7. The costs range from the lowest - that of a surface water intake with little water loss - to an inexpensive borehole water.

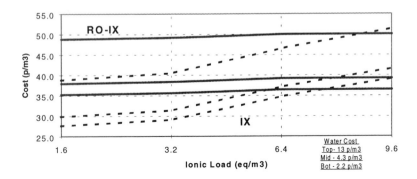

Figure 7 *Cost of treated water for 50 m³/h - cost of water sensitivity*

Over the water cost range taken the breakeven point for a 50 m³/h plant ranges from 7.6-8.7 eq/m³. If, however, mains water is taken (e.g. 50 p/m³) or the cost of effluent treatment is expensive, the costs of RO/IX vs IX increase markedly and no break-even point is found under 10 eq/m³.

Direct comparison of the present results to existing literature data is difficult due to the many different assumptions and parameters used in each study. However, it is possible to make some qualitative comparisons. The base case salinity break-even points of the present study (6.4-8.0 eq/m³) are substantially higher than the 2.6-3 eq/m³ (125-150 ppm as $CaCO_3$) values obtained in previous USA studies[1,4] and goes against the trend of declining salinity break-even point observed in recent years[5]. This shift to higher salinity can be explained by the economic advantages of the packed bed counter-flow regenerated system over co-flow systems and the higher power costs in Europe compared with the USA, both of which are unfavourable to RO. A similar conclusion was made in a European study where a break-even point of ~10 eq/m^3 was found[3].

4 CONCLUSIONS

This economic evaluation considers the major factors contributing to the total cost of treated water by RO/IX and IX alone in Europe. The effect of system size and the latest technology in both resins and membranes has been included. Major conclusions that can be drawn from this study apply to new water treatment plants and are summarised below:-
1. The break-even point above which it is more economical to use RO/IX versus IX alone is 6.3-7.9 eq/m^3 (315-395 ppm as $CaCO_3$ TDS). This is significantly higher than comparable USA studies and reflects developments in packed bed counter-flow regenerated IX systems and higher power costs in Europe.
2. The break-even points derived in this study represents water salinities that are commonly encountered in Europe. For many new plant projects a clear choice of one technology often cannot therefore be made on purely economical grounds but other considerations and local factors must be taken into account, such as water availability, type and disposal regulations, existing installed capital and end-user technology preference or familiarity.
3. Although capital has a significant effect on the total cost of water for all options considered, operating costs represent the major portion at 70-80% of the total.
4. This study considers mainly surface water and low cost discharge of effluent from the water treatment plant into a river. Unless the concentrate from an RO plant can be used elsewhere on site, increasing water costs and disposal costs will very likely penalise RO over ion-exchange.
5. Chemical costs for IX and electrical power costs for RO are the most important operating expenses and those that need to be carefully considered in the decision for a new plant.
6. This study has identified a number of factors that impact the economics and hence market position of IX versus RO technology now and in the future.

Relative familiarity with IX technology combined with the economics shown in this paper could limit the growth of RO technology in industrial water applications. The development of lower energy membrane modules and a trend away from the use of regenerant chemicals could change this scenario.

References

1. See for example, A. F. Ashoff, *UltraPure Water*, July/August 1995, p. 39.
2. VGB-Kraftwerkstechnik GmbH literature, May 1995.
3. K. Grethe and C. Beltle, "Power Station Make-up Water using RO and Ion Exchange for Demineralisation" Steinmuellertagung, 1993.
4. S. Beardsley, S. Coker and S. Whipple, *Watertech Expo '94*, 9 Nov 1994.
5. S. Whipple, E. Ebach and S. Beardsley, *UltraPure Water*, 1987, October.
6. Capital Estimates provided by Satec Ltd., Memcore Ltd., ERG Plc. and Steinmüller GmbH.

THE REVERSIBLE REMOVAL OF NATURALLY OCCURRING ORGANICS USING SODIUM CHLORIDE REGENERATED ION EXCHANGE RESINS : PART 2

M. C. Gottlieb

ResinTech
Cherry Hill, NJ
USA

Ion exchange is the only mechanism involved for the removal of the vast majority of naturally occurring total organic carbon (TOC) substances. The exception is those organics with low molecular weights. The concept of an exclusive ion exchange mechanism where non-ionized organics are not removed by the resin bed, (regardless of the resin's porosity), was proven for the first time by Symons [4].

His work shows that ion exchange kinetics and equilibrium relationships are the important areas to be investigated when studying the reversible exchange. Resin characteristics must also be defined and correlated because they affect the kinetics and equilibrium.

In Part 1 of this series, the major emphasis was on the equilibrium relationships which were used to explain laboratory test results, which indicated the kinetic factors were at least as important. Here, in Part 2, these relationships are explored both with laboratory data and with data from a full scale field trial of a product, SIR-22P designed for exchanging organic ions.

Naturally-occurring organics have very slow diffusion rates. They are weak acids, and therefore, their ionic nature is affected by pH, temperature and concentration. Since they have slow diffusion rates, contact time, especially during regeneration, is known to be an important factor, equally as important as the regeneration dose.

The higher the relative amount of water in the gel phase of the resin, the more rapid its kinetics will be and the lower the relative affinities for the various ions. Both these changes enhance the resin's performance in exchanging and being stripped of the organic ions. All of the various natural organics are similar in that they are based on multiple benzene rings and have significant carboxylic acid functionality.

Calculations based on Symons[4] data show there are, on average, between 1 and 2 carboxylic groups per benzene ring (**Figure 9**). Several other studies of these materials have shown that the carboxylic content of humic substances varies usually between 3 to 12 milliequivalents per dry gram which is similar. We conducted similar tests on a purified tannic acid which was used in our column experiments. This material has a nominal molecular weight of 1,750. Our results were in close agreement with Symons. Our column tests on this material are in agreement with the data from the full scale trial on a natural water.

Symons[4] data clearly shows that the macroporous resins do not perform any better than gel resins in removing these substances. This had also been shown by others [1, 6, 7]. Even for the very high molecular weights (above 10,000) the gel phase water retention was the only reliable predictor of removal performance. The gel phase water retention is controlled by the crosslinker level for a given polymer system.

Sulfate ions being di-valent offer significant competition against the TOC substances for the ion exchange sites. Some of the lower molecular weight organic matter (less than 5,000) has been shown to be less preferred than sulfates and are "dumped" off when runs extend beyond the sulfate exhaustion and reached concentrations higher than the inlet concentrations. A substantial portion of the less than 1,000 MW portion is apparently non-ionic and not removed by resins; this can be removed by activated carbon.

Equilibrium relationships are important to the regeneration process. The potential efficiency of regeneration can be affected greatly by changing the salt concentration which affects the separation factor.

Table 1 shows the selectivity coefficients of the various ions for some gelular anion resins. Those substances with the higher selectivities are more difficult to regenerate from the resin. These resins are all more selective for the chloride ion than for hydroxide, so they are more efficiently regenerated by sodium chloride than by sodium hydroxide. For the standard Type 1's, the relative affinity of the chloride is from 11 (ResinTech SBG1P) to 25 (ResinTech SBG1) times higher than the hydroxide ion.

The separation factor, which is the same as the selectivity coefficient for monovalent exchange, determines the potential efficiency of a regenerant. In the case of multivalent ions, such as sulfates and naturally occurring humic substances, the separation factor, during regeneration, is related to the "apparent selectivity coefficient[8]." The apparent selectivity coefficient varies with the total ionic concentration and resin capacity.

Table 1   *Relative Affinities (OH=1)*

	Type 1 Standard Gel	Type 1 Porous Gel	Type 2 Standard Gel
Lignosulfonate	800	400	120
Benzenesulfonate	500	253	75
Salicylate	450	358	65
Citrate	220	110	23
Iodine	175	97	17
Nitrate	65	42	8
Bromide	50	31.1	6
Chloride	22	11.1	3
Bicarbonate	6	3.6	2.3
Hydroxide	1	1	1

In regenerant concentrations, it is similar to the selectivity coefficients, therefore, the selectivity coefficients can serve as quality indicators for regenerant ions. Aside from hydroxide selectivity, Type 1 and Type 2 resins exhibit similar selectivity relationships between the other ions. Therefore, sodium chloride regeneration of organics laden anion resins is as effective for Type 1 resins as for Type 2. This can best be seen in **Figure 1**

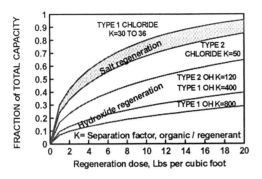

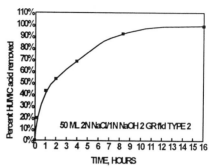

Figure 1[9] *Theoretical regeneration of an organic substance*

Figure 2 *Time vs Humic acid removed*

which shows the theoretical conversion versus regenerant dose levels for chloride and hydroxide regenerations. This gives a good picture of the effect on regeneration efficiency of the relative selectivity. The curves in **Figure 1** are based on the assumption that enough contact time has been given for the regenerant and resin to reach equilibrium. In **Figure 1** we see the porous Type 1 gel resins have the potential to remove TOC substances and be well regenerated with chloride regenerants. However, regenerant contact time is a very important factor in the brine regeneration of an organic laden resin.

When brine injection times of thirty minutes were used for the regeneration of organic-laden resins, it was shown that the same brine could be used over and over, up to 9 times, without loss of capacity or quality in TOC reduction in subsequent cycles. This is due to the slow diffusion rates of the TOC substances, the low rate of build-up and non impairment of the regenerant solution's effectiveness. This shows the regeneration is kinetically limited. It is obvious that much longer relative regeneration times are needed for organic ions.

It has been shown that the equilibration times for aquatic organic matter is best measured in hours[5] compared to minutes for inorganic ions. For example, Wilson[5] studied Type 2 resins that had been fouled. He soaked them in brine/caustic solutions. He studied the amount of organics removed from the resin over time. Over 16 hours was required for what only appeared to be equilibration. It took only 4-6 hours to reach 70-80% of the 16 hour amount. A portion of this data is shown in **Figure 2**.

Using time as a parameter, we should be able to project regenerant effectiveness and predict operating capacities in a similar manner to the capacity versus the regenerant level curves used in inorganic ion exchange. The impact of individual resin characteristics on diffusion rates is quite significant, so it will be necessary to establish separate curves for each type of resin.

Figure 3 shows the cumulative elution of TOC from humic and tannic acid laden SIR-22P. The resins were exhausted in column experiments with solutions of laboratory grade tannic and humic acids. The bed grade tannic and humic acids. The bed volumes were 380 mL and the regenerant flow rate was 5.1 mL per minute in all cases.

There was near linear and rapid rate of removal during the first 100 minutes, (1.3 BV). The elution rate then dropped rapidly by a factor of almost 30 and remained fairly steady over the next several hours. After about 8 hours of continuous elution, the rate of elution had dropped by another 50%. This was generally the same for tannic acid and humic acid, regardless of whether salt or salt and caustic regeneration was used. The

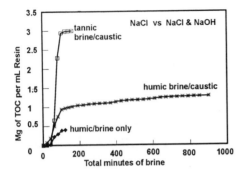

Figure 3 *Regeneration of SIR-22P*

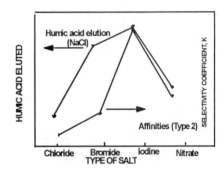

Figure 4 *Humic acid removal &*
selectivity coefficients

elutions were run at 21°C (70°F). No attempt was made to use other temperatures. This is planned for future work.

The rate of TOC elution during the first 100 minutes from a sample of SIR-22P, taken from the field trial at Cape Hatteras, after 49 cycles, averaged between 0.04 and 0.05 mg of TOC as C per mL of resin per minute. Under identical elution conditions, the rate of TOC elution from the SIR-22P resin that had been loaded in our laboratory with tannic acid was 0.045 mg of TOC/mL of resin/min. These values are remarkably similar.

Even through the absolute elution numbers are different for different regenerants and substances, the relative changes are virtually identical in all cases. It is obvious that time was more important than volume in these elution experiments. About 75% of the total elution (16 hour basis) occurs in the first 100 minutes. Elution rate remains fairly linear during this period. It is logical, therefore, to choose this as the regenerant contact time at ambient temperatures.

Figure 4 compares the known relative selectivity coefficients of various anions with the relative organic elution from organically fouled Type 2 resins. Indirectly, this shows the importance of proper concentration control in order to enhance, or in this case, reduce the apparent selectivity of the multivalent TOC substances during regeneration by using high regenerant concentrations.

Loading tests with varying amounts of sulfate ions showed a definite, but not quite straightforward, effect. Sulfate being di-valent is able to compete for the resin sites. As sulfate exchanges onto the resin, the sites available to the TOC ions are reduced which increases the length of the exchange zone. Since the exchange zone is already longer for the organic ions, the result is that the TOC leakage increases before the sulfate does. **Figure 5** and **Figure 6** show this effect.

The Preuss factor named after its discoverer, Albert F. Preuss, by co-worker, Sally Fisher[2], was used to rate the relative fouling resistance of anion resins in the early 50's when researchers were looking for ways to define organic fouling and to rate fouling resistance in strong base anion resins. It was also applied to uranium recovery. The Preuss factor is the molar ratio of gel phase water to ion exchange capacity. It was found to be an excellent predictor of performance within resin types; i.e. Type 1 or Type 2.

At first, in attempts to quantify fouling and organic/resin interactions some researchers tried to equate ion exchange intermolecular distances to equivalent gel phase porosity[3]. It was frustrating to do that because of the wide variation in the shapes of the organic molecules. The Preuss factor describes the relative size of the lake the organics have to move in as they move to and from ion exchange sites, inside the gel phase of the resin.

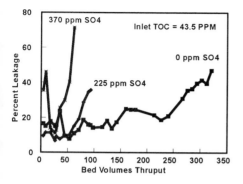

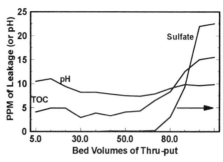

Figure 5 *TOC leakage vs sulfate leakage*

Figure 6 *Leakage from SIR-22P Effect of sulfate breakthrough*

Figure 6 shows the relative order of TOC reduction abilities measured by Symons, compared with the calculated Preuss factors of the resins for the 4 molecular weight ranges that were studied. His tests included gel and macroporous types, styrenics and acrylics. **Figure 6** shows that from below 1,000 to over 10,000 molecular weight, the Preuss factor provides an excellent correlation of increasing absorption capabilities (less leakage). **Table 2** shows the molecular weight of natural organics from two sources.

Table 2 *Molecular Weight Distribution of TOC (Per Cent)*

Molecular Weight		Lake Houston Texas (Symons)	Lake Anna Virginia Tech. (Wiser [10] et. al.)	Tannic Acid Grade ResinTech
+100	K		11.5	
10 - 100	K		10.1	
10 +	K		10.9	
5 - 10	K	28.0		
1 - 10	K		37.6	100%
1 - 5	K	9.5		
100 - 1	K	51.6		
500 - 1	K		17.0	
150 - 500	K	23.8		

The idea that the bigger the lake each ion exchange group has surrounding it, the more effective it is in removing the various ions, holds up well. There can be no doubt that organic substances are considerably larger and diffuse more slowly than the common inorganic ions and require higher column heights or lower flow rates to be exchanged to the same degree as the inorganic ions. If we consider the ratio of gel phase water per ion exchange site, the ion exchange lake, then when these substances fill that lake, diffusion becomes hindered and slowed. The higher the water content of the gel phase, the more of the physical load the resin can handle. Standard grade gel resins offered today, regardless of the supplier, have similar capacity and water retention values. The values used in **Figure 7** are calculated from midpoint values for typical gel resins plus ResinTech SIR-22P, (a special purpose resin intended for exchanging organic ions).

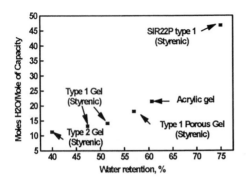

Figure 7 *Water retention vs relative water for strongly basic gel anion resins*

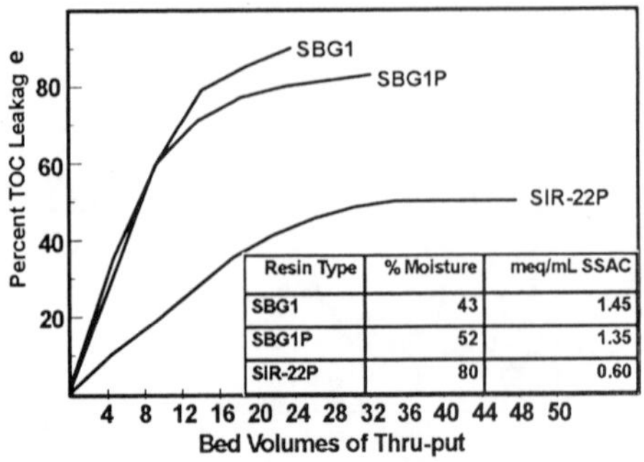

Figure 8 *Comparison of leakage based on % water Humic Acid*

Interestingly, the Type 2 gel resin has the lowest and the acrylic resin has the highest Preuss factors of the standard gel resins. You would, therefore, expect the acrylic resins to be able to remove the most organics and Type 2 to remove the least amount of organics of the common gel resins. Both of these products are acclaimed for good organic fouling resistance. For organic fouling resistance, regeneration efficiency is the key factor, not necessarily removal. The acrylic gel resins do remove the most organics of all the common strong base gel anion resins. Type 2's do not foul because they let a significant portion of the organics slip by.

Figure 8 shows the TOC leakage during the column loading step for SIR-22P and two other styrenic Type 1 gel resins. These resins differ primarily in the crosslinker levels in the polymers. This results in different levels of gel phase porosity which can be seen indirectly by the differences in the water retention values. The columns were loaded at a rate of 0.14 BV per hour. The feed solution contained 43.5 ppm of TOC from purified humic acid dissolved in low TOC demineralized water.

It can be seen that the exchange is limited by diffusion of two related facts. The resins with the highest porosity, i.e. higher water retention, removed the most and lasted the longest. The resins with the highest static capacities removed the least and had the lowest dynamic capacities. It is worth mentioning that increasing the gel phase porosity increases water retention and reduces the total volumetric capacity. It is also apparent that SIR-22P performed significantly better than the standard resins.

Since the ability to elute the organics is a key factor in fouling resistance, so is the relative affinity between the organic ion and the elution ion. Symons[4] determined the carboxylic acid content of the TOC. **Table 3** and **Figure 9** shows the calculated calcium carbonate to TOC ratio from the Symons' data together with the theoretical values for various aromatic carboxylic acids.

The higher molecular weight organics have a lower ratio of acidity. This is consistent with the theory of adding functional groups by biological and photo chemical oxidation that was discussed in Part One of this series. The calcium carbonate equivalency ranged from 0.55 to 0.75 ppm of $CaCO_3$ per ppm of TOC as the molecular weight lowered from above 10,000 to 1,000. The trend appears to not exceed the limiting value of a dicarboxylic acid of approximately 1.05. We can now use these values to estimate the equivalent capacity load of TOC values in the same manner as for inorganic ions. We obtained a value of 0.67 meq/mg of C for tannic acid. This fits well with Symons' data. The humic acid results were much different and not reported because of the lack of confidence in the test results.

Figure 9 *CaCO₃ Equivalency to TOC. Calculated from the carboxylic content*

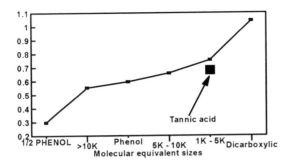

Table 3   Table 4

Table 3	Table 4
Ionic Strengths of TOC	*CAPE HATTERAS DATA*
Calculated from Symons' Data of Carboxylic	
Content of TOC Values in Lake Houston Water	

Table 3

M.W.		meq of COOH per gram of TOC	Equivalency ppm as CaCO₃ per ppm TOC
10	K	11.03	.55
1-10	K	13.13	.66 (.67)*
1-5	K	15.05	.75
< 1	K	ND	

*Calculated from Tech. Grade Tannic acid with nominal MW of 1,750

Table 4

	SIR 22P New	After 49 Cycles	
		*	**
Weak Base Cap. Meq/mL	.03	.05	.05
Strong Base Cap. Meq/mL	.55	.43	.52
Water Retention (Cl) Per cent	.82	78.6	80
Bead Integrity			
Perfect	99		
Cracked	1	*1 hr. contact	
Broken	0	**2 hr. contact	
Size Range Microns	400 - 600	brine caustic	

Table 4 shows that the resin from the field trial retained essentially all its available capacity after 20,000 bed volumes, (150,000 gal./cu.ft.). The inlet water contained an average 320 mg/L of HCO_3, 275 mg/L of Cl, and 35 mg/L of SO_4, all as $CaCO_3$ and 22 mg/L of TOC as C. The effluent TOC remained consistently at, or below, 4 mg/L throughout the service runs. The resin bed contained 1.1 cubic meters, (38 cubic feet).

The flow rate was maintained at 0.14 BV/hr (1.05 gpm/ft³). The service runs were terminated by preset volume at 280 BV in order to insure constant low TOC values at 4 ppm or less. On several occasions the beds were allowed to run until TOC leakage reached 50%, run lengths under these conditions reached over 450 BV.

We have shown that for maximum reduction of organics, it is best to limit the service cycle by sulfate leakage and polish with activated carbon to remove non ionic matter if total TOC removal is required. We recommend a minimum one and a half hour regeneration injection time. The regenerant solution should be a 10% NaCl solution with an additional 2% sodium hydroxide, preferably applied at a temperature of 120°F, but not less than 70°F. If the regenerant and resin are too cold, the necessary contact time will increase significantly. Therefore, it is suggested that a minimum temperature of 20°C up to the design temperature of the equipment or resin be used for best results.

If only partial TOC reduction is needed or maximum throughput is desired, the service cycle can run past the sulfate break. Once sulfate breakthrough occurs, the resin will continue to remove organics. However, all or most of the organics with lower selectivities than sulfate, will be discharged as they are displaced by sulfates and the more strongly held TOC's. During the later stage of the exhaustion cycle, after the sulfate break, the overall removal of organics may drop to as low as 30%. After the sulfate breakthrough, TOC levels will rise and the concentration of the lower molecular weight TOC's can rise above their influent levels.

SUMMARY

The properties that would make an efficient organic scavenger are:

A.   Sufficient gel phase water to physically accommodate the organic matter

B.   Sufficient ion exchange functionality to provide acceptable capacity

C.   Excellent osmotic shock characteristics to withstand the internal stresses during loading and regeneration

The knowledge necessary to design an organic scavenger are:

A.   Complete water analysis including TOC and inorganics

B.   Approximate molecular weight of the TOC substances

C.   Maximum leakage of TOC desired

D.   Maximum amount of organics that can be loaded (equal to the maximum amount eluted-calculated from TOC elution rate and the contact time during regeneration)

A full scale application of the resin cited in this study is discussed in a paper by Meyers[11].

References

1.   R. E. Anderson, *J. Chromatography*, 1980, 201.
2.   S. Fisher, Personal Communication, August, 1993.
3.   R. Kunin, "Ion Exchange Resin", Robert Krieger Publishing Co., 1972.
4.   J. M. Symons, P. L-K. Fu, P. H-S. Kim, *International Water Conference Proceedings Book*, Pittsburgh, PA, October, 1992, 92.
5.   A. L. Wilson, *J. Appl. Chem.*, 1959, **9**, 352.
6.   J. J. Wolff, Dia-prosium Bulletin, 1968, 68.C.2.
7.   J. J. Wolff, Dia-prosium Bulletin, 1968, 68.C.3.A.
8.   M. Gottlieb, F. X. McGarvey and S. Ziarkowski, Liberty Bell Corrosion Course Four, Philadelphia, PA, September, 1980.
9.   M. Gottlieb, *Industrial Water Treatment*, 1995, **27**, No. 3, May, 1995.
10.   S. L. Wiser, Y. H. Lee, C. R. Stroh and C. R. O'Brien, *International Water Conference Proceedings Book*, Pittsburgh, PA, November, 1985, 1.
11.   P. Meyers, International Water Conference Proceedings, IWC95-1, Pittsburgh, PA, Oct., 1995.

IMPROVEMENTS IN POWDERED ION EXCHANGE TECHNOLOGY FOR INDUSTRIAL APPLICATIONS

J. M. Ragosta

Graver Chemical
200 Lake Drive
Glasgow, DE 19702
USA

1 INTRODUCTION

Attendees of this conference are well aware of the benefits and problems associated with the use of ion exchange resins for purification of process streams. While modern ion exchange resins have been greatly improved by their manufacturers, there is still a need for improvements in the use of these products. For specialty applications such as pharmaceutical process, waste water treatment, and chemical processing, improvements are needed.

There are many applications where dissolved solids and suspended solids are present at the same time and where both must be removed. Among the common applications are metal finishing waste water treatment, groundwater treatment, sugar processing, pharmaceutical processing, and many others. In many of these applications, the use of a two step purification process, with filtration followed by ion exchange, is impractical or, at least, very expensive.

Ion exchange has another inherent limitation. To obtain good kinetics, small particle sizes are needed. Unfortunately, these small particles have poor hydraulic performance and use in a column is not practical for process flow rates. While it is possible to use fine mesh adsorbents in a batch adsorption process, the efficiency is poor.

2 DESCRIPTION OF TECHNOLOGY

We have developed a technology which combines filtration with adsorption in a single step. Based on our laboratory work and an understanding of adsorption and filtration, we have concluded that this process can also show better hydraulics than a conventional precoat filter and better adsorption performance than a conventional deep bed adsorption process.

The starting point for understanding this technology is to consider a physical mixture between a filter aid such as fiber or diatomaceous earth and a finely powdered adsorbent. This mixture would functionally act as a mixed filtration / adsorption process. The problem is that this mixture has very poor hydraulics. As shown in Table 1, a physical mixture of filter aid and adsorbent has only slightly better hydraulics than the powdered adsorbent alone. The flow rate and pressure drop through this mixture would be unacceptable.

Table 1   *Hydraulic Performance of Precoats*

Material	Relative Flow Rate
Powdered Ion Exchange Resin	1
Powdered Ion Exchange Resin plus Coarse DE (physical mixture)	2
Flocculated Powdered Ion Exchange Resin (Ecosorb®)	10 - 1,000

The solution to this dilemma is to flocculate the adsorbent and filter aid together.[1] By doing so, the permeability is increased many fold. This allows reasonable flow rates with acceptable surface areas.   Figure 1 shows the mechanism for this behavior—by flocculating the adsorbent and filter aid, the bed porosity is increased dramatically. Since the permeability is related to the fourth power of bed porosity, an increase in porosity has a great effect on flow rate.   Because the kinetics are limited by particle size of the adsorbent, the adsorption kinetics are excellent for all three examples in Figure 1.

Figure 1   *Ecosorb® hydraulic behavior*

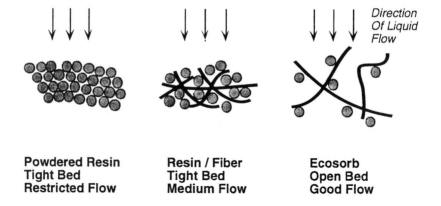

Powdered Resin	Resin / Fiber	Ecosorb
Tight Bed	**Tight Bed**	**Open Bed**
Restricted Flow	**Medium Flow**	**Good Flow**

Direction Of Liquid Flow

This process is very flexible and allows a wide range of filter aids and adsorbents to be used. While there is no reason to believe that many others could not be used, Table 2 lists some of the materials which have been used. In addition to excellent hydraulic properties, the Ecosorb® product is usually easier to handle than the constituents. For example, powdered activated carbon (PAC) is especially difficult to handle because of its tendency to generate dust. Flocculated precoats can be made from PAC which contain enough moisture to prevent dusting. These flocculated precoats can be formulated in a manner which complies with FDA and BATF regulations for food and beverage contact.

Table 2 *Ecosorb® Precoat Compositions*

<u>Adsorbents</u>	<u>Filter Aids</u>
Powdered Activated Carbon	Fiber (all types)
Zeolite Powders	DE
Alumina	
Silica	
Powdered Ion Exchange Resins	
-cation	
-anion	
-chelating	
Proprietary Adsorbents	
Fuller's Earth	

The new product is used in the same equipment as conventional precoats. Laboratory and field experience show that it can be applied to candle filters, plate and frame filters, leaf filters, and filter bags. Surprisingly, the product can be applied to a sand or other media filter, as well. Because of the excellent hydraulics, the material can be applied as a deep bed, with "precoat" beds as deep as 60" being used.[2] This results in what amounts to a deep bed of very fine mesh (as small as 1 micron) powdered adsorbent. In one case, a customer is using powdered activated carbon for a chromatographic purification—with 36" deep beds. Obviously, the pressure drop across a 36" deep bed of normal powdered carbon would be excessive.

3 APPLICATIONS

Because of the unique properties of this product, it both excellent permeability and filtration. For some applications, such as removal of particulates from feed water, the precoat is capable of removing even colloidal species (Figure 2). This is due to the ability of the precoat to adsorb undesirable species as well as to filter them. Even so, we normally recommend prefiltration to remove most solids before treatment with the adsorbent. This will extend run length and improve economics.

Figure 2 *Removal of particulates*

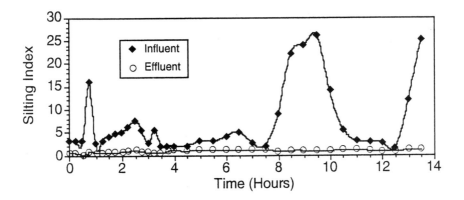

As an adsorbent, this technology allows the use of finely divided adsorbents in column mode rather than the typical batch mode. For example, powdered activated carbon has been used for many years in purification of sugar solutions. Normally, the PAC and a filter aid are mixed with the sugar in a batch mode and then filtered out. This process does not take full advantage of the adsorbent's capabilities. When the same materials are converted to an Ecosorb® formulation and precoated, the adsorbent utilization is greatly improved (Figure 3).

Figure 3 *Batch vs column performance*

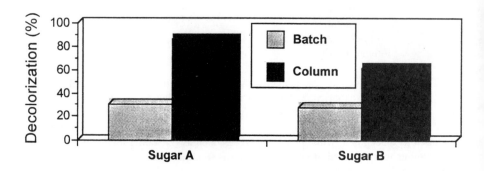

Finally, this new formulation can be used where conventional granular adsorbents (such as ion exchange) are not practical because of slow adsorption kinetics. When the same adsorbent is finely powdered and properly flocculated, the product has hydraulics comparable to bead ion exchange resins, but greatly improved kinetics. Figures 4 and 5 show comparisons between conventional weakly acidic ion exchange resin and an Ecosorb® formulation made from the same product after particle size reduction. In Figure 4, Cytochrome C is removed from a solution containing 900 ppm. The results of stripping Cytochrome C from the loaded adsorbent are shown in Figure 5. The adsorption efficiency and regeneration efficiency are both much greater for the new formulation.

Figure 4 *Cytochrome C recovery*

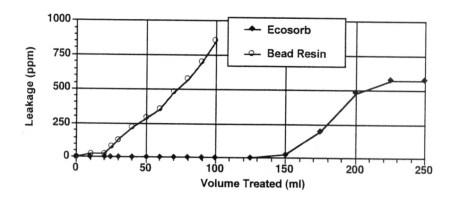

Figure 5 *Cytochrome C regeneration*

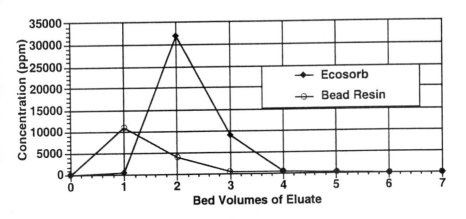

Similarly, Vitamin B12 was recovered from solution as shown in Figure 6. Figure 7 shows the results of regeneration of the adsorbents.

Figure 6 *Vitamin B12 recovery*

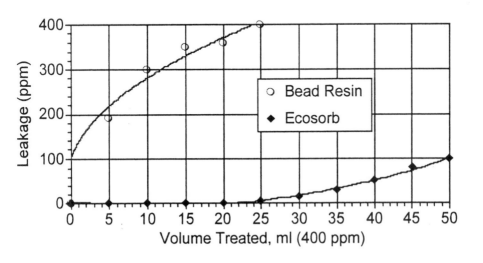

Figure 7 *Vitamin B12 regeneration*

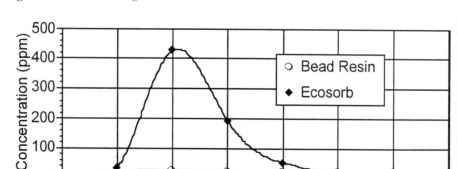

Another common application for Ecosorb® is waste water treatment. In particular, the product is used as a polishing adsorbent after conventional precipitation / clarification processes. In this case, influent metal concentrations are typically around 0.1 to 1 mg / L. Adsorbent performance is shown in Table 3.[3] This performance is considerably better than would be achieved with a bead form ion exchange resin, particularly considering the short bed depth used.

In addition to heavy metal removal from waste water, Ecosorb® has been used to remove phenol and chlorinated organics from water.[3]

Table 3   *Waste Water Treatment Application*

Ion	Influent Concentration (mg / L)	Effluent Concentration (mg / L)
Cd^{2+}	0.18	< 0.05
Cr^{3+}	0.17	< 0.05
Co^{2+}	0.20	< 0.05
Cu^{2+}	0.20	< 0.05
Ni^{2+}	0.20	0.08
Zn^{2+}	0.20	0.05
Pb^{2+}	0.20	< 0.05

Like all technologies, this one has limitations. The major one is that it is most useful in polishing applications where traces of contaminants must be removed. For contaminant concentrations in the low ppm or ppb range, it is especially useful. The product may be useful, however, at high concentration levels when the product being purified is very valuable. Thus, the two major applications are waste water treatment, where low concentrations of contaminants are being removed, and pharmaceutical processing where the purified product is valuable. Sugar processing is also very economical, due to the large amount of labor and materials required to precoat the filters several times per shift with powdered activated carbon and only once with Ecosorb®.

The current flocculated product is normally not regenerated. This limitation accounts for the preference for using this product to remove low concentrations of contaminants as the run length would be short if high concentrations of contaminants were removed. There are, however, applications where the product can be regenerated one or more times. Finally, the product is best used on streams with moderate turbidity. If high turbidity is present, the stream should first be treated with a coarse precoat or non-precoat filter to remove the bulk of the solids with Ecosorb® being used as a polisher.

Even with these limitations, this product has found significant use in applications which require the removal of traces of soluble and insoluble contaminants. Since two unit operations are combined into one, the process is quite inexpensive, especially for high flow rate streams.

There are a large number of potential applications for this product. In general, applications where conventional adsorbents are not efficient enough are ideal candidates. This technology allows the use of finer particles without suffering from excessive pressure drop. A few commercial applications are listed in Table 4.

Table 4   *Some Current Ecosorb® Applications*

Waste Water Polishing
-removal of heavy metals

Sugar Purification
-decolorization
-calcium removal
-odor removal

Pharmaceutical Purification
-decolorization
-removal of specific contaminants

Industrial and Potable Water Pretreatment
-removal of heavy metals
-removal of toxic organics, including halomethanes

4 CONCLUSION

We have developed a unique formulation which allows the use of finely divided ion exchange resins and other adsorbents in deep beds and in thin layers. This formulation provides the benefits of fine mesh resins (fast kinetics, high adsorbent utilization) and bead resins (excellent hydraulics). For polishing applications, the economics are much better than conventional adsorbents, as well.

References

1.   C. J. Halbfoster, U.S. Patent 4,238,334, Dec. 9, 1980; other U.S. patents.
2.   E. Salem, B. L. Libutti, R. Kunin, U.S. Patent 5,022,997, Jun. 11, 1991.
3.   G. Wilber, R. Kunin, *Reactive Polymers*, 1986, **4**, 71.

EFFECT OF THE PROPERTIES OF WEAKLY BASIC ANION EXCHANGERS ON THE DEIONIZATION PROCESS

F. X. McGarvey

Sybron Chemicals Inc.
Birmingham
New Jersey, USA

1 INTRODUCTION

Weakly basic anion exchange resins have been used for strong acid removal, i.e. sulfates, chlorides and nitrates, in standard water treatment. Only a few applications have been developed, where their capacity for weak acid removal has been employed. Kunin and Vassiliou[1] developed a desalting process which employed weakly basic anion exchange in the bicarbonate form to convert chlorides and sulfates to bicarbonate salts which could be removed by weakly acidic cation exchange beds. This process developed in the mid 1960's received considerable attention as the DESAL process. The use of reverse osmosis ended this development.

In recent years Rodgers, Tucker and Mommaerts[2] reported on the use of a weakly basic anion exchanger after a weak acid bed to assist in the deionization of water for the Canadian Oil recovery project at Syncrude Canada Ltd. The main purpose of this resin was to reduce organic fouling and improve the efficiency of salt removal.

2 MECHANISM

Weakly basic anion exchange resins remove acidic substances by interaction which involves their reaction with water. One concept assumes a simple acid adsorption mechanism.

$$RNH_2 + HX \rightarrow RNH_3X$$

Another mechanism involved protonation of the amine with the hydroxyl ion from the disassociation of water formed by an exchange reaction.

$$RNH_2 + OH^- + X^- \rightarrow RNH_3X + OH^-$$

$$OH^- + H^+ \rightarrow H_2O$$

These relationships have been developed quite fully by Kunin[3]. Since the amine carbon dioxide salt is that of a weak acid and a weak base, the interaction with water becomes an important factor. Since water will act as a regenerant to the amine-carbonic acid complex, the width of the exchange point will vary with the temperature of the water. The matter becomes quite complex since a Donnan film would not be pronounced and the kinetics would be solid phase control[4]. From the standpoint of this study the titration curves show an equilibrium result while column tests are non-equilibrium.

This study has approached weak acid removal by a direct measurement of carbon dioxide as part of a two-bed system which is the conventional way that a weak base resin is used in practice as part of a two- or three-bed system with or without a degasification step. Four different weakly basic resins were selected as representative of commercially available resins. Table 1 gives their characteristics using standard test methods. Titration curves were measured by contacting the resin with standard amounts of hydrochloric acid in 0.1N KCl. The results are shown in Figure 1. With the exception of Resin B, there is no evidence of quaternary ammonium strong base groups.

Table 1 *General Properties of the Weakly Basic Anion Exchanger*

Resin	A	B	C	D
Type	Dimethylamino pentamine	Trimethyl amine	Propylamine tetramine	Triethyl pentamine
Structure	Acrylic	Styrene-DVB	Acrylic	Epichlorohydrin
Water Retention, %	58	46	54	60
Total Capacity				
meq/g	10.8	3.4	4.3	11.4
meq/mL	2.9	1.7	1.4	2.4
Strong Base Capacity				
meq/g	1.17	0.92	0.83	0.89
meq/mL	0.31	0.35	0.27	0.23
Effective Size[1], mm	0.27	0.41	0.44	0.60
Uniformity Coefficient[2]	1.65	1.6	1.6	1.6
% Strong Base	10.7	20	19.3	9.6

[1] Effective size is the size at 90% retention on the screens.
[2] Uniformity coefficient is the size at 40% retention divided by the size at 90% retention.

The total capacity of the resin for acid was determined by standard methods[5]. The resin sample is converted to the chloride form with hydrochloric acid and through an anion interchange with nitrate the displaced chloride determined. After this the sample is stripped with caustic and rinsed and then contacted with sodium chloride and the hydroxyl ion generated is titrated with standard sulfuric acid using phenolphthalein as an indicator.

The titration curves shown in Figure 1 were determined by addition of standard amounts of acid to known volumes of weakly basic anion exchanger. The contacts were allowed to reach a constant pH which took about one week in most cases.

It is important to understand that these curves show substantial functionality in the pH range of 4-7 which is the pH range for carbon dioxide coming down the column ahead of the strong acid component. Encouraged by these observations a series of runs was made. Initially a water high in strong acids was selected. The water analysis used in the tests is given in Table 2. Silica was included in this test to determine how this very weakly acidic substance would react in the test conditions.

3 APPLICATION

The laboratory tests shown in the paper were determined under quite standard operating conditions in a two-bed configuration. A large fully regenerated cation exchange bed was used to generate the mineral acidity, free carbon dioxide and silica which was used directly as feed to the beds. Runs B and C were also performed in the same fashion.

The results were quite similar for each resin. Silica was removed initially for a small capacity and was displaced by the carbon dioxide and finally by the chloride and sulfate. The results shown for Figure 2, 3, 4 and 5 were surprisingly similar. Resin B which was prepared with a small amount of trimethyl amine shows the sharpest silica curve but the carbon dioxide curve was similar to the other resins. No attempt was made to calculate a silica balance. However some retention of silica, perhaps as a colloid, is likely. The results for Waters B and C are shown in Figure 6.

The conductivity of effluent was determined as well as free carbon dioxide by titration. For the tests reported in this study, the conductivity gives a method for following the breakthrough of carbon dioxide. In cases where cation leakage may occur, it is unlikely that conductivity can be used effectively. Each installation will require evaluation to determine the proper way to establish carbon dioxide leakage. It is possible that a conductivity probe in the bed could be used to determine breakthrough of carbon dioxide.

Table 2   *Test Water Used for Weak Base Study*

Water	A	B	C
Sulfate, meq/liter	4.36	2.00	0
Chloride, meq/liter	1.80	1.04	0
Alkalinity, meq/liter	1.68	3.00	5.00
Silica, meq/liter	0.28	0	0

4 DISCUSSION

These tests results show that carbon dioxide can be removed efficiently by weak base resins. In fact silica removal has also been observed although it is quite effectively removed by the carbon dioxide. There are several reasons why these findings are of interest from the standpoint of ion exchange plant design. For the most part they are driven by economics.

(1) Since weakly basic anion exchange resins are regenerated at approximately 100% efficiency with caustic, soda ash and ammonium their use can represent a savings in chemicals.

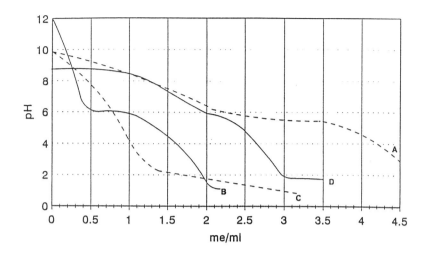

Figure 1 *Titration curves for weakly basic anion exchange resins*

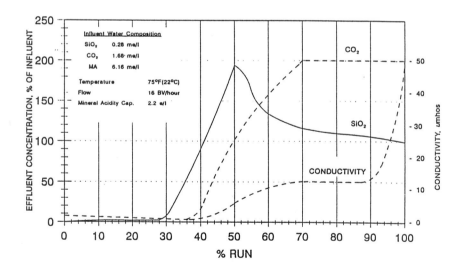

Figure 2 *Typical run Resin A*

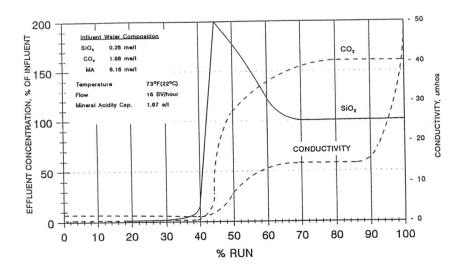

Figure 3 *Typical run Resin B*

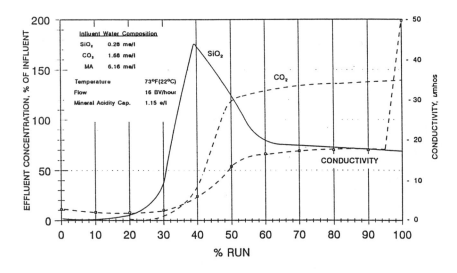

Figure 4 *Typical run Resin C*

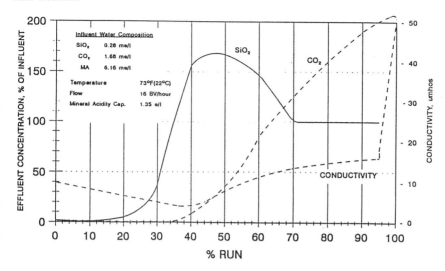

Figure 5 *Typical run Resin D*

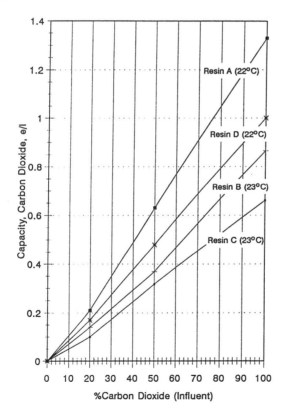

Figure 6 *Relation between CO_2 capacity and influent CO_2 concentration*

(2)   The mechanical removal of carbon dioxide by degasification can be quite expensive since the water must be repressurized.   In addition, the degasification equipment is frequently the source of abnormal contamination due to bacterial substances, fumes, etc.

(3)   The removal of weak acids on the weak base amine exchanger can have some interesting effects on the design of degasification equipment.   If the degasification equipment is installed ahead of the weak base, carbon dioxide load on the weak base will be slight; however, the initial removal of silica followed by dumping as the column is exhausted could have an adverse effect on the strong base unit which is designed so that the silica level is based on the influent composition.   This could result in higher than expected silica leakage during a portion of the run.

(4)   When the degasification step follows the weak base unit, the unit will have a low load of carbon dioxide initially.   When the carbon dioxide breaks through, the concentration of carbon dioxide can double over the remainder of the run.   This increase can overload the degasification equipment particularly if a vacuum process is used.

When neutralization of the regenerants is required, caustic savings may not be achieved since additional base will be required to neutralize the acid.   This is not exactly true since the spent regenerant from the weak base will have neutralization capacity equivalent to the amount of carbon dioxide removed.

(5)   The hydrolysis effect mentioned previously must be recognized.   These test were run at 22-24°C.   Since the dissociation of water is sensitive to temperature it is likely that the capacity for carbon dioxide will be affected by the feed temperature.   Work is needed to determine the magnitude of the thermal effect.

5 ECONOMICS

The value for removal of carbon dioxide on a weakly basic anion exchanger depends on the particular plant and on conditions related to the environment of the plant and the availability of acceptable grade of caustic.   A cost calculation was made for the operation of a three-bed deionization plant.

The following calculation will illustrate a case where caustic consumption is the main factor.   The volume and cost factors are based on treatment of 1000 liters of deionized water with silica removal required.   The water to be fed to the anion portion of the plant has the composition shown in Table 3.

The calculation for a three-bed deionization is summarized in Table 4.

This estimate shows a substantial reduction in strong base resin volume of almost 90% and an increase in the weak base volume of more than 50%.   This ratio of volume is not realistic in a plant where the design would be based on flow limitation.   Likely the strong base would have a volume based on 26 BV per hour and regeneration of the strong base would occur every 4-5 cycles depending on demand.   Of major importance would be the reduction in caustic usage of about 40%.

The caustic usage per 1000 liters for several possible conditions are given in Table 5.

Table 3 *Water Composition for Calculation*

Component	Concentration
Carbon Dioxide	2.0 meq/liter
Chlorides, Sulfates and Nitrates	1.8 meq/liter
Silica	0.2 meq/liter

Table 4 *Calculation Summary*

Case A - Three-bed Demineralization

Three Bed System: cation, weakly basic, strongly basic

System	Conventional		Weak Base Carbon Dioxide Endpoint	
Components	WB	SB	WB	SB
Carbon Dioxide, meq/liter	0	2.0	1.92	0.08
Silica, meq/liter	0	0.20	0	0.20
Strong Acids, meq/liter	1.8	0	1.8	0.00
Capacity, eq/liter	0.870	0.64	0.826	0.757
Loading Volume, equivalents	1.8	2.2	3.72	0.28
Resin Volume, liters	2.07	3.44	4.50	0.37
Regeneration Level,				
gms NaOH/liter	36	80	36	128
gms NaOH	74.5	275	162	47.4
Total Caustic, gms	349.5		209.4	

Table 5 *Caustic Consumption for Various Equipment Options*

System	Endpoint	Grams NaOH/1000 Liters
WB-SB	Conventional - SiO_2 B	349.5
WB-SB	CO_2 Break - SiO_2 B	209.4
Mechanical Degasification		
SB	Silica Break	300
Two-bed		
SA-SB	Silica Break	500

References

1. R. Kunin and B. Vassiliou, *I. & E. C. Process Design and Development*, 1964, **3**, 404.
2. M. Rodgers, B. Tucker and G. Mommaerts, "Ion Exchange Advances" (Proceedings of IEX'92), M. J. Slater, ed., SCI Elsevier, 1992, p. 57.
3. R. Kunin, "Ion Exchange Resins", second edition, reissue 1990, R. E. Kreiger Publishing Company, 1990, p. 55.
4. F. Helfferich, "Ion Exchange", McGraw Hill, USA, 1962, p. 146.
5. G. Simon, "Ion Exchange Training Manual", Van Nostrand Reinhold, 1991, p. 145.

APPLICATION AND PERFORMANCE OF DOW GEL AND PUROLITE SUPERGEL RESINS IN CONDENSATE POLISHING AT TORRENS ISLAND POWER STATION

T. Little

Torrens Island Power Station
Port Adelaide
South Australia

1 INTRODUCTION

This paper details the developments in condensate polishing philosophy at Torrens Island Power Station, dealing specifically with the change from an all macroporous to a gel system after the adoption of an all volatile boiler water treatment. Modifications to the polishing plant resin regeneration procedure are also discussed.

Before the introduction of an all volatile boiler water treatment at Torrens Island, macroporous resins were used in the polishing plant and performed satisfactorily. After the introduction of an all volatile boiler water treatment, polished condensate was no longer of a sufficiently high standard to allow easy compliance with the new stricter guidelines for boiler water contaminants, due to excessive leakage of chloride and sulphate from the polishing plant. The problem was solved by changing to gel resin and altering the polisher resin regeneration program.

2 PLANT DESCRIPTION

2.1 Power Plant

Torrens Island Power Station (TIPS) is a natural gas fired, seaboard power station, located close to Adelaide in South Australia. Owned and operated by the ETSA Corporation, TIPS was built to take advantage of extensive natural gas deposits discovered at Moomba, some 800 km. to the North East. The generating plant consists of two separate stations known as 'A' and 'B' Station respectively, with A Station being the older. Work on the first of the four units in A station began in 1963 and the entire power station project was completed in 1981. 'A' Station has 4 x 120 MW Parsons' triple expansion steam turbines that do not have condensate polishing. 'B' station has 4 x 200 MW Parsons steam turbines, each using 170 kg/sec of 540°C steam at 16.54 MPa(g); full flow condensate polishing is available on all four units.

2.2 Polishing Plant

2.2.1 Units B1 and B2. The Permutit Water Treatment Co. designed and installed the polishing plant on units B1 and B2 and this plant consists of two 100 % duty service vessels with a two vessel resin separation / regeneration plant.

The service vessels operate with a 2:1 cation to anion resin ratio and each of the three resin charges contains 3.68 m³ of cation and 1.84 m³ of anion resin.

Resin transfer is a totally hydraulic process that is easy to use and gives continuous trouble free operation over many cycles of the polisher's operation. A schematic of the B1 and B2 polishing plant area is shown as Figure 1.

2.2.1.1 Operating Conditions B1 and B2. The B1/B2 plant was designed to polish 160 litres/sec of condensate for 150 hours before the cation resin becomes saturated with ammonia and moves into the ammonia cycle. To minimise ammonia usage the plant is operated at 80 litres/sec., half its rated flow rate, except during machine run-ups and when there is a condenser leak. Polisher performance is monitored using conductivity as the measured variable; direct conductivity refers to the conductivity of the raw sample adjusted to 25°C and acid conductivity is the conductivity of the sample after it has been passed through a column of cation resin in the hydrogen form.

Condensate direct conductivity is 4.80 ± 0.25 µs/cm, being derived from 600 ± 50 ppb NH_4OH, approximately 2.5 ppb of $NaCl$ and 12 ppb Carbonic species, condensate acid conductivity is thus maintained at less than 0.11 µs/cm. To minimise sodium ingress into the boiler water, the resin is transferred after 300 hours in service and regenerated, even though the acid conductivity of the treated water may still be acceptable.

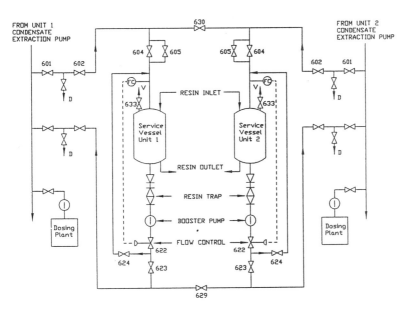

Figure 1 *B1 and B2 condensate polishing plant service area*

2.2.2 Units B3 and B4. The B3 and B4 polishing plant has the same basic layout and operating principles as that of the B1 and B2 units; the plant was designed and installed by Degremont Warman Australia and is the only plant in Australia built by this company. The service vessels are capable of polishing 160 litres/sec. of condensate at 27.3 kg/m² sec (98 m/hr) and operate at approximately 1200 kpa, using 2 m³ of anion and 4 m³ of cation resin. As with the B1 and B2 units, the B3/B4 service condensate flow is 80 kg/sec., half the maximum design level.

A combined hydraulic/pneumatic transfer system is used to move the resin from the service vessel to regeneration area and vice versa. The vessel containing the resin to be transferred is pressurised pneumatically, forcing the resin into a sluicing stream, which entrains and carries it to the other area; the transfer system is very reliable and a complete transfer operation takes approximately 50 minutes. The service vessel lay-out, which is very similar to that of the B1/B2 plant is not shown. A detailed drawing of the B1/B2 polishing plant regeneration area is shown in Figure 2.

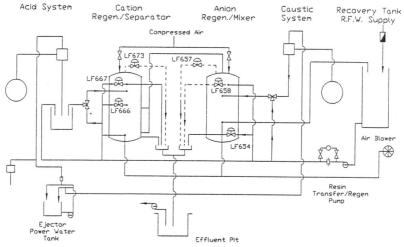

Figure 2 *B1 and B2 polishing plant regeneration area*

2.2.2.1 Operating Conditions B3 and B4. Design regenerant levels for the B3/B4 plant are 70 gms of sulphuric acid and 70 gms of caustic soda per litre for the cation and anion resins respectively; the regenerant levels were set lower than those for the B1/B2 plant during design, reflecting the different resins specified and less conservative operating philosophy of the B3/B4 plant supplier. There is no provision for hot anion resin regeneration. Condensate characteristics are identical to those of the B1/B2 units, as are run times and polisher effluent conductivities.

Polisher run time is nominally 300 hours at 80 kg/sec and over-runs are usually limited to 2 days. Since changing to gel type resin, the direct conductivity and the acid conductivity of the polisher effluent in normal operation will both be 0.055 μs/cm @ 25°C. This compares favourably with the figures of 0.058 μs/cm direct and 0.060 μs/cm acid conductivity when using macroporous resins. See Tables 1 and 4 for details on boiler water limits and resin performance.

3 SPIKE TESTS

To test whether the resins can effectively remove a larger amount of dissolved solids than normally encountered in the condensate spike tests are carried out. These tests are performed on each charge at 2 yearly intervals to track resin condition, and are used to determine when replacement is necessary.

Under normal operating conditions polisher resin has very little work to do due to the low contamination level of the condensate, so a spike test that simulates a condenser leak of approximately 150 litres of seawater per hour is performed. With the turbo generator on-line, and the polisher on full flow, a constant amount of a concentrated brine solution is injected into the condensate to raise its conductivity to 18.0 μs/cm @ 25°C. The acid conductivity of the polisher effluent is monitored until the end-point of 0.08 μs/cm @ 25°C is reached. Since there is twice as much cation resin as anion resin, the anion resin is exhausted first, increasing the acid conductivity of the treated water at the end point. By measuring the quantity of salt absorbed, anion resin capacity can then be determined.

We have found that on installation, anion resin capacity will be close to theoretical and then drop slightly each year, by approximately 0.5 kg NaCl per m^3 resin. After 5 years, a more marked decrease of 2.0 kg NaCl per m^3 resin occurs. New resins will often remove 13.0 kg NaCl per m^3 and are replaced when this falls to less than 9.0 kg NaCl per m^3 of resin.

4 POWER STATION OPERATING HISTORY

The first 2 units in B station, B1 and B2 were commissioned in 1975. At that time, boiler water treatments used high pH (>9.50) regimes and solids, NaOH and in some cases Na_3PO_4 were dosed, along with hydrazine. Due to the strong buffering effect of this chemical treatment, boiler water contamination with NaCl and Na_2SO_4 was permitted to reach levels much higher than is acceptable now, with the more modern all volatile treatments.

4.1 B Station Boiler Tube Failures

B1 boiler had been on line for approximately 5 years before the type of problems that can be encountered when using solid chemical water treatment occurred. In 1980, the first of several tube failures in a high heat flux area of the boiler (inclined tubes in the lower hopper area) was experienced. Upon examination, it became apparent that the tubes had failed due to a condition known as caustic gouging. This type of failure occurs when water, containing dissolved solids -usually caustic water treatment chemicals- is subjected to a heat rate high enough to promote "film boiling". During periods of heat flux so intense that film boiling occurs, water at the tube surface is evaporated to dryness, leaving the dissolved water treatment chemicals as a deposit. An initial low concentration of NaOH, for example 1 ppm, which would give a pH of less than 10.0 and be protective of the tube material, can be deposited on to the tube surface as 100 % NaOH and, due to the high temperature, immediately begin to react with it. These areas become thin, brittle and then fail. A characteristic "gouged" look is evident at the reaction site, as the corrosion products left on the tube surface, being soluble, are dissolved by the water that contacts them during times of lower heat flux.

Boiler water treatment philosophy in TIPS B was not changed after the first tube failures, but the caustic soda concentration of the boiler water was reduced from 2 ppm to 0.5 ppm.

After several more tube failures on the B1 and then B2 boiler in 1980, serious consideration was given to the problem and it was decided to change to an all volatile water treatment, in line with European practice. Refer to Table 1 for details of the changes to the boiler water treatment.

Caustic soda concentration in the boiler water is now kept below 0.1 ppm. Target ammonia concentration is 0.60 ppm, set to give an operating pH of approximately 9.20 and NaCl concentration in the boiler water now does not exceed 0.20 ppm. To maintain this lower chloride level, polished condensate has had to reach and maintain a new quality standard.

At the time of the tube failures, all four B station polishers employed macroporous resins and used the simple regeneration programs installed by the polishing plant manufacturers. This combination of factors lead to low quality polished water (refer Table 4), which had appeared to be acceptable due to its apparently low conductivity. Limits for contaminants in the boiler water were so high (in comparative terms) that blowdowns could be kept within the acceptable limit of 1 per week. When the new lower limit for chloride was implemented with the all volatile treatment, blowdowns became more frequent, being required 2 or three times per week as the polishers could not produce water good enough to satisfy the new operating guidelines.

Table 1 *Boiler Water Quality Parameters for Solid and All Volatile Treatments*

Boiler Water Limits	NaOH ppm	NaCl ppm	NH$_3$ ppm	N$_2$H$_4$ ppb	Direct Cond. µS/cm @ 25°C	Acid Cond. µS/cm @ 25°C
Solid Dosing	< 2.0	< 1.0	0.10	50	18.0 max	7.0 max
All Volatile	< 0.10	< 0.2	0.60	Nil	6.0 max	1.50 max

All four of the polishing plants using macroporous resin ran with typical water quality of 0.058 µS/cm direct and 0.060 µS/cm acid conductivity which proved to be unsatisfactory when an all an volatile boiler water treatment was implemented in 1982.

After consulting the literature in 1982, it became apparent that gel type resins may have been able to help alleviate the problems we were experiencing, due to their lower leakage rates and better regeneration characteristics.

Table 2 *Macroporous Resins Used in Original Charges*

Date Installed	Units	Cation Resin Type	Anion Resin Type
June 1976	B1/B2	Zerolit 525	Zerolit MPF-1
January 1980	B3/B4	Ionac CFP110	Ionac A 641

In August 1983, the Dow Chemical Company wrote and offered to loan us a charge of gel type resin for an operational trial. The offer was accepted and in March 1984, a single mixed bed charge of Dowex gel resin was installed into the B1 polishing plant.

The charge consisted of 3.74 m^3 of Dowex *HGR-W2-H and 1.87 m^3 of Dowex SBR-C-Cl resin.

During a trial period of 6 months from 20/3/84 to 1/10/84, the gel resin was tested thoroughly. Spike tests were carried out and showed that the gel resin did have lower leakage levels and a higher capacity than the macroporous resin. It was obvious that the unit that had the gel charge in service also needed fewer blowdowns than the others.

Conductivity of the polisher treated water was noticeably lower throughout a run, 0.055-0.056 µS/cm direct and 0.055-0.056 µS/cm acid conductivity, and this remained constant to the end of the period in service.

The resins in the B1/B2 plant were due for replacement, so in August 1985 three charges of Dowex *HGR-W2-H and Dowex SBR-C-Cl gel resin were purchased.

Use of the new gel resins resulted in a marked reduction in blowdowns for the B1 and B2 boilers, from twice per week to once per ten days.

In August 1989, the macroporous resins in the B3/B4 polishers were also replaced with Dowex *HGR-W2-H and Dowex SBR-C-Cl gel resin. The number of blowdowns required on the B3/B4 boilers was then also cut by the improved polisher leakage.

B1 and B2 polisher resins were replaced again in November 1993. The new resins are the Supergel type manufactured by Purolite, A400 TL anion resin and SGC 100 *10TLH cation. The Purolite resins have performed satisfactorily, maintaining the low leakage rates required; Supergel resins are also claimed to have improved physical properties that give them an extended operating life when compared with other formulations.

Table 3 *Gel Resins in Current Charges*

Date Installed	Units	Cation Resin Type	Anion Resin Type
November 1985	B1/B2	Dowex *HGR-W2-H	Dowex SBR-C-Cl
August 1987	B3/B4	Dowex *HGR-W2-H	Dowex SBR-C-Cl
November 1993	B1/B2	Purolite SGC 100 *10TLH	Purolite SGA 400 TL

Table 4 *Comparative Performance Data for Macroporous and Gel -Type Resins*

Resin Performance	Direct Cond. µS/ cm @ 25°C	Acid Cond. µS/ cm @ 25°C	NaCl (ppb)	Na₂SO₄ (ppb)	Carbonic Species (ppb)
Condensate	5.00	0.11	2.5	< 0.1	12.0
Macroporous	0.056 - 0.059	0.058 - 0.062	1.50 max	1.0 max	No Data
Gel Type	0.055 - 0.056	0.055 - 0.056	0.50 max	0.20 max	No Data

5 CHANGES TO REGENERATION PROCEDURE

Using gel resin made it easier to achieve lower final rinse conductivities after regenerations and gave better performance through a run, but it was felt that further improvements could be made to final rinse conductivities and regeneration water usage by re-designing the regeneration programs. The original programs had been designed without attempting to minimise either final rinse conductivity or rinse water usage. Testing the conductivity of separate resin rinse stage water, after chemical injection in the original regeneration program, proved that the separate resins had not been washed completely free of regenerant before being mixed and given a final rinse.

To overcome this, the automatic regeneration program was shut down after chemical injection, the separate resins were rinsed manually until the effluent from each vessel, when tested with a portable conductivity meter, was less than 10 µs/cm. The resins were then mixed together and given a final rinse in the usual way.

Due to the time required to perform the new regeneration steps manually, and the difficulties experienced in attempting to re-program the original programmable logic controller, a new one was purchased for the B3/B4 plant in 1992 and a program was written to incorporate the innovations that had been made.

By automating the conductivity controlled separate rinse stage, it was found that the amount of rinse water required to rinse the separate resins to less than 10 µs/cm could be substantially reduced. The new program was written to combine air scouring with the rinsing process and the resins are now successfully cleaned during this rinse stage.

5.1 Minimising Rinse Water

After the chemical injection is complete for both resin types, rather than refilling and rinsing to waste, each bed is drained under pressure so that it is left dry. The vessels are then refilled sufficiently to allow air scouring. After this air scour, the vessels are again emptied under pressure. This process is repeated twice more for the anion resin and three times for the cation; at the end of the final fill and empty stage for each resin, the vessel is refilled to the top and the resins are then rinsed until the rinse water conductivity is less than 10 µs/cm. Rinse conductivity is monitored by a cell in the drain line from each regeneration vessel and when both separate rinse conductivities are less than 10 µs/cm, the resins are mixed together and rinsed; a final rinse conductivity of less than 0.06 µs/cm will be achieved in approximately 10 minutes.

5.2 Resin Clean-Up

The original programs had extensive steps at the *start* for resin clean up using air scouring, but after careful observation it was determined that these steps had very little effect on the cleanliness of resin after regeneration. By trial and error it was discovered that the resin became much cleaner if it was air scoured *after* it had been regenerated, presumably due to the beads swelling when regenerated and cracking the adherent layer of crud. This cleaning process was incorporated into the regeneration procedure and was done initially by operating the plant manually. These cleaning steps are now incorporated into the rinse operation as described above and done automatically.

5.3 Performance Improvement and Cost Savings

When using the original regeneration program, a resin charge could not be regenerated to give a final rinse conductivity less than 0.06 µs/cm and would have used approximately 250 m³ of water. Using the new regeneration method, a charge of resin can be regenerated to the above standard using only approximately 150 m³ of water, this water saving is worth approximately $17,000 per year for the four units. Regeneration time has been reduced and to 5 hours, compared to 7 hours previously.

The new programmer is very reliable and many regenerations are successfully completed with less than 1 hour of supervision by staff members, compared with 4.5 hours required previously; labour savings amount to approximately $10,000 per year.

An overall saving of approximately $27,000/year has been experienced on the B3/B4 plant and it is anticipated that introduction of an automatic version of this regeneration method will produce similar savings on the B1/B2 plant.

The B1/B2 polishing plant is regenerated manually using the new method, producing the same final rinse and on-line results. This has proved to be worthwhile in maintaining boiler water quality and lowering running costs.

6 CONCLUSION

The innovations made to the Torrens Island B Station polishing plant since commissioning in 1975 are:

i.) April 1976, instigated regular spike testing of resins, to monitor anion resin capacity and on-line performance

ii.) March 1984, began 6 months of service and spike testing on loaned charge of gel resin

iii.) August 1985, changed first pair of polisher units from macroporous to gel resin to overcome leakage problems during service runs

iv.) August 1987, changed second pair of units to gel resin after success of first trial

v.) April 1988, altered regeneration program to cut regeneration water usage and produce lower final rinse and service conductivities

vi.) November 1993, purchased Purolite Supergel resins with a view to extending polisher resin life.

The changes instigated at Torrens Island Power Station (TIPS) have significantly lowered the cost of regenerating polisher resins and lowered polished condensate conductivities.

7 FUTURE PLANS

Plans for further work that will be done on the polishing plant at TIPS include:

i.) A new PLC will be fitted to the B1/B2 regeneration plant to automate regenerations and thereby achieve more efficient plant operation.

ii.) Purolite Supergel formulations are expected to be physically stronger and therefore last longer than other gel type resins; the physical properties of the resin will be monitored in service over the next 8 years to determine if this is indeed the case.

References

D. J. O'Sullivan, Experiences with a 'Consep' Condensate Polishing Plant, Presented at the 8th EPRI Workshop on Condensate Polishing, Little Rock Arkansas May 31st to June 2nd 1989.

THE DEVELOPMENT OF ELECTROCHEMICAL ION EXCHANGE FOR THE TREATMENT OF INDUSTRIAL LIQUORS

P. M. Allen, P. G. Griffiths, C. P. Jones, A. R. Junkison and R. I. Taylor

Electrochemical Department
AEA Technology
Harwell OX11 0RA

1 INTRODUCTION

Electrochemical ion exchange (EIX) is an advanced ion exchange process, where an exchange materials has been incorporated into an electrode structure using a suitable binder. Ion exchange is controlled by application of an electrode potential between the EIX electrode and a counter electrode. The combination of the EIX electrode and the counter electrode, in various configurations, comprises the EIX cell.

Historically, applications of EIX have focused on the nuclear industry, where it is important not only to reduce the concentration of radionuclides to below legally defined limits, but also to minimise the overall volume of any wastes produced. More recent developments have concentrated on non-nuclear applications, which include:

1) The removal of toxic, heavy metal ions.
2) The recovery of precious metals.
3) Water deionization.
4) Corrosive anion removal.
5) Nitrate removal.

EIX is unique in that exchange processes are controlled electrochemically. The use of regenerating chemicals is often unnecessary and it is possible to achieve large volume reduction factors (VRF) since elution can often be carried out within a single bed volume. The objective of the series of experiments described in the paper was to demonstrate that significant performance and selectivity improvements could be obtained, for this type of electrochemical system, by increasing the surface area of the ion exchange structures. A very important consideration in defining the method manufacture was that any technique used should be adaptable in future scale operations.

2 THREE DIMENSIONAL EIX

There are several types of EIX system. In conventional EIX, the electrode is made an integral part of the ion exchange matrix by casting the matrix/binder mix onto the mesh electrode. In continuous EIX, the electrode is separated from the ion exchange matrix and the matrix is sealed within the cell in order to provide distinct treatment and electrode compartments. A controlled amount of water is allowed to permeate through the membrane, thus creating a separate concentrated electrolyte stream. In 3d continuous

EIX, the gap between the electrodes is filled with a composite mixture of a high porosity ion exchange medium sandwiched between two low porosity strong acid cation exchange zones. This is schematically illustrated in Figure 1, where, for example, the target species is zinc, adsorbing onto a cation exchange matrix.

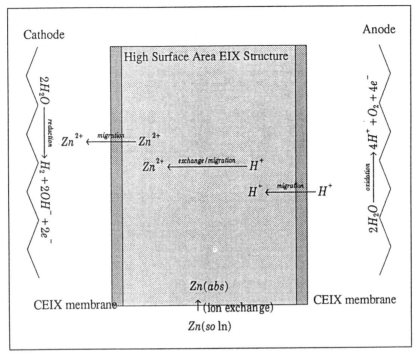

Figure 1 *High surface area 3dEIX removal of zinc*

The feed stream passes up through the high surface area structure. Ion exchange occurs with the target ion (zinc in this case) exchanging for hydrogen ions on the exchanger. Water is oxidised at the anode to produce hydrogen ions and oxygen. The hydrogen ions then migrate towards the cathode and pass through the cation EIX membrane and into the high surface area EIX structure. The bulk of the current flow will be from anode to cathode. The ionic conductivity of the ion exchanger is higher than that of the solution and, therefore, the hydrogen ions migrate through the three dimensional structure displacing other ions from exchange sites. Migration of the ions involves a mechanism of cations hopping from site to site in the direction of the migration (anode to cathode). The hydrogen ions displace any zinc ions from the ion exchange matrix with kinetics dictated by the ion exchange properties of the exchanger.

At the interface between the high surface area structure and the less porous structure that defines the catholyte compartment, the cations (both zinc and hydrogen) will pass into the cation EIX membrane. If the zinc ions are soluble in the catholyte solution, then the concentration of zinc in this compartment solution rises and the zinc plates onto the cathode. If the zinc is not soluble in the catholyte solution, then the zinc will remain adsorbed within the cation EIX membrane. The amount of zinc within the exchanger will increase until the exchange capacity is exceeded, at which point an insoluble salt of zinc (probably carbonate) will form in the pores of the cation EIX membrane.

3 EXPERIMENTAL

The three dimensional structures were produced either by casting a slurry of powdered resin in a polymer binder solution (18.75% styrene butadiene copolymer in trichloroethylene solvent) onto a foam substrate or by spraying a binder/resin mixture onto a thin porous structure. The latter system produced a very open fibrous web like structure. The EIX membranes were manufactured by casting a similar slurry of powdered ion exchanger in a solution of an appropriate binder. The macroporous structures were approximately two centimetres thick after soaking for twenty four hours in deionised water. The cast EIX membranes were approximately five millimetres thick after soaking in water. The macroporous structures produced contained sorbers such as nickel ferrocyanide (designated NiFeCN, ex STMI) in the form of a coated foam, a strong acid cation exchanger (designated PrCH, ex Purolite) foam and a strong acid cation exchanger (PrCH) web. The EIX cell and flow scheme used are illustrated in Figure 2.

The cell contained two EIX microporous membranes, manufactured from the strong acid cation exchanger (PrCH). Each membrane had an area of approximately 50 square centimetres. The central zone, or feed compartment, was filled by the macroporous three dimensional structure. The macroporous structure had a volume of one hundred cubic centimetres.

The performance of the systems was compared by passing a feed stream through the cells. The feed contained 10ppm cobalt (as cobalt nitrate) and 50ppm sodium (as sodium nitrate) in deionised water. The feed was pumped through the cell by means of a Watson Marlow 101UR peristalic pump. The bleeds from the anolyte and catholyte compartments were drawn off using similar pumps. The rate of pumping was such that the electrode compartments were always full of liquor. The cell current used throughout these experiments was 100mA, which corresponds to a current density of $2mA/cm^2$.

4 RESULTS

The performance results, for the various systems, as a function of feed flow rate, are given in Table 1, and illustrated in Figure 3 and Figure 4.

At $1dm^3/h$, the current efficiencies for the four processes are 7.4% continuous cation EIX, 49.8% with the NiFeCN foam, 51.9% with the PrCH foam, and 54.4% with the PrCH web. The total cationic decontamination factors at the same flow rates were 1.12, 3.74, 4.22 and 4.99 respectively. If we define the selectivity of the system as follows:

$$K_s = \frac{(cobalt_r)(sodium_s)}{(cobalt_s)(sodium_r)}$$

Where the species r and s subscripts refer to the metals on the resin and in solution respectively. Then the K_s values for the four systems (again at $1dm^3/h$) were 0.32, 1.8, 1.77 and 4.68. These results imply that the web structure offered the best selectivity, even though the resin used (a strong acid cation resin) is not the most selective for cobalt.

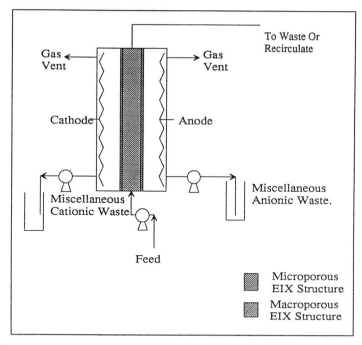

Figure 2 *High surface area EIX equipment schematic. Cell size 26cm high, 9cm wide by 6cm deep*

5 CONCLUSIONS

The experiments have compared the performance of electrochemical ion exchange systems containing different high surface area adsorption structures with one that contained a simple two dimensional membrane. The objective of the experiments was to confirm that the increased surface area component would provide a more effective adsorption route, thus increasing the current efficiency for the electrochemical process and improving the selectivity.

The experimental results indicate that the introduction of the structures improved the system performance, 0.56ppm cobalt in effluent at 1dm³/h with PrCH web structure as compared to 9.6ppm cobalt at the same flow rate without a structure - a nineteen fold increase in the amount of cobalt removed from the stream. The variation in the defined selectivity coefficient between the two systems was 4.68 with the web structure and 0.32 without any structure - a ten fold improvement in selectivity.

The results also indicate that the form of the structure is important. Microscopic examination indicated that the web possessed a more open structure than the foams, and it is not unreasonable to conclude that this in turn would result in an increase in the surface area. It should be stressed that surface area measurements have yet to be performed. There was an even more significant increase in selectivity coefficient, which increased from 1.77 to 4.68 as the structure was changed from foam to web.

To summarise, the results show that significant improvements in the electrochemical ion exchange system performance can be achieved by increasing the surface area of the adsorbing material. The form of the adsorbing structure plays an important role in defining the electrochemical performance and the selectivity of the system.

Table 1   *Performance Tests With and Without High Surface Area Three Dimensional EIX Structures*

No structure			
Flow/(cm^3/h)	Voltage/V	Na/ppm	Co/ppm
2000	5.16	46.8	9.9
1500	5.31	45.0	9.63
1000	5.24	44.0	9.58
500	5.68	38.7	8.68
250	5.79	32.6	7.27
NiFeCN Foam			
Flow/(cm^3/h)	Voltage/V	Na/ppm	Co/ppm
2000	32	19.8	4.4
1500	33.4	14.7	3.04
1000	34.7	14.1	1.79
500	32.5	9.95	0.68
250	20	9.00	0.41
PrCH Foam			
Flow/(cm^3/h)	Voltage/V	Na/ppm	Co/ppm
2000	6.23	-	2.78
1500	6.24	20.8	2.37
1000	6.27	12.5	1.58
500	6.25	10.2	0.72
250	6.32	8.4	0.22
PrCH Web			
Flow/(cm^3/h)	Voltage/V	Na/ppm	Co/ppm
2000	16.1	17.2	1.6
1500	17.8	11.9	1.07
1000	21.7	11.2	0.58
500	29.8	8.92	0.197
250	20	5.8	0.086

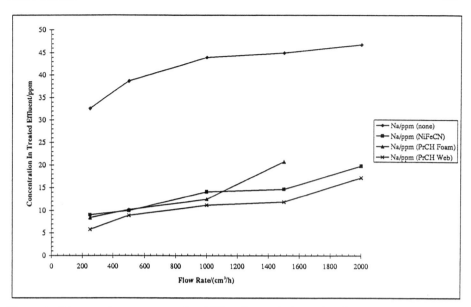

Figure 3 *Flow performance test for EIX cells, variation of sodium concentration in effluent with flow rate. Flow stream 10ppm cobalt (as nitrate), 50ppm sodium (as nitrate), cell current 100mA*

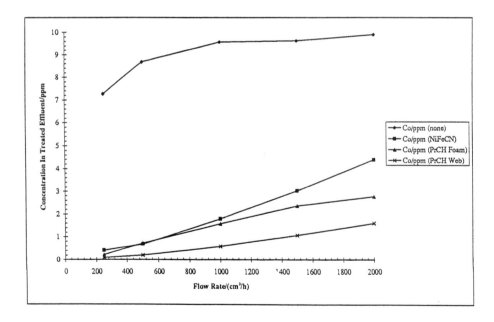

Figure 4 *Flow performance test for EIX cells, variation of cobalt concentration in effluent with flow rate. Flow stream 10ppm cobalt (as nitrate), 50ppm sodium (as nitrate), cell current 100mA*

ROLE OF MORPHOLOGY AND INTERFACIAL CHEMISTRY OF ION EXCHANGER FOR ALUM RECOVERY FROM CLARIFIER SLUDGE

P. Li, A. K. SenGupta, L. Cumbal and S. Gokhale

Environmental Engineering Program
Department of Civil and Environmental Engineering
Lehigh University
Bethlehem, PA 18015, USA

1 INTRODUCTION

Alum is widely used as a coagulant in water treatment plants and ends up in clarifier sludge primarily as insoluble aluminum hydroxide along with suspended solids, natural organic matter and other insoluble impurities. The clarifier sludge is generally discharged into water bodies or disposed of at landfill sites. But the discharge and the disposal of the clarifier sludge are recently being subject to closer scrutiny because of its high aluminum content. Also, the cost of solid waste disposal is increasing very rapidly in every industrialized country. As a result, the possibility of alum recovery from clarifier sludge and subsequent reuse are receiving favorable consideration.[1,2]

The traditional method for alum recovery is acidic extraction using sulfuric acid. The acidic extraction process is based on a simple concept that insoluble aluminum hydroxide in clarifier sludge is dissolved when sulfuric acid is added into the sludge.[1] However, this process is non-selective, i.e., along with alum it recovers all other undesirable substances, namely, natural organic matter, manganese and heavy metals. Understandably, the recovered alum from the acid extraction process may not be acceptable as a coagulant in drinking water treatment plants. The alum recovery process should ideally be a selective one, i.e., the process should reject other desirable constituents present in clarifier sludge.

Recently, thin sheet (0.2-0.5 mm thick) composite ion-exchange materials (CIM) have been synthesized where very fine exchanger beads (less than 100 microns) are embedded or physically trapped in highly porous PTFE (polytetrafluoroethylene).[3] These ion-exchanger encapsulated sheets can be easily introduced into and withdrawn from slurry reactors without any major separation problem and cleaned with water jet (very robust). They also offer excellent kinetics because of fine particle sizes of the enmeshed exchanger beads and porous structure of supporting PTFE. Although PTFE works as the background material, it constitutes less than 20% of the mass of the CIM while ion-exchanger beads comprise over 80%. As a result, the total exchange capacity of the CIM is comparable to that of the spherical ion-exchanger beads. Figure 1 provides a scanning electron microphotograph of the CIM. (On occasion, the CIM has been referred to as "membrane" in the open literature including this paper. However, its physical configuration and properties are significantly different from the traditional ion-exchange membrane used in electrodialysis processes.)

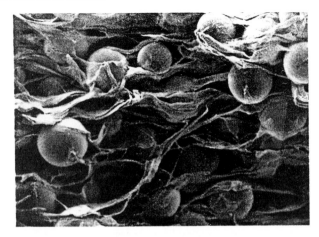

Figure 1 *A scanning electron microphotograph of the composite ion exchange materials (x300)*

2 TWO-STEP ALUM RECOVERY PROCESS

The composite ion-exchange material (CIM) with thin-sheet like physical configuration forms the heart of the two-step alum recovery process. Figure 2 shows a conceptual arrangement showing how the CIM can be used quite conveniently and almost continuously for recovering alum from clarifier sludge. Many other alternate arrangements are also possible. Since, in most water treatment plants, clarifier sludge is discharged only intermittently once or twice a day, the arrangement in Figure 2 appears to be easy-to-implement and operationally simple. Every batch of sludge, after necessary pH adjustment, can be stored in the sludge tank (tank 1) from which aluminum(III) is selectively sorbed by the moving CIM (as a belt after necessary reinforcement) and subsequently regenerated with 5-10% v/v H_2SO_4 and collected as aluminum sulfate in the regeneration tank (tank 1).

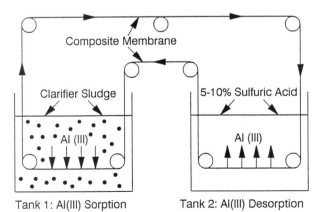

Figure 2 *The conceptual two-step alum recovery process using the CIM*

The proposed process is semi-continuous in operation and involves the following two-steps:

Step I - Selective Sorption of Al(III)

The CIM, when suspended in the clarifier sludge after adjusting its pH to around 3.5, will selectively remove dissolved aluminum(III) from the aqueous phase. Consequently, more aluminum hydroxide precipitates will dissolve and aluminum uptake by the CIM will continue. The two simultaneous reactions occurring in series can be written as follow:

Dissolution:

$$Al(OH)_3(s) + 3H^+ \rightleftharpoons Al^{3+} + 3H_2O \tag{1}$$

CIM uptake:

$$3\overline{R - H} + Al^{3+} \rightleftharpoons \overline{R_3 - Al} + 3H^+ \tag{2}$$

R and overbar denote the composite exchanger phase.

Step II - Desorption of Al(III) or Regeneration of CIM

The second step of the process involves withdrawing the CIM from the acidified clarifier sludge and introducing them into the stirred regeneration tank containing 5-10% sulfuric acid where tiny exchanger-beads in the CIM are effectively regenerated according to the following reaction:

$$2\overline{R_3 - Al} + 3H_2SO_4 \rightleftharpoons 6\overline{R - H} + Al_2(SO_4)_3 \tag{3}$$

The regenerated CIM can now be withdrawn and reintroduced into the clarifier sludge tank.

During the last three years, an extensive number of laboratory experiments were carried out using clarifier sludges from different water treatment plants. The primary objectives of this research were to explore the effectiveness of the proposed two-step process, to assess the selectivity and durability of the new ion exchange materials—CIM, and to evaluate the effectiveness of the recovered alum as a coagulant.

3 EXPERIMENTAL MATERIALS AND METHODS

Composite Ion Exchange Materials (CIM) In this research, two kinds of the CIM were used: the strong acid membrane and the chelating membrane. The membranes used in this study were purchased from Bio-Rad Inc., CA. The commercial name for the strong acid membrane is *AG 50W-X8*, and the commercial name for the chelating membrane is *Chelex 100*. Figure 3 shows functional groups for the spherical beads of these two materials.

Clarifier Sludge One of the sludges used in this study was obtained from the Allentown Water Treatment Plant (AWTP), Pennsylvania. In the AWTP, alum is used to remove the turbidity in the surface water which ranges from 2 to 680 NTU, and alum dosage ranges from 10 to 50 mg/L. The sludge is drained intermittently about two times per week. The primary constituents of the AWTP sludge are given in Table 1.

Simulation of the Two-Step Process - Cyclic Tests The experimental simulation of the two-step process was conducted in one sludge tank and one regeneration tank. In the tests, pieces of the membranes were immersed in the sludge tank for 45 minutes, and they were then withdrawn from the sludge tank and immersed in the regeneration tank for 30

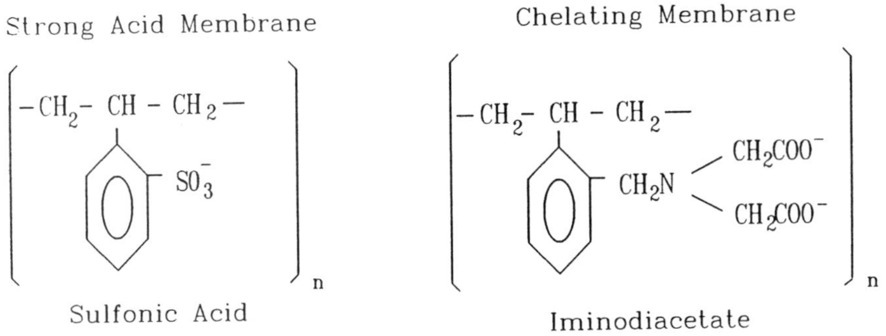

Strong Acid Membrane

Chelating Membrane

Sulfonic Acid

Iminodiacetate

Figure 3 *The functional groups of two composite ion-exchange materials (CIMs)*

Table 1 *Composition of the Sludge Obtained from the Allentown Water Treatment Plant*

ELEMENTS	CONCENTRATION (mg/L)
Al	2400-5600
Fe	240-400
Mn	20-70
Zn	5-15
Cu	1-4
Pb	2-4
Cr	1-2
Dissolved Organic Carbon (DOC)	250-860

Note: Concentrations were measured for acidized sludge solution in which pH was less than 1.0.

minutes. This represented the completion of one sorption-desorption cycle. After one complete cycle, the pieces of the regenerated membranes were ready for the next cycle. The same procedure was used in the entire course of the cyclic tests. In parallel tests, under otherwise identical conditions, intermittent rinsing was introduced in one test, while intermittent rinsing was excluded in the other.

The volume of the sludge used in the tests was 10.0 liters. 4.0 liters of 10% sulfuric acid (10% H_2SO_4) was used as regenerant. The mixing in the sludge tank was attained by a 180W mixer @ 2000 rpm. The mixing in the regeneration tank was achieved by aerating the contents of the tank with nitrogen gas.

Other details about the experimental protocols have been provided by SenGupta and Shi[2], and Li.[4]

Analytical Methods Aluminum analysis was performed by using a spectrophotometer (Perkin-Elmer, Model Lambda 2). The analysis of aluminum employed the eriochrome cyanine R method described in *Standard Methods* (18th edition, 1992)[7]. DOC analysis was conducted by using a TOC analyzer (Dohrmann DC-190). All metals, other than aluminum, were analyzed by an atomic absorption spectrophotometer (Perkin-Elmer, Model 2380). pH was monitored by a pH meter (Fisher Scientific). Turbidity was determined by using a turbidimeter (Hach, Model 2100).

4 RESULTS AND DISCUSSIONS

Cyclic Tests Figure 4 displays Al(III) concentration in recovered alum versus the cycle number for tests with the strong acid membrane and the chelating membrane. In the test with the chelating membrane, Al(III) concentration in recovered alum was 1800 mg/L after 30 cycles; while, in the test with the strong acid membrane, Al(III) concentration in the recovered alum was 4000 mg/L after 30 cycles. In general, the strong acid membrane performed better than the chelating membrane in cyclic tests. Figure 5 presents the composition of recovered alum after 30 cycles with the strong acid membrane (without intermittent rinsing).

The test with the strong acid membrane indicated that the two-step process can recover over eighty percent alum from the sludge without significant problem. For aluminum content of the sludge around or below 2000 mg/L as aluminum(III), such a high recovery can be attained rather easily by using approximately 130 cm² of the strong acid membrane per liter of the sludge. For aluminum content of the sludge well over 2000 mg/L as aluminum, the membrane area and/or the total cycle number per unit volume of the sludge are to be increased.

Figure 6 presents the concentration of dissolved organic carbon (DOC) versus the cycle number for both the tests with and without intermittent rinsing of the strong acid membrane using tap water. The DOC concentration in the test with rinsing was much lower than that in the test without rinsing. These results proved that organic recovery in the cyclic test without rinsing was mainly caused by physical carryover of the solids containing organic matters and not by adsorption. These results are amenable to scientific explanations with the Donnan exclusion principle.[5] Figure 7 provides a schematic illustrating the mechanisms responsible for the exclusion of organics by the strong acid membrane.

Durability Tests Figure 8 shows the results of the durability tests carried out using: the new strong acid membrane; the strong acid membrane after 10 cycles of operation with AWTP sludge; and the strong acid membrane after 100 cycles of operation with AWTP sludge. Note that, after 10 cycles of operation, the composite membrane's aluminum uptake capacity dropped. However, there remains only a marginal difference in the performances between the composite membrane after 10 cycles of operations and the one used after 100 cycles. It may be inferred that any possible fouling of the composite ion-exchange material (CIM) by the sludge medium essentially ceases after a few initial cycles of operation. Similar observations were made in a separate study using the CIM to treat heavy metal-laden sludge.[6]

Recovered Alum as a Coagulant Figure 9 depicts residual turbidity versus alum dosage for three different coagulants during jar tests. All the three coagulants worked satisfactorily; however, the recovered alum performed more efficiently (i.e., required lower dosage) to attain the same turbidity removal compared to the commercial alum.

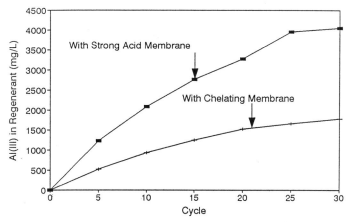

Figure 4 *Al(III) concentration in recovered alum versus the cycle number for tests with the strong acid membrane and the chelating membrane*

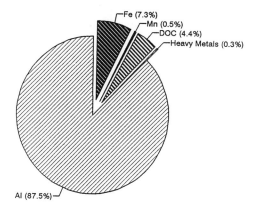

Figure 5 *The composition of recovered alum with the strong acid membrane*

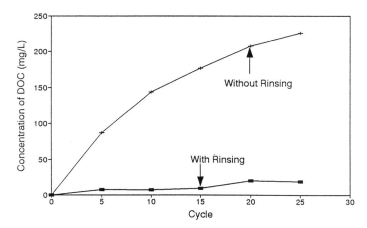

Figure 6 *Dissolved organic carbon (DOC) versus the cycle number for both the tests with and without intermittent rinsing*

Strong Acid Membrane

Figure 7 *A schematic of the mechanisms responsible for the exclusion of organics by the strong acid membrane*

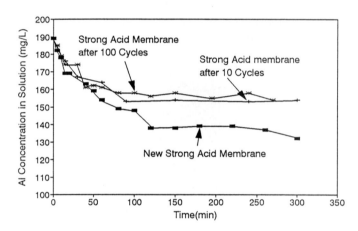

Figure 8 *The results of the durability tests with the new, 10 cycle- and 100 cycle-used strong acid membrane*

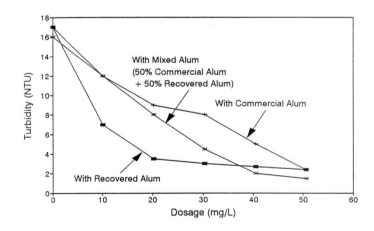

Figure 9 *The results of the tests for evaluating recovered alum as a coagulant.*

5 CONCLUSIONS

The major conclusions of the study can be summarized as follows:

(1)   The two-step process could recover over eighty percent alum from the clarifier sludge;

(2)   The two-step process was quite selective in the sense that recovered alum contained less organic matter and heavy metals than were present in the sludge;

(3)   The use of the strong acid cation exchanger beads within the CIM significantly enhanced the performance of the two-step process;

(4)   The CIM was found quite durable even after two hundred cycles of operation;

(5)   The recovered alum was equally effective as a coagulant, during jar tests, as the commercial alum.

Acknowledgement

The research project was funded by the American Water Works Association Research Foundation (AWWARF).

References

1.   American Water Works Association Research Foundation, "Coagulant Recovery: A Critical Assessment", Denver, CO, 1991.

2.   A. K. SenGupta and B. Shi, *Jour. AWWA*, 1992, **84**, 96.

3.   L. A. Errede et al., *Jour. of Appl. Polym. Sci.*, 1986, **31**, 2721.

4.   P. Li, MS Thesis, Lehigh University, Bethlehem, PA, 1994.

5.   F. Helfferich, "Ion Exchange", Xerox University Microfilms, Ann Arbor, MI, 1961.

6.   S. Sengupta, Ph. D., Dissertation, Lehigh University, Bethlehem, PA, 1993.

7.   APHA, AWWA and WPCF, "Standard Methods for the Examination of Water and Wastewater", 18th Edition, Washington, D. C., 1992.

CHELATED HEAVY METALS: RECOVERY AND SEPARATION FROM LIGANDS

Z. Matejka, H. Parschova and P. Roztocil

Department of Power Engineering
Institute of Chemical Technology
16628 Prague 6
Czech Republic

1 INTRODUCTION

The selective removal of heavy metal cations from waste- or process- streams containing strong anionic complexing agents (citrate, NTA, EDTA) requires special acrylamide resin having oligo(ethyleneamine) moieties (triethylenetetraamine, tetraethylene-pentaamine or pentaethylenehexamine)[1]. The application of ordinary- and also cheap- carboxylic or iminodiacetate cation exchangers is not efficient in the case of NTA and EDTA complexes because of the high stability of anionic metal complexes in the treated solution[2]. Weakly acidic carboxylic cation exchanger is, however, also inefficient on the uptake of metals from citrate solutions.

In this contribution it is shown that, standard carboxylic or iminodiacetate cation exchangers can be efficiently used to remove heavy metals from the above mentioned anionic complexes, provided that suitable oligo(ethyleneamine) ligand is added into the treated solution. The originally present anionic complexes of heavy metals are partially transformed into cationic ethyleneamine-metal complexes (Eq. 1).

$$[NTA\text{-}Me]^- + EA \iff [EA\text{-}Me]^{2+} + NTA \tag{1}$$

Heavy metals present as cationic ethyleneamine complexes will be then removed from solution quantitatively without problems by cation exchangers (carboxylic or IDA-type). In order to keep in the solution the equilibrium state, which is given by the stability constants of the particular metal complexes, anionic metal complexes are gradually and continuously transformed into cationic complexes until the heavy metals are completely removed from the solution by cation exchangers. The breakthrough capacities strongly depend on the rate of metal recomplexation between anionic complexes (NTA, EDTA, citrate) and cationic ethyleneamine complexes. The condition and mechanism of this recomplexation process were studied in this paper.

The separation of heavy metals from EA ligands was tested, IDA-resin and picolylamine type resin compared and suitable operating conditions were determined.

2 EXPERIMENTAL

Resins   - carboxylic cation exchanger (Duolite C 433) Na⁺ form
            - iminodiacetate chelating cation exchanger (Lewatit TP 207) Na⁺ form
            - oligo(ethyleneamine) resins in the free base form: Sumichelate MC 10
                                                         Lewatit E-TEPA
                                                         Lewatit HLH -TETA
            - picolylamine resin (Dowex XFS 4195) - protonated form

Sorption of heavy metals from anionic complexes
         - dynamic column experiments, flow-rate = 8 BV.h⁻¹, pH - 8.5
         - Me^{2+} breakthrough concentration   - 1 mg/L
         - loading solutions : Me^{2+}-1 mM/L
                                citrate or NTA or EDTA - 2 mM/L
                                OEA-ligands (EDA,DETA,TETA,TEPA) - 2mM/L

Separation of metals from EDA
         - EDA - 1 mM/L
         - Me^{2+} - 0.5 mM/L ; pH range 4.0 - 8.5 ; flow-rate = 6 BV.h⁻¹

3 RESULTS AND DISCUSSION

3.1 Sorption of Metals from Citrate Complexes

In the case of Cu^{2+}, not only carboxylic resin but also IDA-resin are not able to take up this metal quantitatively (Figure 1), because citrate complex is stronger than IDA⁻ complex (Table 1), and there is a severe competition of both ligands for Cu^{2+} cation.

By the addition of EDA-ligand into the citrate solution, the new equilibrium will be set up and about 45% of copper is now present as $[Cu-EDA]^{2+}$ cationic complex. Carboxylic resin and IDA-resin are able to take up Cu^{2+} quantitatively and with high breakthrough capacities when EDA ligand was added into the loading citrate solution (Figure 1).

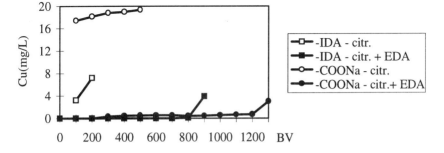

Figure 1 *Sorption of Cu from citrate on carboxylic- and IDA- resin*

Table 1   *Stability of Cu- and Ni- Complexes*

Ligand	$\log K\,(\,Cu^{2+}\,)$	$\log K\,(\,Ni^{2+}\,)$
EDA	10.7	7.5
DETA	16.1	10.6
TETA	20.4	14.1
TEPA	23.1	17.4
EDTA	18.2	18.2
NTA	12.7	11.3
citrate	12.0 - 14.2	6
IDA	10.3	8.2

The sorption of Ni^{2+} is remarkably improved on weakly acidic carboxylic resin (Duolite C 433) and the very high breakthrough capacity is obtained. IDA-resin takes up nickel from citrate solution completely even without the presence of EDA-ligand and therefore no improvement of the sorption ability has been observed for this resin (Figure 2).

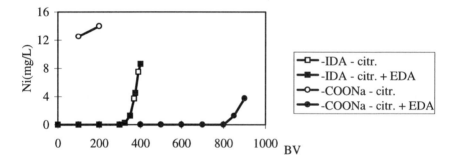

Figure 2   *Sorption of Ni from citrate on carboxylic- and IDA- resin*

3.2   Sorption of Metals from NTA Complexes

The stability of $[Cu\text{-}EDA]^{2+}$ complex is slightly lower than the stability of $[Cu\text{-}NTA]^{-}$ complex (Table 1). Consequently, the equilibrium distribution of Cu^{2+} in solution containing both these ligands (in a molar ratio 1:1) is following : 54.3% of copper remains as $[Cu\text{-}NTA]^{-}$ and 45.7% is transformed to $[Cu\text{-}EDA]^{2+}$ cationic complex.

Carboxylic cation exchanger (Duolite C 433-Na^{+}) removes from this solution only copper from $[Cu\text{-}EDA]^{2+}$ complex because the leakage level is 55% of the inlet Cu-concentration (35 mg/L) and this is the concentration of copper in $[Cu\text{-}NTA]^{-}$ (Figure 3). The possible explanation is that the recomplexation rate $[Cu\text{-}NTA]^{-}$ <=> $[Cu\text{-}EDA]^{2+}$ is rather slow in this system.

IDA-resin (Lewatit TP 207-Na^{+}), on the other hand, takes up copper quantitatively: the high chelating ability of the iminodiacetate functional group is obviously responsible for the remarkable increase in the recomplexation rate, which makes copper the very easily removable cation (Figure 3).

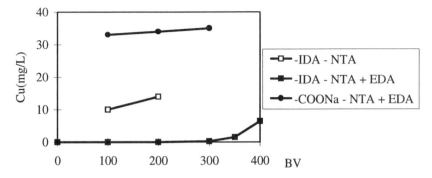

Figure 3 *Sorption of Cu from NTA on carboxylic- and IDA- resin*

The coordination ability of EA-ligands toward nickel are much weaker compared to copper and also the chelating ability of IDA-resin for nickel is substantially lower than for copper (Table 1). In order to achieve the high recomplexation rate, which is required for the quantitative Ni-uptake on IDA-resin it is necessary to add into the treated citrate solution the TETA-ligand (Figure 4).

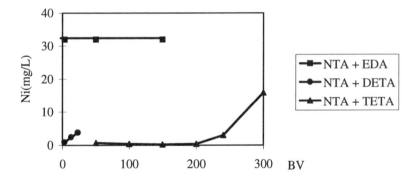

Figure 4 *Sorption of Ni from NTA complex on IDA- resin*

3.3 Sorption of Metals from EDTA Complexes

To achieve the quantitative Cu-uptake on IDA-resin, but with rather low breakthrough capacity, however, it is already sufficient to use DETA-ligand. The equilibrium copper distribution in the loading solution is : 46.9% as $[Cu-DETA]^{2+}$ and 53.1% remains as $[Cu-EDTA]^{2-}$ complex. The application of TETA- or TEPA-ligands will then increase the breakthrough capacity (Figure 5).

In the case of Ni^{2+}, the sorption efficiency of IDA-resin and carboxylic cation exchanger are the same: only partial uptake of metal occurs, corresponding to the amount of metal present in the cationic $[TEPA-Ni]^{2+}$ complex. The stability of $[TEPA-Ni]^{2+}$ complex is only slightly lower than that of $[EDTA-Ni]^{2-}$ complex, but the recomplexation rate in this system is very slow (Figure 5).

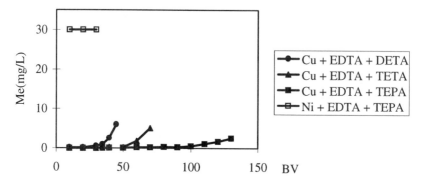

Figure 5 *Sorption of Cu and Ni from EDTA on IDA - resin*

3.4 Separation of Metals from Ethylenediamine Ligand

IDA-resin takes up heavy metals quantitatively, but the separation from EDA is only partial (Figure 6). In the early phase of the sorption run, copper is taken up as $Cu(EDA)_2^{2+}$ cation. As the resin loading by metal continues, the separation degree improves.

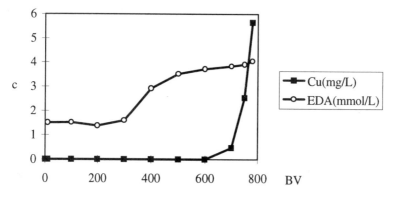

Figure 6 *Sorption of Cu and separation from EDA on IDA- resin*

In a study[1] conducted to select the suitable sorbents for separation of heavy metals from NH_3, the best results were obtained with ion exchangers having OEA functional groups. Heavy metals are taken up by coordination bonds to resin's functional groups, while NH_3 remains completely in solution (Eq. 2).

$$\overline{OEA} + Me(NH_3)_4^{2+} \quad <====> \quad \overline{[OEA\text{-}Me]^{2+}} + 4\,NH_3 \tag{2}$$

The high breakthrough capacities toward heavy metals were obtained.

The separation of heavy metals from EDA complexes is - unlike from NH_3 complexes - much more difficult, because of very high stability constants of these complexes. Several different OEA-resins were tested in our study and only Sumichelate MC-10 was able to take up heavy metals quantitatively (Table 2). The total separation of metals from EDA-ligand is achieved throughout the whole sorption phase (EDA is not taken up at all) for each metal.

EDA-complexes of heavy metals are rather bulky ions compared with $Me(NH_3)_4^{2+}$ complexes, and therefore their diffusion rate inside the resin particle (i.e. the porosity of a resin matrix) will be probably the key factor determining the sorption efficiency of the particular resins. It is also responsible for the big difference in breakthrough capacities observed among resins tested in this study.

Table 2   *Sorption of Metals from EDA- Complexes on Sumichelate MC-10*

pH	Breakthrough Capacity (equiv./L)			
	Cu	Ni	Cd	Zn
4	0.12	0.17	0.3	0.23
5	0.12	0.21	0.21	0.28
6.5	-	-	0.23	0.24
8	0.11	0.11	-	0.30

In order to overcome the adverse effect of the high stabilities of $[EDA\text{-}Me]^{2+}$ complexes on the breakthrough capacity (especially toward copper) the sorption runs were carried out also at the slightly acidic pH range, where metal complexes tend to dissociate and free metal cations can be taken up. But the problem is, Sumichelate MC-10 (and all other OEA-resins as well) has to be applied in the free base form (metal coordination to the protonated ethyleneamine-functional group is not efficient) and the acidic inlet solution is therefore immediately neutralised when contacted with the free base functional group on column operation (Eq. 3)

$$\overline{2(\text{-}OEA)} + Cu^{2+} + EDA.H^+ + H^+ \quad <=====> \quad \overline{2(\text{-}OEA.H^+)} + [Cu\text{-}EDA]^{2+} \tag{3}$$

The effluent pH value lies - regardless of the inlet value - in the range 7-9; the breakthrough capacities are not increased, because metals are again present as EDA-complexes. Therefore no effect of the inlet solution pH value on breakthrough capacity has been observed for Sumichelate MC-10 (Table 2).

The substantial improvement of the breakthrough capacity toward copper is achieved on picolylamine-type resin in the protonated form (Dowex XFS 4195) at solution pH 4.

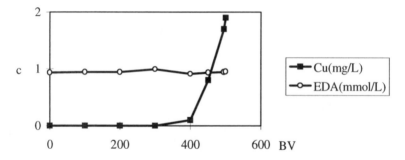

Figure 7 *Sorption of Cu and separation from EDA on Dowex XFS-4195*

The characteristic feature of this type of functional group is its ability to co-ordinate metals even in the protonated form[3]. The pH value of an effluent is then kept constantly in the acidic range 1.8 - 3.0 and copper is sorbed from solution quantitatively as free cation (Eq. 4) with the total separation from EDA-ligand (Figure 7).

$$\overline{\text{-picolyl.H}^+} + Cu^{2+} + H^+ 2(EDA.H^+) <==> \overline{[\text{-picolyl-Cu}]^{2+}} + 2H^+ + 2(EDA.H^+) \qquad (4)$$

Zinc and cadmium are sorbed with lower efficiency because the solution pH value is too low to co-ordinate these metals (Table 3).

Table 3   *Sorption of Metals from EDA- Complexes on Dowex XFS-4195.H⁺*

pH	Breakthrough Capacity (equiv./L)			
	Cu	Ni	Zn	Cd
4	0.40	0.18	0.025	0.025
5	-	0.18	0.05	-
6.5	-	-	0.035	-
8	0.11	-	0.05	-

4 CONCLUSION

1) The presence of EA-type ligands in the loading solution improves the sorption ability of carboxylic- and IDA-resins to take up heavy metals from the strong anionic complexes. Weakly acidic carboxylic resin takes up quantitatively heavy metals from citrate solutions and IDA-resin is able to remove heavy metals from NTA-complexes and Cu^{2+} even from EDTA-complex.

2) The key factor which is responsible for the improvement of the sorption ability of carboxylic- and IDA-resin is the fast recomplexation rate between anionic metal complexes and EA-type cationic metal complexes. The condition of this recomplexation process were determined.

3) The total separation of heavy metals from EDA-ligand is achieved by Sumichelate MC-10 resin in the free base form. Picolylamine-type resins (Dowex XFS 4195) in the protonated form are very suitable for the separation of Cu^{2+} and Ni^{2+} at low pH value.

Acknowledgements

Part of this work was financially supported by Grant No.104/93/0927 from the Grant Agency of Czech Republic for which grateful acknowledgement is made. Thanks are due to Sumitomo Chem. Co. (Osaka, Japan), Bayer AG (Leverkusen, Germany), Dow Chem. Corp. (Midland, MI, USA) and Rohm & Haas Co. (Chauny, France) for the kind supply of resin samples.

References

1. Z. Matejka, Z. Zitkova, R. Weber and K. Novotna, "New Developments in Ion Exchange", Kodansha Ltd., Tokyo, 1991, p. 591.
2. R. Naujocks, *Galvanotechnik*, 1983, **74**, 12, 3.
3. K. C. Jones and R. R. Grinstedt, *Chemistry and Industry*, 1977, August, 6.

ACID DYE REMOVAL FROM TEXTILE INDUSTRY EFFLUENT USING BIOLOGICALLY ACTIVE CARBON ADSORBENT

L. R. Weatherley, G. A. Walker and B. Al-Duri

Department of Chemical Engineering
The Queens University of Belfast
Northern Ireland, BT9 5AG
UK

1 INTRODUCTION

Activated Carbon is one of the most widely used adsorbents in the water industry. Although it is used extensively as a tertiary treatment after biological processing, the combination of biological activity with granular activated carbon (GAC) is a relatively recent development.[1] This may be due to the detrimental effects of clogging by biological growth or the entrainment of pathogenic bacteria in the outlet stream.[2] The term Biological Activated Carbon (BAC), i.e. a combination of bio-degradation and adsorption of a compound simultaneously on activated carbon, was first suggested by Rice et al[3] with various authors confirming its advantages over conventional carbon adsorption. Over a period of time micro-organisms will colonise GAC columns whether it is desired or not, due to the favourable environment provided by its porous external structure which provides shelter from hydraulic shear and may provide a reduction in concentration of toxic compounds due to carbon adsorption.[4] In most cases the biological activity is aerobic and therefore aeration, oxygenation or ozonation stages are therefore needed.

The aim of this work was to establish the effect of biological activity upon the performance of granulated activated carbon for the adsorption of three acid dye pollutants from waste water. Three dyes were chosen: Tectilon Blue 4R-01 an anthraquinone dye, Tectilon Red 2B an azo dye and Tectilon Orange 3G which is a di-azo dye. Although these are relatively non-toxic, quite small concentrations of around 10 ppm are very noticeable and untreated discharge even at this concentration is increasingly unacceptable.

Due to the inherent colour fastness properties of many dyes, they are difficult to remove from waste water by conventional biological processes. However due to the polar nature of their functional groups, dyes are relatively easy to remove by adsorption processes. Several bacterial species are known which degrade aromatic and azo compounds from which the dyes are synthesised and if these species were promoted on GAC it was thought that it may be possible to enhance dye removal by adsorption.

2 ADSORBENT SELECTION

The adsorbent used throughout the study was Filtrasorb F400 activated carbon (Chemviron) and was chosen as a result of preliminary kinetic and equilibrium adsorption studies, comparing dye uptake on peat and lignite with commercially available activated carbons. The results showed that although lignite does adsorb acid dyes, the rate of adsorption is markedly slower than that of F400, with equilibrium isotherms confirming a dye monolayer capacity on F400 of around 15 times greater than that of lignite and 25 times that of peat.

3 BATCH ADSORPTION SYSTEMS

Initially Filtrasorb 400 and the three dyes were contacted in a series of batch experiments to compare equilibrium uptake behaviour and kinetic behaviour for each dye. Full details of the experimental procedures are given elsewhere[5].

3.1 Equilibrium Studies

Adsorption isotherms were determined and the effects of particle size and of dye species in single, binary and ternary solutions were established. In the three single solution isotherm systems results indicated that smaller particle sizes were associated with increased adsorption capacity. This may be due to incomplete saturation, caused by pore blockage. A BET pore size distribution analysis showed that 95% of the pores were less than 10 nm across, see Figure 1. Since the dyes were of molecular diameter of approximately 1-2 Angstrom it may be possible with dye aggregation that blockage of smaller pores could lead to the non availability of active sites. This phenomenon would be exaggerated in larger particle sizes. A further factor is the benificial modification of pore structure during comminution which could explain the decrease in capacity with particle size.

Pore Size Distribution F400 (QUB)

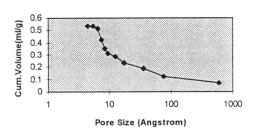

Figure 1 *Pore sized distribution F400*

Each of the isotherms was modelled using Langmuir, Redlich-Peterson and Freundlich analyses (Figure 2) with each of the models giving good agreement. In single dye systems the capacity for each dye varied from 535 mg dye/g carbon for Tectilon Blue, 710 mg/g for Tectilon Red, and 851 mg/g for Tectilon Orange. This apparent order of preference, Tectilon Orange>Tectilon Red>Tectilon Blue was again observed in binary and ternary system measurements.

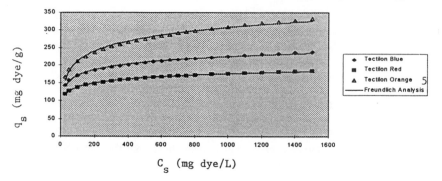

Figure 2 *Freudlich analysis of ternary solution filtrsorb 400 355-500 microns*

3.2 Kinetic Studies

The kinetics of dye uptake were studied with respect to the following variables (i) agitation rate, (ii) initial dye concentration, (iii) carbon concentration, (iv) carbon particle size, (v) dye solution temperature, (vi) pH of the solution, (vii) multicomponent solutions. These experiments were carried out in a 2L stirred tank reactor according to a method described elsewhere.[6] The adsorptions were followed spectrophotometrically.

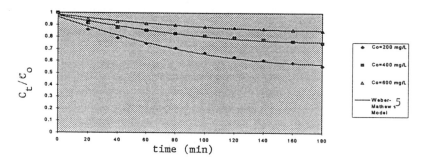

Figure 3 *Effect of initial dye concentration of tectilon red up-take on F400*

The results obtained from these experiments were mathematically modelled using single and double resistance models. Figure 3 shows the effect of dye concentration on the system with an external mass transfer and intra-particle diffusion model based on work by Weber-Mathews[7,8] and good agreement with experimental results over a wide range of concentrations, particle size distributions and agitation rates was obtained.

4 FIXED BED STUDIES

Scale-up for industrial work required data from continuous systems and in almost all cases GAC is used in either fixed or fluidized bed columns. Laboratory experiments were carried out using 25mm and 50mm diameter Perspex fixed bed columns. Bed heights of up to 500mm were used with sample ports located at 50mm intervals along the length of the column. The parameters studied included flow rate, concentration, particle size, bed height, and number of components. Results from these experiments were modelled using the Bed Depth Service Time (BDST) model[9] and the Brauch-Shlunder Column model.[10]

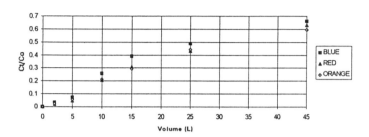

Figure 4 *Breakthrough curve ternary + pollutants*
(F400 5 cm Dia H= 5cm F= 2.5 mL/min Co= 300 mg/L)

A typical breakthrough curve for the ternary system is shown in Figure 4 and a characteristic 'S' shape is shown. Even at this low linear velocity the curve is quite shallow, indicating a large mass transfer zone, which was later confirmed in bed height analysis, see Figure 5. Slow diffusion rates within the pore structure, caused by possible repulsion of the dye functional groups by surface groups on the carbon pore wall may explain this. The effective capacity of the carbon in the column was calculated from constants derived from the BDST model and was found to be approximately 40% of the equilibrium capacity.

The column experiments showed agreement with the batch work with Tectilon Blue breaking through first, see Figure 5, indicating a lower adsorption rate compared to the other dyes although no dye displacement appears to have taken place. The Brauch-Schlunder model accurately described the experimental results, see Figure 5. It was noted that the effective diffusivity values used in the model were approximately 1/10th that of the molecular diffusivity for accurate correlation with experimental results. Intra-particle diffusion appears to be the rate controlling step, indicating the importance of contact time and linear flow rate.

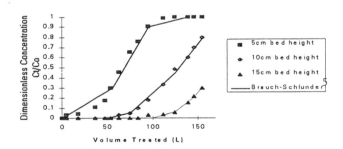

Figure 5 *Breakthrough curve of tectilon red on F400 linear flow= 5 cm/min*
Co= 100 mg/L

5 BIOLOGICALLY ACTIVATED CARBON

Bacteria were chosen which had the ability to utilise the azo dye compounds. *Bacillus Gordonae* (NCIMB 12553), *Bacillus Benzeovarans* (NCIMB 12555) and *Pseudomonas Putida Sp.* (NCIMB10015) are known to oxidise phenol and aromatic ring compounds[11,12,13] and *Flavobacteria Sp.* (NCIMB 9776) were used by Overney (1979)[14] to break the azo bond. These bacteria were also selected because they are relatively innocuous (NCIMB group 1),[15] metabolise under aerobic conditions and operate at 25°C and pH7 which made them suitable for use with the dyehouse effluent of interest here.

These strains were grown on suitable solid and liquid media with growth curves indicating an exponential growth phase from 6 hours incubation to 48 hours incubation, it was cells in this exponential growth phase which were contacted with the dyestuffs in shake flask experiments.

Initial results indicated that the dyes were bio-degradable according to Michaelis-Menten kinetics.

Subsequently a series of experiments was carried out with a stirred tank reactor configuration, volume 3L, in which an effluent of known dye concentration was contacted with; (i) bacteria immobilised on GAC, (ii) bacteria immobilised on sand, (iii) sterile carbon and (iv) free bacterial cells. These experiments were carried out in a water bath at 25°C with each of the bacterial species and a mixed culture containing them all. Aeration of this system proved difficult due to extreme frothing using conventional aerators, however the dissolved oxygen content was maintained at 6-7mg/L by a second impeller designed to entrain air from the liquid surface.

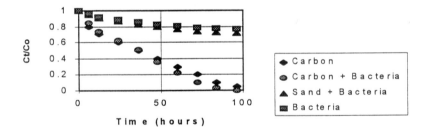

Figure 6 *Colour uptake curve bacillus gordonae & tectilon blue*

The data presented in Figure 6 show that in all four cases colour content was reduced but at different rates.

The contribution of free cells in this batch system was significant but with the addition of inert sand particles, biological activity increased beneficially due to the advantages of cell immobilisation at the external surface of the particles.

The addition of bacteria to the carbon initially reduced the adsorption rate, possibly due to macropores becoming clogged with nutrient broth or biomass. But subsequently the seeded carbon outperformed the sterile carbon in colour reduction. This could have been due to the combined effects of carbon adsorption and bio-degradation or due to carbon regeneration involving extra-cellular enzymes reacting with the adsorbate in the carbon pore structure. The latter is a contentious theory, as some investigators[16] suggest that this would be unlikely due to the high molecular volume of these enzymes compared to the carbon pore size and because of their tendency to be adsorbed.[17]

An alternative theory has been proposed[4] that bioregeneration is caused by desorption of solute due to decreased liquid phase concentration at the particle boundary and subsequent degradation by bacteria attached to the surface of the particle. There is experimental evidence that this is the case[5].

6 BIOLOGICAL ACTIVATED CARBON COLUMNS

The effect of biological activity upon column performance was determined. Fixed bed experiments were conducted with activated carbon seeded with one or more of the bacterial species.

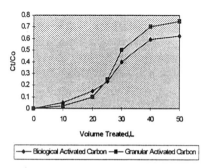

Figure 7 *BAC vs GAC Columns   The effect of pseudomonas putida on tectilon blue adsorption*

Results showed enhanced bed performance over conventional fixed beds although as before an initial acclimatization period was required. Additional nutrients required by the bacteria were fed constantly by a peristaltic pump but dissolved oxygen levels needed for aerobic metabolism could not be maintained in the upflow column system.

Several models have been developed by other investigators[18] to characterize the performance of biologically active adsorption systems.[19,20,21,22] Many of the assumptions inherent in these models make their general application difficult. Work is continuing to see if any of these models could be adapted for the current data.

7 PILOT PLANT

The final objective of the work was to assess the performance of an adsorber at pilot scale treating an actual dyehouse effluent. Preliminary bench-scale fixed bed experiments using a simulated dyehouse effluent containing not only dyes but other pollutants showed that adsorption was effective in reducing C.O.D. but that the dyestuffs broke through earlier in the presence of the other pollutants. On the basis of COD removal the biologically active columns out-performed the conventional adsorbers.

The pilot plant system was designed using a direct scale-up from these bench-scale data on the basis of linear velocity. The pilot plant system consisted of two 2m fixed bed columns, 0.5m diameter, operating in an upflow mode and were able to treat up to 2500L per day of dyehouse effluent. One of the columns was seeded with micro-organisms and initial results showed an enhanced bed performance. Figure 8 shows the predicted and actual outputs for the non-biologically active column plotted as COD against volume of water treated. The apparent increase in observed performance could be explained by unrecorded variations in inlet concentration from the dyehouse and an increase in the amount of fine particles in the column. The predicted breakthrough curve was obtained from a small scale simulation conducted in the 50 mm diameter laboratory scale columns at equivalent flowrate.

Results of these early trials showed that the pilot scale columns operating on the plant had breakthrough time of 100 days for the non-biologically active column and up to 150 days for the biologically active system.

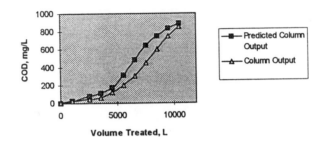

Figure 8 *Pilot Plant COD Column Breakthrough Estimated for C.V. Carpet Effluent COD=1200 mg/L Flow= 45 L/hour, approximately 1000L/day per column*

8 CONCLUSIONS

- GAC F400 has been selected from a range of adsorbents and used successfully to treat dyehouse waste water, reducing colour content and C.O.D.
- Adsorption of the acid dyes Tectilon Blue, Red and Orange on F400 has been demonstrated studied in batch and continuous systems and modelled effectively using Weber-Mathews and Brauch-Schlunder models.
- The three dyes were degraded using several bacterial species with degradation following Michaelis-Menten kinetics. Additional nutrients were required.
- A combination of biological activity and GAC adsorption enhanced carbon performance in dye decoloration and COD reduction of an actual dyehouse effluent.
- Pilot plant and bench scale studies show that by careful selection of bacteria and using adequate nutrient and dissolved oxygen levels, biologically seeded GAC columns can out perform conventional GAC columns.

Acknowledgements

This work was funded by The Department of Education for Northern Ireland and Coates Viyella Ltd.

References

1.   R. G. Rice, and C. M. Robson, "Biological Activated Carbon", Ann Arbor Science, Ann Arbor, Mich.,U.S., 1982.
2.   M. W. LeChevalier et al., *Appl. Env. Micro.*, 1984, **48** (5), 918.
3.   R. G. Rice et al., "Biological Processes in the Treatment of Municipal Water Supplies", Final report submitted to U.S. EPA., Municipal Environmental Research Lab., Cinc., Ohio, U.S., 1980.
4.   K. P. Olmstead and W. J. Weber, *Chem. Eng. Comm.*, 1991, **108**, 113.
5.   G. M. Walker, Ph. D. Thesis, The Queens University of Belfast, 1995.
6.   B. Al-Duri, Ph. D. Thesis, Queen's University of Belfast, 1988.
7.   W. J. Weber and A. P. Mathews, *A. I. Ch. E., Symp. Ser. Water*, 1976, **73** (166), 91.
8.   B. Al-Duri and G. McKay, *Chem. Eng. Sci.*, 1991, **46**, 193.
9.   R. A. Hutchins, *Am. J. Chem. Eng.*, 1973, **80**, 133.
10. V. Brauch and E. U. Schlunder, *Chem. Eng. Sci.*, 1975, **30**, 539.
11. S. M. Sala-Trepat, *Eur. J. Biochem*, 1972, **28**, 347.
12. T. Taniuchi, *J. Biol. Chem.*, 1964, **239** (7), 2204.
13. T. Kappeler et al., *Textile Chem. Color.*, 1978, **10** (8), 1.
14. G. Overney, Dissertation No. 6421, Swiss Federal Institute of Technology, Zurich, 1979.
15. T. R. Dando and J. Young, eds., "Catalogue of Strains", The National Collection of Industrial and Marine Bacteria, Aberdeen, 1990.
16. X. Zhang, *Wat. Res.*, 1991, **25** (2), 165.
17. A. Y. L. Li and F. A. DiGianno, *J. Wat. Poll. Cont. Fed.*, 1983, **55** (2), 392.
18. C. Chin et al., *J. Wat. Poll. Cont. Fed.*, 1973, **45**, 283.
19. W. C. Ying and W. J. Weber, *J. Wat. Poll. Cont. Fed.*, 1979, **51**(11), 2661.
20. G. F. Andrews and C. Tien, *A. I. Ch. E. J.*, 1981, **27**(3), 396.
21. H. T. Chang and B. E. Rittman, *Env. Sci. & Technol.*, 1987, **21**(3), 273.
22. G. E. Speitel and X. J. Zhu, *Env. Eng.*, 1990, **116**(1), 32.

CHEMICAL STRATEGIES FOR PRODUCING ANION EXCHANGERS FROM LIGNOCELLULOSIC RESIDUES TO BE USED FOR REMOVING TEXTILE DYES FROM WASTEWATER

J. A. Laszlo*
USDA-ARS, National Center for Agricultural Utilization Research,
Biomaterials Processing Research, 1815 N. University St.
Peoria, Illinois, USA 61604

I. Šimkovic
Institute of Chemistry, Slovak Academy of Sciences
84238 Bratislava, Slovakia

1 INTRODUCTION

The presence of textile dyes, particularly fiber-reactive dyes, in the wastewater effluents of dyehouse operations is a growing environmental concern.[1] Difficulties arise in effectively and economically treating these effluents because they can contain a great variety of dyes and high concentrations of co-solutes, such as inorganic salts, sizing agents and surfactants.[2] Fiber-reactive dyes, dyes that react to form covalent bonds with fabric, typify the adverse conditions posed. Five to fifty percent of the reactive dye, which is applied at pH 11-12, washes into the wastewater, accompanied by very high concentrations of sodium carbonate, sulfate, and chloride salts. None of the traditional wastewater-treatment methods (aerobic biological treatment, coagulation, membrane filtration, etc.) are effective in decolorizing these effluents in a satisfactorily cost-effective, environmentally-sound manner.

Because most types of textile dyes (acid, reactive, direct, mordant, vat, and azoic) are negatively charged, anion-exchange resins could effectively treat dye-containing wastewaters. Unfortunately, the high cost of commercial anion exchangers and the difficulty of their regeneration after saturation with dye has led the textile industry to dismiss considering the use of ion exchangers.[3,4] However, low-cost anion-exchangers prepared from the lignocellulosic byproducts generated by the processing of agricultural crop products may prove to be quite effective in decolorizing textile wastewaters. These biomass-based exchangers can be disposed of by composting or incineration, thus avoiding the need for regeneration.

Examples of potentially useful agricultural byproducts are sugarcane bagasse, soybean hulls, wheat straw and bran, rice straw and hulls, sugar beet and citrus pulps, and oat hulls. Each of these byproducts are produced in huge quantities (i.e., 10^9 kg/yr), have little or no value, and in some cases may pose a significant disposal problem themselves.[5] All of the byproducts are composed of cellulose, hemicellulose, lignin and acidic polysaccharides, in proportions that vary with byproduct type. Procedures developed to modify and exploit one type of byproduct generally can be applied to all types, although there may be some differences in the properties of the final product.

---

*   Names are necessary to report factually on available data; however, the USDA neither guarantees nor warrants the standard of the product, and the use of the name by USDA implies no approval of the product to the exclusion of others that may also be suitable.

Our goal is to find simple, inexpensive methods to convert agricultural byproducts into anion exchangers. This report discloses four synthetic routes for producing anion exchangers from sugarcane bagasse and wheat straw, and describes some aspects of their dye-binding behavior.

2 EXPERIMENTAL

2.1 Materials

Sugarcane bagasse, obtained from a Louisianna (USA) cane processor, was milled to a size less than 1 mm, then washed and dried as described previously.[6] Wheat straw was prepared as described previously.[7] N-(3-chloro-2-hydroxypropyl) trimethylammonium chloride (CHMAC) was obtained from Dow Chemical Co. (Midland, MI, USA). Epichlorohydrin and Alizarin Red S (approx. 70% dye) were purchased from Aldrich Chemical Co. (Milwaukee, WI, USA). Imidazole was purchased from Sigma Chemical Co. (St. Louis, MO, USA). Aqueous (93% solids) 1,3-bis (3-chloro-2-hydroxypropyl)imidazolium hydrogensulfate (BCHIHS) was obtained from Spolek pro Chemickou a Hutní Výrobu (Ústí nad Labem, Czech Republic). Remazol Brilliant Red F3B (Reactive Red 180) was obtained from Hoechst Celanese (Charlotte, NC, USA). This dye was hydrolyzed at pH 11.6, as described previously.[8]

2.2 Chemical Modification Procedures

Poly(hydroxypropylamine) bagasse (HPA-bagasse) was prepared by reacting bagasse in a closed reaction vessel with various amounts of epichlorohydrin, NH_4OH, NaOH, and water for 24 h at room temperature. The insoluble reaction product was equilibrated with dilute HCl (pH 2.0), then collected by vacuum filtration, washed with water to remove excess HCl, and lyophilized until dry. Imidazole-modified bagasse (IE-bagasse) was prepared similarly to HPA-bagasse, but with the substitution of imidazole for NH_4OH. Bis-(hydroxypropyl)-imidazolium-modified wheat straw (BHPI-wheat straw) was prepared by reacting wheat straw with various quantities of BCHIHS, NaOH and water under reflux. Details of the above reactions will be published elsewhere. Trimethyl-ammoniumhydroxypropyl-modified bagasse (TMAHP-bagasse) was prepared by crosslinking bagasse with epichlorohydrin (6.0 mmol per g of bagasse) and reacting the crosslinked product with CHMAC (4.0 mmol per g of crosslinked bagasse). The product was washed as described above and dried under vacuum at 25°C. Details of the CHMAC/bagasse reaction conditions are given elsewhere.[6]

2.3 Dye-Binding Assay Methods

The dye-binding capacities of the prepared exchangers were determined by equilibrating 1.0 eq/m^3 dye solution for 20 h (unless otherwise noted) at 25°C with a known amount of dry exchanger. The amount of bound dye was calculated from the equilibrated solution residual dye concentration, determined spectrophotometrically. Dye bound is expressed as equivalents of dye bound per kg of dry exchanger (eq/kg). Testing was performed with one dye only present in solution at a time.

Two dyes were employed to measure the dye-binding potential of prepared exchangers. Alizarin Red S (ARS) is a moderately small monovalent anion (Figure 1) belonging to the mordant dye class. Hydrolyzed Remazol Brilliant Red F3B (abbrev. F3B) is a moderately large trivalent anion (Figure 1). As the absorption spectrum of both dyes varies with pH, absorption measurements were made at the respective absorption maxima of the dyes (ARS, 594 nm; F3B, 540 nm) with the dyes diluted into 0.1 kmol/m^3 sodium carbonate (pH 11.6).

Throughout this work the assumption is made that there is a 1:1 stoichiometric association of dye sulfonic acid groups with anion exchange sites within the prepared exchangers. Support for this supposition is given by the observation of such a relationship for dye binding to water-insoluble chitosan[9] and soluble chitosan.[10] However, because of the complex and heterogeneous nature of the exchangers examined herein, this assumed stoichiometry of dye association is provisional.

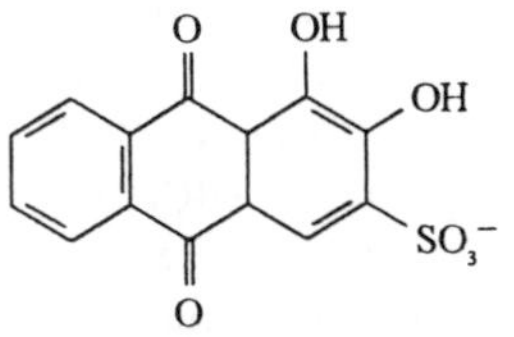

Alizarin Red S

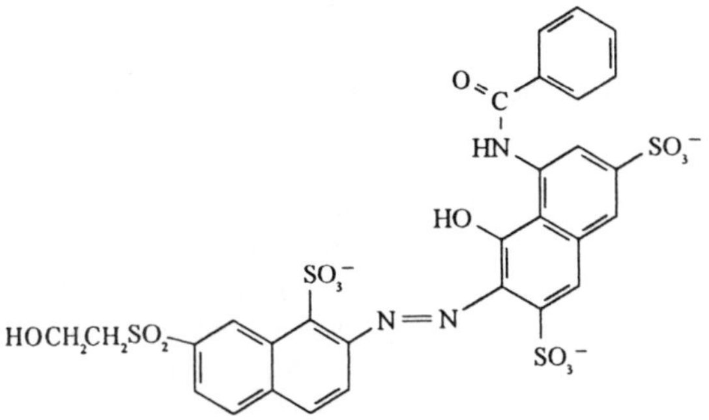

hydrolyzed Brilliant Red F3B

Figure 1 *Structure of dyes*

3 RESULTS AND DISCUSSION

3.1 Ammonia/Epichlorohydrin-Derivatized Bagasse

The simplest chemistry, and perhaps the least expensive method for introducing anion exchange sites into agricultural residues, is to react epichlorohydrin with ammonia in the presence of the lignocellulosic substrate. The resulting water-insoluble poly(hydroxypropylamine) material behaves like a weak anion exchanger. Potentiometric titration in the absence of added salts indicated that HPA-bagasse has an apparent pK_a of about 5.0, but, as with other weak base and weak acid exchangers, electrostatic interactions within the polymer matrix make the pK_a indistinct and highly dependent on solution-phase salt conditions.

The total exchange capacity of HPA-bagasse is readily manipulated by varying the epichlorohydrin : NH_4OH : bagasse ratio in the reaction mixture. HPA-bagasse prepared with a reactants ratio of 0.02:0.02:1 (mol of epichlorohydrin : mol of NH_4OH : g of bagasse) has total anion-exchange capacity of 3.7 eq/kg. However, the dye-binding capacity of this sample is substantially less than that expected based on its total exchange capacity, even at low pH (Table 1). Regardless of reaction conditions used to generate the HPA-bagasse (exchange capacity range: 1.6 to 4.6 eq/kg), the dye-binding capacity of the material (at pH 3.0) is half, or less, of the total exchange capacity. Furthermore, the extent of dye binding was highly dependent on solution pH (Figure 2), which is consistent with the weak-base properties of HPA-bagasse.

The inability to fully exploit the exchange capacity of HPA-bagasse for dye binding is a severe drawback. The reason for the inability of either small or large dyes to saturate the exchange sites of this material is unclear. Steric hindrance effects and hydrogen bonding of amine groups to bagasse hydroxyl groups may explain the low dye-binding capacity. However, exchangers prepared similarly with starch showed a much greater agreement between total exchange capacity and dye-binding capacity.[11] Regardless of the cause for diminished dye-binding capacities, the results indicate that HPA-bagasse is a poor exchanger for dye-removal applications.

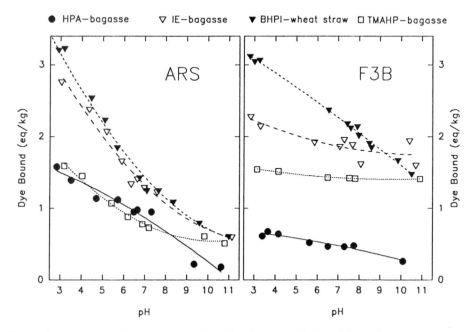

Figure 2 *Variation with pH in extent of dye binding to agricultural byproducts*

Table 1   *Dye-binding Capacity and Total Exchange Capacity of Selected Chemically Modified Agricultural Byproducts*

Material	Exchange Capacity (eq/kg)[a]		
	ARS	F3B	Total
HPA-Bagasse	1.9	1.9	3.7
EI-Bagasse	2.8	2.8	3.1
BCHI-Wheat Straw	2.9	3.1	3.6
TMAHP-Bagasse	1.9	1.8	1.7

[a]   Dye binding determined at pH 3.0. Total exchange capacity determined by potentiometric titration or calculated from nitrogen content.

3.2   Imidazole/Epichlorohydrin-Derivatized Bagasse

Treatment of bagasse with imidazole and epichlorohydrin produces a quaternized, strong anion exchanger (IE-bagasse). This one-step procedure provides both the needed crosslinking as well as the desired anion exchange capacity. Potentiometric titrations of this material were consistent with the presence of only strongly-basic exchange groups ($pK_a > 11$). Like the poly(hydroxypropylamine) exchanger produced using epichlorohydrin and ammonia, IE-bagasse probably consists of large blocks of epichlorohydrin-imidazole chains linked at one or both ends to lignocellulose. IE-bagasse prepared with a reactants ratio of 0.05:0.05:1 (mol of epichlorohydrin : mol of imidazole : g of bagasse) has a total anion-exchange capacity of 3.1 eq/kg. The high yield of insoluble product from this reaction (2 g, starting with 1 g of bagasse) indicates that there was a high degree of productive crosslinking.

At pH 3, the amounts of ARS and F3B bound is nearly equivalent to the total exchange capacity of the material (Table 1). Surprisingly, the extent of dye binding to IE-bagasse is pH dependent (Figure 2). This behavior may result from the association of aromatic imidazolium groups, which retain their positive charge over the entire pH range under consideration, with negatively charged groups present in the bagasse polysaccharides and lignin at high pH. At lower pH, the bagasse polymers lose their charge, permitting increased dye association with the exchanger groups.

3.3   BHPI-Derivatized Wheat Straw

The bis-reactive quaternization reagent BCHIHS introduces quaternary ammonium groups into lignocellulosic substrates with the added potential of producing polymer crosslinks. The derivatization of bagasse with BCHIHS has been described previously.[12] Therefore, wheat straw was examined instead. Reaction of wheat straw with BCHIHS at a ratio of 4 mmol BCHIHS per g of substrate produces an exchanger with a moderately high capacity (1.9 eq/kg) in good yield (1.6 g). As one would expect, higher ratios of BCHIHS : substrate result in higher exchange capacities in the product (data not shown). Unlike the block polymers that are likely formed during the reaction of epichlorohydrin with imidazole (see Section 3.2), the product produced from reaction with BCHIHS has individual bis-(hydroxypropyl)imidazolium adducts introduced into the lignocellulosic polymers of the wheat straw.

At pH 3, ARS and F3B bound to an extent slightly less than the total exchange capacity of the BCHI-wheat straw (Table 1). As was observed with IE-bagasse, dye binding to BHPI-wheat straw is strongly pH dependent (Figure 2), with ARS being more so than F3B.

3.4 TMAHP-Derivatized Bagasse

A final derivatization strategy is to quaternize agricultural byproducts using CHMAC and to crosslink the material with epichlorohydrin in a separate step. In this case it is clear that the quaternary ammonium groups are individually incorporated into the substrate polymers (i.e., no block copolymers formed). Quaternization proceeds with moderately high efficiency (approx. 60% of the applied CHMAC is incorporated into the final product) using 4 mmol of CHMAC per g of substrate in the reaction, producing a product having a total exchange capacity nearly as high as the BCHIHS product produced using the same reactants ratio.

ARS and F3B bound to TMAHP-bagasse very efficiently, utilizing all of the exchange capacity of the material at pH 3 (Table 1). The extent of dye binding varied little as a function of solution pH (Figure 2). Thus, unlike the other chemical modifications described, quaternization with CHMAC produces a dye-binding exchanger with a capacity that is largely independent of pH. This is important because reactive dyes are usually applied to fabric under alkaline conditions.

Table 2 *Potential Advantages and Disadvantages of Various Chemistries for Producing Anion Exchangers from Agricultural Byproducts*

Chemical Reagents	Criteria		
	Cost of Reagents	*Dye-binding Capacity*	*Potential Biodegradability*
Epichlorohydrin/NH₄OH	Low	Low	Low
Epichlorohydrin/Imidazole	Moderate	High	Low
BCHIHS	High	High	Moderate
Epichlorohydrin/CHMAC	Moderate	Moderate	Moderate

4 CONCLUSIONS

Each of the four chemistries described have specific advantages and disadvantages with regard to their use in making ion exchangers out of agricultural byproducts (Table 2). From a practical viewpoint, the important considerations are the cost of reactants, effectiveness of the product, and potential biodegradability. Biodegradability is important because the most (socially) acceptable and environmentally-benign method for disposal of dye-saturated exchanger may be by composting. If it is assumed that the blocks of synthetic polymer present in some of the prepared exchangers (HPA- and EI-bagasse) are highly resistant to biodegradation, then this represents a significant drawback. However, even for the exchangers lacking blocks of synthetic polymer (i.e., BCHI-wheat straw and TMAHP-bagasse), there is no assurance of biodegradability at this time. When cost, effectiveness and degradability are all considered, there is no clear "winner" among the four chemistries. Further work is needed before the best approach is identified for producing dye-binding exchangers from agricultural byproducts.

References

1.   J. Pierce, *J. Soc. Dyers Colour.*, 1994, **110**, 131.
2.   V. M. Correia, T. Stephenson and S. J. Judd, *Environ. Technol.*, 1994, **15**, 917.
3.   P. Cooper, *J. Soc. Dyers. Colour.*, 1993, **109**, 97.
4.   A. Reife, "Encyclopedia of Chemical Technology", Fourth Edition, M. Howe-Grant, ed., John Wiley and Sons, New York, 1993, Vol. 8, p. 753.
5.   M. L. Shuler, "Utilization and Recycle of Agricultural Wastes and Residues", CRC Press, Boca Raton, 1980, p. 23.
6.   J. A. Laszlo, *Textile Chem. Color.*, 1996, in press.
7.   I. Šimkovic, M. Antal and J. Alföldi, *Carbohydr. Polym.*, 1994, **23**, 111.
8.   J. A. Laszlo, *Textile Chem. Color.*, 1995, **27**, 25.
9.   G. G. Maghami and G. A. F. Roberts, *Makromol. Chem.*, 1988, **189**, 2239.
10. F. Delben, P. Gabrielli, R. A. A. Muzzarelli and S. Stefancich, *Carbohydr. Polym.*, 1994, **24**, 25.
11. I. Šimkovic, J. A. Laszlo, and A. R. Thompson, *Carbohydr. Polym.*, 1996, submitted.
12. I. Šimkovic, J. Mlynár, J. Alföldi, and M. M. Micko, *Holzforschung*, 1990, **44**, 113.

SYNTHETIC CLAY ANION EXCHANGERS: THEIR STRUCTURE, MODIFICATION AND APPLICATION IN REMOVING COLOUR AND TOXINS FROM TEXTILE PROCESS WATERS

M. Webb

Crosfield Ltd.
Warrington WA5 1AB
UK

1 INTRODUCTION

Unlike their more naturally-abundant cationic counterparts, anionic clay minerals have received relatively little attention ("cationic" and "anionic" refer to the exchangeable counter-ions of the clay). However, the preparation, structure and applications of synthetic "hydrotalcites" have been the subject of studies and commercial products are available for uses in medical antacid formulations, polymer stabilisation, oil-well drilling muds and as catalysts or catalyst intermediates[1,2]. This paper is concerned with their application as absorbents for contaminants found in textile industry process waters - particularly dyes and organotoxins which are present in effluent streams in significant amounts. It gives an overview of the structure of hydrotalcites, of how the structure of synthetic hydrotalcites has been optimised to increase their performance as absorbents for these contaminants and of their performance in field trials in an effluent treatment system which we are currently commercialising.

2 HYDROTALCITE STRUCTURE

Hydrotalcite is structurally-related to brucite - $Mg(OH)_2$ - a layered structure comprising edge-sharing $Mg(OH)_6$ octahedra which thus form two parallel sheets of $(OH)^-$ ions with Mg^{2+} cations occupying all the octahedral holes between them. In hydrotalcite, the structure is identical except that there has been stoichiometric replacement of some of the Mg^{2+} cations by Al^{3+} cations, the result of which is that the lattice has a nett positive charge. This is compensated for by anions associated with the lattice surface, which are mobile and hence exchangeable.

Hydrotalcite itself has the chemical composition:

$$Mg_6Al_2(OH)_{16}(CO_3).4H_2O$$

It is rare in nature but simple to synthesise by co-precipitating mixed metal solutions - most readily by raising their pH, heating to crystallise, filtering, washing and drying[3]. The resulting structure comprises hydrotalcite platelets aligning themselves in parallel to form small domains of stacked platelets. Between the platelets are the counter-anions (CO_3^{2-} in hydrotalcite itself) and the water of crystallisation. These clay domains are themselves aggregated in a typical product particle and micro/mesopores exist between them.

In addition to the parent hydrotalcite structure, a whole family of what are frequently termed "mixed metal hydroxides (MMH)", "layered double hydroxides (LDH), "hydrotalcitic materials" or simply "hydrotalcites" exists - both in nature and as the result of laboratory synthesis. Some variations on the structural theme are as follows:

* **Mg:Al ratio:** This is 3:1 in the parent material and tends to be a preferred stoichiometry, but it can be varied considerably. One effect of doing so is to change the lattice charge density and thence the anion-exchange capacity of the material.

* **Metals other than Mg and Al:** Analogous structures can be made, some of which occur naturally, where there are other metal(II) cations, e.g. Cu(II), Zn(II), Fe(II), other metal(III) cations, e.g. Fe(III), Cr(III) and mixtures of cations.

* **Anions other than CO_3:** The anions are exchangeable for other inorganic or organic anions and anion exchange has been extensively studied[2]. Synthetic hydrotalcites have anion contents which reflect the environment during precipitation although, because of the preference for carbonate, this will always be present unless carbonate and CO_2 are specifically excluded[2,4].

* **Crystallinity:** Platelets tend to be hexagonal, typically several hundred nm in lateral dimensions for natural hydrotalcite, but crystallite size can be controlled in synthetic materials by the method of preparation and by Ostwald ripening[4].

* **Product particle morphology:** The way in which the individual platelets aggregate within a product particle can be controlled and this affects the porosity characteristics of the product particle and the accessibility of anion-exchange sites.

3 STRUCTURE-PERFORMANCE STUDIES ON DYE ABSORPTION

The textile industry consumes large quantities of water in its processes. It also discharges large quantities of waste water contaminated with a variety of chemical components including salts, raw material debris, processing additives and spent dye. Colour in large volumes of water can be a visible contaminant even at levels well below 1 ppm and measures are in hand to introduce more stringent consent levels to reduce this contamination. However, even with best dyehouse management practice, colour levels in textile effluents can exceed the proposed consent levels by ten times.

Most commonly-used dyes are anionic in nature. The ability of hydrotalcite to absorb such dyes has been recognised for a long time and our early work confirmed that it had a considerable potential as an absorbent in textile effluent treatment. It is highly selective for colour and toxins (the problem contaminants), its absorption kinetics are rapid, it has a high capacity and contaminants can be removed to very low levels at realistic treatment levels. Laboratory trials with real effluents confirmed this promise but we wanted better weight-effectiveness in order to make treatment costs more affordable. To this end we looked at how variations on the compositional and structural themes (see part 2) affected the absorption kinetics and the binding capacity for dyes typifying those found in textile effluent. The highlights of this study are:

3.1 The Effect of Crystallinity

Dye solutions are neither chemically pure nor are they complete molecular dispersions, which complicates absorption/ion exchange studies considerably. However, our early studies showed the following:

(a) The inter-layer spacing in the XRD pattern is unchanged upon dye absorption whereas we would expect replacement of carbonate by a large organic ion to cause a significant change.

(b) The change in pore size distribution (from mercury intrusion and nitrogen BET) upon dye absorption indicates that some absorption occurs in larger mesopores even when capacity, therefore surface area, is available in smaller mesopores. Thermodynamics suggests that absorption should occur into the smallest available pores.

(c) Although some carbonate replacement occurs, it is not stoichiometric.

Our interpretation of these data is that dye absorption on hydrotalcite is by a combination of "colloidal sieving" of incompletely dissolved dye aggregates into mesopores, physical absorption onto mesopore walls and anion exchange, but that, under typical conditions of use, intercalation of dye molecules does **not** occur. Thence we hypothesised that, if absorption is chiefly at crystallite surfaces and not interlayer, we should increase dye binding by reducing crystallite size and thereby increasing the available surface area.

Figure 1 shows a typical improvement in dye absorption kinetics for a low crystallinity material (sample B) compared with well-crystallised material (sample A). This very sensitive kinetics test consists of taking a stirred solution of 0.01 g/L dye (Reactive Red), adding 14 ppm hydrotalcite and removing and filtering samples on a micropore filter at various time intervals. The filtrate colour content is quantified by uv-vis absorbance and expressed here as Absorbance ratio = Absorbance at time t / Initial Absorbance. Because absorption kinetics are sensitive to product particle size, these data are for materials of the same particle size (ca. 6 micron).

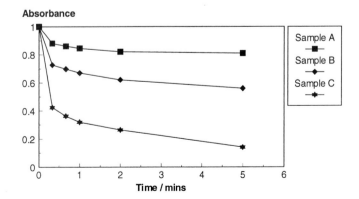

Figure 1 *Kinetics of dye absorption onto different hydrotalcite structures*

Table 1 shows the increase in dye binding capacity of low crystallinity material (sample B compared with sample A). The test consists of adding 0.03 g/L hydrotalcite to a 0.01 g/L dye solution (Procion Turquoise) for 2h then determining the dye uptake of the absorbent by filtration and measurement of the reduction of dye in the solution phase by uv-vis spectrophotometry. This is not a true equilibrium uptake and is also particle size-sensitive to some extent but results correlate well with performance in the field provided that comparisons of materials with the same particle size are made - in this case ca. 6 micron.

Table 1   *Dye Absorption Capacities of Different Hydrotalcite Structures*

Sample ref	Sample description	Capacity / % w/w
A	Typical synthetic HT	8
B	Low crystallinity HT	15
C	Undried, low crystallinity HT	30

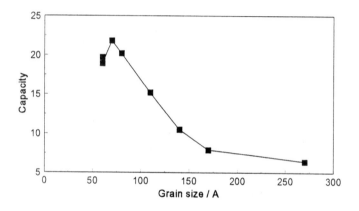

Figure 2   *The effect of crystal grain size on hydrotalcite capacity*

Figure 2 shows the relationship between crystallinity (grain size as determined by line broadening in the x-ray diffraction pattern[6]) and absorption capacity (determined as for Table 1) for a ca. 14 micron material. These results appear to confirm the hypothesis that we need to maximise the available surface area for absorption by reducing the size of the clay crystallites.

3.2   The Effect of Drying

Anyone researching or manufacturing porous materials finds that drying conditions, and particularly drying rate, can affect the pore structure and thence absorption characteristics of the product. However, what we found in this study was a surprisingly large and irreversible loss of performance when the low crystallinity products were dried - whatever the drying conditions.

Figure 1 and Table 1 show the differences in performance between a reslurried filter cake (sample C - typically 20% solids) and the fully dried material (sample B) at constant particle size. Figure 3 shows the results of steadily reducing the solids content of a filter cake (under conditions designed to maintain a constant particle size). This shows that there is a sharp drop in performance once the material is dried below about 15% moisture.

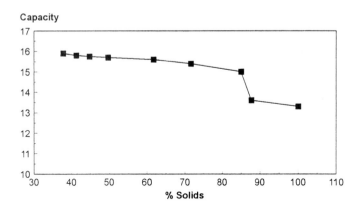

Figure 3 *The effect of removing water from hydrotalcite on its capacity for dye*

Very few techniques are available for determining the structure of porous solids whilst they are in an undried state. One technique which can is thermoporometry - based upon deducing the pore size distribution, from melting point depression, of a material whose pores contain water. Figure 4 shows the results for an undried filter cake compared with the same material dried and then rewetted to the same moisture content. The difference is that the undried material contains a significant volume (ca 0.2 cm^3.g^{-1}) of pores in the 3-10 nm region. Since surface area is dominated by small pores, this loss of porosity in small pores seems to be the explanation of the performance loss on drying.

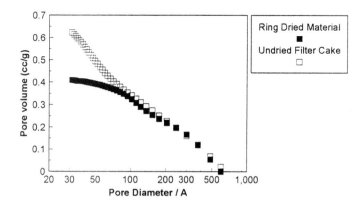

Figure 4 *Pore size distributions of wet samples by thermoporometry*

4 PERFORMANCE IN AN EFFLUENT TREATMENT SYSTEM

We have developed a system[7] based upon the following sequence:

* **Effluent Balancing:** i.e. passing effluent into a tank sufficiently large to smooth out variations in dye concentrations and types.

* **Chemical Treatment:** pH is lowered, hydrotalcite (plus, in some cases, other precipitants/flocculants) is added then pH is rapidly raised. During this step several events occur simultaneously: dye is absorbed upon the hydrotalcite (and co-flocculants if any), other organics present are "de-natured" and co-precipitated and the resulting "flocs" are densified and enlarged to make them more readily-settling.

* **Gravity Settlement:** preferably using a device such as the cross-flow clarifier (accelerated gravity settlement).

This treatment system generates a clear water stream, which is discharged (or re-used, see below) and a contaminated slurry phase which is disposed of (regeneration has not yet been developed).

Tables 2 to 4 show results derived from extensive field trials at textile plants and illustrate the levels of improvement in effluent quality which can be obtained with this system.

4.1   Colour Removal

In a dyeing operation, a very large number of different dyes will be used. The consequence is that the effluent also contains a wide range of colours deriving from an even wider range of molecules. Quantifying the degree of contamination of such a solution is a matter for ongoing debate, but one technique is to measure absorbance at different wavelengths in the visible spectrum. The results in Table 2 are a typical spot check on the effluent during a trial (at a dosage level of 0.1 g/L).

Table 2   *Colour Removal from a Textile Effluent by a Hydrotalcite System*

(Figures quoted are uv absorbance values in a 1cm cell at 100 nm intervals)

	Absorbance at wavelength / nm			
	400	*500*	*600*	*700*
Before treatment	0.59	0.89	0.23	0.11
After treatment	0.025	0.008	0.005	0.000
Proposed Consents (Leicester)	0.181	0.103	0.062	0.041

These results show that considerable colour reductions can be achieved in real life and proposed consent limits comfortably met.

4.2 Pesticide and Mothproofer Removal

Pesticides generally derive from natural fibres and are fairly low molecular weight organics with only slight aqueous solubility. We have found that hydrotalcite can remove a wide range of organochlorine and organophosphorous pesticides - including the frequently-encountered sheep dip chemicals propetamphos and diazenon - and narrow spectrum pesticides, such as the mothproofers permethrin and cyfluthrin.

Table 3 *Mothproofer Removal by Hydrotalcite*

	Permethrin	Cyfluthrin
Before treatment	31	4
After treatment	<0.27	<0.1
Consent limits	2.5 to 8	

Table 4 *Pesticide Removal by Hydrotalcite*

	Propetamphos	Diazenon
Before treatment	26	1.5
After treatment	0.9	0.1

Figures are parts per billion (ppb); dosage at 0.2 g/L for Tables 3 and 4.

4.3 COD Reduction

A further benefit of the treatment system is that, during the chemical treatment stage, other organic contaminants are co-flocculated and subsequently removed by separation, thus reducing the effluent COD substantially. Experience on wholly cotton dyeing effluents suggests that an average COD reduction of up to 75% can be achieved when using a typical 0.1 g /L dosage (another cost saving).

4.4 Water Re-use

The treated effluent produced from the above system has a reduced COD, is substantially free of colour and could be re-used in a number of dye-house processes. The commercial and environmental benefits of water re-use are at least as important as the removal of contaminants: less water needs to be bought into the plant, less water is discharged as effluent from the plant therefore treatment charges are lower, heat is recycled along with the water (because of the relatively short treatment times involved) and chemicals in the effluent e.g. scouring aids are also recycled. Computation for a representative UK textile site of treatment costs as a function of level of water re-use shows that, once re-use exceeds 15%, the nett result is operating cost **savings** rather than treatment costs incurred. With the current state of technology, we would anticipate that up to 50% water re-use should be possible.

5 CONCLUSIONS

Hydrotalcites are effective absorbents for colours and organotoxins from textile effluents. Their performance has been considerably improved by reducing crystallinity and by using them without drying, both of which increase the available surface area for absorption. A system for using them in effluent treatment, consisting of absorption followed by solid-liquid separation, has enabled proposed discharge consents to be met on colour and organotoxins and gives the additional benefit of reducing the total COD. Colour reduction occurs to such an extent that some water re-use is possible which can considerably reduce the cost of effluent treatment and, in some instances, enable actual cost savings to be made from treating effluent.

References

1. W. T. Reichle, *Chemtech*, 1986, 58.
2. S. Miyata, *Clays and Clay Minerals*, 1983, **31**, 305.
3. W. T. Reichle, *Solid State Ionics*, 1986, 135.
4. S. Miyata, *Clays and Clay Minerals*, 1980, **28**, 50.
5. Concannon et al, US Patent 5,159,456.
6. E. F. Kaelble, "Handbook of X-rays", McGraw-Hill, 1967.
7. P Cooper, ed., "Colour in Dyehouse Effluent", Society of Dyers and Colourists, 1995, Chapter 9.

THE NOVEL ABSORBER EVALUATION CLUB - A REVIEW OF RECENT STUDIES

E. W. Hooper and P. R. Kavanagh

AEA Technology
Harwell Laboratory
Harwell, Didcot
Oxon OX11 ORA

1 INTRODUCTION

The separation processes of ion exchange and sorption have been widely used for many years throughout the nuclear industry to decontaminate aqueous waste streams. The two processes are analogous, both removing ions from solution either by exchange of a counter ion (ion exchange) or by direct physical removal (sorption). Standard formulations of sorbents are used worldwide to treat waste streams, but as new materials are manufactured and discharge authorisations are lowered, it becomes increasingly difficult for operators to adhere to the governing principles of ALARP (*as low as reasonably practicable*) and BATNEEC (*best available technology not entailing excessive cost*). The objective of the Novel Absorber Evaluation Club is to remain aware of new sorbent developments and produce decontamination data on these products for direct comparison with standard 'benchmark' materials.

The Novel Absorber Evaluation Club (NAEC) was inaugurated in 1988 by AEA Technology on behalf of the UK Nuclear Industry as part of an overall requirement to minimise any environmental impact resulting from operations.

Membership of the NAEC is open to all parties with an interest in the decontamination of liquid radioactive effluents. The work programme of the NAEC is guided by the membership with a remit to provide quantitative experimental information of the ability of new and novel absorbent materials to remove radionuclides from solution.

Manufacturers and suppliers from industry and research are invited to submit materials for testing. Members prioritise the testing which comprises an initial screening of the material in a series of batch contacts with up to five standard NAEC reference waste streams. Further evaluation at suppliers request may include flow through packed beds or use as a finely divided addition in combination with membrane filtration. Test conditions may be varied by change of pH (in the range 3 - 11) and additions of organic contaminants or complexants.

With a membership covering the spectrum of the nuclear industry, the NAEC is in a unique position as an independent test facility with input from acknowledged experts in the field. Present members include British Nuclear Fuels plc, Nuclear Electric plc, Ministry of Defence (represented by the Atomic Weapons Establishment, Aldermaston and Director General Submarines, Bath) and UKAEA Government Division. A parallel contract also exists with the Her Majesty's Inspectorate of Pollution for full exchange of information.

Features of the Club programme include:

- Independent test facility for manufacturers of sorbents

- Utilisation and access to a database of results and reports generated since 1988

- Use of full radioactive facilities, experimental techniques and operator expertise

- NAEC methods and reference waste streams internationally recognised by IAEA

- Reference waste streams cover the range of radionuclides expected to be found in virtually any nuclear stream

- Worldwide contacts through Club experience, membership and exchange of information with research centres

- Capability of tailoring an absorber recipe to optimise the decontamination of specified waste streams

This paper will detail the tests carried out since 1991 and summarise the results obtained for the sorption of a range of elements including Mn, Co, Cr, Zn, Ru, Cs, Sr, Cd, Hg, Ag, Fe, Sb, Tc and alpha emitters (Pu + Am).

2 TEST DETAILS

2.1 Reference Waste Streams

Five reference waste streams are used in the test programme, known as NAEC (Novel Absorber Evaluation Club) S1, S2, S3, S4 or S5.
All waste streams are 0.05M in sodium nitrate to reduce peptisation of the sorbent.

NAEC S1   Contains ^{137}Cs, ^{60}Co, ^{65}Zn, ^{51}Cr, ^{59}Fe, ^{54}Mn, ^{125}Sb, ^{106}Ru, ^{203}Hg, ^{109}Cd, ^{110m}Ag and ^{144}Ce at the 100Bq/mL level.

This stream is analysed by gamma spectrometry with a limit of detection of 1Bq/mL or less i.e. a decontamination factor of 100 or more can be detected.

NAEC S2   contains ^{99}Tc as the pertechnetate (TcO_4^-) at the 100 Bq/mL level.

Analysis is by beta scintillation counting.

NAEC S3   contains $^{239}Pu(IV)$ at 2 Bq/mL, ^{241}Am at 1 Bq/mL and ^{90}Sr at 5 Bq/mL

Analysis is by both alpha and beta scintillation counting.

NAEC S4   is NAEC S1 plus 0.25 g/L ethylene diamine tetra-acetic acid added as sodium form, and 0.15 g/L citric acid.

Analysis is by gamma spectrometry.

NAEC S5   Contains ^{239}Pu(IV), ^{241}Am and ^{237}Np (V) at 1 Bq/mL plus 1g/L NaHCO$_3$ to produce anionic carbonato species in alkaline solution.

Analysis is by alpha scintillation counting.

The test solutions are adjusted to the required pH value and then kept for at least 24 hours at room temperature before use so as to allow equilibration of ionic species.

The waste streams used in the test will depend on the type of sorbent being examined. NAEC S1, S3 and S4 will be generally used for cation absorbers and S2, S4 and S5 for anion absorbers.

2.2 Sorbent Conditioning

Prior to use, sorbents are conditioned by washing with water adjusted to the experiment pH value using sodium hydroxide solution or dilute nitric acid. The washings are continued until the pH of the wash remains at its original value for 2-3 hours or preferably overnight.

The absorber is used in wet condition after decanting the wash liquor and removing excess moisture with a cellulose tissue.

2.3 Batch Contact Experiments

1 mL portions of the conditioned absorber are measured into 100 mL borosilicate conical flasks using a hypodermic syringe which has had the conical end removed to provide a 'full-bore' syringe. 50 mL of the reference waste stream are added and the flask sealed with a polythene stopper before placing in an agitated thermostatted water bath at 25°C. 1.5 mL portions of the liquid are removed after 1, 2, 4, 6 and 24 hours, centrifuged and then 1 mL removed for counting.

A control experiment, i.e. no sorbent present, is included to determine the extent of 'plate-out' on the walls of the container.

All experiments are performed in duplicate.

The analytical results are presented as Bq/mL and also as a calculated decontamination factor with reference to the original waste stream activity level:

$$\text{Decontamination Factor (DF)} = \frac{\text{activity per mL in feed}}{\text{activity per mL in sample}}$$

No correction is made for the volume changes resulting from sampling.

2.4   Column Experiments

Glass columns of 1 cm bore are packed with 5 mL absorber. 500 mL of water adjusted to the test pH are washed through each column to condition the absorber. Each column is then allowed to stand for a further 5 days, the pH of conditioned effluent being checked daily and adjusted if necessary. The columns are operated at a throughput of 10 bed volumes per hour (approximately 400 mL per day) using reference streams NAEC S1, S2 and S3, as appropriate. The pooled effluent is sampled daily for analysis.

3 ABSORBERS TESTED

The absorbers tested by the NAEC have been provided by the following organisations (number of samples in brackets):

Duratek Corporation	(2)	USA
Toray Industries	(3)	Japan
Universal Chemicals	(2)	UK
Recherche Appliquee du Nord	(1)	France
MEL Chemicals	(2)	UK
Institute of Nuclear Chemistry	(4)	Poland
University of Reading	(8)	UK
Whatman Paper Ltd	(2)	UK
Tate & Lyle	(1)	UK
Rohm & Haas	(2)	USA
Termoxid Company	(3)	Russia
Technical University of Prague	(5)	Czech Republic
Allied Colloids	(1)	UK
BASF	(1)	Germany
Comenius University	(2)	Slovak Republic
Union Carbide		USA

The particular absorbers supplied by each organisation are identified in Table 1.

Absorbers to be tested in the future have been obtained from:

Eichrom	USA
Degussa	Germany
Sutcliffe Speakman	UK
Chemviron	UK

4 RESULTS

The results of tests carried out during 1988-1991 were reported at IEX '92[1].

Table 1 shows the results obtained with reference waste streams NAEC S1, S2 and S3 during the period 1992 to the time of writing. It should be noted that, for ease of scanning, the table shows either decontamination factor for each nuclide after 24 hours contact or mean effluent nuclide DF in the case of column experiments. These are indicated by either 'b' or 'c' respectively in column 2 of the table. In all cases, a > sign indicates that the nuclide was removed to below the limit of detection of the analytical method used.

All the results in Table 1 were obtained at a solution pH of 9 except where stated.

Table 2 lists the Decontamination Factors obtained for selected absorbers after 24 hours contact with NAEC S1 at pH values of 3, 6 and 11.

Table 3 shows the decontamination factors obtained with four absorbers contacted with NAEC S5 at pH values of 7, 9 and 11.

5 SUMMARY

The results presented in this paper show only a fraction of the data collected by the Novel Absorber Evaluation Club. A regularly updated Data Base provides Members of the Club with relevant information on sorbents for the decontamination of radioactive waste streams.

The information obtained from the Club's Test Programme is also of value in designing treatment processes for the clean-up of industrial wastes to meet discharge authorisations, recover metal values and to allow water recycle.

References

1.   E. W. Hooper, "Ion Exchange Advances" (Proceedings of IEX'92), M. J. Slater, ed., SCI Elsevier, 1992, p. 310.

Table 1   *Decontamination Factors after 24 hour Contact*

Absorber	batch/ column	Cr-51	Mn-54	Co-60	Zn-65	Ru-106	Cd-109	Sb-125	Cs-137	Hg-203	Fe-59	Tc-99	Sr-90	α
Clinoptilolite	b	11.4	6.0	9.1	25.6	>50	8.5	1.1	63	1.8	--	1.0	19.0	14.9
AW 500	b	12.2	49.6	37.6	>60	>60	>100	1.1	66	2.7	--	1.0	7.1	8.7
IRN 77/78 (pH 7)	b	--	2.5	2.8	>5	--	--	9.1	2.0	--	--	--	2.5	3.0
MEL Chemicals														
ZrP/ZrOH (9:1)	c	13.8	3.3	18.6	95.3	132	--	47.6	33.2	28.8	85	--	--	--
Inst. Nuclear Chemistry, Warsaw														
NiFC composite	b	4.1	2.4	4.1	13.1	3.4	--	1.1	35.4	--	--	--	--	--
CoFC composite	b	2.6	4.3	4.0	14.7	3.8	2.9	1.0	30.2	1.4	--	--	--	--
Polyan composite	b	1.9	5.7	5.4	12.2	3.2	3.0	1.0	1.0	1.4	--	--	2.3	2.8
TiFC composite (pH 6)	b	4.2	9.0	6.2	10.0	3.6	--	1.1	24.8	--	--	--	--	--
University of Reading														
Na Bentonite	b	1.6	1.8	1.8	2.7	1.4	--	1.0	1.3	--	>150	--	--	7.3
Ca Bentonite	b	1.7	2.1	2.1	3.3	2.6	--	1.0	1.2	--	6.0	--	--	13.6
Tin Hydrogen Phosphate	b	2.3	10.9	10.0	13.1	4.9	--	1.1	1.2	--	>150	--	--	2.1
Phosphatoantimonic acid	b	1.3	1.4	1.4	2.5	1.8	1.3	1.1	1.2	1.3	2.2	--	--	--
Oxide Y (pH 6)	b	3.4	5.7	7.4	11.1	4.8	--	1.2	1.1	--	15.4	--	31.5	54.5
LDH (pH 6)	b	1.9	4.5	4.5	4.0	2.5	2.8	1.0	1.1	1.6	3.1	--	1.6	--
Whatman Paper Ltd														
Partisil 4002	b	--	1.2	1.3	1.2	1.3	1.5	1.2	1.3	4.8	2.0	3.6	2.4	>150
Partisil 6002	b	--	1.1	1.0	1.0	1.0	1.1	1.2	1.2	2.8	1.4	2.0	1.6	15
Tate & Lyle														
Brimac 216 (milled)	b	1.8	8.5	7.4	>206	>254	--	1.2	1.0	--	--	--	1.0	>26
Brimac 216 (granular)	c	>123	36.4	21.2	>206	>254	--	1.9	1.1	--	--	1.1	3.2	80
Rohm & Haas														
Amberjet 1200	b	4.1	484	127	133	4.0	11.0	1.1	2.9	8.6	58.2	--	43.1	>1.5
Amberjet 4200	b	--	--	--	--	--	--	--	35.2	--	--	27.7	1.1	35.2

Table 1 (cont) Decontamination Factors after 24 hour Contact

Absorber	batch/column	Cr-51	Mn-54	Co-60	Zn-65	Ru-106	Cd-109	Sb-125	Ag-110	Cs-137	Hg-203	Fe-59	Tc-99	Sr-90	α
Termoxid Company															
Termoxid 3A	b	80.8	71.5	38.7	69.2	3.8	18.2	--	55.4	>800	5.1	36.7	--	3.1	2.7
Termoxid 5	b	>194	>930	>890	180	>204	>128	--	144	1.9	95.6	110	1.0	25	107
Termoxid 231	b	--	--	--	--	--	--	--	--	--	--	--	1.3	8.1	>107
Technical University, Prague															
NiFC-PAN	b	2.3	1.9	2.0	--	3.0	>77	--	1.2	>1260	1.3	2.3	--	1.6	15
TiO-PAN	b	>312	1.2	1.2	--	>218	24.1	--	1.6	1.0	1.2	9.8	--	5.4	>110
NaTiO-PAN	b	1.9	3.4	1.4	--	2.7	2.3	--	3.5	12.0	17.5	13.1	--	8.0	2.0
MnO-PAN	b	1.6	2.7	3.6	--	2.8	10.4	--	1.6	3.9	4.3	3.0	--	8.4	2.3
Nm-PAN	b	2.2	2.7	2.7	--	2.9	>77	--	1.1	>1260	1.6	3.0	--	57	22
University of Reading, II															
Kanemite	b	2.0	6.7	6.1	68	1.9	10.6	--	1.4	3.7	4.8	--	1.1	4.6	1.9
HDS	b	--	--	--	--	--	--	--	--	--	--	--	1.3	1.6	86.7
LDH-X	b	--	--	--	--	--	--	--	--	--	--	--	1.2	--	--
Allied Colloids															
Beringite (fine)	b	2.4	3.6	2.4	5.8	2.9	3.8	1.3	1.4	2.8	2.2	--	--	1.1	>106
Beringite (coarse)	c	>107	>1450	573	>730	12.4	11.0	1.2	>97	2.7	112	>378	--	--	--
BASF															
Divergan HM	b	8.2	17.4	25.1	31.3	1.3	>69	--	>280	1.2	>65	3.8	--		
Comenius University, Bratislava															
Nalsit G	b	11.3	5.8	5.5	5.5	2.1	4.4	--	5.2	6.1	2.5	11.6	--	7.3	2.4
Y-sit (coarse)	b	11.6	2.7	3.0	3.4	1.9	3.7	--	3.7	8.0	1.6	17.3	--	5.9	1.7
Y-sit (fine)	b	9.6	6.0	8.1	13.9	2.3	18.6	--	108	22.6	3.0	9.1	--	32.8	11.5

Table 2 *Effect of pH on the Decontamination Factor - NAEC S1, 24 hr Contact*

Absorber →	ZrP			ZrOH			Goethite			HTiO			NiFC			BRIMAC		
pH →	3	6	11	3	6	11	3	6	11	3	6	11	3	6	11	3	6	11
Cs-137	>200	9	>200	1	1	1.7	1	1	1	1	1.6	1.2	15	127	1.2	1.7	1.7	1.4
Fe-59	31	13	13	17	740	>8000	5	>1500	>8000	19	>1500	22	6	493	>8000	2	>1500	180
Co-60	231	290	7	27	290	23	1	>870	38	21	>870	15	9	290	140	132	435	58
Ag-110m	38	1.6	77	1.3	101	>2710	1	77	>2710	1.4	>1000	>2710	>114	10	1	1	101	12
Sb-125	1.3	1	-	29	>275	-	10	>275	-	1.5	1	-	1.3	1	-	1.2	1.8	-
Cr-51	4	5	3	6	>275	15	2	>275	6	4	>275	13	3	22	7	1.2	>275	18
Mn-54	116	>970	8	12	>970	55	1	>970	60	36	>970	17	7	23	86	3	>970	66
Zn-65	48	57	-	21	194	-	1	>970	-	16	243	-	6	323	-	22	323	-
Hg-203	1	3	4	2	10	214	1.3	5	65	11	25	17	5	>234	2	1	69	>904
Cd-109	8	6	22	5	16	54	1.2	49	>313	6	54	45	5	>70	>313	9	>70	>313
Ru-106	7	2.5	3	11	>110	44	2.5	>110	40	9	9	4	3	5	>886	>257	17	>886

Table 3 *Decontamination Factors obtained with NAEC-S5, Batch Contacts, 24 hours*

Sorbent	pH 7	pH 9	pH11
Milled Brimac 216	>216	216	>216
HTiO	4.5	6.4	9.6
Goethite	>258	>258	>258
Termoxid 5	2.4	2.8	3.1

REMOVAL OF COPPER FROM DILUTE AQUEOUS SOLUTIONS BY ADSORBENT AND ION EXCHANGE MATERIALS

M. Ulmanu, T. Segarceanu, C. Vasiliu and I. Anger

Reseach Institute for Non-Ferrous and Rare Metals
Bucharest
Romania

1 INTRODUCTION

The increase in industrialisation and urbanisation poses new and varied ecological problems and as a result the preservation of the environment becomes more and more important. Water in particular is very exposed to the danger of pollution from the large quantities of effluent that industry daily discharges into rivers, sewage works, etc. with the ions of non-ferrous heavy metals such as copper, zinc, cadmium and lead posing special problems due to their high toxicity. Several methods are available for the removal of such ions from waste waters, including precipitation and resin ion exchange which are widely used in industry. In recent years other processes have been developed which offer certain advantages, among these being the use of compounds with adsorbent and ion exchange properties derived from natural materials such as: clays, Zeolites, ashes, cellulose and agricultural wastes[1-7]. The advantages of these materials are they are cheap and non-polluting and the process is relatively easy to operate and control. This paper describes the results obtained for the removal of copper from dilute aqueous solutions containing between 66 - 240 mg/L copper by 12 different materials having both adsorbent and ion exchange properties. The influence of contact time and the nature of the anionic species on the efficiency of copper removal has been studied and adsorption capacities determined.

2 EXPERIMENTAL

2.1 Materials

Experiments were performed using synthetic solutions containing copper ions prepared from copper acetate, chloride, nitrate and sulphate (Merck quality) using the following materials: silica gel, activated carbon, kieselguhr and kaolin, all Merck quality; Zeolites ZSM-5 and Y; bentonite (major component montmorillonite); diatomite (major component illite); fly ash, (a mixture of quartz, mullite, albite, hematite and magnetite); saw-dust and a residual ash from the food industry. The latter consisted of potassium and sodium aluminosilicates, calcium, potassium and magnesium carbonates and small quantities of silica.

3 SORPTION STUDIES

3.1 Optimum Contact Time

Equal amounts of adsorbent (2.000 g) were contacted with a stirred copper sulphate solution (100 mL) containing 180 mg Cu/L at 25°C for between 1.0 and 3.5 hours. At defined intervals the solid was removed by filtration, the pH of the filtrate measured and the copper content determined by atomic absorption spectrophotometry. The results are given in Table 1, Figure 1, showing the quantity of by copper retained by the materials under test, and Figure 2, the efficiency of copper both as a function of contact time. The best results were obtained when using the residual food ash, (1 hour contact); bentonite and Zeolite ZSM-5, (3 hours contact).

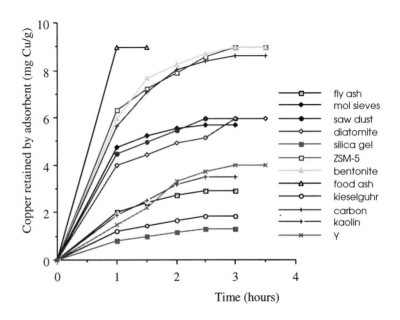

Figure 1 *Variation of copper removal from a copper sulphate solution (180 mg/L) with contact time*

3.2 Influence of Copper Ion Concentration on Copper Removal

A series of experiments were carried out under the same conditions as above with aqueous solutions containing 66, 104, 180 and 240 mg Cu/L using the optimum contact times for each adsorbent. The results are shown in Table 2 and Figure 3, showing the variation of copper retained by the absorbent as a function of copper concentration in solution, and Figure 4 the related removal efficiency.

The adsorbent materials fall into several groups which reflect their overall capacity for the metal. Thus for fly ash, saw dust, diatomite and silica gel there is a decreasing percentage of copper retained as the feed concentration increases. In the case of kaolin,

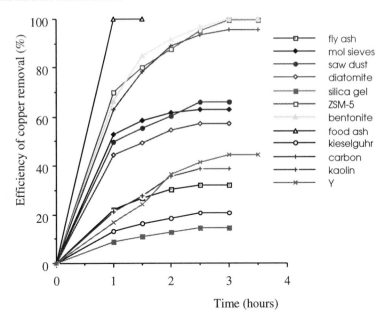

Figure 2 *Variation of copper removal efficiency with contact time*

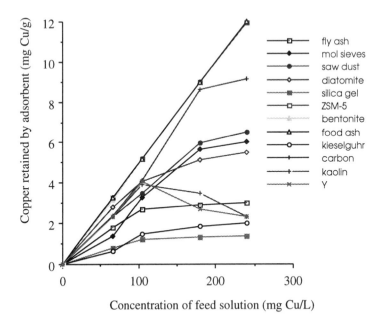

Figure 3 *Variation of copper removal as a function of the concentration of the copper feed solution*

Table 1   *Influence of Contact Time on the Removal of Copper by Materials under Test from an Aqueous Copper Sulphate Solution (180 mg/L)*

Material	1.0 mg/g	1.0 %	1.5 mg/g	1.5 %	2.0 mg/g	2.0 %	2.5 mg/g	2.5 %	3.0 mg/g	3.0 %	3.5 mg/g	3.5 %	pH
Fly ash	2.00	22.20	2.40	26.66	2.75	30.55	2.90	32.22	2.90	32.22	-	-	5.40
Molecular Sieves	4.75	52.75	5.26	58.44	5.55	61.66	5.68	63.11	5.68	63.11	-	-	6.09
Saw dust	4.48	49.77	4.98	55.33	5.45	60.55	5.95	66.11	5.95	66.11	-	-	4.73
Diatomite	4.00	44.44	4.45	49.44	4.95	54.44	5.15	57.22	5.15	57.22	-	-	6.23
Silica gel	0.80	8.88	1.00	11.11	1.15	12.77	1.30	14.44	1.30	14.44	-	-	4.25
Zeolite ZSM-5	6.31	70.11	7.21	80.11	7.91	87.88	8.55	95.00	8.97	99.66	8.97	99.66	7.60
Bentonite	5.97	66.33	7.66	85.11	8.25	91.66	8.68	96.44	8.99	99.88	8.99	99.88	7.63
Food ash	8.99	99.88	8.99	99.88	-	-	-	-	-	-	-	-	10.67
Kieselguhr	1.20	13.33	1.45	16.11	1.65	18.33	1.85	20.55	1.85	20.55	-	-	4.95
Activated carbon	5.66	62.82	7.07	78.55	8.01	89.00	8.39	93.22	8.60	95.55	8.60	95.55	4.20
Kaolin	1.90	21.11	2.50	27.77	3.20	32.55	3.50	38.88	3.50	38.88	-	-	4.80
Zeolite Y	1.50	16.66	2.20	24.44	3.30	36.66	3.71	41.22	4.00	44.44	4.00	44.44	5.25

Contact time (hours)

mg/g = mass of copper removed/g material and is given by: $x/m = (C_i - C_f)/10m$; where C_i = initial concentration of copper (mg/L), C_f = final concentration of copper (mg/L), m = mass of material (g) and 10 = dilution factor. % = efficiency of copper removal given by $E(\%) = ((C_i - C_f)/C_i).100$;

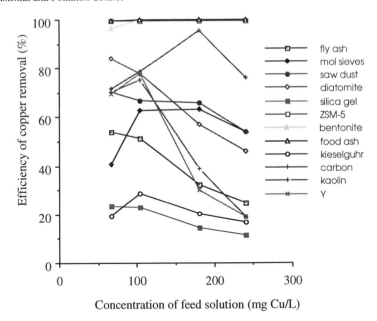

Figure 4 V*ariation of efficiency of copper removal as a function of feed concentration*

and Zeolite Y there is an initial increase followed by a subsequent decrease in retained copper after the 104 mg/L solution, with similar behaviour for active carbon and molecular sieves with a reduction in removal efficiency following the 180 mg/L copper solution. Finally for Zeolite ZSM-5, bentonite and the food ash there is a direct correlation between copper retained and feed concentration with no evidence of an approach to the capacity of the material.

3.3 Influence of the Anionic Species Present in Solution

To determine if the nature of the anionic species in solution has any influence on the absorption of copper a series of experiments were performed at constant copper concentration and contact time but varying the anions in the series sulphate, chloride, nitrate and acetate. The results, Table 3, indicate that the nature of the anion does influence the copper uptake with a variation between the maximum and minimum amounts of copper removed between 0.03 - 1.35 mg Cu (0.9 - 41.5%). The greatest difference is found for those materials which display the poorest copper removal, i.e. fly ash (33.3%); molecular sieve (34.2%); silica gel (41.5%); kaolin (40.9%) and Zeolite Y (30.3%). For these materials the largest amount of copper removed is found when the anion is chloride. Four materials fall in the range 20 - 26% with saw dust and active carbon favouring chloride while diatomite and kieselguhr favour nitrate. Finally the three materials, bentonite, food ash and Zeolite ZSM-5 achieve over 95% removal for all anions and show very little variation. In most cases the lowest value of copper removal is observed when acetate is the accompanying ion.

Table 2 *Influence of Copper Concentration on the Uptake of Copper by the Materials under Test*

Materials	Copper concn. 66 mg/L mg/g	%	Copper concn. 104 mg/L mg/g	%	Copper concn. 180 mg/L mg/g	%	Copper concn. 240 mg/L mg/g	%
Fly ash	1.78	53.93	2.67	51.34	2.90	32.22	3.00	25.00
Molecular Sieves	1.35	40.90	3.27	62.88	5.68	63.11	6.00	50.00
Saw dust	2.32	70.30	3.48	66.92	5.95	66.11	6.50	54.16
Diatomite	2.78	84.24	4.05	77.88	5.15	57.22	5.50	45.83
Silica gel	0.78	23.63	1.2	23.00	1.30	14.44	1.4	11.66
Zeolite ZSM-5	3.28	99.39	5.18	99.61	8.97	99.66	11.97	99.75
Bentonite	3.18	99.36	5.19	99.80	8.99	99.88	11.98	99.83
Food ash	3.29	99.66	5.19	99.80	9.00	99.99	12.00	99.99
Kieselguhr	0.66	19.69	1.50	28.84	1.85	20.55	2.00	16.66
Activated Carbon	2.36	71.51	4.10	78.84	8.60	95.55	9.13	76.08
Kaolin	2.32	70.30	3.90	75.00	3.50	38.88	2.30	19.16
Zeolite Y	2.30	69.69	4.05	77.88	2.70	30.00	2.30	19.16

Table 3 Influence of the Nature of the Anion on the Uptake of Copper from Solution (66 mg/L) retained by the Materials under Test

Material	Anion present in Copper Solution								$Q_{max} - Q_{min}$ (mg Cu / g material)	
	sulphate		chloride		nitrate		acetate			
	mg/g	%	mg/g	%	mg/g	%	mg/g	%	mg	%
Fly ash	1.78	53.93	2.15	65.15	1.05	31.81	1.20	36.26	1.10	33.33
Molecular Sieves	1.35	40.90	2.25	68.18	1.80	54.54	1.12	33.93	1.13	34.24
Saw dust	2.32	70.30	2.31	70.00	2.45	74.24	1.77	53.63	0.68	20.60
Diatomite	2.78	84.29	2.96	89.54	2.46	74.24	2.24	67.87	0.71	21.66
Silica gel	0.78	23.63	1.70	51.51	0.79	23.93	0.33	10.00	1.37	41.51
Zeolite ZSM-5	3.28	99.39	3.29	99.60	3.28	99.39	3.25	98.48	0.04	1.12
Bentonite	3.18	96.36	3.30	99.96	3.30	99.96	3.27	99.06	0.14	4.21
Food ash	3.29	99.66	3.30	99.96	3.30	99.96	3.27	99.06	0.03	0.90
Kieselguhr	0.66	19.69	1.05	31.80	1.35	40.90	0.69	20.90	0.66	20.00
Activated carbon	2.36	71.51	2.87	86.96	2.05	62.12	2.00	60.60	0.87	26.36
Kaolin	2.32	70.30	2.40	72.72	1.05	31.80	2.25	68.18	1.35	40.90
Zeolite	2.30	69.69	2.30	69.69	1.85	56.06	1.30	39.39	1.00	30.30

3.4 Determination of Absorption Capacity

The removal of copper in the systems studied can occur by a number of mechanisms, e.g. adsorption; ion exchange with the alkali, alkaline earth, and hydrogen ions within the material; and by precipitation of copper hydroxide. The formation of copper hydroxide is indicated by the variation of the solution pH on contact with the adsorbents (Table 1). These results show values above pH 7 for the Zeolite ZSM-5 (7.60), bentonite (7.63), and the food ash, strongly alkaline at pH 10.67. Other systems were close to pH 6: molecular sieves (6.09) and diatomite (6.23); or weakly acidic: activated carbon (4.20), silica gel (4.25), saw dust (4.73), kaolin (4.80), kieselguhr (4.95), Zeolite Y (5.25) and fly ash (5.40). These values are only indicative of an approximate pH value as it has been shown to vary both with contact time and concentration of the initial feed solution. These observations are the subject of another paper.

Therefore because of the complexity of these systems the term 'absorption capacity' is defined as the total amount of copper removed from solution by the adsorbing material. The 'absorption capacities' for copper on the Zeolite ZSM-5, bentonite and food ash were determined using a copper sulphate solution containing 350 mg Cu/L. For the first two materials, one gram was contacted with the above solution (100 mL) for 3 hours, while the food ash (1g) was contacted with the copper solution (500 mL) for 1 hour. The results obtained are given in Table 4.

Material	Copper content of solution (mg)		Copper removed by material (mg/g)
	initial solution	*final solution*	
Zeolite ZSM-5	35.00	21.48	13.52
Bentonite	35.00	16.90	18.10
Food ash	175.00	43.00	132.00

Table 4   *Adsorption Capacities of Zeolite ZSM-5, Bentonite and Residual Food Ash*

4 CONCLUSIONS

This paper describes some experiments to find a suitable cheap material for the removal of metal ions from dilute waste waters such as may be found in association with the extractive metal industry and to produce an effluent with a residual metal concentration which is below the maximum accepted concentration allowable by Romanian regulations. Of the twelve materials studied those which exhibit the best performance are the Zeolite ZSM-5, bentonite and a residual ash obtained from the food industry. Of these three the residual food ash showed an adsorption capacity of 132 mg Cu/g adsorbent nearly ten times that of the other materials. The ash is very readily available and cheap so should prove to be both an economic and environmentally acceptable material for this process.

Acknowledgements

The authors would like to thank Professor Michael Cox, University of Hertfordshire, for his interest and for help in preparing this paper.

References

1. T. Viraraghvan and A. Kapoor, *Appl. Clay. Sci.,* 1994, **9(1)**, 31.
2. M. J. Zamzow and J. E. Murphy, *Sep. Sci. and Tech.*, 1992, **27(14)**, 1969.
3. K-O. Subeha, R. C. Cheeseman and R. Perry, *J. Chem. Technol. Biotechnol.,* 1994, **59(2)**, 121.
4. T. Viraraghvan and F. M. Dronanraju, *Water Pollut. Res. J. Can.*, 1993, **28(2)**, 369.
5. A. Wilazak and Th. M. Keinath, *Water Environ. Res.*, 1993, **65(3)**, 238.
6. S. R. Shukla and V. D. Sukhardande, *J. Appl. Polym. Sci.*, 1990, **41**, 2655.
7. G. Vazquez, G. Antorrena, J. Gonzales and M. D. Doval, *Bioresource Technol.*, 1994, **48**, 251.

SELECTIVE RECOVERY OF URANIUM FROM DILUTE AQUEOUS SOLUTIONS : DOWEX 21K AND URANYL-TEMPLATED GEL-FILLED SORBENTS

M. Chanda and G. L. Rempel

Department of Chemical Engineering
University of Waterloo
Waterloo, Ontario
Canada N2L 3G1

1 INTRODUCTION

Most researches in ion-exchange development aim at enhancing functional properties such as sorption selectivity, capacity, and rate behavior which influence the applications of the resins. Working with poly(4-vinyl pyridine), Nishide et al[1] demonstrated the possibility of introducing memory effect, and thus increasing the selectivity, by templating with a chosen metal before crosslinking, though the effect was significant only for certain metal ions. Thus among Cu(II), Co(II), Zn(II), and Cd(II), the maximum memory effect was observed for Cu(II).

On the other hand, the problems of non-attainment of theoretical capacity of resins at equilibrium and slow rate of attainment of equilibrium sorption are often related to inaccessibility of sorption sites in the interior of resin beads. Though a significant improvement in both these respects results from preparing the sorbent as a thin layer on high surface area substrate, their application on a large scale for metal recovery and purification becomes uneconomic because of the requirement of large volume to make up for the reduced overall capacity of the sorbent. This advantage as well as the deficiency of gel-coated sorbents and the possibility of imparting a memory effect by templating led us to the development of a templated gel-filling (TGF) process[2] to make a Cu(II)-selective sorbent of high capacity and fast kinetics. With suitable modification, the process has now been used to make a uranium-selective TGF sorbent which shows significantly higher uranium capacity and kinetics in U/Fe dilute solution as compared to the commercial resin Dowex 21K, though its uranium selectivity relative to iron(III) is not as high as that of Dowex 21K. The results of the study are presented in this paper. Also reported, briefly, is the sorption selectivity that results from templating PEI with several other metal ions.

2 EXPERIMENTAL SECTION

2.1 Preparation of Sorbents

A commercial branched polyethyleneimine (PEI) (Aldrich Cat. No. 18,197-8) of average molecular weight 50,000-60,000 was used for making templated sorbents. For preliminary investigation of the effect of templating, a number of sorbents, PEI(M).XG, were prepared by simply complexing PEI with different metals (M) as template, followed by crosslinking(X) the remaining free amine sites by reaction with glutaraldehyde(G) at room temperature and subsequent removal of the template ion by leaching with acid. By this process, however, the sorbents are obtained only in a powder form.

The TGF process was therefore used for making a gel-type granular sorbent with UO_2^{2+} as template ion. The process is illustrated schematically in Figure 1. In a typical procedure, 100 g of silica gel (14-28 Tyler mesh) was soaked in 200 mL of 2% (w/v) UO_2SO_4 solution and evaporated to dryness on a water bath with continuous mixing (I). PEI (20 g), available as 50% aqueous solution was dissolved in 400 mL n-butanol (II). (I) was added to (II) and shaken on gyrotory shaker for 6 h. The solution was decanted off and the gel was washed several times with n-butanol. The gel-coated solid was then dispersed in 400 cm³ of 10% (w/v) glutaraldehyde solution in n-butanol and the mixture was vigorously agitated at room temperature for 10 h. The PEI-UO_2^{2+} complex gel-coated on silica and crosslinked by reaction with glutaraldehyde in this way was then treated successively with 2N H_2SO_4 and 2N NaOH to leach out UO_2^{2+} and SiO_2, respectively, leaving behind porous, hollow shells of templated and crosslinked PEI. The shells were next impregnated by soaking in a small volume of 20% (w/v) UO_2SO_4 solution for a few hours and then shaken on a gyrotory shaker at room temperature in a solution of PEI (80 g, 50% aq. solution) in n-butanol (800 mL) for 10 h. This resulted in filling of the hollow shells with a gel of PEI-UO_2^{2+} which was then crosslinked by treating with 300 mL 5% (w/v) solution of 1,4-dibromo-2-butene in n-butanol at 110°C for 12 h on a gyrotory platform shaker fitted with mantle heater and reflux condenser. The crosslinked and UO_2^{2+}-templated PEI granular gel was quaternized by heating under reflux at 90°C in a mixture of 100 mL $(CH_3)_2SO_4$ and 200 cm³ CH_3CN for 12 h on a gyrotory platform shaker. The granular sorbent was then washed successively with 1.5M Na_2CO_3 and 1.0M Na_2SO_4 and thoroughly washed with water. The resulting granular sorbent, designated TGF-PEI(U).XD(Q), was stored as wet solid (75% w/w moisture). The N content of the dried sorbent was 19.8% (w/w).

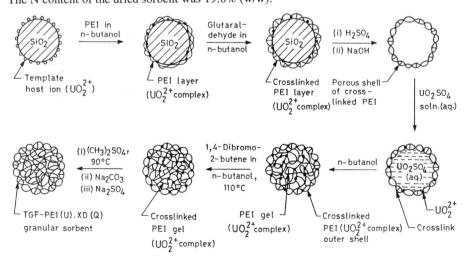

Figure 1 *Schematic of templated gel-filling process for making gel-type granular sorbent from polyethyleneimine using UO_2^{2+} as template*

2.2 Sorption Experiments

For the measurement of sorption, small scale dynamic contacts between resin and metal solution of specified compositions were effected in tightly stoppered flasks at 27°C on a wrist-action shaker. The extent of sorption was calculated from the residual concentration of the sorbate in the equilibrated solution. For equilibrium sorptions shaking was done for 20h.

3 RESULTS AND DISCUSSION

3.1 Template Effect on Selectivity

The selectivity of the sorbents for metal M_1 relative to metal M_2 was determined by measuring equilibrium sorptions in binary mixtures of equimolar M_1 and M_2 and calculating the separation factor $\alpha\,^{M_1}_{M_2}$ from

$$\alpha\,^{M_1}_{M_2} = \frac{x^*_{M_1} \times C^*_{M_2}}{x^*_{M_2} \times C^*_{M_1}} \qquad (1)$$

where $x^*_{M_1}$ and $x^*_{M_2}$ represent equilibrium sorptions (mmol/g dry resin) of M_1 and M_2; $C^*_{M_1}$ and $C^*_{M_2}$ are the equilibrium concentrations (mmol/L) of M_1 and M_2 in the presence of the sorbent. The results are summarized in Table 1.

Table 1   *Selectivity of Templated Resins*

Sorbent	Separation Factor[a]			
	$\alpha\,^{Cu}_{Ni}$	$\alpha\,^{Cu}_{Cd}$	$\alpha\,^{Ni}_{Cd}$	$\alpha\,^{U}_{Fe}$
PEI.XG	8.0	3.6	1.4	0.8
PEI(Cu).XG	9.5	7.9	-	-
PEI(Ni).XG	7.6	-	4.2	-
PEI(Cd).XG	-	2.5	0.9	-
PEI(Fe).XG	-	-	-	0.5
PEI(U).XG	-	-	-	1.6
PEI(U).XG(Q)	-	-	-	2.5
TGF-PEI(U).XD(Q)	-	-	-	4.2
Dowex 21K	-	-	-	10.0

[a]   Measured in binary mixtures of equimolar (10 mmol/L) solution of each metal at pH 5, except U/Fe solution which is equimolar with 1 mmol/L of each of UO_2^{2+} and Fe(III) at pH 3.0.

A comparison of the results in Table 1 shows that templating of PEI with a metal before crosslinking generally increases the selectivity for the same metal. However, this effect is more pronounced where there is sizeable difference between two metal ions and the smaller ion is used as the template, as can be seen for Cu/Cd and Ni/Cd systems.

Templating with UO_2^{2+} improves the $\alpha\,^{U}_{Fe}$ value of PEI and it is further augmented by quaternization, as can be seen by comparison of the $\alpha\,^{U}_{Fe}$ values for PEI.XG, PEI(U).XG, and PEI(U).XG(Q). It may be noted that the reaction of PEI with glutaraldehyde introduces OH groups which tend to increase[3] Fe(III) sorption. Therefore, for making uranyl-templated PEI resin by the TGF process, 1,4-dibromo-2-butene was used as the crosslinking agent. This product, further quaternized and designated as TGF-PEI(U).XD(Q), has significantly higher α^U_{Fe} than PEI.XG. Though Dowex 21K has an even higher value of α^U_{Fe}, its U sorption at low substrate concentrations is, however, very low.

3.2 Effect of pH on Sorption Capacity

The effect of pH on the uranium sorption of the two sorbents is compared in Figure 2(a). On both the sorbents a sorption peak is observed at pH ~ 2. Since at pH ~ 2 uranium exists predominantly as undissociated (neutral) uranium sulfate[4] (Figure 2b), the sorption may be described as taking place by the reaction

$$\overline{(QN^+)_2\ SO_4^{2-}} + UO_2SO_4 \Leftrightarrow \overline{(QN^+)_2\left[UO_2(SO_4)_2\right]^{2-}} \tag{2}$$

where QN^+ represents quaternized nitrogen in the sorbent. This mechanism is also favored by the fact that undissociated UO_2SO_4 being uncharged, is able to enter the resin freely without hindrance by the Donnan effect.

The uranium sorption on both the sorbents increases again with increasing pH at pH > 3. The increase, however, is more marked for the TGF sorbent. This may be explained by the fact that this sorbent has many free amine sites (which were held by the template ion during quaternization), besides QN^+ sites, and so can pick up UO_2^{2+} which are available in significant concentrations at pH > 3.

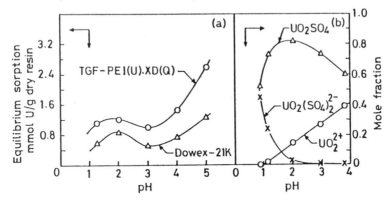

Figure 2 *(a) Effect of pH on equilibrium sorption of UO_2^{2+} on TGF-PEI(U).XD(Q) and Dowex 21K in H_2SO_4 media. Initial concentration of UO_2SO_4: 5.0 mmol/L; resin loading: 4.0-8.0 g (wet)/L; temperature 30°C. (b) Mole fraction distribution of uranyl species in solution at different pH values*

3.3 Sorption Isotherm

All the equilibrium sorption data on the TGF sorbent [Figure 3(a)] fitted well to both Langmuir and Freundlich isotherms. On the other hand, the sorption data on Dowex 21K

[Figure 3(b)] did not fit to the Langmuir isotherm, though they fitted well to the Freundlich isotherm. Defining the parameters A_s and K_b as the saturation sorption capacity (mmol metal/g dry resin) and the sorption binding constant (L/mol), respectively, the Langmuir isotherm is written as

$$x_A^* = \frac{\left(10^{-3}\, K_b\right) A_s\, C_A^*}{1 + \left(10^{-3}\, K_b\right) C_A^*}$$

(3)

where x_A^* is the equilibrium sorption (mmol metal/g dry resin) and C_A^* is the equilibrium sorbate concentration (mmol/L).

The Freundlich isotherm is written as

$$x_A^* = p\left(C_A^*\right)^q$$

(4)

where p and q are parameters, and x_A^* and C_A^* defined as above.

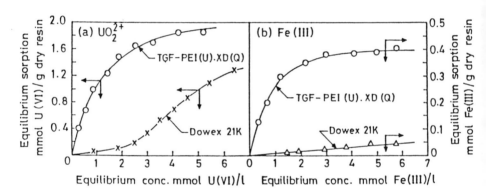

Figure 3 *Sorption isotherms for (a) UO_2^{2+}/SO_4^{2-} (pH 4.0) and (b) $Fe(III)/SO_4^{2-}$ (pH 2.0) on TGF-PEI(U).XD(Q) and Dowex 21K resins. Resin loading 4.0 g (wet)/L; temperature 30°C*

The values of A_s, K_b, p, and q determined by least squares fit of sorption data in Figures 3(a) and 3(b) are presented in Table 2. Both U(VI) and Fe(III) have much stronger sorption on the TGF sorbent than on Dowex 21K, as shown by the much smaller values of the Freundlich parameter q for the TGF sorbent, as compared to Dowex 21K.

Table 2 *Langmuir and Freundlich Isotherm Parameters for UO_2^{2+} (pH 4.0) and Fe(III) (pH 2.0) Sorption*

Sorbate/Sorbent	Langmuir Isotherm			Freundlich Isotherm		
	A_s /	K_b /	Corr.	p	q	Corr.
	mmol g^{-1}	L mol^{-1}	Coeff.			Coeff.
UO_2^{2+}/TGF-PEI(U).XD(Q)	2.44	741	0.999	0.92	0.53	0.979
Fe(III)/TGF-PEI(U).XD(Q)	0.48	1119	0.998	0.22	0.49	0.969
UO_2^{2+}/Dowex 21K	-	-	-	0.021	2.38	0.990

3.4 Effect of Common Anions

The equilibrium sorptions of U(VI) on the TGF sorbent and Dowex 21K in the presence of varying concentrations of NaCl and Na_2SO_4 are plotted in Figure 4. The uranium preference in these plots has been defined by

$$\text{U preference} = \frac{\text{equilibrium sorption of U in presence of salt}}{\text{equilibrium sorption of U in absence of salt}} \qquad (5)$$

The drastic fall in U sorption on Dowex 21K in the presence of Cl^- and significant increase in the presence of SO_4^{2-} are both explained by the fact that the sorption on Dowex 21K takes place mostly by ion exchange with SO_4^{2-} promoting the formation of the ionic species $[UO_2(SO_4)_3]^{4-}$. In sharp contrast, the effect of anions is much less for the TGF sorbent on which the U sorption takes place both by ion exchange and chelation.

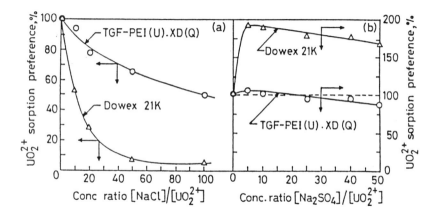

Figure 4 *Sorption preference for UO_2^{2+} in the presence of (a) NaCl and (b) Na_2SO_4 on TGF-PEI(U).XD(Q) and Dowex 21K. Initial concentration of UO_2^{2+} 5.0 mmol/L; resin loading 6.0 g (wet)/L; pH 4.0; temperature 30°C*

3.5 Kinetic Considerations

The rate of attainment of equilibrium sorption on the TGF sorbent (Figure 5a) is seen to be nearly independent of the sorbate concentration, suggesting that an ordinary particle diffusion control (pdc) model may be applicable[5]. On the other hand, the external solution concentration is seen to have a significant effect on Dowex 21K sorption kinetics (Figure 5b). Such a concentration effect is not consistent with the ordinary pdc model, but is in accord with the shell-core or ash-layer diffusion model. With the assumption that the resin is spherical and that the bulk solution concentration remains essentially constant ("infinite solution volume" or ISV condition), the relation for fractional attainment of equilibrium[5]

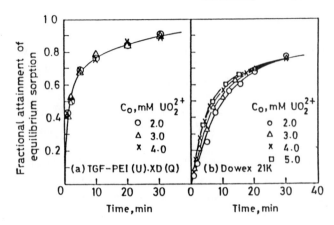

Figure 5 *Rate of sorption of UO_2^{2+} on (a) TGF-PEI(U).XD(Q) of bead size (diameter) 0.71-1.52 mm and (b) Dowex 21K of bead size 0.40-0.75 mm in UO_2SO_4 solutions of different concentrations, C_o at pH 4.0. Resin loading 2.0 g (wet)/L; temperature 30°C; vigorous agitation*

$$F = 1 - \frac{6}{\pi^2} \sum_{n=1}^{\infty} n^{-2} \exp\left(-\pi^2 n^2 \overline{D} t / \overline{r}^2\right) \tag{6}$$

applies to the range of particle diffusion. To a good approximation, Equation (6) can be simplified to

$$F \cong \left[1 - \exp\left(-\pi^2 \overline{D} t / \overline{r}^2\right)\right]^{1/2} \tag{7}$$

from which the particle diffusion coefficient $\overline{D}$ is approximately obtained as

$$\overline{D} = -0.233 \, \overline{r}^2 \, t^{-1} \, \log\left[1 - F^2\right] \tag{8}$$

Since in the present work, $\omega \ll 1$, where ω is the equivalent ratio[6], the rate data in Figure 5(a) could be applied to Equation (8). This yielded nearly constant $\overline{D}$ values of 1.0-1.2 x 10^{-6} cm²/s in the initial conversion range up to 60%.

Under ISV conditions, the model applied to ion-exchange sorption with shell-core behavior can be written in terms of fractional attainment of equilibrium (F) as[5]

$$\left[1 - 3\left(1 - F\right)^{2/3} + 2\left(1 - F\right)\right] = \left[\frac{6\left(\lambda \overline{D}\right)C_o}{\overline{C_r}\overline{r}^2}\right] t \tag{9}$$

where $\overline{C}_r$ is the resin phase concentration or resin capacity, and λ is the molar distribution coefficient defined as $\lambda = \overline{C}_o / C_o$ in which $\overline{C}_o$ is the concentration of the sorbate in the resin shell at the surface corresponding to the external solution concentration C_o. These may be related by the Glueckauf expression[7]

$$\overline{C}_o = kC_o^{2-z} \tag{10}$$

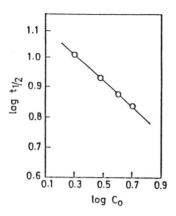

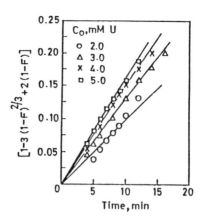

Figure 6 *Logarithmic plot of $t_{1/2}$ (min) of sorption vs. initial concentration, C_o (mmol/L), of solution for Dowex 21K [Figure 5(b)]*

Figure 7 *Test of mathematical model [Equation (9)] for the kinetic sorption data on Dowex 21K [Figure 5(b)]*

where k and z are characteristic constants. A logarithmic plot of $t_{1/2}$ vs. C_o shown in Figure 6 for the experimental data in Figure 5(b) is linear, and regression gives the following equation (corr. coeff. 0.999)

$$t_{1/2} = 14.03 \times C_o^{-0.45} \qquad (11)$$

thus yielding for the Glueckauf parameter z a value of 1.55. Therefore, λ can be related to C_o by $\lambda = k / C_o^{0.55}$ which shows that λ varies significantly with the external solution concentration C_o.

The graphical correlation, in Figure 7, of t vs. $\left[1 - 3(1 - F)^{2/3} + 2(1 - F)\right]$ gives straight lines passing through the origin, a good confirmation of the assumed model. From the slopes of the straight lines, $\lambda \overline{D}$ values were calculated. The values of $\lambda \overline{D} \times 10^6$ (cm²/s) so obtained are noted below, along with the external solution concentration C_o (mmol U/L) given in parentheses: 1.4 (2.0), 1.2 (3.0), 1.0 (4.0), and 0.8 (5.0). An increase in $\lambda \overline{D}$ values at lower solution concentrations would be expected in view of the dependence of λ on solution concentration, as noted above. Since usually $\lambda \gg 1$, it can be concluded that $\overline{D} < 10^{-6}$ cm²/s. The particle diffusivity of the TGF sorbent is thus greater than that of Dowex 21K.

4 CONCLUSIONS

By a new process of templated gel-filling (TGF), a gel-type granular sorbent has been prepared using polyethyleneimine as the chelating resin, UO_2^{2+} as the template ion, 1,4-dibromo-2-butene as the crosslinking agent, and $(CH_3)_2SO_4$ as the quaternizing agent.

Designated as TGF-PEI(U).XD(Q), this new sorbent has higher U selectivity (α_{Fe}^{U} = 4.2) than the base resin PEI (α_{Fe}^{U} = 0.8). Though the commercial resin Dowex 21K, used for U recovery, has still higher U selectivity (α_{Fe}^{U} = 10), its very low U uptake, especially at low substrate concentrations, is a drawback. Unlike Dowex 21K which exhibits very low U sorption and practically no Fe sorption in dilute ($\leq$ 1 mM) U/Fe solution, the TGF sorbent shows high U sorption accompanied by much smaller but still substantial Fe sorption under the same conditions. With selective stripping of the sorbed U with Na_2CO_3 or $(NH_4)_2CO_3$, the TGF sorbent thus offers the possibility of recovering from dilute U streams iron-free U concentrates of much higher concentration than what is possible with Dowex 21K. An added advantage of the TGF sorbent is its significantly faster kinetics than Dowex 21K.

References

1.   H. Nishide, J. Deguchi and E. Tsuchida, *J. Polym. Sci., Polym. Chem. Ed.*, 1977, **15**, 3023.
2.   M. Chanda and G. L. Rempel, *Ind. Eng. Chem. Res.*, 1995 (August).
3.   M. Chanda and G. L. Rempel, *React. Polym.*, 1995, **25**, 25.
4.   M. Chanda and G. L. Rempel, *React. Polym.*, 1992, **18**, 141.
5.   F. Helfferich, "Ion Exchange", McGraw-Hill, New York, 1962.
6.   M. Tetenbaum and H. P. Gregor, *J. Phys. Chem.*, 1954, **58**, 1156.
7.   E. Glueckauf, *Proc. Roy. Soc., London, Ser. A*, 1962, **268**, 350.

SEPARATION OF GADOLINIUM FROM URANIUM IN A CHROMATOGRAPHIC COLUMN WITH TBP ADSORBED ON XAD-BEADS AS A STATIONARY PHASE

D. Bjernedal and M. Rasimus

Process Development and Conversion
Nuclear Fuel Division
ABB Atom AB
Västerås, Sweden

1 INTRODUCTION

In Västerås, Sweden ABB has a nuclear fuel factory manufacturing fuel elements for light water BWR's and PWR's.

The first part of the factory was built in 1968 and was complemented with a conversion plant in 1976. In 1986, the uranium recovery plant was built and in this plant the ASATUR process was introduced. (ASATUR = ASEA-ATOM's process for Uranium Recovery). ASATUR is a chromatographic process with TBP absorbed on XAD-beads as the stationary phase and uranium in nitric acid solution as the feed solution.

Since 1986 about 20 tons of contaminated uranium have been recovered, most of it contaminated scrap from vacuum systems. Some of the uranium is recovered from the more heavily contaminated sludge generated in the water cleaning system and ashes from combustion of burnable waste.

In 1994 tests were done with uranium contaminated with up to 3% gadolinium. Since then about 1 ton uranium has been recovered from this type of material and has been reused as nuclear grade uranium oxide.

2 PROCESS EQUIPMENT

The uranium recovery plant consists of two different systems, one for calcination and leaching of the contaminated uranium, one for separation of impurities and uranium in solution.

All process equipment is constructed for corrosion resistance against TBP, nitric acid and hydro fluoric acid. This means that nearly all surfaces in contact with the nitric acid solution or the TBP-bearing beads is coated with fluorine containing polymers. In some cases, other resistant polymers have been chosen.

The system for calcination and leaching consists of one pusher-type furnace, two leaching vessels with heat exchanger, one pump for recirculation of leaching slurry, one filterpress, three storage vessels for UNH-feed for the separation system and one pump for mixing UNH-feed and for feeding the separation system. All equipment has dimensions or built-in neutron absorbers that make the equipment critically safe for handling of uranium enriched up to 5% U-235.

The system for separation consists of four chromatographic columns, - volume about 700 litres - twelve storage vessels for raffinate, eluate and washing solutions - diluted nitric acid - with volume 700 litres, two 25 litres columns for separation of TBP from eluate, one 250 litres column for end treatment of raffinate and one 2500 litres storage tank for raffinate with uranium content less than 5 mg/L.

3 PROCESS DESCRIPTION

Contaminated uranium is dissolved in diluted nitric acid. Solutions from several leaching batches are mixed together to give a feed solution that is about 100 g/L in uranium and about 3M in free nitric acid. The solution is pumped through two chromatographic columns, the second column being a back-up column, with a stationary phase consisting of TBP-bearing XAD-beads. The feeding is done with a flowrate of approximately 200 litres per hour and during approximately 5 hours.

The beads are made from a copolymer of divinylbenzene and styrene with a high degree of cross-linking. The area is about 600 m^2/g beads. The particle size for the beads is greater than 0.25 mm and less than 1.2 mm. The extracting agent is loaded onto the adsorbent by contacting the beads with an excess of the reagent for several weeks. The beads are then loaded to the columns from the top and the excess of reagent is drained through nozzles in the bottom of the column. The beads can approximately carry about 0.5 kg TBP/litres beads indicating that one column has nearly 250 - 300 kg TBP impregnated. When feeding with solution 100 g/L in uranium and 3M in free nitric acid it is possible to reach a concentration of uranium in the TBP of about 350 g uranium/L TBP. This means that one column can extract between 87.5 and 105 kg uranium. Because of the band-broading when washing with 500 litres fresh 3M nitric acid, flowrate 200 litres per hour, about 90% of the uranium is recovered with small amounts of impurities.

The uranium is eluted with deionised water with a flowrate of approximately 200 litres per hour. The time for the elution step is about 15 hours including about 7 hours extra rinsing to minimise uranium tailing to the raffinate in the next separation run. This extra quantity of fluid is then reused in the beginning of the next elution. Before sampling, the eluates are treated in two 25-litres columns with XAD-beads to minimise TBP contamination in the further processing of the eluates. With the two columns in series the TBP content can be reduced from 200 - 300 mg/L to less than 10 mg TBP/L. The volume of the eluate is about 1.5 - 2.0 m^3 with uranium content of 40 to 50 g/L. This means that the capacity for the separation process is about 100 kg uranium for 24 hours. The pH of the solution is about 1 and the solution usually has a very low content of impurities. After analyses and acceptance for further processing, the uranium is fed back to the main uranium flow by being a part of the start solution for the AUC-precipitation in the conversion shop. In the AUC-precipitation, most of the remaining impurities are complexed with ammonia and carbonate as well as fluoride when UF$_6$ is in the uranium feed for the AUC-precipitation.

The uranium content in the raffinates, after TBP-extraction, are about 100 - 200 mg/L. Before end treatment of the raffinate the uranium content must be reduced below 5 mg/L. This is done in a 250-litres chromatographic column with a mixture 1:1 of TBP/D2EHPA as the extracting agent. Before treatment with lime the TBP/D2EHPA dissolved in the raffinate is reduced in a 25-litres column filled with XAD-beads. In a final stage, the raffinate is treated with lime and the resulting solution is dumped to the municipal waste water treatment. The lime sludge is so far dumped at the SAKAB dump site.

The XAD-beads used as adsorbent for leaking extracting agent, can be reused in the chromatographic columns after preparation with more extracting agent. XAD-beads from the chromatographic columns that have lost their TBP-bearing capacity are eluted to the lowest possible level of uranium and then destructed by combustion in air. For scrubbing off of the phosphoric acids, formed during thermal decomposition of the extracting agents, the XAD-beads are mixed with lime - which works as a dry scrubbing agent - before combustion. The uranium in the ashes is then dissolved in nitric acid solution and once again the impurities are separated from the uranium.

A flow diagram for the process is shown in Figure 1.

Flow diagram for uranium recovery

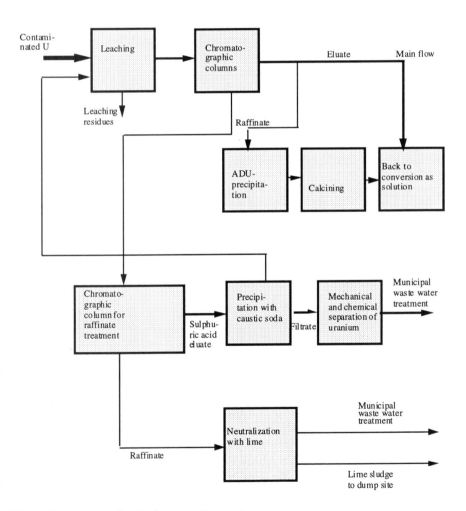

Figure 1 *Principal sketch of process for uranium recovery*

4 PROCESS DESCRIPTION FOR URANIUM-GADOLINIUM SEPARATION

The goal for the process development was to minimise the gadolinium content in the eluates to less than 100 µg/g uranium. This was achieved by feeding with 70% of the uranium capacity for one column, slightly modifying the free nitric acid in the feed solution from 3 M to about 2.5 M, keeping the wash solution to 3 M free nitric acid and extension of the washing step to 600 litres instead of 500 litres. This means that the losses of uranium were greater because of extended band broading. To prevent great losses, the feeding, washing and elution steps are performed in two columns in series. The mass balance for a recovery cycle with gadolinium contaminated uranium is presented in Table 1.

Table 1   *Mass Balance Data for One Recovery Cycle with Gadolinium Contaminated Uranium*

Solution input	Solution output
700 L UNH with U = 100 g/L HNO_3 = 2.5 M	350 L reuse in next elution 350 L raffinate to neutralization
600 L washing solution HNO_3 = 50 L/ h Deionised water = 150 L/ h	600 L raffinate to neutralization 300 L raffinate reused in leaching
1400 L reused fluid	1500 L Eluate, U = 45 g/L
1400 L deionised water	1000 L reused in next elution
Solution input 4100 L	Solution output 4100 L

Uranium input	Uranium output
70 kg U in UNH	1500 L eluate with U = 45 g/L = 67.5 kg U
	950 L raffinate with U = 0.2 g/L = 0.2 kg U
	300 L raffinate with U = 3 g/L = 0.9 kg U
	1350 L solution for reuse with U = 1 g/L = 1.35 kg U
	Uranium output ~70 kg

For eluates that are less than 100 µg Gd/g U, the main part of the uranium content is precipitated with ammonia to ammoniumdiuranate, ADU. The precipitation is done in such a way that a separation of uranium and gadolinium is obtained. This is achieved with careful control of the pH at the end of the precipitation. The optimal pH is about 4 - 4.5 and at that pH range, 98 - 99% of the uranium is precipitated with only 10% or less of the gadolinium coprecipitated.

Through heat treatment of the ADU, the uranium is converted to U_3O_8 and the uranium is dissolved in nitric acid to form uranylnitratehexahydrate (UNH). In the ammoniumuranylcarbonate (AUC) precipitation, the UNH-solution is used as the uranium source instead of UF_6. In the AUC-precipitation, a maximum 10% of the gadolinium is coprecipitated with the uranium. The AUC-precipitate is then heat treated to produce a nuclear grade uranium oxide powder for fuel pellet production. The gadolinium contaminated filtrates from the precipitations are treated in the uranium recovery shop to eliminate the risk of cross contamination of nuclear grade uranium dioxide powders produced in the conversion shop.

5 RESULTS

During August 1994 - April 1995 about 1 ton of uranium has been recovered as nuclear grade uranium oxide. The gadolinium content in the eluates has been between 20 to 100 µg/g uranium. With start concentrations of about 30000 µg/g, separation factors between 300 to 1500 were achieved in the column separation and with two precipitations, the uranium can be used as nuclear grade uranium oxide with a gadolinium content less than 1 µg/g uranium.

ADSORPTION AND ELUTION BEHAVIOR OF PLATINUM-GROUP METALS IN NITRIC ACID MEDIUM
A Study on the Application of a Newly Developed Ion-Exchange Process to Spent-Nuclear-Fuel Reprocessing

Y. -Z. Wei, M. Kumagai and Y. Takashima
Institute of Research and Innovation, 1201, Takada, Kashiwa, 277 Japan

K. Takeda, Shibaura Institute of Technology, Japan

M. Asou, T. Namba and K. Suzuki, Tokyo Electric Power Co. Inc., Japan

A. Maekawa and S. Ohe, The Kansai Electric Power Co Inc., Japan

1 INTRODUCTION

With its attractive advantages such as simple separation procedures, organic solvent free and compacted equipment, ion-exchange has extremely large potential in the application of reprocessing spent-nuclear-fuel. In order to develop an advanced ion-exchange process to recover uranium, plutonium and other valuable fission products (FP) such as platinum-group metals from the spent-fuel of light water reactors (LWR), we have prepared a new type of anion-exchanger characterized by high mechanical strength, a rapid ion-exchange rate and excellent radiation-resistance. The adsorption and elution behavior of U, Pu and some non-adsorptive FP elements such as Cs, Sr, Nd in hydrochloric acid medium has been studied by numerical simulation, and the separability between U and these FP elements has also been investigated experimentally using this novel anion-exchanger.[1] On the other hand, the separation procedures might become relatively simple in nitric acid medium because most of the FP elements show no or only weak adsorbability onto anion-exchangers.

Spent LWR-fuel typically contains 94-95% (by weight) U, 1.0-1.5% Pu and 4-5% FP elements including $1.9 \times 10^{-1}\%$ Ru, $3.2 \times 10^{-2}\%$ Rh and $8.5 \times 10^{-2}\%$ Pd.[2] In the nitric acid solution of spent LWR-fuel, the second triad of group VIII, i.e. Ru, Rh and Pd, are expected to be the main adsorptive elements which would interfere with the separation of U and Pu. In this study, the adsorption experiments of these platinum-group metals in nitric acid solution by using the new type anion-exchanger were carried out, and the adsorption behavior was examined by comparing with a conventional anion-exchanger. Furthermore, the selective elution of the metals was investigated by using nitric acid and complexing agents, and the separation probabilities of these metals with U were estimated.

2 EXPERIMENTAL

2.1 Materials

The anion-exchanger named AR-01, with the resin embedded in porous silica spherical particles of about 50 μm in diameter, was used. This anion-exchanger contains strong-base quaternary and weak-base tertiary benzimidazole groups which act as the exchange sites. Table 1 shows the properties and the functional group structures of the anion-exchanger. The preparation method and the detail properties of the anion-exchanger have been given in References 3 and 4. In order to examine the ion-exchange behavior of Pd(II), several conventional anion-exchanger shown in Table 2 were also used.

The nitric acid solutions of Pd(II) and Rh(III) were prepared by dissolving the nitrate salts, $Pd(NO_3)_2$ and $Rh(NO_3)_3 \cdot 2H_2O$ into HNO_3 aqueous solutions, respectively. The nitric acid solutions of Rh(III) were made-up from a $RuNO(NO_3)_3$ solution containing 1.5% Ru.

Table 1   *Properties of AR-01 Anion-Exchanger*

	Tertiary Benzimidazole	Quaternary Benzimidazole
Ionic Group		
Matrix Structure		PS-MR
Resin Content (wt%)		24.4
Carrier Bead		Porous SiO_2
Total Capacity (eq/kg-resin)		3.4
Quaternary Capacity (eq/kg-resin)		2.0
Bead Diameter (μm)		40 - 60

Table 2   *Properties of the Conventional Anion-Exchangers*

Exchanger	Matrix Structure	Ionic Group	Capacity (eq/kg)	Bead Diameter (μm)
IRA-400	PS-Gel	$-N(CH_3)_3^+$	3.7	400 - 530
IRA-900	PS-MP	$-N(CH_3)_3^+$	4.2	450 - 550
IRA-93ZU	PS-MR	$-N(CH_3)_2$	4.4	410 - 510

2.2   Experiment Procedures

All the adsorption and elution experiments were carried out at a constant temperature of 333 K. In the batch adsorption experiments, about 1 g of anion-exchanger was equilibrated with HNO_3 solutions of concentrations ranging from 0.1 to 9.0 kmol/m^3 prior to contact with the anion-exchanger. The anion-exchanger and 20 cm^3 of a platinum-group metal containing solution contained in a glass flask was set in the water-bath maintained at 333 K and was shaken mechanically. Then the solution was separated by filtration, the metal concentration in the solution was measured by ICP spectroscopy (Shimadzu ICP-1000III). The anion-exchanger was washed with a dilute HNO_3 solution as well as pure water and then dried at 393 K for 10 hours. The distribution coefficient, Kd, defined as the ratio of the metal concentration in the resin of the ion-exchanger and that in the solution, was calculated.

Nitric acid solutions and the complexing agent solutions of thiourea (Tu) or ammonia (NH_3) were used as the eluents of Pd(II). The AR-01 anion-exchanger with adsorbed Pd(II) was transferred to the glass flask containing 20 cm^3 of an eluent solution, and was shaken mechanically. The elution rate was calculated from the concentration of Pd in the solution.

Separation experiments for a similar nitric acid solution of 10 typical FP elements including the platinum-group metals were conducted using the glass column with 10 mm internal diameter and 200 mm long. The ion-exchanger was packed to the column in slurry state, the volume of the ion-exchanger bed was about 12 cm^3. The feed and eluent solutions were fed to the column maintained at 333 K in a constant flow rate of 1.1 x 10^{-3} m/s, the effluents from the column were recovered by fractional collectors of 10 cm^3.

3 RESULTS AND DISCUSSION

3.1 Adsorption

Figure 1 shows the time evolution of the adsorption of the platinum-group metals from nitric acid solutions onto AR-01 anion-exchanger. The adsorption of Ru(III) reached an equilibrium state at about 20 minutes with Kd values of 60-65. Rh(III) showed negligibly slight adsorption. On the other hand, Pd(II) showed significantly strong adsorption onto the anion-exchanger, the adsorption was very slow and still continued after 2 hours.

Figure 2 shows the distribution coefficients of the metals in nitric acid solutions with concentrations from 0.1 to 9.0 $kmol/m^3$ using AR-01. Rh(III) showed only very weak adsorption throughout the HNO_3 concentration range. Adsorption of Ru(III) is small in dilute HNO_3 and increases to a maximum at a HNO_3 concentration of about 1.0 $kmol/m^3$ and then decreases with increasing HNO_3 concentration. In the case of Pd(II), strong adsorption was observed at all the concentrations studied and the Kd value decreased with increasing HNO_3 concentration. Here the adsorption behavior shown by Pd(II) was characterized by the significantly strong adsorption at diluted HNO_3 solutions with Kd more than 1000.

Figure 3 shows the time evolution of Pd(II) adsorption from nitric acid solutions onto AR-01 as well as the several conventional anion-exchangers, IRA-400, IRA-900 and IRA-93ZU. The results indicated that the adsorption of Pd(II) onto these conventional anion-exchangers was very rapid and reached an equilibrium state within 30 minutes with the Kd values of 15-20. In other words, the unique adsorption behavior shown by Pd(II) onto AR-01 anion-exchanger was not observed in these conventional anion-exchangers.

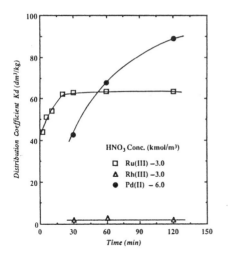

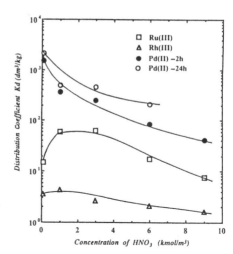

Figure 1 *Time evolution of adsorption of platinum metals from nitric acid solution (AR-01)*

Figure 2 *Adsorption of platinum metals from nitric acid solution (AR-01)*

The ion-exchange reaction of anionic complexes in nitrate media is generally expressed as

$$\nu_A (R^+ \cdot NO_3^-) + A^{-\nu_A} = (\nu_A R^+ \cdot A^{-\nu_A}) + \nu_A NO_3^- \tag{1}$$

where R^+, A, ν_A denote the fixed-ionic-group, the counter-ion (anionic complex) and the charge number of the counter-ion, respectively. In many case of anion-exchange of complexes, maximum adsorption is observed in a certain ligand concentration based on the equilibria of complex formation and the competitive adsorption of anionic ligand.[5,6]

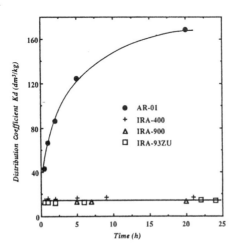

Figure 3 *Time evolution of Pd(II) adsorption onto various anion-exchangers (6.0 kmol/m³ - HNO₃)*

From the results shown in Figures 1 and 2, it was considered that Ru(III) adsorbed onto the anion-exchanger as anionic nitrosylnitrato-complexes such as $RuNO(NO_3)_4^-$ and $RuNO(NO_3)_5^{2-}$, which are formed in HNO_3 solution.[7]

However, from the results shown in Figures 1-3, it is considered that the adsorption of Pd(II) onto AR-01 anion-exchanger is not according to the general anion-exchange reaction as shown by Equation (1). The unique adsorption of Pd(II) was assumed to result from complex (such as $RN:Pd(NO_3)_3$) formation between Pd(II) and the nitrogen atom that possesses lone-pair electrons at the fixed-ionic-group (benzimidazole) of AR-01 ion-exchanger. The complex formation reactions are supposed as follows:

$$RN^+ \cdot NO_3^- + Pd^{2+} + 2NO_3^- = RN^+ : Pd(NO_3)_3^- \tag{2}$$

$$RN^+ \cdot NO_3^- + PdNO_3^+ + NO_3^- = RN^+ : Pd(NO_3)_3^- \tag{3}$$

$$RN^+ \cdot NO_3^- + Pd(NO_3)_2 = RN^+ : Pd(NO_3)_3^- \tag{4}$$

The adsorption of Pd(II) onto AR-01 ion-exchanger accompanied by complex formation was also recognized in hydrochloric acid solution.[8]

Only a few investigations have been performed on the anion-exchange of platinum-group metals in nitrate media until now. In early studies, it was reported that Pd(II) was adsorbed onto the strong-base anion-exchangers such as Dowex-1 with Kd values smaller than 100.[9,10] Recently, the dependence of Pd(II)-adsorption behavior on the structure of anion-exchanger has also been reported by Liebmann et al.[11]

3.2 Elution and Separation

Prior to the batch elution experiment, adsorption of Pd(II) using AR-01 had been performed at 6.0 kmol/m³ HNO_3 solution for 24 hours. This resulted in the adsorption of Pd to a level of 50-60 g/kg-resin in the anion-exchanger. Figure 4 shows the time evolution of Pd-elution by HNO_3 solutions at various concentrations. It was found that almost no Pd adsorbed in the ion-exchanger was eluted when HNO_3 concentration is smaller than 6.0 kmol/m³. In the case of 14.0 kmol/m³ HNO_3, near 40% elution was obtained, however the elution is quite slow. These results agreed with the dependence of Pd(II)-adsorption on HNO_3 concentration as shown in Figure 2.

Figure 5 shows the results of the batch elution using complexing agent solutions of ammonia and thiourea. The Pd(II) was rapidly eluted using 0.5 kmol/m³ Tu contained in 0.1 kmol/m³ HNO_3 solution with the elution rate over 85% in 15 minutes. However, in the case of 0.5 kmol/m³ Tu - 6.0 kmol/m³ HNO_3, the elution rate of Pd dropped to under 10. This was considered to be due to the decomposition of thiourea through redox reactions with the relatively concentrated HNO_3. As shown in the figure, the Pd was also eluted effectively by 1.0 kmol/m³ HN_3 solution, but the elution is slower than using thiourea.

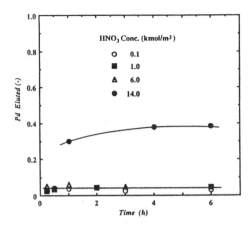

Figure 4 *Time evolution of Pd(II) elution with nitric acid solutions (AR-01)*

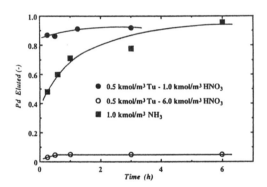

Figure 5 *Time evolution of Pd(II) elution with complexing agents (AR-01)*

Thiourea and NH_3 both have strong complexation affinity with transition metals through coordinate bonds, because they contain nitrogen atoms which possess lone-pair electrons. The elution of Pd(II) by these complexing agents is considered to result from ligand-exchange reactions between the Pd(II) complex in the anion-exchanger and these agents as follows:

$$RN^+ : Pd(NO_3)_3^- + 4NH_3 = RN^+ \cdot NO_3^- + Pd(NH_3)_4^{2+} + 2NO_3^- \qquad (5)$$

$$RN^+ : Pd(NO_3)_3^- + 4Tu = RN^+ \cdot NO_3^- + Pd(Tu)_4^{2+} + 2NO_3^- \qquad (6)$$

$$RN^+ : Pd(NO_3)_3^- + 2Tu + 2H^+ = RN^+ \cdot NO_3^- + Pd(NO_3)_2(TuH)_2^{2+} \qquad (7)$$

In order to investigate the elution behavior of platinum-group metals in an ion-exchange column, separation experiments using simulated nitric acid solutions containing 10 representative FP elements were carried out. The solution feed is in accord with a proposed separation process for U recovery from spent LWR-fuel solutions. Figure 6 shows a typical result from the column experiments. The feed solution contains 2-6 mol/m^3 of the FP elements and 6.0 kmol/m^3 HNO$_3$. As shown in the figure, the elements of Cs, Sr, Y, Nd, Mo, Zr and Rh showed no adsorption and were washed out with the feed solution and the 6.0 kmol/m^3 HNO$_3$ washing solution. A part of Pd(II) was strongly adsorbed by the ion-exchanger and was effectively eluted by thiourea. Since U(IV) is adsorbed by the anion-exchanger as anionic nitrato-complexes, it is predicted that the U(VI) adsorbed in the anion-exchanger is eluted by a dilute HNO$_3$ solution due to the decrease of ligand concentration. From the Pd(II) elution curve shown in Figure 6, near complete separation between Pd and U is expected to be achieved. On the other hand, Ru(III) showed a quite complicated elution behavior and its separation from U needs to be examined further.

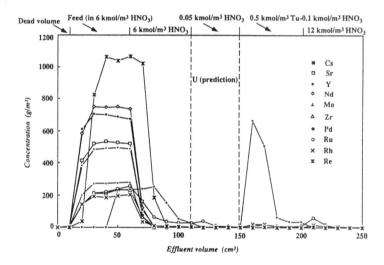

Figure 6 *Elution curves by column experiment for a simulant nitric acid solution of typical FP elements (AR-01)*

4 CONCLUSION

The adsorption and elution behavior of the platinum-group metals in nitric acid solution has been studied using the new type anion-exchanger, AR-01, and the separation probabilities of these metals with U were discussed. The results are summarized as follows:

1) Rh(III) showed only slight adsorption onto the anion-exchanger. Rh(III) adsorbed as nitrosylnitrato-complexes and showed an adsorption peak at about 1.0 kmol/m^3 HNO$_3$ with Kd values of 60-65.

2) Pd(II) showed strong adsorption, the adsorption was very slow and increased with reducing HNO$_3$ concentration. This unique behavior was considered to result from the complex formation between Pd(II) and the nitrogen atom in the fixed-ionic-group of the anion-exchanger.

3) Pd can be satisfactorily eluted by complexing agents such as thiourea and ammonia through ligand-exchange reactions.

4) In column experiments, Pd was strongly adsorbed by the anion-exchanger and was effectively eluted by thiourea. It is estimated that near complete separation between Pd and U can be achieved. Ru showed a quite complicated elution behavior and its separation from U needs to be examined further.

Nomenclature

A	counter ion (anionic complex)	
FP	fission product	
Kd	distribution coefficient	dm³/kg
R	fixed ionic group of ion-exchanger	
v_A	charge number of counter ion	

References

1. M. Kumagai et al., *Proc. of IMechE 1992*, Manchester, Nov. 17-18, 1992, p. 113.
2. M. Benedict, T. H. Pigford and H. W. Levi, "Nuclear Chemical Engineering", McGraw-Hill, New York, 1981, p. 388.
3. K. Takeda, M. Akiyama and T. Yamamizu, *Reactive Polymers*, 1985, **3**, 173.
4. K. Takeda, F. Kawakami and M. Sasaki, *J. Macromol. Sci.-Chem.*, 1986, **A23[9]**, 1137.
5. F. Helfferich, "Ion Exchange", McGraw-Hill, New York, 1962, p. 205.
6. K. A. Kraus and F. Nelson, *Proc. Intern. Conf. on Peaceful Uses of Atomic Energy*, Geneva, 1956, **7**, 113.
7. J. M. Fletcher et al., *J. Inorg. Nucl. Chem.*, 1959, **12**, 154.
8. Y. -Z. Wei et al., *Proc. of Intern. Conf. on Ion-Exchange*, Takamatsu, Dec. 4-6, 1995, I-A3, p. 19.
9. J. P. Faris and R. F. Buchanan, *U. S. Atomic Energy Comm. Rept.*, 1963, **ANL-6811**, 185.
10. F. Ichikawa, S. Urano and H. Imai, *Bull. Chem. Soc. Japan*, 1961, **34(7)**, 952.
11. R. Liebmann and G. Pfrepper, *Kernenergie*, 1984, **27(1)**, 29.

NOVEL ANION EXCHANGE RESINS WITH THERMAL STABILITY: SYNTHESIS AND CHARACTERISTICS

H. Kubota, K. Yano, S. Sawada, Y. Aosaki, J. Watanabe, T. Usui, S. Ono, T. Shoda and K. Okazaki
Mitsubishi Chemical Corporation, Yokohama Research Center
1000 Kamoshida-cho, Aoba-ku, Yokohama, 227 Japan

M. Tomoi, Yokohama National University, Yokohama, 240 Japan

M. Shindo and T. Onozuka, Tohoku Electric Power Co. Inc., Aoba-ku, Sendai, 981 Japan

1 INTRODUCTION

Over fifty years have passed since styrene-based, anion exchange resins were first reported. Currently, and throughout their history, most of the commercially significant anion exchange resins have had polystyrene-divinylbenzene(DVB) matrices, and have conventionally been functionalized with dimethylamine (for weakly basic), trimethylamine (for type I, highly basic) or dimethylhydroxyethylamine (for type II, strongly basic). Traditionally, in order to facilitate the introduction of the tertiary (weak base) or the quaternary (highly or strongly basic) functionality, using the above amines, the polymer matrix is first chloromethylated, so as to insert into the structure a highly reactive site, to which the amine can be attached (i.e. to provide the "key" to such functionalization). There are however, some serious disadvantages to this traditional approach in the synthesis of anion exchange resins: 1) the quaternary ammonium functionality in the benzyl position is relatively unstable in the OH-form (i.e. under basic conditions); 2) the chloromethylation reaction is usually carried out with either chloromethyl methyl ether (CME), or bischloromethylether (BCME) both of which (especially BCME) are highly carcinogenic; and 3) aminolysis of the chloromethyl group may at times be incomplete giving rise to possible chloride leakage.

Today, many styrene-based strongly basic anion exchange resins have been employed for water treatment. There are, however, many disadvantages in conventional resins: (1) decrease of salt splitting capacity due to poor durability, (2) trimethylamine odor and organic contamination of cation exchange resins in Mixed Beds, based on leakage of amines derived from the degradation of the quaternary ammonium functionality. It is well known that quaternized trimethylammonium (type I) and dimethylethanolammonium (type II) functionality on the benzyl position are very easy to eliminate when in the OH-form. These problems arise from the instability of the quaternary ammonium functionality of styrene-based resins, because the acidity of the hydrogen on methylene chain in conventional resins being high due to the existence of the benzene ring. Therefore, we felt it is doubtful whether it is possible to resolve the problem just by chemical modification of the ammonium functionality on the benzyl position.

We should also be very aware to reduce the leakage of organic impurities from the resins themselves. It is clear that the demands on water quality are still increasing, and it has been suggested that it could even become the limiting factor in the production of items in almost every field of advanced technology. The objective of this research, is to develop a resin answering these increased quality requirements.

One way to dramatically improve the stability of the quaternary ammonium functionality is not to modify chloromethylated polystyrenes with various amines, but to separate the quaternary ammonium functionality from the benzene ring on the polyethylene backbone with a chemically stable spacer-arm.

Two families of spacer-arms considered chemically stable, have been incorporated in these new resins, to separate the positive charge from the benzene ring. One of these is an alkyl type (Figure 1) and the other an ether type (Figure 2).

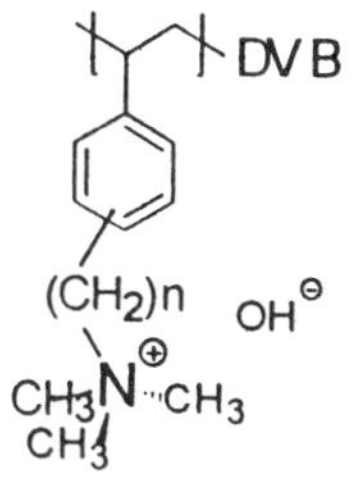

Figure 1 *Novel resins with alkyl spacer-arms* 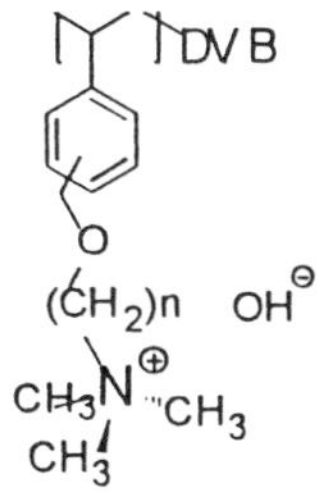   **Figure 2** *Novel resins with ether spacer-arms*

The novel, basic anion exchange resins with spacer-arms, were hypothesized to be a good basis for new resins that could overcome the problems outlined above. The novel anion exchange resins might be able to innovate processing and usage of anion exchange resins.

We believe that the novel resins with spacer-arms, would prove to be suitable for not only demineralization of water, but a wide range of applications. This paper discusses the preparation of these novel resins, properties, and their potential uses as anion exchange resins.

2 EXPERIMENTAL SECTION

2.1 Materials

Chloromethylstyrene (CMS) monomer (pure 96%) was purchased from Tokyo Kasei Industries Ltd. DVB (Divinylbenzene) (pure 80%, balance of monomer is ethylvinylbenzene) was donated by Nippon Steel Chemical Co. Ltd. AIBN (azobisisobutyronitrile, Wako Pure Chemical Ind. Ltd.) was used as an initiator without further purification.

A series of ω-bromoalkylstyrenes was prepared by the coupling of the corresponding Grignard reagents with excess 1,ω-dibromoalkanes in 45% to 84% yield. ω-Bromoalkoxymethylstyrenes were prepared by the Williamson ether synthesis of vinylbenzylalcohol with excess 1,ω-dibromoalkanes in sodium hydroxide solution in good yield (76% to 92%). These resulting monomers were distilled *in vacu* and stored in a refrigerator until use.

2.2 Resins

All ion exchange resins were commercially available. DIAION® SA10A, SA12A (type I resin), DIAION® SA20A, SA21A (type II resin) and DIAION® SK1B (cation exchange resin) were available from Mitsubishi Chemical Corporation (Tokyo). Amberlite® IRA-400 (type I) was available from Sigma. The novel resins with alkyl spacer-arms were named "CnA", $(St-(CH_2)_n-N^+R_3)$ and those with ether spacer-arm were named "CnE", $(St-CH_2-O-(CH_2)_n-N^+R_3)$.

2.3 Analyses

The TOC analyser used was a Shimadzu TOC-5000. ^{1}H-NMR was run in $CDCl_3$ at 300MHz on a Varian Unity-300.

2.4 Preparation of Novel Resins with Alkyl/Ether Spacer-arms

A general method of synthesizing quaternary ammonium resins with spacers was defined.   Polymer beads were first produced by oil-in-water type suspension polymerization. Each was prepared separately at room temperature under nitrogen. The continuous phase (1.5 L water and stabilizer) was placed in the reaction vessel which was fitted with a Teflon stir-paddle. The monomer phase (425 g of ω-bromoalkylstyrenes/ ω-bromoalkoxymethylstyrenes, 30 g of DVB (80% purity) and 1.50 g of AIBN initiator) was poured into the continuous phase. The suspension was stirred and heated to 80°C, at which temperature it was maintained for 8 hours polymerisation. The beads were isolated and washed with methanol and demineralized water; they were spherical and transparent. The resulting beads were aminated with 30% trimethylamine aqueous solution.   The aminolysis was essentially complete and yielded quaternary ammonium resins. Using the beads prepared above, novel type II resins with spacer-arms were also prepared, by aminating with dimethylethanolamine at 50°C for 6 hr. The products were washed with demineralized water.

2.5 Measurement of Ion Exchange Capacity

Ten mL of resin (Cl^- form) were packed in a column, and 250 mL of 4% aqueous solution of NaOH was passed through the column. The effluent was recovered, and titrated with 1N HCl. Following measurement of neutral salt splitting capacity, weakly basic anion exchange capacity was measured. One hundred mL of 0.1N HCl solution was passed through in the column, and subsequently, 60 mL of methanol was passed through to rinse the resin. The effluent was recovered, and titrated with 1N NaOH.

2.6 Evaluation of Resins

2.6.1 Thermal Stability (Static Test). All resins used were completely converted to the hydroxide form with 2N NaOH. One hundred mL of completely regenerated anion exchange resins (OH form) were also poured into individual sealed glass autoclave tubes. The resins were held at 70°C to 140°C for 30 days or 90 days in demineralized water or ethyleneglycol (EG) under nitrogen.   After exposure, neutral salt splitting capacities, weakly basic anion exchange capacities of the resins, moisture content, pH and TOC's of the supernates were measured. Degradation products of the resins were identified by GC. On the basis of the experiments, percentage of residual salt splitting capacity was calculated. Percentage of remaining salt splitting capacity = total salt splitting capacity (per unit volume) of recovered resin after test ÷ total salt splitting capacity (per unit volume) of resin prepared before test.

2.6.2 Thermal Stability (Dynamic Test). For these studies, 50 mL of anion exchange resins were converted from Cl form to OH form. Resins in the Cl form were completely regenerated with 2N NaOH. Then, the regenerated resin (OH form) was packed in a stainless steel column. UCP was run upwards through the column (40 cm x 30 mm i.d.) by SV 100 (5.0 L/hr) for 90°C for 30 days (3.6m^3/M). After experiments, ion exchange capacity and moisture content of the recovered resins were measured.

2.6.3 Durability of the Resins (Field Test). To demonstrate long life expectancy of the novel resins, durability experiments were carried out in demineralization columns at our Kurosaki plant. Seita River water was used as the raw industrial water. The raw river water was pretreated by coagulation, flocculation, sand filtration and disinfection. This pre-treated water supplies the numerous production units of our Kurosaki plant with a total of 10,000 cu. meters of deionized water per day. The novel resins, together with DIAION® SA10A, and SA12A as references (i.e. baseline) materials, were all suspended in plastic weave bags, in the anion columns of the deionizers, and maintained at room temperature from Nov. 20, 1994. These anion columns were regenerated with 100g 100% NaOH/L-resin at 40°C, once every 30 days. After regeneration, the resins were washed with 4% NaCl and 1N NaOH solution, ion exchange capacities and moisture content of the resins were measured.

2.6.4 Leachability of Resins. For these studies, 80 mL of each anion exchange resins were converted from Cl form to OH form. Resins in the Cl form were regenerated with 1N NaOH (200g/L-resin) at SV of 4L/hr. Then, the OH ions were eluted with 160 mL of UCP (2BV). UCP was passed through column (50 cm x 14 mm i.d.) by SV 20 at 25°C or 70°C; electrical conductivity, pH and TOCs of the effluent were measured. UCP (DO < 5 ppb, TOC < 0.5 ppb) was passed through up to 100 BV (8.0 L/80 mL-resin).

2.6.5 Chemical Stability. To examine the chemical stability of these resins, 25 mL of each resin was maintained at 30°C for 30 days in 1N/25% NaOH (OH form) solution, 1N/35% HCl solution (Cl form), 1% H_2O_2 (OH form and Cl form) solution and demineralized water (OH form and Cl form). After treatment, TOCs of the supernate and total ion exchange capacities of the resins were measured.

2.6.6 Stability for γ-ray Exposure. For each of the resins tested, separate 70 mL wet sample aliquots of Cl and of OH form resin were placed in Erlenmeyer-flasks, and irradiated from a ^{60}Co source. After exposure, each sample was washed with deionized water, and the ion exchange capacity and the moisture content of each was measured.

3 RESULTS AND DISCUSSION

3.1 Overview

There are significant differences between the polymer structures of the novel resins and those of conventional styrene-based resins, most notably the existence of alkyl or ether spacer-arms functionality in the former. Consequently, there are many differences in behavior and physical characteristics; as summarised in Table 1.

Table 1   *Characteristics of the Novel Resins*

Property		Alkyl type	Ether type	Conventional
Salt splitting capacity	(meq/mL)	1.1-1.4	1.0-1.2	1.3-1.4
	(meq/g)	3.5-3.8	3.0-3.3	3.6-4.1
Effective salt splitting cap.	(meq/mL)		0.88	0.90
Thermal stability (type I)				
	Static Test (°C)	90	90-100	70
	Dynamic Test (°C)	80-90	70-80	55
Durability (field test)		now testing	high durability	gradual degradation
Thermal stability for type II resin		metastable	metastable	unstable acetaldehyde odor
Leachability	TOC (of resin)	less leachable	less leachable	leachable
	TOX (remaining halogens)	less leachable	less leachable	leachable
Silica removal		high capacity		standard
Basicity of functionality		high	high	standard
Chemical stability		very stable	less stable for oxidants	standard
Odor (in the OH form)		no odor	no odor	amine odor
Stability for γ-ray exposure		excellent	good	standard

Thus, various performance enhancements were envisioned, primarily due to the incorporation of spacer-arm functionality, and it was hoped that the novel resins would prove to be suitable for a wide range of applications.

3.2   Thermal Stability (Static Experiments)

Two series of novel resins were prepared to investigate the factors affecting thermal stability. Figure 3 and Figure 4 showed the results of selected experiments (autoclave tube test) of thermal stability of styrene-based conventional resin (alkyl spacer, n=1), novel resins with alkyl spacer and ether spacer-arm.

As can be seen from Figure 3, in two families of resins with spacer-arms, as the number of methylene chain increased, thermal stability of resins (OH form) was found to be improved, although with two exceptions (alkyl spacer-arm n=2, ether spacer-arm type n=3). When comparing styrene-DVB based products, the thermal stability of OH-form resins with longer spacer-arms was found to be considerably higher than that of conventional resins, such as, DIAION® SA10A, DIAION® SA12A and Amberlite® IRA-400.

The thermal stability of these novel resins with alkyl spacer-arm, was also evaluated in 100% EG, and in 60% aqueous EG, maintaining the sample aliquots at 80°C over a 30-day period. After the treatment, percentage of remaining salt splitting capacity was 89% for C4A and 41% for DIAION® SA10A in 100% EG, respectively. It was demonstrated that the novel resins with alkyl spacer-arm were extremely stable, even in organic solvent with high nucleophilicity. Furthermore, it was expected, that these novel resins with longer spacer-arms might prove to be stable at temperature in excess of 140°C in the OH form (highly basic conditions).

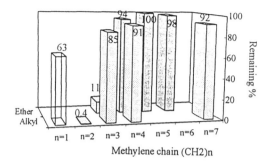

Figure 3 *Thermal stability vs. methylene chain (100°C, 30 days)*

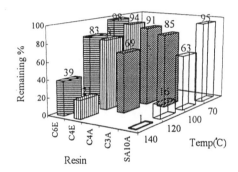

Figure 4 *Thermal stability of various resins (30 days)*

As can be seen from the above results, the best way to overcome the instability of quaternary ammonium functionality in conventional resins was to incorporate spacer-arms, such as alkyl spacer-arms or ether spacer-arms. The separation between positive charge and benzene ring with spacer-arm greatly enhanced the stability of the quaternary ammonium functionality at higher temperatures. The distance between the benzene ring and the positive charge was found to be significant in improving the thermal stability of the quaternary ammonium functionality. As a result of the separation, thermal stability, stability for γ-ray exposure, excellent durability, high regeneration efficiency and leachability of resins were found to be superior to conventional styrene-based resins. These improvements all are results of the greater stability of the quaternary ammonium functionality.

However, it is expected that neutral salt splitting capacity per unit weight of resin decreases when the number of methylene groups in a chain increases. Considering this correlation, the novel resins with alkyl spacer-arms in which the number of methylene chain is n=3 to n=6 which is the side-chain recommended as ideal, for the intended improvement. When comparing with styrene-based resins, alkyl spacer resin with ethylene side-chain (n=2) was extremely unstable under high-temperature (thermal) conditions. In the case of the ethylene chain (as just mentioned) containing two ethylenic links, the typical E1cB reaction (the Hoffmann degradation) was found to occur. Because of the presence of a β-hydrogen on the alkylene spacer, this side-chain exhibited a tendency to degrade under basic conditions (i.e. in the OH form).

As regards resins containing ether-type spacers, it was found that as the number of methylene links increased, thermal stability improved. With the propylene links (for n=3) the presence of the oxygen atom also led to the Hoffmann degradation. However, with a butylene-linked chain (n=4), it was supposed, that the oxygen atom could stabilize the amine functionality, owing to intra-molecular (six-member structure) hydrogen bonding between the oxygen atom, and the positive charge of the neighboring hydrogen on the α-carbon.

In strongly basic anion exchange resins with spacer-arms, degradation of the quaternary ammonium functionality is suggested by three different paths. These could therefore be: a) E2 elimination (the Hoffmann degradation) with production of olefin and tertiary amine; b) S_N2 reaction, resulting in production of an alcohol, and tertiary amine; and c) production of a tertiary amine by elimination of a methyl group on the quaternary amine functionality.

Table 2   *Ratio of Loss of Salt Splitting Capacity/Production of Weakly Basic Capacity*

Resin	Remaining salt splitting cap. %	Loss of salt splitting cap. (meq/g) / Production of weakly basic cap. (meq/g)
C 6 E	98	0.07 / 0.07
C 4 E	94	0.34 / 0.18
C 4 A	91	0.28 / 0.19
C 3 A	85	0.50 / 0.34
DIAION® SA12A	66	1.13 / 0.43
DIAION® SA10A	63	1.25 / 0.49
C 3 E	10	2.95 / 0.20
C 2 A	0.4	3.69 / 0.13

Table 2 shows the effect of exposing the various resin samples to a temperature of 100°C for a period of 30 days, in terms of loss of salt splitting capacity (and corresponding increase in weak base functionality). In general, loss of salt splitting capacity was found to be almost inversely proportional to the thermal stability of the quaternary ammonium functionality. Resulting weakly basic capacity, however, was found to be constant in the range of 0.1 meq/g to 0.5 meq/g. Therefore, production of weakly basic anion exchange resin was independent of the degradation of quaternary ammonium functionality. Consequently, it was proposed that the separation by the 'spacer(s)' inhibited the quaternary ammonium functionality from degrading under high temperatures. In particular, based on the results of CP-MAS NMR analysis, the E2 elimination reaction (the Hoffmann degradation) was found to be inhibited.

3.3   Thermal Stability of Novel Type II Resins

From a practical standpoint, type II resins are useful as well as type I resins, but are less stable at elevated temperature. It is commonly believed, that the use of type II resins (in the OH form) must be restricted to a maximum of around 40°C. Yet, type II resins are very desirable by virtue of their high regeneration efficiency, high operating (breakthrough) capacity, and excellent resistance to organic fouling. And now it may also be possible to add thermal stability to this list of advantages. We termed novel type II resins with alkyl spacer-arm CnA II (Figure 5). It was expected, that CnA II would have low leachables, combined with high regeneration efficiency owing to high flexibility and stability of the functional group.

Table 3 showed that the novel type II with spacer-arm (C4A II) was thermally stable as well as the novel type I resins. It is found that spacer-arms inhibited the quaternary ammonium functionality from degradation. It was suggested that the decrease of acidity of neighboring hydrogen reduced the severity of degradation of the quaternary ammonium functionality. Moreover, the novel type II resin was found to yield lower leachables than conventional type II resin even at elevated temperature (Table 5).

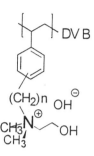

Figure 5 *Novel type II resins*

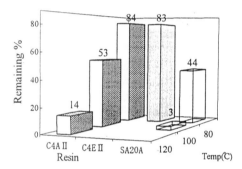

Figure 6 *Thermal stability of some type II anion exchange resins*

3.4 Thermal Stability (Dynamic Experiments)

Figure 7 shows the thermal stability of anion exchange resins. As shown in Figure 4, the novel resins were appreciably more stable than conventional styrene-based resin, such as SA10A. It has always been commonly believed, that type I resins in the OH form, were restricted to a maximum of around 60°C. On the basis of the results illustrated in Figure 6, our new resins with alkyl/ether spacer-arms, were found to withstand operation at 80-90°C, whereas the conventional type I resins are commonly used at around 55°C.

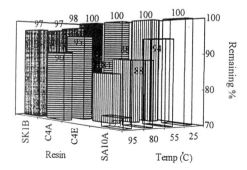

Figure 7 *Thermal stability of the resins (30 days)*

3.5 Durability of the Resins

These novel resins with thermal stability were also expected to exhibit long life expectancy. Durability experiments have been performed in operational demineralizer vessels (5 m³) at our Kurosaki factory plant since Nov. 20, 1994.

Table 3 *Durability of Anion Exchange Resins tested at Kurosaki Plant*

Date Period	Resin	C 4 E Low WR	DIAION® SA10A	C 4 E High WR	DIAION® SA12A
11/94	Salt splitting cap.	1.20 / 3.20	1.42 / 3.80	0.96 / 3.28	1.34 / 4.24
At start	Water retention (%)	45.7	45.1	54.7	54.1
2/95	Salt splitting cap.	1.10 / 2.89	1.30 / 3.37	0.91 / 2.89	1.21 / 3.73
After 3 months	Water retention (%)	45.0	45.3	53.5	53.2
5/95	Salt splitting cap.	1.11 / 2.91	1.24 / 3.21	0.92 / 2.90	1.18 / 3.57
After 6 months	Water retention (%)	44.3	44.2	53.2	52.3
6/95	Salt splitting cap.	1.10 / 2.87	1.20 / 3.12	0.92 / 2.89	1.15 / 3.51
After 7.5 months	Water retention (%)	44.3	43.4	53.3	52.7

Salt splitting capacity: left meq/mL   right meq/g

It is well known that conventional resins exhibited a tendency to degrade, even under normal (conventional) use conditions. The decrease of salt splitting capacity suggested that conventional resins showed degradation of the quaternary functionality at room temperature. As expected, and based on the results of the thermal stability tests described in 3.4 and Figure 6 above, the novel resins were found to resist "leakage" of the functional groups.

However, we felt it was a drawback, that these novel resins showed lower salt splitting capacities, than the comparing conventional resin types, with the ion exchange capacity per unit weight decreasing, as the number of methylene links in the spacer arms increased. Consequently, we devoted considerable time to further improve these new resin types, especially with regard to increasing the salt splitting capacity. At the time of writing, the next generation of these new resin types is in the final stages of development and testing, exhibiting *dramatically high* salt splitting capacities, of the order of 1.6-1.9 meq/mL (4.5-5.2 meq/g).

3.6 Leachability of Resins

Ultrapure water is used in a wide variety of industries with major users being the power and semiconductor industries. Water quality demands for fossil fuel fired power stations and for nuclear power stations with once-through steam generators, are now capable of being met. As shown in Table 4, effluent from the novel resin (C4A) is of an extremely high quality. All these results are due to the effect of the stability of the spacer-arm reinforced side chains supporting the quaternary amine functional groups of these novel resins. Based on these results (see Table 4 and Table 5), it is hoped, that these new types of resins will prove suitable for a wide range of applications, such as the production of ultrapure water for a variety of purposes, as well as for other process uses.

Table 4   *Leakage from Type I Resins*

Resin	C 4A				DIAION® SA10A				Amberlite® IRA-400	
Column temp (°C)	25		70		25		70		25	
Rinsing volume (BV)	Cond µS/cm	TOC (ppb)	Cond µS/cm	TOC (ppb)	Cond µS/cm	TOC (ppb)	Cond µS/cm	TOC (ppb)	Cond µS/cm	TOC (ppb)
Before regeneration	5.1	111	11.8	526	6.1	3510	15.1	5790	7.6	2070
20	0.30	135	1.8	775	3.3	2750	25.8	4120	4.3	2000
60	0.11	108	1.5	601	2.3	1620	18.7	2360	2.5	1250
100	0.08	90	1.3	478	1.8	1120	11.0	1440		

Table 5   *Leakage from Type II Resins*

Resin	C 4A II				DIAION® SA21A			
Column temp (°C)	25		70		25		70	
Rinsing volume (BV)	Cond (µS/cm)	TOC (ppb)	Cond µS/cm	TOC (ppb)	Cond µS/cm	TOC (ppb)	Cond µS/cm	TOC (ppb)
Before regeneration	0.77	38	23.3	696	0.74	6160	14.6	1140
20	0.32	135	1.6	2340	1.2	1570	12.7	4074
60	0.21	130	1.6	2390	0.94	680	12.7	3990
100	0.16	108	1.6	2160	0.69	356	12.0	3970

Resin volume; 80 mL (Cl form), Regenerant: 1N NaOH, Column temp. 25°C / 70°C
Regeneration level; 200g NaOH/ L-R, Effluent: UCP, Flowrate: SV20

3.7   Chemical Stability

Chemical stability of anion exchange resins was estimated from the results of TOCs derived from the leakage of the resins. With regard to the novel resins with alkyl spacer-arms, Table 6 shows that in no case was there any loss of ion exchange capacity, and the resins were stable under various chemical conditions. Furthermore, novel resins with alkyl spacer-arm in the OH form were less leachable than conventional resins such as DIAION® SA10A, SA12A and Amberlite® IRA-400. As a result, no trimethylamine odor, in the OH form, was leached from the novel resins.

On the other hand, novel resins with ether spacer-arms were chemically less stable under oxidizing conditions, when compared with conventional resins, because the ether bond showed a tendency to cleave with dissolved oxygen and oxidants. This was the only drawback of the novel resins with ether spacer-arm.

Table 6   *Chemical Stability of Novel Resins in terms of ppm TOC of Supernate (Static Tests)*

Solution used (30 days at 30°C)	C 3 A	C 4 A	C 4 E	DIAION® SA10A	DIAION® SA12A
Demineralized water (Cl/OH)	5 / 20	7 /15	not tested.	14 /82	6 / 84
1N / 25% - NaOH (OH)	28 / not tested.	32 / 112	55 / 386	81 / 245	93 / 189
1N / 35% - HCl (Cl)	15 / not tested.	17 / 34	76 / not tested.	53 / 96	41 / 77
1% - H_2O_2 (Cl/OH)	12 / 66	16 / 84	74 / 890	14 / 71	4 / 70

30 days exposure at 30°C in the OH form or Cl form

3.8 Stability for γ-Ray Exposure

Anion exchange resins in the OH form or Cl form were irradiated with ^{60}Co γ-rays. The effect of γ-ray exposure on the anion exchange capacity is shown in Table 7. Loss of salt splitting capacity was observed at doses of up to 0.1 MGy. A marked decrease in salt splitting capacity was observed in the range of 0.1 MGy to 1 MGy. The novel resin with alkyl spacer-arm was more stable than conventional resins. Resins in the Cl form were generally more resistant to γ-ray exposure than those on the OH form. The net effect, however, was found to be a decrease in salt splitting capacity. Linear relation was observed between thermal stability and stability for γ-ray exposure.

Table 7 *Residual Salt Splitting Capacity at a Dose of 1 MGy (meq/g)*

Resin	C 4 A		C 4 E		DIAION® SA10A	
Salt form	OH	Cl	OH	Cl	OH	Cl
Residual % capacity	83	88	73	78	60	72

4 SUMMARY

This presentation represents **a true "first"** in recent anion exchange resin development, achieved by Mitsubishi Chemical Corporation. Novel, strong base anion exchange resins, with much greater than usual thermal stability, are still based on styrene-DVB matrices, but contrary to the conventional (traditional) structures, the side chain supporting the quaternary amine functional groups, contains **"spacer arms"** of alkyl, or ether types. The properties of these new resins differ from those of conventional resins in many interesting ways. One was particularly noteworthy: separation between the positive charge and the benzene ring with spacer-arms greatly enhances the stability of the quaternary ammonium functionality of the resins. As a result, thermal stability, stability for γ-rays irradiation, excellent durability, high regeneration efficiency, low resin leachability and extremely high silica removal capacity are superior to conventional styrene-based quaternary ammonium resins. Furthermore, compared to conventional styrene-based type II anion exchange resin, the novel type II resins exhibit an appreciably higher thermal stability. Finally, continuing development of additional versions of these new resin types with "spacer-arms" are resulting in strong base resins with dramatically higher salt splitting capacities, of the order of 1.6-1.9 meq/mL. These novel resins based on new concepts will greatly innovate current processes and usage of anion exchange resins.

Acknowledgements

The authors gratefully acknowledge the assistance of Mr A. Nishimura (Manager), together with that of Mr T. Morita. They would also like to recognize the valuable contribution made by Mr K. Matsuoka, in implementing the durability tests at our Kurosaki Plant.

OXIDATION RESISTANCE OF CATION EXCHANGE RESINS

J. A. Dale and J. Irving

Purolite International Ltd.
Pontyclun
Mid Glamorgan CF7 8YL

1 INTRODUCTION

Ion Exchange Resins with resistance to strong oxidising agents has been a major objective for many years. Poor resistance has influenced the performance of synthetic ion exchange resins since their inception. This problem has been addressed regularly over the years. The influences on the rate of oxidation are complex, and many of the factors which affect the resin structure and the functional groups have been clarified.

Some improvement in oxidation resistance was obtained with higher crosslinked cation resins, which can withstand attack for a longer period before their performance suffers significantly.

2 OXIDATION RESISTANCE

Resins with modified matrices[1] have been proposed with partial success. These generally are effective for a limited period because the oxidising agent attacks the modifying group in a sacrificial manner rather than the polymer backbone. Success is limited, eventually the oxidation reverts to that of the unmodified resin matrix.

Resins with good oxidation resistance can also suffer other disadvantages. For instance, a more highly crosslinked resin shows slower diffusion, and is generally more difficult to regenerate. It is also more prone to fouling with high molecular weight organics and also with multi-valent ions. It has been shown[2] that hydrated ions such as hydrogen and hydroxide, have to shed hydration shells to be accommodated in highly crosslinked resins. The resulting increase in charge density allows these aggressive ions to attack the functional group causing less thermal stability. Hence, the extended life obtained from increased oxidation resistance may be curtailed by other factors.

3 MECHANISMS OF OXIDATIVE ATTACK

The chemistry of oxidative attack is now well known[3,4]. Many papers[5,6] have shown that the mechanisms can be complex. Hydrogen peroxide, ozone, free chlorine, hypochlorite, chlorine dioxide, chromic, nitric, and peracetic acids, have been studied by a number of workers; all cause resin degradation, especially in the presence of catalysts: UV light, radiation, transition metals. To a lesser extent atmospheric oxygen or chlorate also cause problems, particularly at higher temperatures.

Hydrogen peroxide is a very useful model for studies. Firstly, it does not add anything other than oxygen to the structure (unlike chlorine or hypochlorite which can chlorinate the matrix, or chromate which may become irreversibly bound to the resin). Secondly, it is fairly easily handled. Nitric acid on the other hand can produce run-away exotherms.

One of the most important variables affecting the rate of degradation is the presence of transition metals on the resin or in solution. Of these, iron is by far the most important[4,5,6] because it catalyses the decomposition of hydroperoxides. It is also ubiquitous, and can be expected to be present in any normal ion exchange process, unless special effort is made to remove it. The elimination of iron from systems with hydrogen peroxide or atmospheric oxygen can reduce degradation by orders of magnitude. Copper is also important but less powerful and less common. Clearly, all tests on oxidation resistance must be carried-out at controlled metal concentrations. Hydrogen peroxide concentration, temperature and time need to be specified.

Other variables need to be eliminated or controlled:

3.1   UV light (including daylight) in presence of air (oxygen)

3.2   Radiation

3.3   Resin pH - resins in H^+ form are as a rule more readily oxidised.

In industrial applications these conditions have to be taken into consideration when discussing the operating performance of a resin and/or equipment. The presence of oxygen in condensate polishing has been shown to be detrimental to thermal stability[7]. For this reason hydrazine has historically been used. More recently, health hazards associated with hydrazine together with changes in materials of construction and the water conditioning chemistry have prompted some industrial power plants to operate oxygenated systems. The full effects on consequent oxidative degradation still have to be evaluated.

Chromium/chromate can act both in a redox/oxidation mode by a subtlety different mechanism in which the oxidation is self catalysing, so it is useful to compare this oxidation process also.

4 INFLUENCE ON OPERATING PERFORMANCE

Increased moisture retention gives rise to many problems: loss of physical strength, irreversible resin swelling causing impaction, increased mechanical stress, channelling, and poor distribution of regenerant. Any or all of these can result in severe resin physical breakdown. Removal of the excess resin produced by the irreversible expansion may be necessary. This reduces the capacity of the resin bed (fewer active groups). Also changes in resin selectivity, can produce a lower throughput because of increased leakage/premature end-point.

Linear polymer leachables can foul resin beds down-stream[8,9]. This is particularly important in the operation of mixed beds[10-13]. The effect of carboxylic groups on rinse performance is well known, as is the effect of low molecular weight polyamines (possible degradation products of anion resins) on the kinetics of cation resins[14,15]. An oxidatively resistant resin would have decisive advantages in ultra-pure water production.

The ultra-pure water industry is ever mindful of any impurities in the system, particularly TOC, which can effect the quality of production e.g. silicon chips. As the techniques for determination of TOC, have improved, so the full effects of trace leachables have been clearly recognised. Leachables also act as nutrients for bacteria and algae. Regrettably, oxidising agents are the best means of controlling biological growth. Ozone and UV light are frequently used as part of the integrated and complex system needed for ultra-pure water. Incidences of both ozone and UV light causing resin degradation are well known.

5 NEW DEVELOPMENTS

Purolite International Ltd. has developed a new cation exchange resin type with real oxidation resistance. Comparison with the previous modifications, which showed certain advances over conventional, more highly crosslinked, strong acid cation resins, is dramatic. The resin structure may undergo oxidation, but the performance behaviour itself shows little degradation (criteria 1-6 given below). A standard oxidation test, using specific amounts of 1% hydrogen peroxide and iron at 40°C shows striking differences from previous resins.

5.1 Changes in Moisture Content

These are relatively small. Table 1 compares the moisture retention before and after oxidation of the new product, D2908, with a number of standard cation resins including a newer modified product with partial resistance. All these are severely degraded, without exception (See Figure 1).

5.2 Increase in Polymer Leachables

These are negligible when compared with standard resins (see Table 2).

5.3 The Products have Extremely High Breaking Weights

Table 3 gives a comparison of breaking weights before and after oxidation.

5.4 Extent of Oxidation

It is evident that the oxidation itself cannot be obviated. Also, the new material is not precisely comparable with conventional resins which lose up to 10% of their mass for an increase in moisture of 10%. Many of the carboxylic groups generated are lost in the leachables whereas the new structure is essentially intact, and the products of oxidation are retained, bound to the resin structure.

5.5 Loss of Functional Groups

When considering cation resins, the loss of functional groups is more or less pro-rata with loss in resin mass, so point 5.4 applies.

Table 1 *Oxidation Resistance using 1% w/v Hydrogen Peroxide*

Temperature	40°C	Moisture %				
Iron Content	500ppm	Days				
RESIN		0	1	2	3	5
D2908	VALUE:	56.9	57.4	58.6	60.2	65.5
H form	NETT D:		0.5	1.7	3.3	8.6
5%DVB	VALUE:	65.1	74.3	92.9	sample dissolved	
SP	NETT D:		9.2	27.8		
8%DVB	VALUE:	56.3	62.2	70.0	80.1	sample
Gel H form	NETT D:		5.9	13.7	23.8	dissolved
8%DVB	VALUE:	49.6	73.3	90.4	sample dissolved	
Gel Na form	NETT D:		23.7	40.8		
8%DVB	VALUE:	64.5	68.2	75.4	92.9	-
MP	NETT D:		3.7	10.9	28.4	
12%DVB	VALUE:	55.9	58.2	60.5	70.0	-
MP	NETT D:		2.3	4.6	14.1	

Table 2 *Polymer Leachables*

RESIN	BEFORE ppm	AFTER ppm
D2908	3.2	2.9
8%DVB GEL H FORM	2.9	> 100

After 24 Hour Oxidation Test
48 Hour TOC Test

Table 3 *Breaking Weights*

RESIN	BEFORE g	AFTER g
D2908	1184	946
8%DVB GEL H FORM	600	300

After 24 Hour Oxidation Test

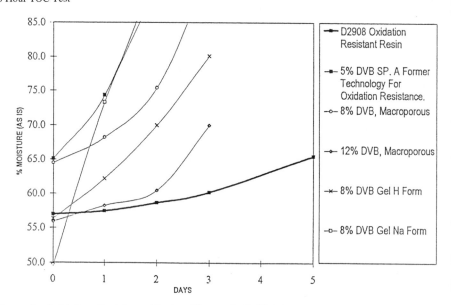

Figure 1 *Oxidation Resistance Tests - Change in % Moisture*

5.6 Resin Selectivity

Both values and trends differ from those of conventional resin. The effect of oxidation has been investigated for sodium/hydrogen exchange only (See Figures 2 and 3).

A second test using 5% w/v chromic acid/10% sulphuric acid mixture at 90°C was also briefly investigated. The dramatic difference between the conventional resin and the new resin was evident for this oxidative medium also. The results are given in Table 4.

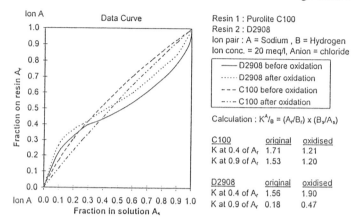

Resin 1 : Purolite C100
Resin 2 : D2908
Ion pair : A = Sodium , B = Hydrogen
Ion conc. = 20 meq/l, Anion = chloride

—— D2908 before oxidation
······ D2908 after oxidation
— — C100 before oxidation
— ·· — C100 after oxidation

Calculation : $K^A/_B = (A_r/B_r) \times (B_s/A_s)$

C100	original	oxidised
K at 0.4 of A_r	1.71	1.21
K at 0.9 of A_r	1.53	1.20

D2908	original	oxidised
K at 0.4 of A_r	1.56	1.90
K at 0.9 of A_r	0.18	0.47

Figure 2 *Comparison of sodium/hydrogen selectivity*

Table 4 *Oxidation Resistance in 5% Chromic Acid*

TEMPERATURE: 90°C		D2908		8% X-LINK GEL	
		Before	After	Before	After
% Moisture	%	57.4	60.5	54.8	sample
Difference			3.1		dissolved
% Dry Matter	%	42.6	39.5	45.2	0.00

Mass (4g Orig Dry Wt)	g	9.39	8.97	8.85	sample
Mass Change	g		-0.42		dissolved
Mass Change	%		-4.47		
Mass Dry Matter	g	4.00	3.54	4.00	0.00
Mass Dry Matter Change	g		-0.46		-
Mass Dry Matter Change	%		-11.42		-

6 RESIN SELECTIVITY

Resin selectivity was determined essentially by the technique outlined by Gregor[16], Bonner and Smith[17] using concepts and notation developed later by Anderson[18].

20 mL samples of resin were equilibrated with solutions containing various mole fractions of ion pairs. The resins were then analysed to determine the fraction of each ion loaded.

The resin loading data was then used to plot comparative selectivity curves Figures 2-5. Figure 2 gives a selectivity comparison of fresh resins with the oxidised counterparts. It is most interesting that oxidation changes selectivity in the opposite direction to that for standard resin: as it degrades it behaves like a more highly crosslinked material.

The data for sodium/hydrogen exchange compares D2908 with Purolite C-100, for chloride anion (Figure 2) and sulphate (Figure 3). Both resins are marginally more selective for sodium where the co-ion is sulphate. Purolite C-100 was in line with 8% crosslinked resins given in Helfferich "Ion Exchange"[19].

The new oxidation resistant resin, shows considerable differences in selectivity. Where the fraction of sodium is < 0.25 in solution, the novel resin has higher selectivity for sodium; above this level the selectivity is significantly lower. In certain situations this difference may offer significant advantages. It is more difficult to predict kinetic and fouling behaviour, but the fact that the sites have the selectivity of a more easily-regenerated or easily-cleaned, lower crosslinked resin, offers considerable incentive to study these. The combination of high selectivity for a limited number of sites, together with a tough oxidation resistant structure which does not produce leachables as a result of oxidative breakdown, makes this resin potentially useful for condensate polishing where iron levels build up to produce an oxidation prone environment.

For this reason, sodium/ammonia selectivity was studied (see Figure 4). Again the data for Purolite C-100 are in line with published values. The data for the new resin are interesting; for a mole fraction high in ammonia, the resin is more selective for sodium. However when using limited amounts of regenerant counterflow, as the fraction of ammonia falls, so the resin becomes more selective for ammonia. This resin may therefore be used to advantage to remove trace levels of sodium from solution. It would be expected that the higher hydration of hydrogen ion will be kinetically more efficient.

The calcium/sodium equilibria are remarkably similar, Figure 5. However because of concentration differences in the resin phase, the numerical values for K vary. It remains to be seen how well the D2908 will perform as a softening resin. In any case, the resin is unlikely to be cost effective for most softening applications.

7 THERMAL STABILITY

The thermal properties of this novel material has not been evaluated so far. However the selectivity data would suggest that these products could offer some useful advantages here also. The resins contain a substantial majority of sites which are less selective for hydrogen ion than the conventional 8% crosslinked resins. Conventional resins with a higher moisture content (with lower crosslinking) are found to be more thermally stable. This has been attributed to a lower charge density of the more highly hydrated hydrogen ion, present in a lower crosslinked product. Ions of higher charge density attack the sulphonic acid functional group causing thermal desulphonation. Meyers and Boyd[20] showed that the higher the crosslinking the larger the Na/H selectivity coefficient. It follows that the higher the charge density of the hydrogen ion the slower the diffusion and the more relatively selective becomes the resin. The converse therefore should apply.

The D2908 has sites with a larger range of selectivities than conventional resins, however the majority of sites are of lower selectivity when compared to an 8% cross-linked resin and would therefore appear to be more highly hydrated. Therefore, the majority of sites should be more thermally stable. This will be experimentally investigated.

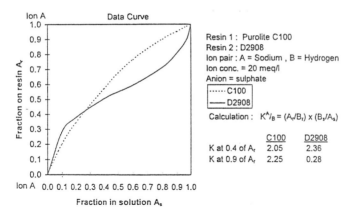

Figure 3 *Sodium hydrogen selectivity (anion = sulphate)*

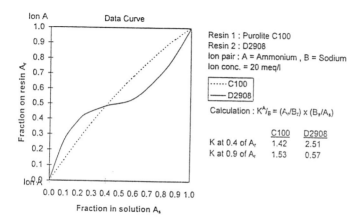

Figure 4 *Ammonium sodium selectivity*

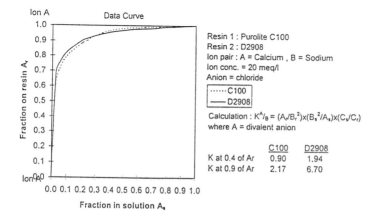

Figure 5 *Calcium sodium selectivity*

Acknowledgements

Thanks to Andreas Gotthardt who carried out the selectivity determinations, to Graham Crooks for testing the oxidation resistance, to Steve Plant who prepared the resin samples and to J. R. Millar who provided much historic information on various characteristics of existent resins.
We are obliged to Purolite International Ltd., for permission to present this paper.

References

1. Y. R. Dhingra, P. R. Van Tol, US Patent 5,302,623, assigned to the Dow Chemical Co.
2. B. Chu, D. C. Whitney, R. M. Diamond, *J. Inorg. Nucl. Chem.*, 1962, **24**, 1405.
3. J. R. Stalbush, R. M. Strom, *ReactivePolymers*, 1990, **13**, 233.
4. G. J. Moody, J. D. R Thomas, *Laboratory Practice*, 1971, **21(9)**, 632.
5. L. Goldring, "Theory & Practice of Ion Exchange", SCI 1976, 7.1.
6. C. Inami, et al, *Pittsburgh Water Conf.*, 1991, IWC-91-34, 266.
7. K. Dorfner, Ion Exchangers 350.
8. S. Fisher, G. Otten, *Pittsburgh Water Conf.*, 1981.
9. S. Fisher, G. Otten, *Pittsburgh Water Conf.*, 1986.
10. D. C. Auerswald, "Ion Exchange for Industry", M. Streat, ed., SCI, Ellis Horwood Ltd., 1988, p. 11.
11. J. R. Stalhlbush, et al., *Pittsburgh Water Conf.*, 1987, IWC-87-47.
12. K. Wieck-Hansen, "Ion Exchange Advances", M. J. Salter, ed., SCI Elsevier Applied Science, 1992, p. 128.
13. K. Daucik, *Ultra Pure Water 5*, SCI, 1994.
14. H. Small, *J. Amer. Chem. Soc.*, 1968, **90**, 2217.
15. W. E. Bornak, J. W. Griffin, *Ultra Pure Water,* 1987, September.
16. H. P. Gregor, J. Belle, R. A. Marcus, *J. Amer. Chem. Soc.*, 1955, **77**, 2713.
17. O. D. Bonner, I. I. Smith, *J. Phys. Chem.*, 1957, **61**, 362.
18. R. A. Anderson, *A. I. Ch. E. Symposium Series*, 1973, **152**, 236.
19. F. G. Helfferich, "Ion Exchange", McGraw-Hill, 1962, 151.
20. G. E. Myers, G. E. Boyd, *J. Phys. Chem.*, 1956, **60**, 521.

NOVEL BIPOLAR ION-EXCHANGE RESINS - THEIR PREPARATION AND CHARACTERISATION

U. Schumacher, S. Semmelbeck and M. Grote

Department of Applied Chemistry
University of Paderborn
D-33095 Paderborn
Germany

1 INTRODUCTION

A special group of bipolar ion exchangers are the so-called snake-cage exchangers containing linear polyanions or polycations in their network[1]. In contrast to well-known amphoteric resins, the electric charge of the counterions can be compensated because their active groups are not anchored in the matrix and have a higher mobility. The forming of inner salt structures makes snake-cage exchangers able to sorb anions and cations simultaneously. Generally their regeneration can easily be carried out by means of water, but as a great disadvantage this type of exchanger cannot be synthesised reproducibly. The products exhibit a not clearly defined mixture of different functional groups which leads to a strong limitation in technical application. Consequently there is a great interest in developing bipolar ion exchangers having a well-defined structure giving rise to stoichiometric interaction.

The monofunctional exchanger P-TD seems to be a suitable starting point. Grote at al.[2,3] have already studied this weakly basic, precious metal and nitrate-selective anion exchanger in detail which contains the tetrazolium cation as effective polar group. Attaching the S-atom of dehydrodithizone to chloromethylated polystyrene seems to be an easy way to synthesise P-TD.

Our aim of preparing bipolar ion exchangers with well-defined steric arrangement of the exchanging groups ought to be reached by introducing carboxylic or sulphonic acid groups into the starting unit dehydrodithizone. Furthermore, the effects on the snake-cage properties have to be worked out with special respect to the influence of the bipolar structure.

2 SYNTHESIS AND CHARACTERISATION

2.1 Preparation of Structurally Defined Bipolar Ion Exchange Resins (P-TD Structures)

2.1.1 Basic Routes. The preparation of ion exchangers with well-defined structure and steric arrangement of the exchanging groups follows two pathways in principle (see Figure 2.1). If the bipolar dehydrodithizone derivatives (precursor) are attached to chloromethylated polystyrene, the P-TD-type-exchangers will be obtained directly (pathway A). Alternatively, immobilisation of analogously substituted dithizone and subsequent oxidation is a further possibility to get the bipolar ion exchanger via an indirect way (pathway B).

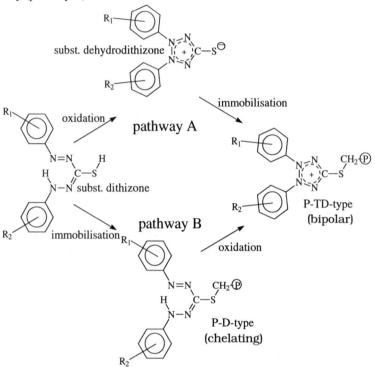

Figure 2.1   *Synthetic pathways A and B to bipolar ion exchange resins (P-TD-type)*
(P: polymeric matrix; R_1, R_2: acidic groups)

2.1.2 Preparation of Acidic Dithizone Derivatives. Figure 2.2 gives a survey of the synthesised compounds which were needed in order to prepare the bipolar ion exchangers.

symbol	R_1	R_2
D-oC$_2$	o-COOH	o-COOH
D-pC$_2$	p-COOH	p-COOH
D-oC	o-COOH	H
D-mC	m-COOH	H
D-pC	p-COOH	H
D-oS	o-SO$_3$H	H

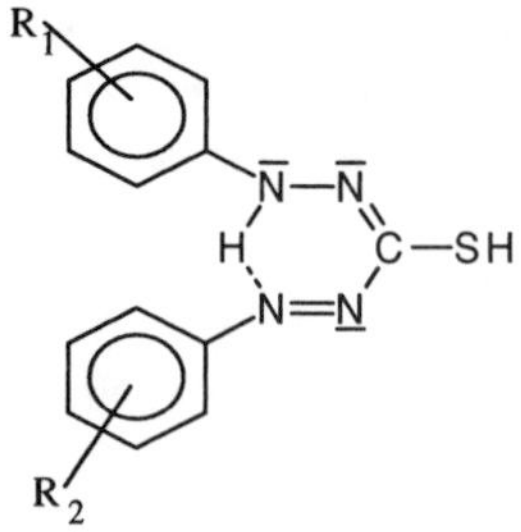

Figure 2.2   *Synthesised starting compounds (acidic dithizone derivatives)*

Because of decomposition during synthesis the use of meta- and para-sulphonic acid derivatives has not been applied any further.

The purity of the starting compounds shown in Figure 2.2 was confirmed directly after preparation by spot analysis (formazans show characteristic purple colour with sulphuric acid) and thin-layer chromatography. IR- and NMR-spectroscopy in connection with elementary analysis confirmed the structure of the desired compounds. However, difficulties in oxidising the dithizone derivatives to dehydrodithizones led to predominant application of pathway B (see Figure 2.1).

2.1.3 Immobilisation via Pathway B. The immobilisation of the starting compounds shown in Figure 2.2 was carried out directly by suspending a batch of chloromethylated polystyrene in a dimethylformamide solution of the appropriate dithizone derivative. Besides the nucleophilic substitution resulting sulphur bonded dithizone groups (P-D-type resins), the carboxyl-substituted starting compounds showed esterification with the reactive polymer. In the case of ortho-substituted compounds this secondary reaction could be avoided by using catalytic amounts of potassium iodide (Finkelstein reaction[5]) instead of the base 1,8-diazabicyclo[5,4,0]- undec-7-en (DBU)[4]. It was observed that in addition to the dithizone component also, the polymer matrix has a distinct influence on the course of the immobilisation. While 1-(2'-sulphophenyl)-5-phenyl-3-thiocarbazone (D-oS) could only be attached to fine grain reactive gel-type polymers the corresponding carboxyl-substituted compound 1-(2'-carboxyphenyl)-5-phenyl-3-thiocarbazone (D-oC) preferred reaction with macroporous chloromethylated polystyrene (Figure 2.2). The effects observed may be caused by different microenvironments within the reactive polymers, thus, different interactions of the reactive sites with the dissolved reactants may occur[6]. In case of fibrous materials, the mode of interactions should differ from those in resin particles. A higher accessibility of reactive groups bound onto a fiber surface was already found when reacting graft polypropylene supported chloromethyl polystyrene (20-40µm fiber, SMOP-3, 1mmol/g dry resin; SMOPTECH, Finland) with carboxy-dithizons, e.g. D-pC (Figure 2.1). The resulting materials are still under study.

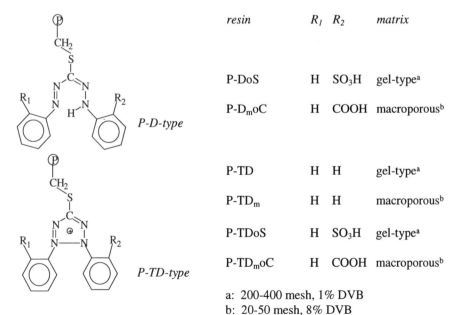

resin	R_1	R_2	matrix
P-DoS	H	SO$_3$H	gel-type[a]
P-D$_m$oC	H	COOH	macroporous[b]
P-TD	H	H	gel-type[a]
P-TD$_m$	H	H	macroporous[b]
P-TDoS	H	SO$_3$H	gel-type[a]
P-TD$_m$oC	H	COOH	macroporous[b]

a: 200-400 mesh, 1% DVB
b: 20-50 mesh, 8% DVB

Figure 2.3   *Investigated resins*

2.1.4 Oxidation. In comparison to the oxidation of the non-immobilised dithizone derivatives, it was much easier to convert the polymer-coupled dithizones. The oxidation of the chelating formazan-groups to the bipolar tetrazolium group succeeded by using N-bromosuccinimide or bromine water in aqueous solution, yielding the P-TD-type polymers.

2.2 Characterisation of Intermediates and Final Products

The monosubstituted intermediates P-DoS (sulphonic acid group in ortho-position) and P-D$_m$oC (carboxylic acid group in ortho-position) were characterised by means of elementary analysis and IR-spectroscopy in comparison to free model compounds.

The structure of the bipolar ion exchange resins P-TDoS and P-TD$_m$oC was confirmed by the same methods. In all cases, the experimental results correspond to the theoretically expected data with respect to the error limits. Figure 2.3 summarises the designation of the prepared ion exchangers and the structure of their functional groups.

3 DISCUSSION OF PROPERTIES

Further characterisation of the exchange resins was done by acid-base titration, electrical conductivity measurements and sorption studies using solutions of strong electrolytes and of heavy metal salts respectively.

3.1 Acid-Base Titration

The two bipolar resins and their intermediates (H$^+$-form) were titrated against sodium hydroxide solution by the miniaturised *direct method*[8] to determine the amount of active acidic groups. For this purpose after every addition of the titrant (0.12 M NaOH in 1M NaCl) to the suspended resin portion (0.1 g in 1 M NaCl) the pH value was measured in 24h - intervals. The capacities values Q(H$^+$) obtained are shown in Table 3.1.

Table 3.1   *Specification of Ion Exchangers and Capacity Values (see Figure 2.3)*

resin type	symbol	moisture content %	theor. capacity Q_0[a) [meq/g,dry resin]	capacity $Q(H^+)$[b) [meq/g,dry resin]
chelating	P-DoS	42	0.55	0.55
(intermediates)	P-D$_m$oC	38	1.05	1.00
bipolar	P-TDoS	35	0.40	0.13
	P-TD$_m$oC	30	0.70	not to determine
weakly basic	P-TD[c)	55	1.10	-
	P-TD$_m$[c)	55	2.0	-

a) determined by elementary analysis   b) determined by pH-titration
c) resins prepared according to Grote et al.[2,3]

The values of the H^+-capacities of both chelating intermediates (P-D-type) correspond satisfactoriliy to the theoretical capacities (Q_0) determined by elementary analysis. As Figure 3.1 shows, P-TDoS exhibits a large slope at the turning point. This is a typical behaviour for a strong acidic cation exchanger. However, the H^+-capacity of 0.13 meq/g dry resin, lies 67% below the value expected from the elementary analysis (Table 3.1). This result is a first experimental hint to the formation of stable inner salt structures.

By contrast, it is not possible to determine capacity values by pH-titration for P-TD$_m$oC because it does not exhibit a clearly defined turning point. This behaviour is not typical for weakly acidic ion exchange resins[8], however, it was found to resemble strongly the snake-cage polyelectrolyte Dowex 11A8, which was titrated under the same conditions.

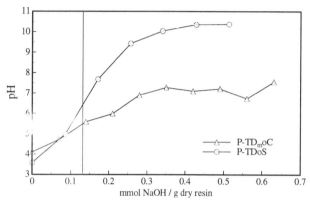

Figure 3.1 *Neutralisation curve of bipolar resins P-TDoS and P-TD$_m$oC (The vertical line indicates the turning point of the P-TDoS curve)*

The existence of inner salt structures can also be concluded from the additionally determined moisture content. The contents of the bipolar exchange resins lie significantly below the ones of the non-oxidised P-D intermediates (Table 3.1), although the oxidation of the non-charged chelating groups to the tetrazolium cations introduces additional polar, i.e. hydrophilic, groups into the polymeric system.

3.2 Conductivity Measurements

The results of the pH-titration and moisture content determinations of both bipolar resins gave hints to the existence of inner salt structures. Nevertheless it was only possible to classify the strength of intramolecular interactions for the resin P-TDoS. In order to get more information about the resin P-TD$_m$oC supplementary conductivity measurements using a measuring device according to Sauer et al[9] were carried out. The electrical conductivity of packed beds containing the bipolar exchanger P-TD$_m$oC equilibrated with aqueous solution of NaCl electrolyte is distinctly lower (specific conductivity of the resin material $\kappa = 80$ μS/cm) than that for the anion exchanger P-TD$_m$ ($\kappa = 365\mu$S/cm). Knowledge of the electrical conductivity of packed beds containing ion exchangers in contact with solutions of electrolytes is of interest as a means of ascertaining the mobility of ions in the exchanger phase. According to the existence and also the stability of the inner salts, there are two counterions per each bipolar functional group which may contribute to the conductivity. Hence, the drastically reduced conductivity of the bipolar polymer P-TD$_m$oC indicates also the formation of stable inner salts.

3.3 Sorption Properties

3.2.1 Electrolyte Sorption under Static Conditions. Not unexpectedly, P-TDoS behaves in sorption experiments (e.g. 100 mg resin, 30 mL 0.002 M LiNO$_3$ solution) like a strongly acidic cation exchanger, with a very low capacity (e.g. Q(Li$^+$) = 0.1 meq/g, dry resin) corresponding quite well to the titration data. As a consequence, the resin is not able to exchange anions like nitrate, sulphate and chloride. The strong shielding effect of the negatively charged sulfonic group on the quaternary cation may be responsible for the total exclusion of anions.

By contrast, P-TD$_m$oC exhibits a pH-dependent sorption behaviour which is not surprising because this bipolar exchanger contains weakly acidic carboxyl groups (see Figure 3.2).

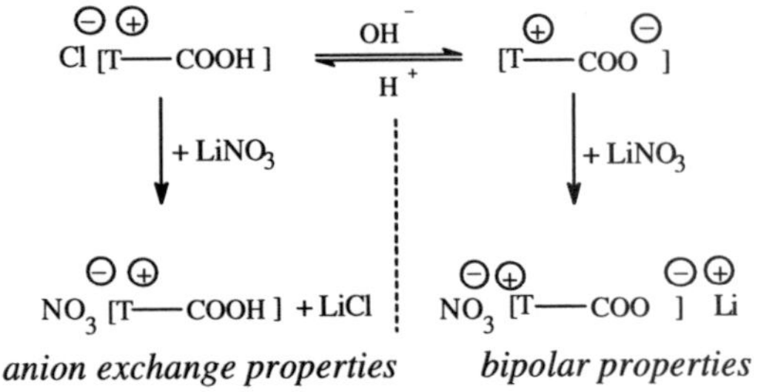

Figure 3.2   *Dissociation equilibrium and sorption properties of P-TD$_m$oC*
[T$^+$]: functional tetrazolium cation

In acidic solution P-TD$_m$oC behaves like a weakly basic anion exchanger (Q(NO$_3^-$) = 0.11meq/g dry resin). In neutral solutions, a weak sorption of lithium and nitrate ions (electrolyte sorption, Q(Li$^+$, NO$_3^-$) = 0.05 meq/g dry resin) was observed. The assumption of stable inner salt structures gives a suitable explanation for the hindered formation of free carboxyl groups and, hence, an exchange effect which is not very pronounced.

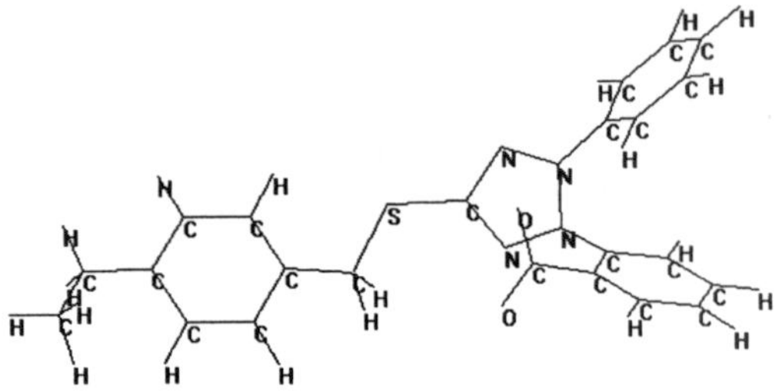

segment of resin matrix   functional group

Figure 3.3   *Quantum mechanically calculated structure of P-TD$_m$oC*
(calculated by "hyper chem" program, gas phase)

The formation of intramolecular salts requires well-defined distances between anionic and cationic charge carriers. Quantum mechanical treatment (Figure 3.3) reveals that an intramolecular inner salt structure can only be formed, if the acidic respectively anionic group is in the ortho-position of the phenyl ring attached to the quaternary heterocycle.

3.2.2 Precious Metal Sorption. The bipolar exchanger P-TDoS shows a rather unusual sorption behaviour towards precious and base metals. Great differences compared to the well investigated resin P-TD can be observed[2]. While P-TD is able to sorb several anionic precious metal chloro complexes, like e.g. Au(III), Pd(II), Pt(V) and traces of base metals. P-TDoS exclusively sorbs only Au (III). Furthermore, in contrast to typical anion exchangers the loading capacity increases with increasing hydrochloric acid concentration (see Figure 3.4).

Another striking non-typical property is the desorption of Au (III) simply by means of water. Ionic bonded precious metal complexes can normally be eluted only by an excess of displacing anions, such as perchloric acid or chelating agents, such as thiourea[3]. Hence, another sorption mechanism is indicated in this case. There are reasons to assume that predominantly tetrachlorogold acid is weakly adsorbed by the bipolar functional groups. The increasing sorption capacity of P-TDoS in hydrochloric acid solution supports this assumption.

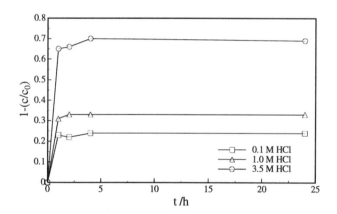

Figure 3.4 *Sorption of Au(III) by P-TDoS as function of hydrochloric acid concentration and shaking time*
(100mg resin; 20 mL 0.001 M Au (III) in hydrochloric acid)

4 SUMMARY

The preparation of bipolar exchangers based on functional tetrazolium groups revealed a lot of preparative difficulties. However, it was possible to carry out successfully the synthesis of P-TDoS, a gel-type monosulphonic acid, and P-TD$_m$oC a macroporous monocarboxylic acid. The polymers were obtained by introduction of acidic dithizone derivatives into a chloromethylated polymer matrix and subsequent oxidation to the corresponding bipolar heterocyclic groups (pathway B). The exchangers were structurally characterised by elementary analysis and IR-spectroscopy. Acid-base titration, electrical conductivity measurements and sorption experiments characterised the basic properties of these resins. In general, the introduction of acidic groups into the non-substituted cationic heterocycle of the P-TD resin drastically changes the original anion-exchange properties of the tetrazolium ring. Some of the discovered particularities are:

The capacity values calculated from acid-base titration lie distinctly below the data obtained from elementary analysis. Furthermore, sorption data and conductivity measurements show the restricted accessibility of free anionic and cationic groups due to the formation of inner salts.

Additionally a highly selective sorption of Au(III) from concentrated hydrochloric acid solution was observed. The sorption can be explained by weak adsorptive interactions between tetrachlorogold acid and the polymer which leads to easy desorption by means of water.

Taking all this facts into consideration, stable inner salt structure must be assumed to dominate the behaviour of the bipolar resins. The steric arrangement of the ortho-sulphonic or carboxylic acid groups and the adjacent tetrazolium cations seems to be rigidly fixed. Consequently the inner salt cannot be opened easily by action of counterions.

Further attempts are in progress to synthesise bipolar functionalities containing a more flexible structure, e.g., by introducing alkylcarboxylic acid groups into the tetrazolium system.

Acknowledgements

We are grateful for the financial support by the "Deutsche Forschungsgemeinschaft" (DFG).

References

1.   F. Wolf, H. Mlytz, *J. Chromatogr.*, 1968, **34**, 59.
2.   M. Grote, A. Kettrup, *Anal. Chim. Acta*, 1985, 175, 223.
3.   M. Grote, M. Sandrock, A. Kettrup, *Reactive Polymers,* 1990, **13**, 267.
4.   C. Heidelberger, A. Guggisberg, *Helv. Chim. Acta*, 1981, **64**, 399.
5.   H. R. Christen, "Organische Chemie", Salle und Sauerländer, Frankfurt, 1985, p. 454.
6.   J. M. J. Frechet, G. D. Darling, *Pure & Appl. Chem.*, 1988, **60**, 353.
7.   M. H. Kotze, F. L. D. Cloete, "Ion Exchange Advances" (Proceedings of IEX'92), M. J. Slater, ed., SCI Elsevier, 1992, p. 366.
8.   M. Grote, P. Schildmann-Humberg, *Ind. Eng. Chem. Res.*, 1995, **34**, 2712.
9.   M. C. Sauer, P. F. Southwick, K. S. Spiegler, M. R. Willy, *Ind. Eng. Chem.*, 1955, 2187.

STRUCTURE AND ADSORPTION PROPERTIES OF HYPERCROSSLINKED POLYSTYRENE SORBENTS

V. A. Davankov, M. P. Tsyurupa, O. G. Tarabaeva and A. S. Shabaeva

Nesmeyanov-Institute of Organo-Element Compounds
Russian Academy of Sciences
117813 Moscow
Russian Federation

1 INTRODUCTION

Currently ion exchange and adsorption technologies compete successfully with traditional methods of purification and isolation of organic and inorganic substances. The main areas of adsorption applications involve waste water treatment, food, textile, chemicals and oil-processing industries, preconcentration of organic pollutants from air and water for their subsequent determination by means of chromatographic techniques, etc. Adsorbents to be used must meet certain specific requirements including the combination of high adsorption capacity with the ease of regeneration, good kinetic properties, mechanical stability and osmotic resistance during many adsorption-regeneration cycles.

In several respects activated carbons meet the above requirements. However, their application is well known to be preferable at very low concentrations of adsorbate since the regeneration of activated carbons is a very energy-consuming and expensive process, accounting in a number of cases for almost 70 % of all operating expenses for their use.

Porous polymeric adsorbents, as a rule, can be regenerated under milder conditions. Nevertheless, certain disadvantages are peculiar to all known polymeric adsorbents including the macroporous styrene-divinylbenzene (DVB) copolymers which are of major practical importance. Here, the adsorption is largely restricted to the surface of the pore walls, resulting in the adsorption capacity often insufficient for favorable practical applications.

With a few representative examples, this paper describes the adsorption properties of hypercrosslinked polystyrene sorbents, the first member of a novel type of porous adsorbents.

2 SYNTHESIS AND STRUCTURE OF HYPERCROSSLINKED SORBENTS

Macroporous styrene-DVB copolymers are commonly prepared by copolymerization of monomers in the presence of an inert diluent. Their porous structure, resulting from a phase separation during the synthesis, comprises of a highly crosslinked, dense and almost impermeable polymeric phase with voids filled with air or any liquid. To preclude the formation of impermeable polymeric regions in the beads of the adsorbent, another synthetic approach is needed.

The principle of the synthesis of hypercrosslinked polystyrene sorbents consists in crosslinking pre-formed styrene copolymers (with a small content of DVB) in a swollen state by means of a large amount of a bifunctional crosslinking agent which must form bridges of restricted conformational mobility.[1] Bis-chloromethyl derivatives of benzene or biphenyl, dimethyl formal and monochlorodimethyl ether are suitable crosslinking agents reacting with polystyrene in ethylene dichloride in the presence of a Friedel-Crafts catalyst. In the case of monochlorodimethyl ether, the reaction proceeds through the intermediate stage of chloromethylation followed by the formation of rigid diphenylmethane type bridges between two carbon chains:

Theoretically, 0.5 mole of bifunctional agent is sufficient for binding all polystyrene phenyl rings to arrive at a network having the crosslinking degree of 100 %. In reality, however, a certain portion of phenyl rings remains unreacted whereas an equivalent fraction of aromatic rings receives three substituents. By using more than 0.5 mole of the cross-agent, the ratio of three- to mono-substituted benzene rings can be further enhanced implying that the density of crosslinking can easily exceed that characterized by the above formal value of 100 %.

Two distinguishing features are characteristic of the hypercrosslinked polystyrene networks.[2]

First, in spite of the extremely dense crosslinking, the hypercrosslinked networks are capable of strong swelling with thermodynamically good solvents such as toluene or ethylene dichloride. Even at the crosslinking degree as high as 100 %, the material behaves like conventional styrene copolymers containing just 2 to 3 % DVB. More importantly, the hypercrosslinked networks swell to the same high extent with methanol and heptane, typical precipitating media for linear polystyrene. Swelling with water is markedly lower. Nevertheless, the non-polar network increases its volume in aqueous media by a factor of 2 (provided the starting copolymer contains less than 0.3 % DVB). This unusual property can find an explanation[2] in that the dry hypercrosslinked network is strongly strained because of the contradiction between the natural tendency to form a dense polymeric phase and the large number of rigid bridges preventing the total collapse of the spacious network (during the removal of the synthesis solvent). Due to the inner strains, even a weaker solvation of the network with non-solvents causes a noticeable expansion of the polymer.

Second, because of the loose packing of polymeric chains, the bulk density of hypercrosslinked polystyrene amounts to about 0.7 g/cm^3 only (compared to the value of 1.04 to 1.06 g/cm^3 for conventional styrene polymers). That is why the hypercrosslinked polymer absorbs high quantities of inert gases at 77 K, as if the material exposes an inner surface area of 1000-1800 m²/g. It should be emphasized here that no real interface is available in the homogeneous, transparent, single-phase hypercrosslinked network, and so we deal with a special kind of "porosity" of polymeric materials.

Both the absence of densely packed impermeable domains and the peculiar tendency of relaxation of inner strains by swelling and volume expansion, account for an exceptionally high adsorption activity of hypercrosslinked polystyrene with respect to numerous organic substances.

Let us consider the properties of two typical batches of the hypercrosslinked polystyrene adsorbents. The first, Styrosorb 2, represents one of beaded laboratory samples the inner surface of which usually varies from 400 to 1500 m²/g depending on conditions of preparation. These polymers are arbitrarily referred to as microporous sorbents because the pore diameter amounts to 2-3 nm. The second batch is represented by sorbents exhibiting similar values of specific surface, but having a bimodal pore size distribution in the range of 2-3 nm and 30 or 100 nm. Currently these sorbents are manufactured by Purolite International Ltd. (Pontyclun, UK) with a trade mark "Macronet Hypersol", MN-series.

3 SORPTION OF VAPOURS OF ORGANIC SUBSTANCES

Figure 1 demonstrates a series of sorption isotherms for vapours of n-heptane, acetone, methanol, perfluorooctane and water on Styrosorb 2. At low and moderate relative pressures, Styrosorb better absorbed vapours of hydrocarbons and acetone, rather than that of polar methanol. However, closer to saturation pressures, the absorption capacities approached the same value of about 1.2 cm³/g irrespective of the thermodynamic affinity of the solvents to polystyrene. The above value exceeded significantly the value of the total pore volume of the dry sorbent, 0.22 cm³/g, implying a strong swelling of the polymer with the vapours sorbed. The uptake of perfluorooctane vapors was markedly lower, about 0.6 cm³/g. No adsorption of water vapour takes place. Therefore, moisture does not exert any negative effect on sorption of organic vapors from gaseous phases.

The product MN-100 has an enhanced crosslinking density. It also differs from Styrosorb 2 in that it contains about 0.4 mmol/g of polar amino groups. This material is less hydrophobic and efficiently absorbs polar compounds even at low relative pressures and elevated temperatures (Figure 1), the ultimate capacity being as high as 0.76 cm³/g. It is not surprising that the sorption capacity for non-polar pentane is reduced in comparison to the sorption of pentane or hexane on Styrosorb 2. Due to the fact that the variation in the bead volume of MN-100 is very small, sorption isotherms for pentane on MN-100 are completely reversible, contrary to the situation with Styrosorb 2.

A particularly important moment is the rate of approach to sorption equilibrium. Owing to the strong swelling of Styrosorb 2, the diffusion coefficients for hydrocarbons on this material prove to depend upon the degree of pore filling. Sorption of n-hexane or n-pentane is a slow process, at the beginning, but then it accelerates with the effective diffusion coefficients, D_e, increasing by three orders of magnitude (Figure 2). On the contrary, the desorption proceeds rapid and easily up to very small residual content of the sorbate. Desorption in vacuum requires about 100 seconds.

When percolating a mixture of 15 % n-pentane vapour with 85 % air through a large column packed with beaded Styrosorb, the breakthrough curves of the sorbate were observed to be steep and symmetric testifying to the formation of a rather narrow sorption zone.

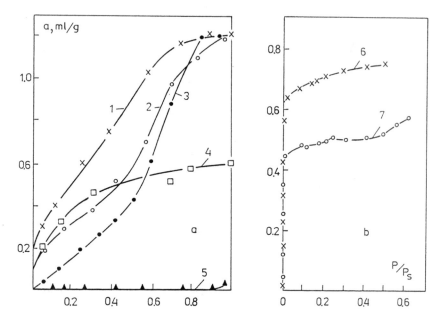

Figure 1 *Adsorption isotherms for (1) n-hexane, (2) acetone, (3) methanol, (4) perfluorooctane, (5) water, (6) ethyl acetate and (7) n-pentane on (a) Styrosorb 2 and (b) MN-100. Temperature, 293 K, with the exception of ethyl acetate (6) where it is 343 K*

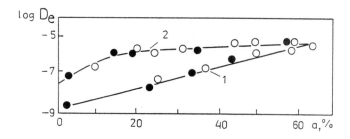

Figure 2 *Dependence of adsorption (1) and desorption (2) effective diffusion coefficients of n-hexane (open circles) and n-pentane (dark circles) on the degree of pore filling (weight percent) for Styrosorb 2 at 293 K*

With the concentration of pentane vapors raised to 50 %, the situation became much more complicated. To obtain more information on the process, 7 sensors were mounted inside the column. Their response to the concentration of pentane and temperature of the sorbent bed in each of the 7 sections are presented in Figure 3. It follows from the data collected that the absorption is a strongly exothermic process. There is a temperature wave, about 40 K high, when the sorption zone reaches the first check point in the column. Naturally, the heat evolves at the sorption site, mainly at the front of the sorption zone. With the subsequent portions of air and pentane arriving after the sorption zone, the packing cools down slowly. Being transported further with the carrier gas, the front of the heat wave obviously moves ahead of the front of the sorption zone. Therefore, every subsequent section of the packing becomes pre-heated long before it is reached by the sorbate. When examined at check points more distant from the column inlet, the heat waves arise at higher temperature levels. However, since the sorption capacity decreases with increased temperature, the heat waves become more and more dispersed. This is not the case with the sorption zone whose front, in the 7 column positions examined, was observed to gradually sharpen, due to the temperature enhanced mass transfer rate. The only unfavorable tendency observed is that the sharp rise in sorbate concentration is always followed by a long and flat part of the sorption plot caused by a slow decrease of the temperature of the packing layer which is situated between the column inlet and each of the corresponding concentration test positions. At the column outlet, only 80 % of the total theoretical sorption capacity was observed to be used up rapidly, whereas the

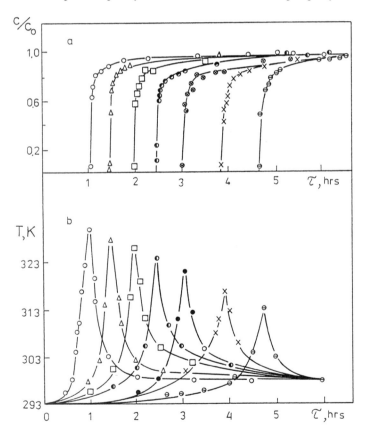

Figure 3 *Pentane breakthrough curves (a) and thermograms (b) for different sections of sorbent bed. Column, 0.5 x 0.02 m, ID; pentane concentration, 52.9 %-wt*

remaining 20 % of the capacity required many additional hours and a complete cooling of the whole packing.

Two important conclusions follow from the above experiments. The first relates to sorption processes with high concentrations of sorbate in its initial mixtures with the carrier gas. Here, the column packing will require a special heat exchanger, if the sorption capacity of the packing is expected to be fully exploited. This is not needed for gas mixtures with low concentrations of sorbate. In this case, the sorption process can safely be started immediately after a possible heat regeneration of the column, since the excessive carrier gas would certainly reduce the temperature of the whole packing long before the sorption front arrives at the end of the column.

The regeneration of Styrosorb in the above column was performed by percolating steam of 378 K at a rate of 0.03 m/sec. Over 90 % of the hydrocarbon absorbed could be recovered within 5-10 minutes. The regenerated sorbent retained no more than 5 % water which did not affect negatively the subsequent sorption cycle. 100 sorption - desorption cycles were performed without any decrease in the sorption capacity of the column or mechanical destruction of sorbent beads.

Although the adsorption of organic substances onto the hypercrosslinked sorbents is accompanied by strong heat generation, no oxidation of the adsorbates is catalyzed. Thus, heating MN-100 and MN-150 samples with absorbed cyclohexanone (about 0.3 g per 1 g of the polymer) did not stimulate formation of 1,2-cyclohexanedione within 360 hours at 358 K (as determined by means of gas chromatography). Contrary to this, activated carbon SKT-6A (Russia) produced 2 % of the above by-product in 48 hours at the same temperature. The amount of dione rose to 20 % during an exposure for 144 hours. Oxidation of cyclohexanone adsorbed on SKT-6A slowly proceeds at room temperature, as well.

In general, the adsorption activity of Styrosorbs with respect to vapours of organic compounds can be estimated easily and rapidly by means of gas chromatography. Table 1 shows specific retention volumes, V_g, for a set of analytes at 423 K obtained on a glass column 300 x 1.5 mm ID packed with Styrosorb 2. The retention at 293 K was calculated by extrapolation of ln V_g versus 1/T.

Table 1   *Specific Retention Volumes,* V_g, *of Organic Compounds*

Compound	$V_g^{423}/mL\ g^{-1}$	$V_g^{293}/L\ g^{-1}$
n-Butane	20	18
n-Hexane	220	340
Ethanol	20	6
iso-Propanol	70	70
Acetonitrile	30	36
Nitromethane	60	34
Benzene	260	300
Methylene dichloride	50	21
Chloroform	160	50
Ethylene dichloride	200	220
CF_2ClBr	10	5
$C_2F_4Br_2$	40	-

Styrosorb 2 is a very weakly specific adsorbent. Although the retention is mainly governed by dispersion interactions and the π-electrons of benzene do not contribute significantly to adsorption (the values of V_g for benzene and hexane are similar), there exists a definite correlation between V_g and polarization of molecules. Both the retention volumes and initial adsorption heats increase on introducing voluminous atoms into a sorbate molecule. The latter fact is responsible for higher values of adsorption heats of halogen containing hydrocarbons in comparison with those of alkanes. The adsorption heats rise in the series $CH_2Cl_2 > CHCl_3 > CCl_4$ (41, 46 and 52 kJ/mol, respectively). Generally, the initial adsorption heats on Styrosorbs are lower than those on activated carbons, which explains the easier regeneration of the hypercrosslinked polystyrene type sorbents.

4 ADSORPTION OF HEAVY METALS IONS ON NEUTRAL SORBENTS

It is quite natural that the hypercrosslinked polystyrenes absorb large amounts of organic substances from aqueous media.[3] Surprising is the fact that mercury (II) ions, as well, readily adsorb on these polymers containing no polar functional groups. With aqueous acetate buffers and nitric acid solutions, the sorption capacity for mercury was observed to increase proportionally to the value of inner surface area of the polymer (Figure 4) amounting to 440 mg/g for the sample with specific surface of 1500 m²/g. Interestingly, mercury sorption is a slow process. Desorption of mercury takes place with solutions

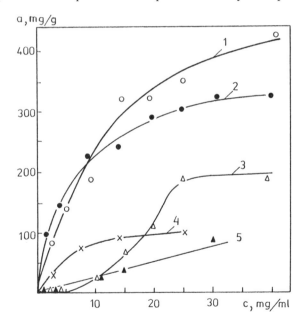

Figure 4 *Sorption of ions of Hg (1,2,4), Ag (3) and Pb (5) from acetate buffer (1,2,4,5) and nitric acid (3) solutions at pH 4. Surface area of Styrosorb 2 samples, 1500 (1,3,5), 1000 (2) and 400 m /g (4)*

of hydrogen chloride or ethylenediamine tetraacetate, but not with fresh portions of acetate or nitrate buffers. HCl and EDTA, however, failed to destroy the chemical C-Hg bond formed in a conventional mercuration reaction of the polymer in glacial acetic acid. These findings suggest that the absorption is caused by a complexation of mercury ions with two or more benzene rings of the polystyrene sorbent. In this case, the slow kinetics would result from a slow conformational rearrangement of the network in accordance with requirements of the metal ion coordination sphere.

Beside mercury(II), ions of Ag^+, Pb^{2+} and Bi^{3+} were found to absorb into Styrosorbs, though with smaller sorption capacities (Figure 4). No absorption was detected for ions of Ni^{2+}, Cd^{2+}, Co^{2+}, Cu^{2+} or Fe^{3+}.

Acknwledgements

The research described in this publication was made possible in part by Grant NM4Z300 from the International Science Foundation and Russian Government. The authors acknowledge "Purolite International, Ltd.", Pontyclun, UK, for support and cooperation in developing new types of sorbents.

References

1.   S. V. Rogozhin, V. A. Davankov and M. P. Tsyurupa, USSR Patent 299165 (1969), USA Patent 3729457, UK Patent 1315214.
2.   V. A. Davankov, M. P. Tsyurupa, *React. Polym.*, 1990, **13**, 27.
3.   M. P. Tsyurupa, L. A. Maslova, A. I. Andreeva and V. A. Davankov, *React. Polym.*, 1995, **25**, 69.

ON THE RELIABILITY OF CAPACITY DETERMINATIONS IN THE MICROSCALE - MINIATURIZED METHODS OF MOISTURE-CONTENT DETERMINATION AND pH-TITRATION

M. Grote and P. Schildmann - Humberg

Department of Applied Chemistry
University of Paderborn
D-33095 Paderborn
Germany

1 INTRODUCTION

The characterization of resins - in the routine work or in the development of novel ion exchange materials - requires the determination of total number of exchanging groups (exchange capacity value Q_O). Usually the exchange capacity value Q_O is expressed as "milliequivalents of exchangeable ion per gram of *dry* resin". As a consequence, the determination of the moisture content has to supplement the various standard procedures which are based on titration techniques. A number of different techniques are established, such as the determination developed by Kunin and Myers[1]. Moreover, methods of pH-titration, e.g., the *direct method*[2,3,4] and the *batch method*[4,5,6] allow examination of the acidity or basicity of ion exchangers. All these procedures are usually applied to macro amounts of resin (about 5g to 20 g). In our study, attempts were made to adjust these titration methods and the determination of the moisture content as well to the microscale using much smaller quantities of ion-exchange resins (Table 1). By following these miniaturized procedures, material and time could be saved, however, the question arises if the quality of the analytical data will be deteriorated.

Table 1 *Properties of Ion-Exchange Resins Investigated*

Designation (Manufacturer)	Resin Type (Polystyrene DVB) Crosslinkage	Particle Size, Moisture Content	Total Exchange Capacity Q_O [meq/g, dry resin][a]
Ion Exchanger I (Merck)	strongly acidic 8 % DVB	20 - 50 mesh 45 - 55 %	4.0 - 4.9
Chelex 100 (Bio-Rad)	weakly acidic 1 % DVB	100 - 200 mesh 68 - 76 %	1.6 - 2.3
AG 1-X2 (Bio-Rad)	strongly basic 2 % DVB	200 - 400 mesh 70 - 78 %	4.1 - 5.6
Ion Exchanger II (Merck)	weakly basic 5 % DVB	20 - 50 mesh 35 - 45 %	4.3 - 5.1

[a]   Calculated by using data of the manufacturer

1.1 Determination of The Moisture Content

The validity of the determination of moisture contents strongly influences the reliability of the capacity data Q_O. As microsized resin samples (≤ 100 mg) were analyzed the precision and the accuracy of the analytical values obtained are governed by the representativity of the individual resin sample, respectively by the number of particles of the sample[7]. Accordingly, requirements such as uniformity and constancy of the particle size fraction sampled are more difficult to fulfil compared with macro sized weighed batches. The particle size of the resins investigated in our study covers ranges of 200-400 mesh and 20-50 mesh. As presented by Table 1, the polymers selected were strongly and weakly basic or acidic commercially available ion exchangers.

1.1.1 Pretreatment and Sampling. The determination of the moisture content of ion exchangers should be performed on an "air dry" basis. That means that the polymers sampled for water determination should have acquired a constant water content, e.g., during storage at the atmosphere for "several days"[5]. As it is difficult in practice to meet the requirement, a significant factor of uncertainty will be usually provided. Instead of the time consuming procedure, we developed in our study a quick and simple filtration technique to bring the water content of micro samples of resins to some reproducible level[8].

An amount of 3g or less of the appropriate preconditioned ion-exchange resin (H^+ form or OH^- form, respectively) was poured into a glass filter crucible and washed with distilled water pH neutral. Then the resin batch was sucked dry by means of a water jet pump just until no more drops of the filtrate appeared at the outlet. Afterwards, pumping was allowed to continue for further 2 minutes. Finally samples were precisely weighed for analysis (portions of about 100 mg or less).

The resin batch pretreated in such a way was always divided in order to weigh a series of samples for the determination of the moisture content and the remaining part at the same time for pH-titration. In the case of basic ion exchangers, the OH^- form of the resin was converted to the chloride - form by 1 M NaCl solution to avoid decomposition on drying respectively heating.

1.1.2 Indirect Determination. The most common method for determining the water content of resin samples involves oven drying of a weighed sample and gravimetric determination of the evolved water from the loss in weight. In general, oven drying is applied at 110-115°C or lower[3,6]. In order to minimize decomposition of the resin, it is possible to reduce the internal pressure so that a lower temperature can be applied[4]. The drying process continues until the weight of the sample has become constant at the conditions chosen. Drying "overnight" is a common instruction[3].

In our study, the reliability of a certain drying procedure was tested applying micro amounts (~100 mg) of ion exchangers (Ion Exchanger I, AG 1-X2). A set of ten samples of each resin, pretreated as already described, were weighed and placed in a vacuum oven. The oven was evacuated over 2h by a water jet pump, then the final pressure was maintained. Simultaneously, the internal temperature was adjusted to 80°C. The recommended time of drying ("overnight") was interpreted as a minimum drying interval of 20 h. Subsequently, the drying process was interrupted in order to determine the loss in weight of the individual samples. The whole procedure was repeated three times until a total drying period of about 133 h (~ 5 days) was reached. The "moisture contents" obtained for the different series of drying are and listed in Table 2.

Table 2 *Repeatability of the Gravimetric Determination of the Moisture Content (% w/w) as Function of Time of Drying*
(Ten 100 mg-samples of each resin; 80°C in vacuum)

time of drying / h	20	40	66	133
Ion Exchanger I				
mean / %	51.70	55.49	56.53	55.63
standard deviation, s	0.19	0.28	0.40	0.39
AG 1-X2				
mean / %	76.01	76.79	76.93	76.08
standard deviation, s	0.19	0.24	0.25	0.38

The data presented reveal two distinct effects. First, a period of about 20 h appears not to be sufficient to achieve a constant weight of the sample. In particular, the acidic resin (Ion Exchanger I) requires a second period of drying to achieve the state aimed, whereas the effect observed for the basic AG 1-X2 is less pronounced. Secondly, the standard deviations of the values determined, increase with an extending time of drying.

The evaluation of the gravimetrical data is based on the assumption that this loss represents exclusively the mass of water in the resin. However, while heated in an oven, several decomposition processes, e.g., desulfonation of sulfonic acid resins or hydrolysis of basic polymers, may occur in addition to the volatilization of water. Thermoanalytical experiments have demonstrated the complexity of these processes[8]. Evidently, the increased level of "moisture content" and the steady increase of standard deviations determined in our series of measurements (Table 2) may be caused by additional decomposition reactions. Consequently, it seems to be impossible to determine a definite "true", "accurate" value of the water content by means of a common heating procedure.

It is still recommended to apply the common "over-night" drying procedure (20 h interval, in vacuum) in order to achieve a sufficient precision of the water determination and to minimize the expenditure of effort.

1.1.3 Chemical Determination. Alternatively, the most important method for the chemical determination of water involves the use of Karl Fischer solution, which is a relatively specific reagent for water. The reagent is composed of methanol, sulfur dioxide, iodine, and an organic base RN (formula (1)) which is nowadays preferably imidazole instead of pyridine (e.g., Hydranal® - Riedel de Haën). Upon addition of water to this reagent, the following reactions occur [9, 10].

$$CH_3OH + SO_2 + RN \rightarrow [RNH]SO_3CH_3 \tag{1}$$

$$[RNH]SO_3CH_3 + 2\,RN + H_2O + I_2 \rightarrow [RNH]SO_4CH_3 + 2\,[RNH]I \tag{2}$$

The reaction step (2) involves the consummation of water and, simultaneously, the oxidation of sulfur dioxide by iodine. If a coulometric measuring device is employed to determine the end point of the titration, iodine can be generated in situ electrochemically (Figure 1). However, difficulties are encountered in the analysis of sorbed moisture. For this purpose a preceding external extraction is required. The integrated system of the "Karl-Fischer-Titrator Aqua 20.00" combines both the extraction step and the computer-controlled microcoulometric titration technique[9].

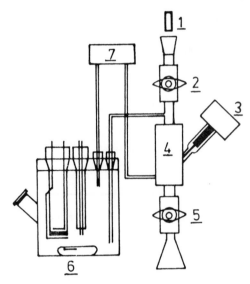

Figure 1 *Apparatus for water determination in solid (and liquid) samples*
"Karl-Fischer-Titrator Aqua 20.00" (according to [9]): 1 glass-container, sample,
2,5 inlet and output valve, 3 magnetic closure, 4 heating chamber,
6 coulometric measuring cell, 7 pump (extracting gas circuit)

As illustrated by Figure 1, the water evolved from the micro sized sample (~10-20 mg) placed in a heating chamber is transferred by a stream of an inert gas into the coulometric measuring cell which contains the absorbing Karl-Fischer reagent.

Such a coulometric determination can be followed from the beginning of the titration until the endpoint by monitoring the course of the "indicator" voltage, expressed in mV (Figure 2). This value indicates the actual concentration of iodine in the Karl-Fischer reagent. Figure 2 demonstrates that in the case of a micro sized sample (10 mg) of Ion Exchanger II, the complete measurement was already finished within 10 minutes at a heating temperature of 150°C. Furthermore, the release of water from the resin sample can be also continuously controlled as a function of time until the final analytical value (43.9 % water) is reached.

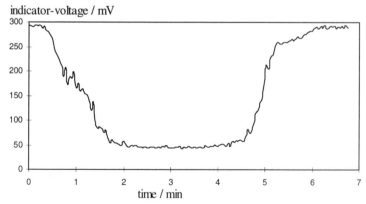

Figure 2 *Determination of moisture content by coulometric Karl-Fischer titration :
indicator voltage as a function of time* (10 mg of Ion Exchanger II, 150°C)

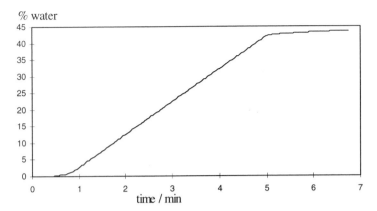

Figure 3 *Water release from Ion Exchanger II at 150°C as a function of time*
(10 mg sample, Karl-Fischer-Titrator Aqua 20.00)

Compared to the analytical data obtained by the common gravimetric procedure the drastically reduced sample size used in the Karl-Fischer method caused a more pronounced deterioration of the precision (Table 2). However, the determinations carried out with the Karl-Fischer-Titrator led to moisture content values which correspond quite well to the data obtained by the drying procedure. Due to its high rapidity, the micromethod based on the Karl-Fischer titration seems to be particularly attractive.

Table 2 *Comparison of the Microdetermination of Moisture Contents of Ion Exchange Resins Determined by Different Methods[b]*
(Ten 10 mg samples of each resin)

Resin	[a]*Moisture content* *Value / %*	[b]*KF - Titration* *(150 °C, 15 min)* *Value / %*		*Drying Procedure* *(80°C in Vacuum, 20h)* *Value / %*	
		mean	s	mean	s
Ion Exchanger I	45 - 55	51.48	2.44	51.05	0.74
Ion Exchanger II	35 - 45	43.88	1.62	42.42	0.93
AG1-X2	70 - 78	76.62	3.82	75.60	0.54
Chelex 100	68 - 76	54.87	1.56	54.17	1.07

a   According to manufacturer
b   Karl-Fischer Titrator AQUA 20.00

1.2   Scaling Down of the Titration Methods

1.2.1. Direct Titration. The direct titration is favourably applied to highly ionized resins. To perform this method, usually a portion of 5g of the preconditioned exchanger is suspended in a sodium chloride solution. Subsequently the titrant is added in small, definite volumes by stirring. After each addition, the solution is allowed to come to equilibrium, before the pH value is measured. We have found that in fact the direct titration can be performed on the microscale by using only one portion of 100 mg of resin. For example, the titration of the strongly acidic resin Ion Exchanger I is described:

One weighed 100 mg-sample (H^+ form) was filled into a polyethylene bottle and suspended in 20 mL of 1 M NaCl solution. The bottle was closed and then the suspension was mechanically shaken over a period of 24 h. When shaking was interrupted, the resin particles were allowed to settle and the pH value of the supernatant aqueous phase was determined by a glass electrode. The pH value was measured in 24h intervals, after every addition of the titrant (0.2 M NaOH in 1 M NaCl) in portions of 50 μL. Finally the pH values are plotted vs. volume of titrant added.

The curve of such a direct titration is shown in Figure 4. Obviously, the shapes of the titration curve resemble the typical curves of a monofunctional strongly acidic resin which are usually obtained with a significant higher amount of polymer. Regarding the aspect of material saving, the miniaturized direct method is particularly attractive if only minimum amounts of novel polymers are available[11].

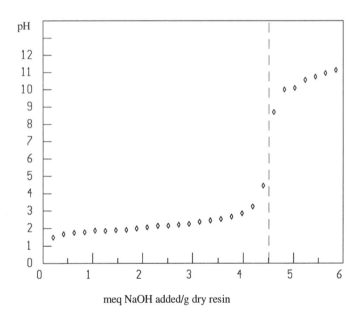

Figure 4   *Direct microtitration of Ion Exchanger I (strongly acidic)*
(100 mg sample, Q_0 = 4.6 meq/g)

1.2.2 The Batch Method. The alternative batch method can be used for any type of ion exchanger including the weakly dissociated[3,4]. The key aspect is the adding of different volumes of the titrant to a number of individually weighed samples. Experimentally, the titration of Ion Exchanger I was carried out as follows:

A total number of twenty screw-capped polyethylene bottles are each filled with 100 mg samples of preconditioned (H+ form) and dried cation exchanger. At the same time portions of the resin were sampled to determine the moisture content (see 1.1.2). Successively decreasing volumes (increments of 2 mL or less) of 1 M NaCl solution and successively increasing amounts (increments of 2 mL or less) of 0.01 M NaOH and 0.1 M NaOH, respectively, in 1 M NaCl were added to each of the bottles so that a constant concentration of chloride was maintained. The bottles were closed and equilibrated on a shaking machine. After a shaking time, varying between 24 h and 30 days, the pH values in the individual bottles were measured and evaluated graphically.

We observed that the equivalence point determined by the batch procedure corresponds to a slightly higher consumed quantity of alkali hydroxide compared to the result of the direct titration (4.6 meq/g, Figure 4). Consequently, a higher capacity value of $Q_0 = 4.7$ meq/g was calculated. A similar relation was found for the weakly acidic, chelating resin Chelex 100 (direct titration: 1.9; batch method: 2.2 meq/g). The adjustment of the different pH-titration methods to basic ion exchangers in the micro scale was also successful in the case of the strongly basic resin AG 1-X2. The data obtained from the two alternative techniques differ analogously in such a manner as has been already described for the acidic polymers (1.9 and 2.2 meq/g resp.).

The weakly basic anion exchanger (Ion Exchanger II) behaves quite different as the changes in pH near the equivalence point are less pronounced during the direct titration

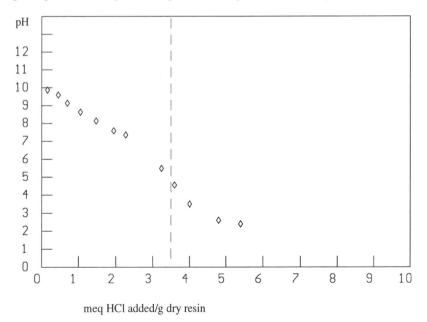

meq HCl added/g dry resin

Figure 5 *Micro batch titration at 30°C of the weakly basic Ion Exchanger II*
(48 h equilibration, 100 mg sample, $Q_0 = 3.5$ meq/g)

(3.4 meq/g) than during the batch technique (3.8 meq/g). 30 days were necessary to obtain an acceptable titration profile of the weak base ion exchanger even in presence of sodium chloride. However, the equilibration can be drastically accelerated by a simple experimental variation introduced by Vyas and Kapadia[12]. The suspension which contains the polymer, dissolved electrolyte and the added titrant is to equilibrate at 30°C. Under such conditions, a reaction time of only 48 h is sufficient to yield an excellent titration curve (Figure 5).

1.3   Reliability of Microdeterminations

Standard deviations were estimated for two series of measurements applying the different titration techniques and the determination of the moisture content by oven drying (100 mg samples) to the Ion Exchanger I. The capacity data obtained for five complete batch procedures yielded a mean of 4.89 meq/g and a standard deviation of s = 0.11 (s_{rel} = 2.25 %). The same number of direct titrations gave a similar precision for the mean value 4.53 meq/g (s = 0.10, s_{rel} = 2.20 %) whereas the repeatability for the weak base resins (s~ 0.2) is comparatively lower. The values determined correspond satisfactorily with the data of the manufacturers (Table 1) and data obtained by other methods, e.g. the standard method of Kunin[8].

The results of our study show, that the miniaturized techniques of titration and determinations of the water content offer a sufficient reliability. Hence, by means of the proposed methods, material and time will be saved compared to the standard macro procedures.

Acknowledgements

The authors thank Analysentechnik Beringer (Mainz), Dr. I. Yenmez, GAP Engineering (Bochum) and Dr. M. Hahn, ECH Elektrochemie Halle GmbH, for providing the Karl-Fischer -Titrator Aqua 20.00 and helpful discussions.

References

1.   R. Kunin, R. J. Myers, "Ion Exchange Resins", Wiley and Sons, New York, 1949.
2.   Grießbach, "Über die Herstellung und Anwendung neuerer Austauschadsorbentien, insbesondere auf Harzbasis", *Beihefte zur Zeitschrift des Vereins Deutscher Chemiker,* Nr. 31, Verlag Chemie, Berlin, 1939.
3.   R. Kunin, "Ion Exchange Resins", Krieger, Huntington, NewYork, 1972, p. 337.
4.   M. Marhol, Ion Exchangers in Analytical Chemistry, in: "Wilson and Wilson's Comprehensive Analytical Chemistry", G. Svehla, ed., Elsevier, Amsterdam, 1982, p. 94.
5.   N. E. Topp, K. W. Pepper, *J. Chem. Soc.,* 1949, 3299.
6.   F. Helfferich, "Ionenaustauscher", Verlag Chemie: Weinheim, 1959, Vol. 1, Chap. 4.
7.   "Analytikum", K. Doerffel, R. Geyer, H. Müller, Deutscher Verlag für Grundstoff-industrie, Leipzig, 1994, Chap. 10.
8.   M. Grote, P. Schildmann-Humberg, *Ind. Eng. Chem. Res.*, 1995, **34**, 2712.
9.   J. Bortlisz, K. Rüther, H. Matschiner, M. Hahn, *Acta Hydrochim. Hydrobiol.,* 1995, **23**, 104.
10. G. Wünsch, A. Seubert, *Fres. Z. Anal. Chem.*, 1989, **334**, 16.
11. U. Schumacher, M. Grote, University of Paderborn, to be published.
12. M. V. Vyas, R. N. Kapadia, *Indian J. Technol.*, 1980, **18**, 411.

REFINEMENTS TO THE *HIAC* TECHNIQUE FOR MEASUREMENT OF PARTICLE SIZE OF ION EXCHANGE RESINS

L. S. Golden

Purolite International Ltd.
Pontyclun
Mid Glamorgan CF7 8YL

1 INTRODUCTION

The particle size distribution has an important influence on the performance of an ion exchange resin in many applications, and therefore accurate measurement of this parameter is essential, both for production and quality control. Accurate particle size measurement has assumed even greater importance over the last few years with the development of the so-called "uniform" particle size resins.

There are only a limited number of methods available for determining the particle size of ion exchange resins, and some of these do not offer sufficient accuracy. Traditionally, analyses were made using standard laboratory sieves, on either wet or dried product. Whilst dry screening is easier, this method can really be discounted, since resins are normally used wet and therefore a second factor, the size change between wet and dry particles is introduced. Since this in turn depends on the chemical properties of the resin, in particular the moisture retention, such a technique cannot sensibly be used as an accurate determination of particle size. The author showed[1] that even wet sieving is beset by inaccuracies, not the least of which is the tolerance of +/- 6% allowed on the nominal size in any given sieve. Figure 1 shows a photograph of a 60 B.S. mesh sieve, with aperture sizes measured in mm from an enlargement (approx. x35). These clearly show

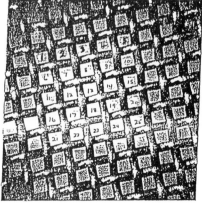

(1) 10.0 x 8.3	(10) 9.4 x 8.8	(19) 10.0 x 9.2
(2) 9.5 x 8.5	(11) 10.4 x 8.8	(20) 9.3 x 9.2
(3) 9.7 x 8.5	(12) 9.9 x 9.2	(21) 10.0 x 9.2
(4) 9.8 x 8.6	(13) 10.1 x 9.4	(22) 10.2 x 9.3
(5) 9.3 x 8.5	(14) 10.0 x 9.3	(23) 10.4 x 9.3
(6) 10.0 x 9.1	(15) 9.7 x 8.7	(24) 10.2 x 9.2
(7) 10.0 x 9.5	(16) 10.2 x 9.6	(25) 10.0 x 9.0
(8) 9.8 x 9.5	(17) 10.8 x 9.5	
(9) 10.2 x 9.0	(18) 9.7 x 9.7	

effective mean aperture	: 9.07 mm
standard deviation	: 0.39 mm = 4.3%
range of sizes	: 230 μm - 270 μm

Figure 1 *Apertures (mm) measured in 60 B.S. mesh (250μm) sieve*

the variations in the size of the openings, and when related back to actual dimensions represent a range of at least +/- 20μm about the mean.

The author presented an instrumental method for measuring the particle size of ion exchange resins in 1977[2] using the HIAC Particle Size Analyser. This was a commercially available instrument, designed primarily for counting particles dispersed in oil, but its operation could be adapted to size resin particles.

Most instrumental techniques have been designed to size particles ranging from the sub-micron up to about 100 - 200 microns. Additionally, a suitable analytical method must be capable of making the measurement whilst the resin particle is fully wetted by water. Instruments such as the Coulter Counter, although able to cope with the 200 to 1500 micron resin bead sizes usually encountered with ion exchange resins, depend on electrical conductivity to measure the size. Since both resin and the salt solution in which the particles would have to be suspended are conductive, this method is not practical. The only other potentially suitable techniques available are optical scanning microscopes and laser counters.

Since the 1977 publication, most resin manufacturers and larger users have adopted the HIAC (or equivalent) for measuring particle size, and the method is now established as an ASTM. However, there are drawbacks and problems even with this method, which the author has sought to overcome, and the most recent of these are described in this paper.

2 PRINCIPLE OF OPERATION

The HIAC operates on a principle of light extinction. A beam of parallel light is projected through a pair of optical windows, between which the particles pass. Light emerging from the second window is detected by a photodiode (Figure 2A). When a particle passes

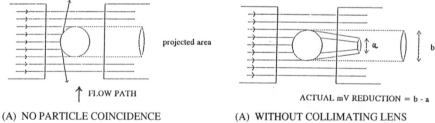

(A) NO PARTICLE COINCIDENCE   (A) WITHOUT COLLIMATING LENS

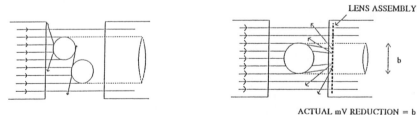

(B) WITH PARTICLE COINCIDENCE   (B) WITH COLLIMATING LENS

Figure 2 *Particle flow through sensor*   **Figure 3** *Effect of collimating lens*

through the cell ('sensor'), it projects a shadow onto the photodiode, the output of which is correspondingly reduced. The extent of this reduction is proportional to the projected cross-sectional area of the particle. In the case of spherical beads, which is the usual form of most ion exchange resins, this is easily related to the bead diameter. The instrument is calibrated using a series of particles of known diameters.

In the earlier papers, the author reported two major pitfalls in the technique, of which the user must be aware and which must be allowed for if accurate analyses are to be obtained. Since the analysis is based on the proportion of light extinguished, it follows that only one particle should be in the sensing zone at any one time, or else the equipment will count more than one smaller particle as one larger particle (Figure 2B). This can best be prevented by ensuring that the dilution of particles in the suspension passing through the cell is sufficient that the statistical chance of more than one particle being present in a sensor volume is minimised.

The second potential problem is associated with the sensor optics, namely that the sensor must have a collimating lens not only before the sensing zone, but also after it. Ion exchange resin particles are somewhat unusual compared with the types of particles normally analysed by these instruments, in that they are swollen by water which is 'chemically' bound into the matrix, and can be suspended in excess water without changing this internal water content. This property actually makes the resin bead behave as a small lens as it passes through the sensor, the focal length of which varies with the moisture retention of the resin (Figure 3A). This results in more light falling on the detector than the theory dictates, making the particle appear smaller that it really is. This can only be eliminated by use of a second collimating lens after the sensing cell (Figure 3B). For sensors without this second collimating lens, the user is forced to calibrate the sensor over a range of resin moisture contents as well as particle sizes. Since the manufacturers do not appear to automatically provide this second collimating lens, the potential user is advised to check this carefully before placing an order.

The magnitude of this effect is clearly illustrated in Figure 4, which shows the apparent size distributions of three different resins which in reality have very similar narrow size distributions. The effect of moisture content of the resin is very clearly demonstrated by using two strong acid cation resins, differing only in their moisture content due to different levels of cross-linking in the matrix. The third example of a macroporous cation resin shows the effect of measuring optically opaque beads compared with the transparent gel beads of the other two samples. Figure 4A shows the true distributions of the 3 samples, using a 600 micron sensor with a collimating lens, whilst Figure 4B shows the false distribution obtained with the same materials using a 600 micron sensor without collimating lens, calibrated for gel resins with moisture retention approximately 45%. This clearly shows how increasing moisture content of the resin reduces the *apparent* size distribution of the particles.

3 TYPE OF INSTRUMENT

The original equipment on which the technique was developed was a 12-channel machine. This means that the output from the photodiode, after electronic amplification, incremented one of twelve channels, depending on the signal size. The thresholds of these channels could be varied at will using thumbnail rotary switches. Consequently, the instrument could be used to scan any required size range, however narrow, but on any one run would only provide twelve data points. Of course, by multiple runs, a greater number

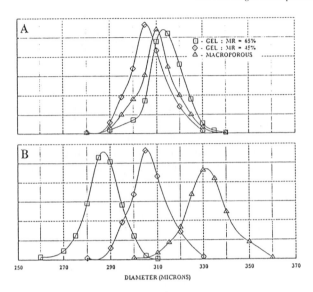

Figure 4 *A - With collimating lens   B - Without collimating lens*

of points could be obtained for a given analysis.

Later designs of equipment dispensed with this 'user interface' for setting channel thresholds, instead providing a greater number of channels (in some cases 32) but fixed in a geometric progression. Since again the instrument is primarily designed for measuring smaller (< 200 micron) particles, the geometric steps very much favoured the smaller particle sizes, with only a few being available in the primary range of interest to ion exchange resin users. Today, the equipment has reverted to user changeable channel thresholds, but, since these require delicate adjustment with a screwdriver, it is not really practicable to make changes with any frequency.

The increasing number of 'laser' analysers suffer from a similar problem. Again, being designed with fine powders in mind, they have the same preponderance of channel thresholds below 100 microns and very few up to 1500 microns or more.

Due to these limitations, the author was not prepared to replace the 12-channel HIAC with any of the current instrumentation available. In discussions with the local service engineer who builds his own particle counters, it was agreed that an instrument would be built to the required specification. This is the 'OMEGA' particle size analyser, which is now available commercially.

This instrument has no predefined channel thresholds, and instead completely scans the entire sensor range in approximately one micron intervals (at the lower end of the sensor range, the interval is three to five channels per micron, at the higher end is two or three microns per channel). All the count information is transferred instantaneously to a PC, and processed so that the data can be displayed. As a particle is counted, it appears on the computer display (graduated in microns). As a result, the displayed distribution is updated particle by particle as the sample is actually passing through the sensor, over the 1 - 2 minutes of a typical analysis run. The computer program requires two correlating runs, and the envelope of the previous run remains on the screen during the second run for operator comparison (Figure 5).

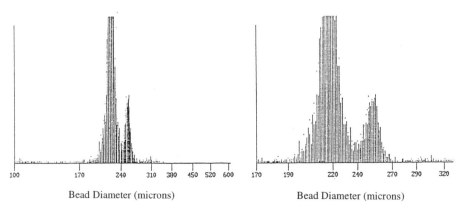

Bead Diameter (microns)   Bead Diameter (microns)

Figure 5 *Screen display from 'Omega' when at full scan and narrowed scan*

The large number of data points (more than 600) would be rather cumbersome to handle, so the program breaks the data down into convenient blocks for analysis. However, the user has control over the range of these blocks and the screen display range, so the distribution over any size range can be checked. The entire raw data can be stored on hard disk and recalled for subsequent analysis, so review of the distribution over any desired range is available at any time. Distributions of up to four different samples can also be displayed on the screen at the same time, for comparison purposes.

4 RESOLUTION

The prime concern regarding the equipment limitations is that of resolution - in other words, whether the data will show sufficient detail of the true size distribution of the sample being measured. Table 1 shows the data provided by a manufacturer of Laser equipment on a resin sample in the size range 300 - 1200 microns. In this case, the 8 or 9 data points is sufficient to give a reasonable analysis, the resolution being approximately that obtainable with sieves, but unfortunately the majority of available channels provide no data at all about the particle distribution.

The problem arises with narrower grade samples, as the data in Table 2 clearly illustrates. Of the 32 channels of information provided by the laser analyser, only three of these are of any value, and all that can be said of the sample is that its distribution lies between 183.4 and 272.3 microns. Sieving would actually give one extra data point, but tells nothing more about the absolute range.

An analysis in 20 micron steps carried out with the OMEGA shows that the distribution actually lies between 200 and 280 microns, but with only a little more detail.

The distribution plots of these analyses are very interesting (Figure 6). The minimal number of data points in the laser analysis actually makes the distribution look broader than it really is. The one extra data point from the sieve analysis gives a distribution very similar to that obtained from the instrumental analysis carried out at exactly 20 micron intervals. In all three cases, a bell-shaped nearly 'normal' distribution is obtained.

During the OMEGA analysis, the real time display of data showed the binodal distribution (see Figure 5), although the way in which the channels had been arbitrarily set for analysis did not clearly reveal it. However, re-analysis of the stored raw data using 10

Table 1 *Data Output from a Modern Laser Analyser on Standard Grade Resin*

Size μm	Cum.% Below	Size μm	Cum.% Below	Size μm	Cum.% Below	Size μm	Cum.% Below
4.9	0.0	23.0	0.0	108.6	0.0	513.6	9.9
5.9	0.0	27.9	0.0	131.9	0.0	623.7	24.4
7.2	0.0	33.9	0.0	160.2	0.0	757.4	48.8
8.7	0.0	41.1	0.0	194.5	0.0	919.7	79.0
10.6	0.0	50.0	0.0	236.2	0.0	1116.9	95.2
12.8	0.0	60.7	0.0	286.8	0.0	1356.3	99.6
15.6	0.0	73.7	0.0	348.3	0.1	1647.0	100.0
18.9	0.0	89.4	0.0	423.0	2.8	2000.0	100.0

Table 2 *Data from Various Sources on the Same Narrow Particle Size Range Product*

LASER		SIEVE		20μm COUNT		10μm COUNT	
Size μm	Cum.% Below	Size μm	Cum.% Below	Size μm	Cum.% Below	Size μm	Cum.% Below
↑	0.0					190	0.0
101.4	0.0			180	0.0	200	0.2
123.6	0.0	149	0.0	200	0.2	210	3.0
150.6	0.0	177	0.0	220	25.1	220	25.1
183.4	0.0	210	3.0	240	70.5	230	63.3
223.5	42.0	250	80.3	260	95.9	240	70.5
272.3	100.0	297	100.0	280	100.0	250	80.3
331.8	100.0	354	100.0	300	100.0	260	95.9
↓	100.0					270	98.2
						280	100.0

instead of 20 micron intervals made a remarkable transformation, and it can be seen that in fact the distribution is binodal, with two separate peaks at about 220 and 250 microns. The binodal distribution is completely missed in the previous cases. The differences between these distributions are more clearly illustrated in the 'cumulative' curves (Figure 7).

This very clearly demonstrates the importance of instrument choice for measurement of any but 'standard grade' resin samples.

5 DEFINITION

This term is used to represent the accuracy with which the HIAC-type counter will define a given distribution, and is critically influenced by the design of the sensor. Sensors are available in a range of nominal sizes. This refers to the maximum orifice size of the sensor, most of which have a dynamic range of x60. For example, a 600 micron sensor will measure in the range 10 - 600 microns.

In a previous paper[1], the author showed that it was undesirable to use a sensor with a maximum orifice size very much larger than that of the sample. Thus, a 600 micron sensor would be more appropriate for measuring particles in the size range 200 - 400 microns than a 1000 micron sensor, and a 1000 micron sensor would be more appropriate

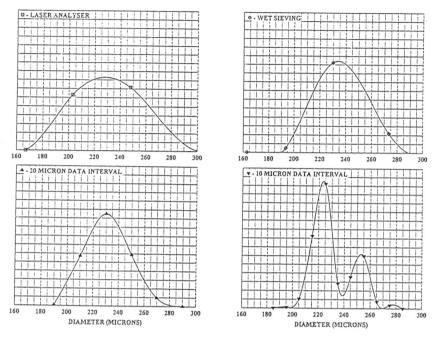

Figure 6 *Distribution plots of same sample by different analytical techniques*

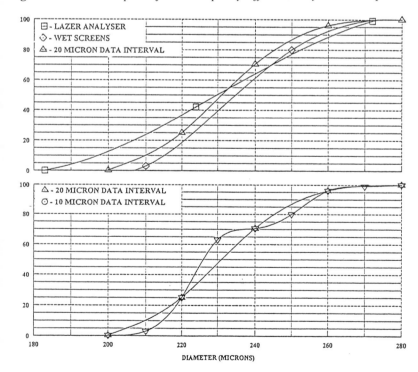

Figure 7 *Cumulative % below size - comparison of distribution data*

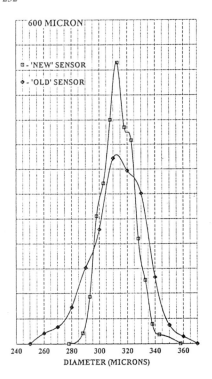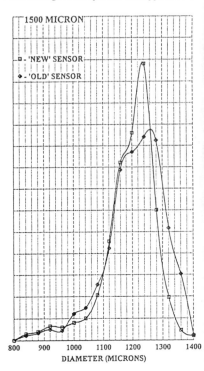

Figure 8 *Comparison of distributions using 600 and 1500 micron sensors of old and new design*

for measuring particles in the range 400 - 800 microns than a 1500 sensor.

The reason behind this is that particle coincidence in the sensor must be minimised, as already discussed. Even with a very dilute sample, it is still possible for two particles to overlap as they pass through the sensing zone, although the closer the size of the particle is to the maximum orifice size, the less the likelihood of such overlap taking place. Thus, if the sample is 'too fine' for the sensor, the resulting distribution will be biased slightly to the coarser end and spread out a little.

HIAC sensors are designed so that, although the distance between the cell windows is precisely that of the nominal aperture, the length of the aperture is much greater, so that on viewing the sensor, the aperture is seen to be a rectangular slit. A new source of sensors has been found, wherein the design is such that the aperture is of square cross-section. (As an incidental, this source can provide sensors of any given aperture size, whereas availability of HIAC sensors is either 600 or 2500 micron, the latter really being too large for even standard resin size distributions.)

The increase in definition using the square sectioned sensor is apparent from the distribution curves shown in Figure 8. This compares distributions for the same set of particles measured with the same nominal aperture size sensors, the 'old' type being of the original rectangular cross-section, and the 'new' being of the square type. The improved definition of the 'new' design is clearly visible.

6 CONCLUSIONS

This work has shown the importance of the instrument being capable of user selection of analytical ranges and of correct sensor geometry. Consequently, equipment must be selected with care to ensure that it has sufficient versatility to carry out the measurements with sufficient resolution on the samples of interest, and more importantly that it does not in itself introduce inaccuracies due to optics which are not ideal for the analysis.

Light extinction particle counters remain the most accurate method for sizing ion exchange resins. The instrument described herein, which could be regarded as a second generation of such counters, appears to satisfactorily meet the criteria.

References

1. L. S. Golden, "Ion Exchange for Industry", M. Streat, ed., SCI, Ellis Horwood Ltd., 1988.
2. L. S. Golden, *Particle Size Analysis Conference*, University of Salford, September 1977.

HARMONISATION OF ION EXCHANGE FORMULATIONS AND NOMENCLATURE

J. Lehto and R. Harjula

Laboratory of Radiochemistry
Department of Chemistry, P.O.Box 55
FIN-00014 Helsinki University
Finland

1 INTRODUCTION

Nomenclature in various branches of ion exchange research varies widely. Same quantities may be named in different ways, which hinders communication between various researchers.[1] Also serious misuse and misunderstanding is involved in the use of many basic definitions. Harmonisation of nomenclature would certainly be of help in this situation. But even careful definition of all the quantities in scientific papers and presentations, which is very often ignored, would reduce this communication problem to a matter of interpretation or translation. Recommendations on ion exchange nomenclature were issued by IUPAC in 1972.[2] Practically no updating of these recommendations has taken place in more recent IUPAC publications.[3,4] This may be one major reason for the fact that these recommendations are so often ignored in the present literature.

Mathematical formulation of ion exchange equilibrium is also a matter of large variation. In general, for various types of ion exchange materials, different types of formulations have been developed. It is common to express the equilibrium in organic resins as "dissociation constant", in chelating resins as "complexation constant" and in zeolites as "selectivity coefficient". Various concentration scales, which are used in these formulations makes mutual understanding even worse. In most cases, these formulations are interconvertible and the equilibrium quantities obtained by their use can be thus compared, even though it is rather laborious. However, considering real applications of different types of ion exchange materials, it is more or less impossible to evaluate their performances based on literature data. Use of uniform basic quantities could considerably reduce the need for laborious and time consuming comparative experiments with ion exchange materials in industrial applications.

To discuss the problems described above, an international workshop on "Uniform and Reliable Nomenclature, Formulations and Experimentation for Ion Exchange" was organised in Helsinki, Finland, in 1994.[5] There were in total, nineteen participants in this workshop from Japan, China, Russia, Belorussia, Hungary, Finland, Germany, UK and USA. It was concluded that updated recommendations for ion exchange nomenclature should be prepared and suggestions were made on how to start such a process. In addition, some tentative suggestions were made to update some of the more essential nomenclature, such as ion exchanger and selectivity coefficient. The outcome of the workshop, including these suggestions, has been published in a special issue of Reactive and Functional Polymers[5]. The ideas of the workshop were also discussed at the

Ion-Ex'95 conference in Wrexham, Wales in 1995.[6] To be successful, the process of updating and revising ion exchange nomenclature and formulations should have the support and guidance from the ion exchange community at large and that any new recommendations should have widespread acceptance prior to publication.

2 CONVENTIONAL NOMENCLATURE AND RECOMMENDATIONS

IUPAC recommendations from 1972 cover a large number of essential terms for ion exchange and it is obvious that the situation in ion exchange literature would be much less confusing if these recommendations had been followed. However, many terms in these recommendations are so broadly defined that a large number of interpretations are possible, or a wide variety of numerical values can be obtained for a given parameter, such as selectivity coefficient. In the following, conventions and recommendations for some of the more essential terms, e.g. selectivity coefficient, ion exchange, and ion exchange capacity, are discussed with a view to demonstrate the confusion that arises from the diverse conventions.

2.1 Ion Exchange and Ion Exchanger

The IUPAC recommendation for an ion exchanger, "A solid or liquid, inorganic or organic, containing ions, exchangeable with others of the same sign of charge present in a solution in which the exchanger is considered to be insoluble" is probably an appropriate definition for organic ion exchange resins. However, for inorganic ion exchangers, this definition is rather ambiguous and confusion may also arise e.g. between substitution reactions and true ion exchange reactions. For instance, when a solution of $CoCl_2$ is contacted with manganese sulphide, transfer of ions (Mn^{2+} and Co^{2+}) between solution and solid takes place and two distinctive solid phases (MnS, CoS) will be present in the system. This is a conventional substitution reaction, which can be written as

$$MnS + Co^{2+} \rightleftharpoons CoS + Mn^{2+}$$

The equilibrium of this reaction can be easily rationalized by using the solubility products of the two metal sulphides. When a zeolite in sodium form is contacted with a solution of KCl, a true ion exchange reaction takes place. At equilibrium, potassium and sodium ions are present in the zeolite framework but no separate solid compounds co-exist. The situation would be similar in a ion exchange resin. Considering the rapid development of inorganic ion exchange materials during the past decades, it is desirable that a more unambiguous definition should be derived.

2.2 Ion Exchange Capacities

Three distinctive static capacities are recommended by IUPAC. Theoretical specific capacity and volume capacity are quantities based on the number of milliequivalents of "ionogenic groups" per gramme of dry ion exchanger or per gramme of swollen ion exchanger, respectively. Practical specific capacity is a quantity based on the number of millimoles of ions taken up per gramme of dry ion exchanger under specified conditions. Theoretical specific capacity is constant for a given ion. Practical specific capacity, on the other hand, depends strongly on the experimental conditions for a given ion and can have values ranging from zero to the value of the theoretical capacity. These terms are, however, very seldomly used. Instead, terms like "ion exchange capacity", "cation exchange capacity", "X"-ion capacity (e.g. Ca-capacity) " are used. Furthermore, varying definitions are used for these terms and various experimental conditions are used to determine the values of these "capacities".

In addition to these terms, it is often necessary to use terms such as "maximum capacity", "maximum ion uptake" or "saturation capacity". In several types of exchangers, a proportion of "ionogenic groups" or "exchange sites" are unavailable for ion exchange, e.g. due to steric and ion-sieve effects. The proportion of these inactive sites depends on the size and chemical nature of the in-going ion. If properly determined, these "maximum capacities" are constant for a given ion and given exchanger.

It is obvious that the existence of so many "capacities" creates confusion in the ion exchange literature, especially if the terms used are not clearly defined. This confusion is likely to have a harmful effect on the reliability of basic ion exchange data.

2.3 Reaction Equation

Various formulations have been developed for the reaction equation of ion exchange. Considering binary exchange reaction between ions A (charge z_A) and B (charge z_B) two types of basic formulations are in common use:

$$z_A \, \overline{B}^{\, z_B} + z_B \, A^{\, z_A} \longleftrightarrow z_A \, B^{\, z_B} + z_B \, \overline{A}^{\, z_A} \tag{1}$$

$$\frac{1}{z_B} \, \overline{B}^{\, z_B} + \frac{1}{z_A} \, A^{\, z_A} \longleftrightarrow \frac{1}{z_B} \, B^{\, z_B} + \frac{1}{z_A} \, \overline{A}^{\, z_A} \tag{2}$$

In addition to overbars, various other symbols are used to denote the ion in the exchanger. No recommendations have been given on how to formulate an ion exchange reaction. However, the IUPAC definition of selectivity coefficient is consistent with reaction equation (1) above (see later 2.5).

2.4 Ion Exchange Equilibria

IUPAC recommendations for equilibrium parameters include <u>selectivity coefficient</u>, <u>corrected selectivity coefficient</u> and various distribution ratios or coefficients. Only the former two will be discussed here. In general, the recommendation for the <u>selectivity coefficient</u> has the following form:

$$k_{A/B} = \frac{\left[\overline{A}\right]^{z_B} [B]^{z_A}}{[A]^{z_B} \left[\overline{B}\right]^{z_A}} \tag{3}$$

where square brackets refer to ion concentrations. <u>Corrected selectivity coefficient</u> is obtained when concentrations of external solution are replaced by activities. In the IUPAC recommendations, it is also noted that "for exchanges involving counter-ions differing in their charges, the numerical value of $k_{A/B}$ depends on the choice of the concentration in the ion exchanger and the solution (molar scale, molal scale, mole fraction scale etc.). Concentration units must be clearly stated in exchange of ions of differing charges".

A lot of freedom is given for the measurement of ion exchange selectivity and this has resulted in a multitude of selectivity coefficients with differing numerical values. Most commonly, molar or molal concentration scales are used. In dilute solution both concentration scales give practically identical value for the selectivity coefficient. When molal scale is used, a molal selectivity coefficient $k^m_{A/B}$ is obtained:

$$k^m_{A/B} = \frac{\overline{m}_A^{Z_B} m_B^{Z_A}}{\overline{m}_B^{Z_A} m_A^{Z_B}} \qquad (4)$$

Also formulations that involve mixed concentration scales, molar scale for solution and rational scales for exchanger are very popular. Mole fractions were introduced e.g. by Högfeldt et al.[7] and correspondingly the corrected selectivity coefficient can be written as

$$k^x_{A/B} = \frac{\overline{X}_A^{Z_B} m_B^{Z_A}}{\overline{X}_B^{Z_A} m_A^{Z_B}} \qquad (5)$$

Equivalent fractions are used in the formulation introduced by Gaines and Thomas,[8] i.e.

$$k^E_{A/B} = \frac{\overline{E}_A^{Z_B} m_B^{Z_A}}{\overline{E}_B^{Z_A} m_A^{Z_B}} \qquad (6)$$

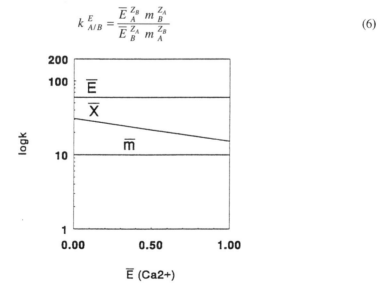

$$\overline{E} \,(Ca2+)$$

Figure 1 *Selectivity coefficients $k^m_{A/B}$, $k^x_{A/B}$ and $k^E_{A/B}$ for Ca^{2+}/Na^+ exchange in a hypothetical ion exchanger for which $k^m_{A/B}$ is constant at 10. Maximum Ca^{2+} loading 3.0 meq/g, total ion concentration in solution: 0.1 eq/L*

Figure 1 demonstrates the great variance arising from the different exchanger phase concentration scales in the numerical values of different selectivity coefficients. Values of $k^m_{A/B}$, $k^x_{A/B}$ and $k^E_{A/B}$ for Ca^{2+}/Na^+ are shown for a hypothetical ion exchanger (maximum Ca^{2+} loading 3 meq/g), which has been given a constant value of 10 for $k^m_{Ca/Na}$. The relationships between the selectivity coefficients are:

$$k^E_{A/B} = k^m_{A/B} \frac{Z_A^{Z_B}}{Z_B^{Z_A}} Q^{(Z_A - Z_B)} \tag{7}$$

$$k^x_{A/B} = k^m_{A/B} \frac{Z_A^{(Z_B - Z_A)}}{Z_B^{(Z_A - Z_B)}} \left[Z_A - (Z_B - Z_A)\overline{E}_A\right] Q^{(Z_A - Z_B)} \tag{8}$$

In addition to the above selectivity coefficients, a rational selectivity coefficient may be used:[9]

$$k^N_{A/B} = \frac{\overline{E}_A^{Z_B} E_B^{Z_A}}{\overline{E}_B^{Z_A} E_A^{Z_B}} \tag{9}$$

This rational selectivity coefficient depends on the exchange capacity Q of the exchanger and total ion concentration (N_t, eq/L) in solution, reflecting the electroselectivity effect, *i.e.*

$$k^N_{A/B} = k^m_{A/B} Q^{(Z_A - Z_B)} (N_t)^{(Z_B - Z_A)} \tag{10}$$

In ion exchange experiments, total ion concentration may be varied at will and the numerical value of $k^N_{A/B}$ can become very large in dilute solutions although the selectivity, as measured by the other selectivity coefficients, is low. For instance, in the example of Figure 1, the value of $k^N_{A/B}$ would be 300.

For application oriented research on ion exchange, it is obvious that it would be very useful if agreement could be reached on the recommended concentration scale. This would make the comparison of different exchange materials much more straightforward. People who are involved in the development and employment of new materials for ion exchange are not all specialists in ion exchange. Chemists synthesising new materials would appreciate a common measure that could be used to compare new materials that they are developing to the older ones prepared by other chemists. Sales staff would be happy to advertise the new materials with selectivity data that are comparable with the competitors' products, without carrying out "suspicious" conversions between the different selectivity measures.

3 TENTATIVE SUGGESTIONS FOR UPDATING NOMENCLATURE

The following tentative suggestions were made in the Helsinki Workshop for the definitions of some of the more essential terms in ion exchange.

3.1 Ion Exchange and Ion Exchanger

"**Ion exchange** is the equivalent exchange of ions between two or more ionized species located in different phases, at least one of which is an ion exchanger, without the formation of new types of chemical bonds."

"**Ion exchanger** is a phase containing osmotically inactive insoluble carrier of the electrical charge (matrix)."

Osmotically inactive means that the carrier cannot migrate from the phase where it is located to another phase. For example, sulphonic acid groups ($-SO_3H$) cannot migrate from polystyrene(PS)-divinylbenzene(DVB) framework into the solution phase. Thus PS-DVB-SO_3H is an ion exchanger, but $BaSO_4$ is not, since SO_4^{2-} can migrate into the solution phase due to dissolution.

3.2 Static Ion Exchange Capacities

"**Ion exchange capacity** is millimoles or milliequivalents of ionizable groups or exchange sites of unit charge per gram of dry exchanger. "

Ion exchange capacity can be determined unambiguously by determining analytically the composition of the ion exchanger. In this way the number of ionizable groups, such as $-SO_3H$ in sulphonic acid resins, or the exchange sites, such as hypothetical AlO_2^- sites in zeolites, can be obtained. In most inorganic ion exchangers, there are no distinctive ionizable groups. For example, in zeolites the negative charge created by Al^{3+} for Si^{4+} substitutions is more or less uniformly distributed to the oxygen atoms in both SiO_2 and AlO_2 in the framework. The term "exchange site" was chosen to describe this kind of cases, even though it is not very unambiguous either. The ionic form, to which the ion exchange capacity refers, should be given, since it affects the numerical capacity value.

"**Loading** is the total amount of ions expressed in milliequivalents or millimoles taken up per unit mass or unit volume of the exchanger under specified conditions, which should be always given. "

Loading refers always to the specific experimental conditions under which the ion uptake value was determined. Loading can of course be equal to the ion exchange capacity. For example, with strongly acidic and basic ion exchange resins maximum loading most probably equals the ion exchange capacity. Maximum loading can exceed ion exchange capacity, for example, due to electrolyte sorption, or it can be lower than the ion exchange capacity due to steric and ion sieve effects.

As an example zeolite Y in sodium-form, $Na_{56}(AlO_2)_{56}(SiO_2)_{136} \cdot 237H_2O$, is treated as follows: The "molecular weight" of the nonhydrous Y is 12013 g/mol and thus the ion exchange capacity 56 equivalents/mol x 1/(12013 g/mol) = 4.66 meq/g. Ten out of fifty six sodium ions are inaccessible to calcium ions and, therefore, the maximum loading in calcium exchange in sodium-form zeolite Y is 46/56 x 4.66 meq/g = 3.83 meq/g. In any other ion exchange experiment, the loading is between 0 and 3.83 meq/g, and in case there is electrolyte sorption, even higher values can be obtained.

To obtain commonly accepted definitions for "capacities", experimental procedures, "standard methods", may be needed for their determination. At present, standard methods (ASTM, DIN) exist only for the capacity determination of organic ion exchange resins. Reliable determination of ion exchange capacity may, however, be difficult, and several methods (titration, ion uptake determinations, exchanger composition determination) are usually needed to obtain the correct value. It is also obvious that one single standard methods cannot be suitable for all ion exchange materials.[10]

3.3   Reaction Equation

In the discussions, it was concluded that both common formulations of Eq. 1 and Eq. 2 are appropriate. It was noted however, that there is a clear distinction between the formulations. Equation 1 refers to the exchange of $z_A z_B$ equivalents of ions whereas Eq. 2 always refers to the exchange of one equivalent of ions.

3.4   Ion Exchange Equilibria

In the tentative suggestions, the terms describing ion exchange equilibrium parameter were divided into two groups:

3.4.1 Equilibrium parameters

Thermodynamic equilibrium constant:

$$K = \frac{\overline{a}_A^x\, a_B^y}{\overline{a}_B^y\, a_A^x} \tag{11}$$

Exponent x is either z_B (Eq. 1) or $1/z_A$ (Eq. 2) and y either z_A (Eq. 1) or $1/z_B$ (Eq. 2).

Corrected equilibrium coefficient:

$$K = \frac{\overline{C}_A^x\, a_B^y}{\overline{C}_B^y\, a_A^x} \tag{12}$$

where C_A and C_B are exchanger phase concentrations. Molarity, molality or full mole fractions were the preferred concentration units. Full mole fractions means that also the water content of the exchanger, and e.g. the sorbed electrolyte, should be considered.

Equilibrium coefficient:

$$K = \frac{\overline{C}_A^x\, C_B^y}{\overline{C}_B^y\, C_A^x} \tag{13}$$

where C_A and C_B are solution phase concentrations. Molality, molarity and mole fractions were the preferred concentration units.

3.4.2 Selectivity parameters

Separation factor:

$$\alpha = \frac{\overline{C}_A\, C_B}{\overline{C}_B\, C_A} \tag{14}$$

Selectivity coefficient:

$$k = \frac{\overline{E}_A^x\, E_B^y}{\overline{E}_B^y\, E_A^x} \tag{15}$$

There is a clear distinction between the two groups of parameters. Separation factor (Eq. 14) and selectivity coefficient (Eq. 15) depend strongly on the total concentration of the exchanging ions in the solution, at given C_A or E_A, in case that their charges are not equal (electroselectivity effect). Equilibrium parameters of Eqs. 12-13 do not reflect this effect and are almost independent of the total solution concentration.

4 FURTHER WAYS TO PROCEED IN THE HARMONISATION PROCESS

A harmonisation process of the ion exchange nomenclature requires guidance and support from the ion exchange community at large. It was suggested in the Helsinki Workshop that further workshops on this topic need to be arranged as part of international ion exchange conferences. Establishing a working committee at a later stage could very valuable in promoting harmonisation. Also international and national academic and industrial organisations would be needed to support this harmonisation process. A short-term goal could perhaps be to work out a booklet/article/book chapter containing the most important definitions and formulations, as well as presenting guidance of proper experimentation in ion exchange studies.

Acknowledgement

R. Harjula had a presentation "Harmonisation of Ion Exchange Formulations and Nomenclature: What Could Be Done?" in the ION-EX'95 Conference in Wrexham, Wales, September 10-14, 1995 and a paper by the same title will be published in the proceedings on this conference which paper contains essentially the same information as this one.

References

1. R. P. Townsend, in "Ion Exchange Procesess: Advances and Applications", A. Dyer, M. J. Hudson and P. A. Williams, eds., Royal Society of Chemistry, Bath, 1993, p. 3.
2. Recommendations on Ion Exchange Nomenclature, *Pure Appl. Chem.*, 1972, **29**, 619.
3. V. Gold, K. L. Loening, A. D. McNaught and P. Sehmi, "Compendium of Chemical Terminology, IUPAC Recommendations", Blackwell Scientific Publications, Oxford, 1987.
4. Nomenclature for Chromatography, IUPAC Recommendations, *Pure Appl. Chem.*, 1993, **65**, 819.
5. J. Lehto and R. Harjula (eds.), *React. Funct. Polymers*, Special Issue, 1995, **27**(2).
6. R. Harjula and J. Lehto, in *Proceedings of the ION-EX'95 Conference*, Wrexham, Wales, September 10-14, 1995 (in press).
7. E. Högfeldt, E. Ekedahl and L. G. Sillen, *Acta Chim. Scand.*, 1950, **4**, 404.
8. G. L. Gaines and H. C. Thomas, *J. Chem. Phys.*, 1953, **21**, 714.
9. F. Helfferich, "Ion Exchange", McGraw-Hill, New York, 1962.
10. J. Lehto and R. Harjula, *React. Funct. Polymers*, 1995, **27**, 121.

COMPUTER MODEL OF SULPHONATED STYRENE-DIVINYLBENZENE CO-POLYMERS

V. S. Soldatov, V. I. Gogolinskii, V. M. Zelenkovskii and A. L. Pushkarchuk

Institute of Physical - Organic Chemistry
Academy of Sciences of Belarus
13 Surganov st.
Minsk 220072
Belarus

1 INTRODUCTION

A model defining mathematical form of dependencies "additive property - ionic composition" for ion exchange systems have been recently suggested[1,2]. In particular, it allows to express in meaningful mathematical form the equilibrium coefficients as a function of equivalent fraction of a counter-ion in the ion exchanger. The main statement of this model is a suggestion that due to structural irregularity of statistical polymer matrix the exchange sites can have different number of nearest neighbours and, therefore, to exist in different energy states. This causes deviation of ion exchanger properties from that of statistical mixtures and influences their dependencies of degree of ion exchange. One of the main parameters of the model is the probability of existence of (P_i) of different number ($i = 0, 1, 2, ... $ n) of neighbouring exchange sites in a sphere of interaction around a central exchange site. These parameters are present in the "property - composition" equation following from the model

$$Y = \sum_{i=0,n} P_i \sum_{j=0,i} \frac{i!}{(i-j)!j!} \cdot y\,(i-j, j) \cdot \left(1 - \overline{X}\right)^{(i-j)} \overline{X}^{\,j} \qquad (1)$$

where Y is some additive property, y constants of the method, $\overline{X}$ is the mole fraction of exchange sites occupied by the entering counter-ion, i total number of neighbours, j number of neighbours occupied by the entering counter-ion.

This equation is a polynomial of power i relative to $\overline{X}$. Therefore, it is possible to obtain some information of the ion exchanger structure from the power of polynomial describing dependence of the equilibrium coefficient of $\overline{X}$. It has been shown that correct description of selectivity of sulphostyrene ion exchanger toward inorganic ions in a wide range of cross-linkage can be achieved if $i = 2$ for resins with low cross-linkage $i = 3$ for that with %DVB>8. A method for evaluation of the probabilities P_i was briefly described[1-3] but the properties of sulphonated co-polymers of styrene and divinylbenzene have not been described in terms of the model in detail. Determination of the i and P_i is a key problem in practical use of this equation, since except direct presence, the first value affects the energy factor y.

The P_i values are structural parameters of polyelectrolytes and can be evaluated if the structure of a polymer is known.

It is assumed that a statistical spatial polymer matrix can be represented by a set of elementary fragments with known structure. They can be randomly integrated into a matrix and some property of the integrated system can be found as a sum of products of the fragment properties and the probabilities of their existence.

The aim of the present paper is to evaluate the P_i for representative elementary fragments of sulphonated co-polymers of styrene and divinylbenzene. It is assumed that fragments responsible for complexity of the $Y = f(\overline{X})$ function can be identified with this knowledge.

1.1 Elementary Fragments

The following three types of fragments have been selected to represent the polymer structure: linear chain, H-shaped fragment and T-shaped fragment. Any other structures can be obtained as some junctions of these fragments. Nevertheless, if the size of the fragment is fixed then small loops should be also considered. They are not discussed here and will be a subject of a special paper.

The choice of size of the elementary fragments is important. We accepted it as a junction of 8 or 9 monomeric units, since with larger fragments it would be impossible to represent the structure of highly cross-linked ion exchangers. The smaller fragments do not have statistical meaning.

Structure of the elementary fragments can be different.

"Head to tail" (*ht*) junctions are predominant[4], but there is no convincing argument to discriminate the *hh* to *tt* junctions. If the polymerisation process is statistical, then their fraction in the total structure of the polymer is 0.25 for each. We name *hh* or *tt* junction irregularities. The linear chains with one irregularity in the middle was considered. Sulphonic groups in the phenyl rings was assumed to be situated only in para-position relative to the chain.

We distinguish four varieties of the H- and T-fragments: the ones with para- and meta- DVB, each with sulphonated and non-sulphonated cross-agent residuals. Only *ht*-junctions were considered.

The T-fragment structure depends on the type of initiator used in polymerisation. We have considered structures formed when benzoyl peroxide was used.

1.2 Method for the Probabilities Evaluation

The centre sulphur atom of one of the a groups was chosen as a central exchange site. Frequencies of finding the centres of the other sulphur atoms belonging to the fragment in a volume surrounded by radii of different length in the range 5-12 Å have been calculated. The same procedure was performed with all sulphur atoms in the fragment except the ones situated on the ends of each fragments (6 for linear chain, 4 for the H-fragment and 5 for the T-fragment). This corresponds to maximum cross-linkage of 25 mole percent.

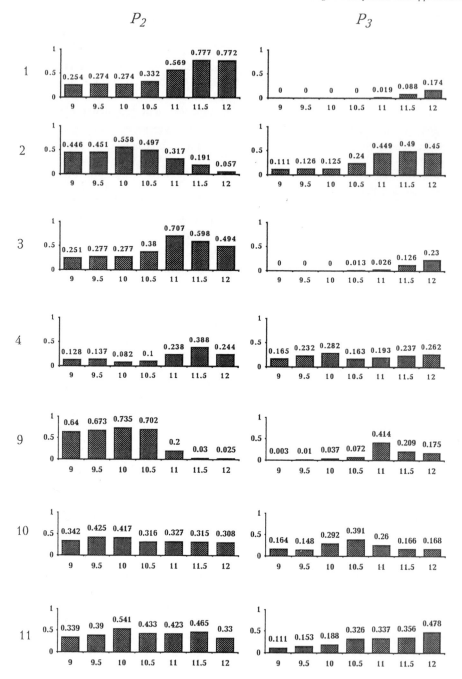

Figure 1 P_2 and P_3 at different R (Å) for structures given in Figure 2 under identical numbers

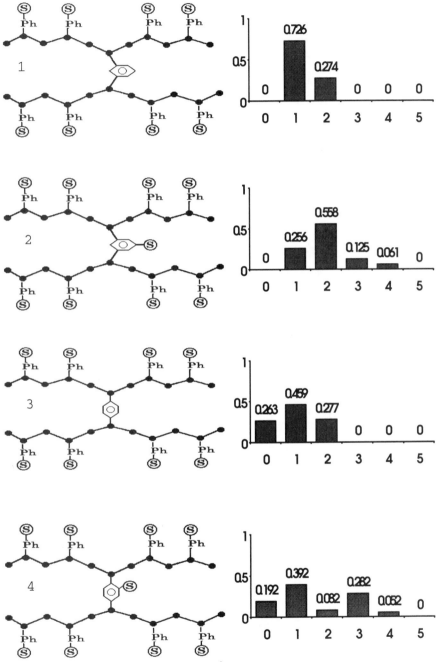

Figure 2a *Structure and P_i for R = 10 Å for the H-fragments, Ph denotes phenyl ring, (S) - sulphonic group*

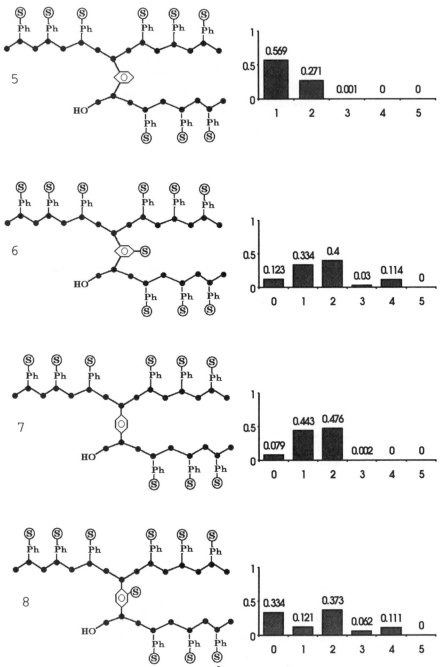

Figure 2b   *Structure and P_i for $R = 10$ Å for the T-fragments, Ph denotes phenyl ring,*
ⓢ *- sulphonic group*

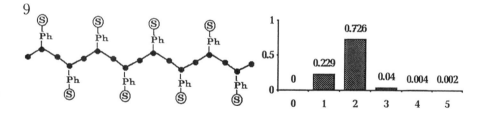

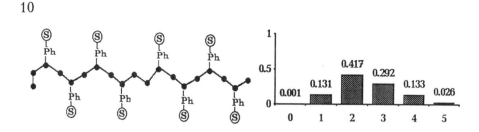

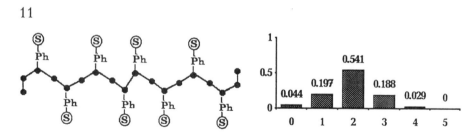

Figure 2c   *Structure and P_i for $R = 10$ Å for the linear fragments, Ph denotes phenyl ring, (S) - sulphonic group*

Geometric size and structure of the fragment was varied in a wide range by variation of the distance of each functional group from the central sulphonic group or the centre of the cross-agent residual. The angle between the sub-fragment linear chains joined by the cross-bridge in the H- and T-fragments was also varied in the mentioned cases. In each case more than 200 conformations were examined. From all considered conformations only those with low energies were taken for further consideration whose introduction into the total polymer structure could be meaningful accounting to their Boltzman's population. The number of meaningful conformation was usually about 20.

The first step of these calculations was computing the total energy of the fragment at small variations of the initially chosen distance to search for the energy minimum. The energy appeared to be highly sensitive to the distance value. The distance corresponding to the energy minimum (the equilibrium value) could be different from the initially chosen one by several tenths of angstrom. It was taken for the frequency calculations. The structure and energy characteristics of different conformers have been computed in molecular mechanics approximation using force field MM2[5]. Finally, the probabilities of the i number of neighbours in a fragment, approximately equal to the frequencies, were found by summing the values for all considered microstates (different distances and all sulphur atoms) accounting Boltzman's population of the conformations.

2 RESULTS

An important problem with the considered model is the estimation of radius of sphere (R), where the neighbours are situated (radius of interaction). It was estimated as 10 Å for our case[1,2].

The results of calculations are illustrated by histograms in Figures 1-2. The following conclusions can be drawn from these data.

As seen from Figure 1 the two neighbours probability, P_2, is high in all cases at R around 10 Å. It means that sulphostyrene ion exchanger of any cross-linkage are described by Equation 1 of not lower than the 2nd power.

The probabilities of different number of neighbours at R = 10 Å are given in Figure 2. The probability of three neighbours, P_3, is negligibly small in a regular (*ht*) linear chain and the knots with non-sulphonated cross-agent.

The P_3 becomes significant at R = 10 Å for linear chains with irregularities and for the knots with sulphonated cross-agent. This indicates that these fragments are responsible for complexity of the $Y = f(\overline{X})$ function. Increasing cross-linkage and polymerisation conditions favouring formation of *hh* and *tt* junctions should cause complicated shape of the $Y = f(\overline{X})$ curves expressed by polynomial of the 3rd or 4th power.

The T-fragment have lower probabilities of $i > 2$ than the H-fragments.

Difference between para and meta DVB is noticeable. It is clearer seen for the H-fragments. It is not substantial for R = 10 Å, but increases with increasing R. Therefore a greater difference in the behaviour of the resins with the p- and m-DVB can be expected for large organic ions.

3 DISCUSSION

The results reported in this paper are part of research aiming to find a general way for describing ion exchange systems with the model suggested earlier[1,2]. The main practical problem is independent estimation of the constants of equation $Y = f(\overline{X})$. Here we have suggested a method for estimation, or, at least, quantitative interpretation, of one groups of the constants, P_i. The method suggested can be applied for the other polymers with spatial matrix. Unsolved remain problems of estimation of the two other groups of constants: the probabilities of different fragments existence and the energy terms y. The work in this direction is in progress. Nevertheless, already obtained results give answers to some old questions of theory of ion exchange: what behaviour can be expected from statistically regular mono-functional ion exchange resins and what is the nature of their ion exchange sites irregularity: why selectivity parameters dependencies of resin loading is more complicated for higher cross-linkage. It becomes understandable why $Y = f(\overline{X})$

dependencies remain simple for high cross-linkage cases when cross-agent does not contain functional groups: sulphostyrene ion exchangers obtained by polymerisation of ethers of p-vinyl sulphonic acid[6], co-polymers of acrylic acids and DVB[7]. It is also evident that a drop in the exchange capacity of ion exchangers should lead to simplification of the $Y = f(\overline{X})$ down to $Y = $ const ($i = 0$, zeroth degree polynomial), corresponding to ideal exchange. Value $i = 0$ is also possible if the interaction radius is smaller than 7 Å. It is the case when ion exchange selectivity is controlled by strong short distance interactions and electrostatic energy change is comparably small. This may be realised in the case of strong polarisation interaction.

Acknowledgements

This work was supported by the foundation for fundamental research of the Republic of Belarus, Grant No. F43 - 119.

References

1. V. S. Soldatov, In "Ion Exchange Advances", (Proceedings of IEX'92), M. J. Slater, ed., Elsevier Applied Science, London, 1992, p. 159.
2. V. S. Soldatov, *Reactive Polymers*, 1993, **19**, 105.
3. V. S. Soldatov, V. I. Gogolinskii, V. M. Zelenkovskii, A. L. Pushkarchuk, *DAN Rossii (Russia)*, 1994, **337**, 105.
4. F. A. Bovey, "High Resolution NMR of Macromolecules", Acad. Press, London, 1972.
5. U. Burket, N. L. Allinger, "Molecular Mechanics", American Chemical Society, Washington, D. C., 1982.
6. J. N. Spinner, J. Ciric, W. F. Graydon, *Canad. J. Chem.*, 1954, **32**, 143.
7. V. S. Soldatov, A. I. Pokrovskaya, R. V. Martzinkevich, *J. Phys. Chem. (USSR)*, 1967, **41**, 1098.

DOUBLE-SELECTIVITY EQUILIBRIUM MODEL FOR ISOTHERMS EXHIBITING A SINGLE INFLEXION POINT

O. J. Bricio, J. Coca and H. Sastre

Chemical Engineering Department
University of Oviedo
33071 Oviedo
Spain

1 INTRODUCTION

Ion-exchange equilibrium, as a reversible reaction that takes place at constant total ionic concentration, is commonly well defined by a simple equilibrium isotherm with just one adjustable parameter: the selectivity coefficient. Isotherms based on the constant-selectivity model are always concave for favourable equilibrium and convex for unfavourable equilibrium. However, sometimes the experimental isotherms show an inflexion point, changing their behaviour from favourable to unfavourable equilibrium. These data, most of the times may by well fitted to a mathematical function that is the result of adding two constant-selectivity isotherms.

2 EXPERIMENTAL

The **equilibrium isotherms** for the exchange of hydrogen by some light metallic cations on Amberlite 200 (macroporous styrene-divinylbenzene resin)[1] were measured by means of batch experiments in which the resin is equilibrated with a solution of a given concentration. Experiments involving high solution concentrations were developed under *infinite solution volume* conditions so the resin had to be regenerated and the eluent analysed.[2] On the contrary, experiments at low concentrations were carried out under *finite solution volume* conditions and the concentration in both phases could be inferred from changes in solution composition.[3,4]

3 RESULTS

Ion exchange isotherms for the exchange of some light cations with hydrogen on Amberlite 200, were measured as part of a greater project.[5] Interest was focused on the influence of concentration, co-ion and temperature on the equilibria. Table 1 shows the experiments carried out to determine if there was some influence of those parameters within the operation range. Some of the isotherms are shown in Figures 1 to 5.

From the experimental data it can be seen that temperature influence on equilibrium is negligible for the system studied. Both isotherms, measured at 20 and 60 °C, are within

Table 1 *Systems Studied*

	Counter-ions	Co-ion	Normality	T (°C)	Analyzed phase	Analyzed ion
1	K⁺- H⁺	Cl⁻	1.2	20	resin (eluted)	K⁺
2	K⁺- H⁺	Cl⁻	1.2	60	resin (eluted)	K⁺ , H⁺
3	K⁺- H⁺	Cl⁻	0.01	20	solution	H⁺
4	K⁺- H⁺	Cl⁻	0.01	60	solution	H⁺
5	K⁺- H⁺	SO₄²⁻	1.2	20	resin (titrated)	H⁺
6	K⁺- NH₄⁺	SO₄²⁻	1.2	20	resin (eluted)	K⁺
7	Na⁺- H⁺	Cl⁻	1.2	20	resin (titrated)	H⁺
8	Ca²⁺- H⁺	Cl⁻	1.2	20	resin (titrated)	H⁺
9	Mg²⁺- H⁺	Cl⁻	1.2	20	resin (titrated)	H⁺

the experimental error of each other. No clear influence of concentration was established in either, and so, further experiments were not done.

The nature of the co-ion present in solution seems to have a strong influence on the isotherm. This is not really true. The isotherms were plotted as a function of the concentration of the metallic cation in solution, but this was calculated by subtraction of the amount of metal inside the resin from the total amount of metal added at the beginning of the experiment. This is not the real concentration of the cation in solution at equilibria. It must be considered that there are some other competitive equilibria in solution when working with sulphate ions. In fact, sulphate is the less common ion in a mixture of

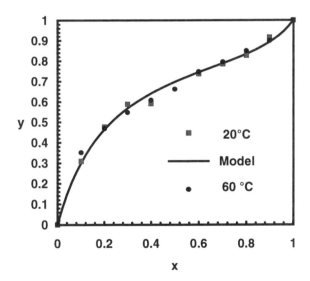

Figure 1 *Equilibrium isotherms for potassium - hydrogen exchange with chloride as co-ion for two temperatures. Total concentration is 1.2 N*

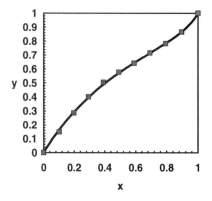

Figure 2 *Equilibrium isotherm for sodium - hydrogen exchange with chloride as co-ion at 1.2 N and 20 °C*

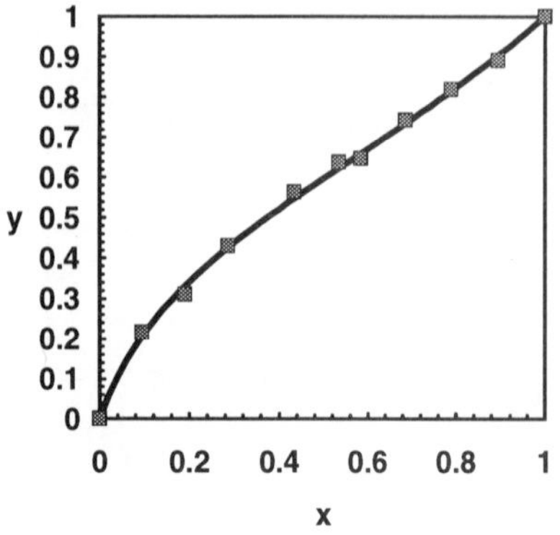

Figure 3 *Equilibrium isotherm for calcium - hydrogen exchange with chloride as co-ion at 1.2 N and 20 °C*

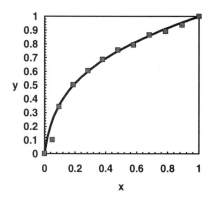

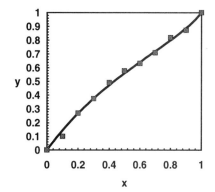

Figure 4 *Equilibrium isotherm for magnesium - hydrogen exchange with chloride as co-ion at 1.2 N and 20 °C*

Figure 5 *Equilibrium isotherm for potassium - ammonium exchange with sulphate as co-ion at 1.2 N and 20 °C*

sulphuric acid and potassium sulphate (1.2 N) due to the formation of hydrogensulfate ions and ion-pairing with potassium. When those equilibriums are removed from the concentrations in solution, leaving only concentrations of free ions, both isotherms, for chloride and sulphate, become the same one as predicted by the Donnan law. According to this, there could not be any influence of the co-ion because it is excluded from the inside of the resin matrix: it is simply not present where equilibrium takes place. Violations occur at high concentration, and it was proved that in these extreme conditions, the whole macroporous structure of the resin is invaded by the external solution, but it is still possible that penetration stopped before invading the microporous structure.[6]

Isotherms were fitted without success to the constant selectivity coefficient model. The inflexion point can not be predicted by this model. Experimental data show a different trend from the predictions of the model, and the fitting residues, with a standard error of 5.3%, are above the experimental error (estimated as less than 3% of the scale).

The mathematical form of the BET equation was applied to fit the data. It was taken as an empiric approximation,[7] because it is not possible to assume the presence of multilayers of ions bonded to the functional groups of the resins. Although, it allows a good fit of the data (standard error of 2.3% of the scale), it is not a good model for predicting the behaviour of the resin in other conditions different from those tested.

3.1. Double Selectivity Model

The inflexion point of the equilibrium isotherms measured for Amberlite 200 has been reported in some other **styrene-DVB resins**.[8-10] It has been explained as a responsibility of the unknown activity coefficients in resin, but it is difficult to verify this experimentally. In this work, we have developed a **model** from the facts known about styrene-DVB resins:

- Average selectivity increases with the degree of cross-linking.

- Selectivity falls as the concentration of the preferred ion increases (Figure 6).

- Cross-linking increases as an effect of the addition of DVB to the polymerization mixture.

- Highly cross-linked resins show a higher degree of heterogeneity than the less cross-linked resins.[11] The environment of the ionogenic groups varies at microscopic scale and, therefore, selectivity of a resin particle may be different from point to point.

- An increasing cross-linking degree leads to a reduction of distances between polymeric chains, thus intensifying ionic interactions inside the resin and, probably, the selectivity for ions.

- There is no reason for the sulphonic polystyrene having the same selectivity than the sulphonated divinylbenzene. Polystyrene is sulphonated preferentially at *-para* position whereas divinylbenzene has this position blocked.

With this knowledge, the following hypotheses have been formulated:

- The effect of simultaneous equilibrium of both sulphonated groups, styrene and divinylbenzene can be differentiated.

- The ratio of both exchange groups is given by the molar percentage of DVB present in the resin (degree of cross-linking).

- An increasing degree of cross-linking means an increasing selectivity.

- Ionic distribution inside the resin is governed by the simultaneous equilibrium with both groups. The general ion-exchange equilibrium theory is verified at microscopic scale.

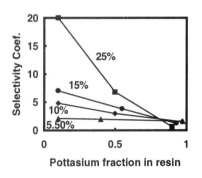

Figure 6 *Evolution of the selectivity coefficient with resin load and the degree of cross-linking for potassium - hydrogen exchange*

The amount of counter-ions inside the resin is distributed between both selective groups or exchange areas:

$$y = X y_1 + (1 - X) y_2 \tag{1}$$

where X is hypothetically the degree of cross-linking.

The rational selectivity coefficient of each phase will be:

$$K_{Bi}^A = \frac{y_i^{z_j} (1-x)^{z_i}}{(1-y_i)^{z_i} x^{z_j}} \tag{2}$$

therefore, for the exchange of single-charged ions, the following equation should describe the equilibrium isotherm:

$$y = X \frac{K_{B1}^A x}{1 + (K_{B1}^A - 1) x} + (1 - X) \frac{K_{B2}^A x}{1 + (K_{B2}^A - 1) x} \tag{3}$$

When exchanging double-charged ions with single-charged ions, the explicit equation is more complex:

$$y = X \frac{2K_{B1}^A + \frac{(1-x)^2}{x} - \sqrt{\left(2K_{B1}^A + \frac{(1-x)^2}{x}\right)^2 - 4K_{B1}^{A\,2}}}{2K_{B1}^A} + (1-X) \frac{2K_{B2}^A + \frac{(1-x)^2}{x} - \sqrt{\left(2K_{B2}^A + \frac{(1-x)^2}{x}\right)^2 - 4K_{B2}^{A\,2}}}{2K_{B2}^A} \tag{4}$$

For higher order cases, it is preferable to solve the implicit equations system.

This model has been proved quite satisfactory for the systems studied. From the pure mathematical point of view, fitting is excellent even as an empiric equation for those systems not reduced to free ionic concentrations. Standard error is 2.1 % of full scale and residues do not show unusual tendencies (Figure 7).

Further checking of the model has been carried out by fitting more complete data from Reichenberg for potassium - hydrogen exchange on styrene-DVB resins.[8,9] The fit is quite good and the general behaviour is as predicted by the model (see Figures 8 and 9).

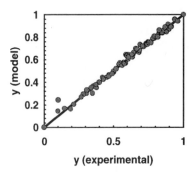

Figure 7 *Data fittings to the double selectivity coefficient model*

The larger exchange area prefers the potassium over hydrogen while the situation reverses for the other area, thus explaining the selectivity reversal that takes place on highly cross-linked resins as they are loaded until they are nearly saturated. This behaviour can be seen in Figure 10 for Amberlite 200. Selectivity decreases as the ionic fraction of the metal in the resin increases because the most selective sites are already occupied. Other resins with a lower degree of cross-linking show an almost constant selectivity coefficient.

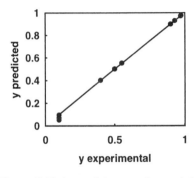

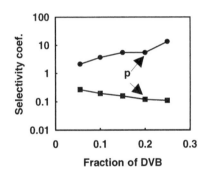

Figure 8 *Fittings of the experimental data compiled by Reichenberg for potassium - hydrogen exchange in styrene-DVB resins, to the double selectivity model*

Figure 9 *Pairs of selectivity coefficients as a function of cross-linking in styrene-DVB resins. Point "p" corresponds to Amberlite 200*

Table 2 *Fitting Parameters of the Isotherms for the Double Selectivity Model*

	Counter-ions	Co-ion	Normality	T (°C)	K_{B1}^{A}	K_{B2}^{A}
1	K^+- H^+	Cl^-	1.2	20	0.14 ± 0.04	5.3 ± 0.3
2	K^+- H^+	Cl^-	1.2	60	0.14 ± 0.04	5.3 ± 0.4
3	K^+- H^+	Cl^-	0.01	20	0.13 ± 0.04	6.3 ± 0.4
4	K^+- H^+	Cl^-	0.01	60	0.057 ± 0.019	5.0 ± 0.3
5	K^+- H^+ *	SO_4^{2-}	1.2	20	0.38 ± 0.06	6.7 ± 0.3
6	K^+- NH_4^+ *	SO_4^{2-}	1.2	20	0.22 ± 0.06	1.90 ± 0.12
7	Na^+- H^+	Cl^-	1.2	20	0.134 ± 0.016	2.25 ± 0.05
8	Ca^{2+}- H^+	Cl^-	1.2	20	0.14 ± 0.04	3.9 ± 0.3
9	Mg^{2+}- H^+	Cl^-	1.2	20	2.4 ± 1.4	9.2 ± 1.5
10	K^+- H^+	corrected	1.2	20	0.11 ± 0.02	3.8 ± 0.2
11	K^+- NH_4^+	corrected	1.2	20	0.16 ± 0.04	4.9 ± 0.2

*   Other parallel equilibria in solution not corrected. Model equation used as empiric approximation.

Nomenclature

z_i	Ionic charge
K_{Bi}^{A}	Selectivity coefficient based on equivalents
X	Fractional cross-linking
x	Equivalent ionic fraction in solution
y	Equivalent ionic fraction in resin

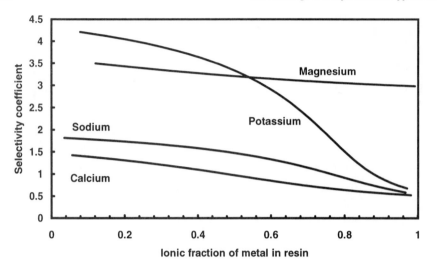

Figure 10 *Evolution of selectivity coefficients of different cations with respect to hydrogen as functions of the load of the resin (Amberlite 200, 20% of cross-linking, total ionic concentration in solution: 1.2 N)*

References

1. Anon., "Amberlite IR-200 Engineering Bulletin. Hydrogen Cycle - Co-current Sulphuric Acid Regeneration", Rohm & Haas, European Region, IX 200/EB/1, 1977.
2. O. Bricio, Ph. D. Thesis, University of Oviedo (Spain), December 1995.
3. A. de Lucas, J. Zarca and P. Cañizares, *Anales de Química*, 1990, **86**, 875.
4. A. L. Liapis, and D. W. T. Rippin, *Chem. Eng. Sci.*, 1977, **32**, 619.
5. O. Bricio, J. Coca and H. Sastre, Abstracts of 6th Mediterranean Congress on Chemical Engineering, Barcelona 1993, p. 57.
6. O. Bricio, J. Coca and H. Sastre, Abstracts of XV Reunión Iberica de Adsorción, Oviedo 1991, p. 83.
7. A. E. Rodrigues, "Percolation Processes, Theory and Applications", A. E. Rodrigues and D. Tondeur, eds., NATO ASI, Ser. E 33, 1983, p. 43.
8. F. G. Helfferich, *React. Polym.*, 1990, **12**, 95.
9. D. Reichenberg, "Ion Exchangers in Organic and Biochemistry", Interscience Publishers, New York, 1957, p. 66.
10. D. Reichenberg, "Ion Exchange", J. Marinsky, ed., Marcel Dekker Inc., New York, 1966, Vol 1, p. 227.
11. L. S. Goldring, "Ion Exchange", J. Marinsky, ed., Marcel Dekker Inc., New York, 1966, Vol 1, p. 205.

A HETEROGENEOUS MODEL FOR THE SIMULATION OF ION EXCHANGE EQUILIBRIA

S. Melis and G. Cao

Dipartimento di Ingegneria Chimica e Materiali
Universita' di Cagliari
09123 Cagliari
Italy

1 INTRODUCTION

Various separation processes are based on the use of ion exchange resins[1]. In particular, the optimal design and operation of the corresponding ion exchange units require the accurate simulation of multicomponent equilibria of the involved chemical species. A number of studies have been devoted to the representation of uptake equilibria of various species by cation exchange resins as a function of solution composition[2]. The models proposed in the literature can be divided in two main groups. Models of the first group describe the ion exchange process in terms of the law of mass action[3,4,5], while those of the second group treat ion exchange as a phase equilibrium[6]. An intermediate approach has been recently proposed[7]. It consists of applying the mass action law under the assumption that both the fluid and the solid phase behave ideally, while accounting for the heterogeneity of functional groups. In other words, in this approach it is assumed that the effect of mixture nonidealities is smaller than that due to the resin heterogeneity and can be neglected. This introduces the idea of accounting for the heterogeneous nature of the solid in the framework of the equilibrium models based on the law of mass action by considering the existence of a statistical distribution of energies of the functional groups, approximated through a discrete distribution of only two types of functional groups. This model, referred to as the Heterogeneous Mass Action Model (HMAM), is able to take the dependence of the exchange process upon solution normality as well as the selectivity changes with the resin composition into account.

It should be noted that the HMAM reduces to the constant selectivity treatment of the model of the first group when neglecting the heterogeneity of the resin functional groups, thus considering groups of only one type. In addition, since the HMAM assumes ideality in both phases, multicomponent systems containing N_c counterions can be described on the basis of only the $(N_c - 1)$ binary equilibria of each exchangeable species with a reference counterion, i.e. the triangle rule holds. In other words, there is no need to study all possible equilibria between each couple of counterions, i.e. $N_c(N_c - 1)/2$ as in the case of nonideal models of the first group. When compared with the model of the second group, the HMAM leads to the same equations as that of Myers and Byington[6] in the case of ions with equal valence, while they are different in the case of ions with higher valence.

The reliability of the HMAM in predicting the equilibrium behavior of multicomponent systems from the knowledge of binary experimental data, has been tested by comparison with experimental results available in the literature for inorganic species[7] as well as data obtained in our laboratory for the case of organic compounds, i.e. amino acids[8].

In this work, we first analyze the validity of the two-sites assumption introduced by Melis et al.[7] for the simulation of the resin heterogeneity. In particular, we compare this assumption with a more complex description of the real energy distribution of the functional groups, where the resin is characterized by three and four types of functional groups, respectively. Then, since the formulation of HMAM with only two types of functional groups appears to be the most convenient, its capability in describing binary uptake data and in predicting the behavior of the third binary equilibrium according the triangle rule is further tested by comparison with suitable experimental data available in the literature. Moreover, by analysing the effect of resin cross-linking degree on model parameters, we propose a qualitative explanation of the origin of resin heterogeneity in terms of inhomogeneities of the resin structure.

2 THE HETEROGENEOUS MASS ACTION LAW MODEL

Let us consider a system containing N_c counterions and select one of them, say the n-th, as reference one. In order to describe the behavior of such a system, we can consider the $(N_c - 1)$ independent exchange reactions between the reference counterion ($S_n^{Z_n+}$) and each one of the other system components ($S_i^{Z_i+}$):

$$z_n S_{i,s}^{Z_i+} + z_i S_{n,r}^{Z_n+} \leftrightarrow z_n S_{i,r}^{Z_i+} + z_i S_{n,s}^{Z_n+} \qquad i = 1, N_c - 1 \qquad (1)$$

where z_i is the charge of the counterion $S_i^{Z_i+}$ and the subscripts r and s refer to the resin and the solution phase, respectively.

In order to describe the exchange process represented by eq 1 we consider the model proposed by Melis et al.[7], where both phases are assumed to be ideal while heterogeneity is accounted for through a given distribution of the standard free-energy change associated to the ion exchange process, ΔG°. This can be approximated by considering the existence of N_f equally abundant different types of functional groups. Hence, for the generic i-th equilibrium reaction (1) occurring on functional groups of type j, the equilibrium constant $K_{j,i}$ is written as:

$$K_{j,i} = \frac{\left(q_{j,i}\right)^{Z_n} \left(C_n\right)^{Z_i}}{\left(C_i\right)^{Z_n} \left(q_{j,n}\right)^{Z_i}} \qquad j = 1, N_f \ ; \qquad i = 1, N_c - 1$$

$$(2)$$

where C_i and $q_{j,i}$ represent the concentrations of the species $S_i^{Z_i+}$ in the solution and on functional groups of type j, respectively. By introducing the ionic fractions in solution and on functional groups of type j

$$x_i = \frac{z_i C_i}{\sum_{k=1}^{N_c} z_k C_k} = \frac{z_i C_i}{N} \qquad ; \quad y_{j,i} = \frac{z_i q_{j,i}}{\sum_{k=1}^{N_c} z_k q_{j,k}} = \frac{z_i q_{j,i}}{Q_j} \qquad (3)$$

where N and Q_j are the solution normality and the exchange capacity of functional groups of type j, respectively, we can express eq 2 in terms of ionic fractions:

$$K_{j,i} = \frac{\left(y_{j,i}\right)^{Z_n}\left(x_n\right)^{Z_i}}{\left(x_i\right)^{Z_n}\left(y_{j,n}\right)^{Z_i}}\left(\frac{Q_j}{N}\right)^{Z_n - Z_i} \qquad j = 1, N_f \quad ; \qquad i = 1, N_c - 1 \qquad (4)$$

Equations (4), in addition to the congruence conditions for the solution phase and for functional groups of type j (i.e. the sum of the ionic fractions of all species equal to unity) can be used to determine the unknown ionic fractions $y_{j,i}$, once the equilibrium constants, $K_{j,i}$, the value of Q_j and the composition of the solution are known. The overall composition of the resin phase can now be evaluated by averaging the ionic fractions (or, alternatively, by summing up the resin loadings $q_{j,i}$) on each type of functional groups:

$$y_i = \sum_{j=1}^{N_f} p_j\, y_{j,i} = \frac{z_i\, q_i}{Q} \qquad\qquad i = 1, N_c \qquad (5)$$

where p_j is the probability that the exchange occurs on functional groups of type j, which, assuming equally abundant types of functional groups, results equal to:

$$p_j = \frac{1}{N_f} \qquad\qquad j = 1, N_f \qquad (6)$$

As discussed in Melis et al.[7], the parameters $K_{j,i}$ can be directly related to the average value and the variance of the standard free-energy distribution associated to the exchange reaction (1):

$$\overline{\Delta G_i^o} = \sum_{j=1}^{N_f} p_j\, \Delta G_{j,i}^o = \sum_{j=1}^{N_f} p_j\left[-RT \ln\left(K_{j,i}\right)\right] \qquad (7)$$

$$\sigma_i^2 = \sum_{j=1}^{N_f} p_j\left(\Delta G_{j,i}^o - \overline{\Delta G_i^o}\right)^2 \qquad (8)$$

However, in order to avoid the temperature dependence of $\overline{\Delta G_i^o}$ and σ_i^2, it may be convenient to define also an average equilibrium constant and a heterogeneity parameter:

$$\overline{K_i} = \exp\left[-\overline{\Delta G_i^o}\,/\,RT\right] = \prod_{j=1}^{N_f}\left(K_{j,i}\right)^{1/N_f} \qquad (9)$$

$$\gamma_i = \exp\left(-\sigma_i\,/\,RT\right) \qquad (10)$$

hence, in the following, we will refer to $\overline{K_i}$ and γ_i as the model parameters.

3 RESULTS AND DISCUSSION

3.1 The N_f-sites Approximation

In order to verify the validity of assuming the existence of only two types of functional groups considered in the original formulation of the HMAM, we compare its results with those corresponding to the more general case, i.e. the resin is characterized by three and four types of functional groups, respectively. To this aim, the equilibrium behavior of four binary systems available in the literature[5,9,10] is taken into account. In addition, in order to appreciate the importance of heterogeneity, we consider also the case of a resin with only one type of functional group. It is apparent from eq 4 that the number of model parameters $K_{j,i}$ is equal to the number of functional groups which characterize the resin. However, in order to make a clearer comparison, we take advantage of eqs 9 and 10, thus considering $\overline{K_i}$ and γ_i as the model parameters, whose value can be estimated by fitting the experimental data through a non linear least square procedure. As it may be seen from the values of the average relative errors reported in Table 1, the multiple-sites approximations allow for a quite satisfactory correlation of the experimental results while this is not the case when a single site is considered. Moreover, it may be seen from the same table that the parameters $\overline{K_i}$ and γ_i are rather similar when considering the existence of more than one type of functional group. This clearly indicates the validity of the two-site approximation since it combines a great flexibility in correlating experimental data to a great simplicity, since it involves only two adjustable parameters per counterion.

Table 1   *Values of Adjustable Parameters and Corresponding Average Percentage Errors in Reproducing Binary Equilibrium Data for the N_f-sites Approximation*

System		$N_f = 1$	$N_f = 2$	$N_f = 3$	$N_f = 4$
	$\overline{K_i}$	2.37	2.107	2.133	2.128
K+/H+	γ_i	–	2.418	2.355	2.387
	ε	6.43	2.200	2.123	2.104
	$\overline{K_i}$	0.98	0.59	0.626	0.628
Cl-/OH-	γ_i	–	5.17	4.741	4.772
	ε	20.3	2.4	0.67	0.67
	$\overline{K_i}$	10.56	10.4	10.377	10.357
Ca++/H+	γ_i	–	1.62	1.63	1.64
	ε	1.91	1.9	1.88	1.88
	$\overline{K_i}$	5.26	4.76	4.834	4.885
Mg++/H+	γ_i	–	2.73	2.705	2.727
	ε	3.25	1.4	1.28	1.24

3.2   Comparison with Experimental Data

The capability of the HMAM along with the two-site approximation, which is considered in the following, is further studied by comparison with experimental data of the binary systems Ag+/H+ and Cu++/Ag+ investigated by Dranoff and Lapidus[11]. The comparison between HMAM results and experimental data is shown in Figure 1a and 1b, where the estimated value of $\overline{K_i}$ and γ_i are also reported. For both systems, the agreement with experimental data is satisfactory. The reliability of HMAM in predicting binary experimental equilibrium data according with the triangle rule, i.e. knowing the equilibrium behavior of only two of the possible binary systems, is then tested. For this we analyze the experimental equilibrium data of the binary system Cu^{++}/H^+ reported by the same authors[11]. By using the same parameter values estimated above for the systems Ag^+/H^+ and Cu^{++}/Ag^+, we compare experimental data and model predictions as shown in Figure 2. It is seen that the agreement is satisfactory (average relative error $\varepsilon = 2.3\%$).

3.3 Effect of Cross-linking Degree on Model Parameters

In the following, we analyse the effect of the cross-linking degree on the binary equilibrium between monovalent, divalent, and one monovalent and one divalent ions. The model parameters are estimated against suitable experimental data available in the literature and are reported, as described in the following, as a function of the cross-linking degree.

For the case of equilibria involving monovalent ions we consider the systems Rb^+/H^+, Tl^+/H^+, and Cs^+/H^+ reported by Bonner[12] and K^+/H^+ (data from de Lucas Martinez et al.[5]) for various values of the degree of cross-linking. It is apparent from Figures 3a to 3d that $\overline{K_i}$ and the heterogeneity parameter, γ_i increase together with the cross-linking degree. In particular, $\overline{K_i}$ and γ_i were invariably found to increase linearly for all systems involving monovalent ions. It is however difficult to explain this behavior, since on the basis of the ideal Donnan theory[13], it is expected that $\overline{K_i}$ has an exponential dependence on the swelling pressure, which in turn linearly depends on the cross-linking degree.

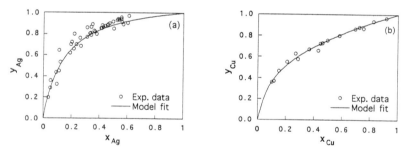

Figure 1 *Comparison between experimental data and model fit: (a) Ag^+/H^+, $\overline{K_i} = 7.58$, $\gamma = 1$; (b) Cu^{++}/Ag^+, $\overline{K_i} = 0.48$, $\gamma = 1.97$*

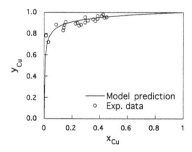

Figure 2 *Comparison between experimental data and model prediction: Cu^{++}/H^+*

When considering the case of equilibria of divalent ions, i.e. Cd^{++}/Ca^{++} and Cu^{++}/Mg^{++} (data from Bonner and Smith[14]), a definite trend of model parameters as a function of the cross-linking degree cannot be established, as may be seen in Figures 4a to 4d. However, it remains evident that $\overline{K}_i$ and γ_i still increases for increasing values of the cross-linking degree.

This behavior was also found for the equilibrium between one mono- and one divalent ion, i.e. the system Zn^{++}/H^+ investigated by Boyd et al.[15]. As shown in Figure 5, $\overline{K}_i$ and γ_i display an approximately exponential dependence upon the cross-linking degree. It is worth mentioning that this behavior, as well as the one corresponding to divalent ions, has to be verified with a larger amount of experimental data.

It is worth noting that the increase in the average value of the equilibrium constant of the HMAM with increasing the cross-linking degree is consistent with previous results reported in the literature[12,14]. In fact, by experimentally investigating several binary ion exchange equilibria on different resins with similar structure but with a different degree of cross-linking, it was found that highly cross-linked resins are more selective than weakly cross-linked ones. This behavior is in particular consistent with the ideal Donnan theory[13], which predicts that highly cross-linked resins are characterized by larger swelling pressures, which in turn result in an increase of the equilibrium constant. In addition, the increase in the heterogeneity parameter of the HMAM for increasing values of the cross-linking degree was somehow expected. In fact, highly cross-linked resins appear to be inhomogeneous both in the cross-link density[16] and pore size[17]. This may result in wider distributions of the swelling pressure and therefore in larger variation of the equilibrium constant.

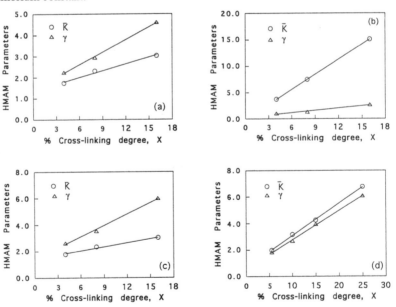

Figure 3 *Effect of cross-linking on model parameters: (a) Rb^+/H^+; (b) Tl^+/H^+; (c) Cs^+/H^+; (d) K^+/H^+*

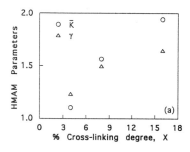 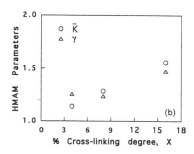

Figure 4 *Effect of cross-linking on model parameters: (a) Cd^{++}/Ca^{++}; (b) Cu^{++}/Mg^{++}*

4 CONCLUDING REMARKS

In the present work we discussed the validity of the two-sites approximation introduced in the formulation of the HMAM by Melis et al.[7] for the simulation of resin heterogeneity. The results showed that this approximation represents the best compromise between model flexibility and the minimum number of model parameters to be obtained by direct comparison with experimental data.

Subsequently, we further tested the capability of the HMAM, along with the formulation which consider the presence of only two types of functional groups, in describing binary uptake data and in predicting the behavior of the third binary equilibrium according to the triangle rule by comparison with suitable experimental data available in the literature.

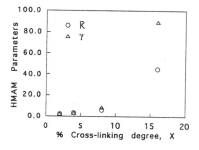

Figure 5 *Effect of cross-linking on model parameters for the system Zn^{++}/H^+*

Finally, we investigated which resin characteristic may be responsible for resin heterogeneity. In particular, we investigated the effect of cross-linking which, according to the ideal Donnan theory, affects the ion exchange equilibria. The results obtained show that both the average equilibrium constant, $\overline{K}_i$ and the heterogeneity parameter, γ_i increase together with the degree of cross-linking. Moreover, it was observed that a linear relationship appears to exist when considering the exchange between monovalent ions.

However, when considering ion exchange equilibria involving counterions of higher valences, such a dependence was not found, maybe owing to the superimposition of other phenomena. It is worth noting that, although the model parameters are not quantitatively correlated with the degree of cross-linking, the results reported in this work may contribute, together with other results recently presented in the literature[18], to establish a general trend in ion exchange equilibria, with the final goal of developing a rational methodology for the estimation of equilibrium based upon a knowledge of resin characteristics and molecular properties.

Nomenclature

C_i	concentration of the ion $S_i^{Z_i+}$ in the solution	kmol/m^3
$K_{j,i}$	equilibrium constant of the reaction between counterion $S_i^{Z_i+}$ and reference counterion occurring on functional groups of type j	
$\overline{K_i}$	average equilibrium constant of the reaction between counterion $S_i^{Z_i+}$ and reference counterion	
N	solution normality	kmol/m^3
N_c	number of counterions present in the system	
N_f	number of types of functional groups	
p_j	probability that the exchange reaction occurs on functional groups of type j	
$q_{j,i}$	concentration of the ion $S_i^{Z_i+}$ on functional groups of type j	kmol/m^3
Q	total exchange capacity	kmol/m^3
Q_j	total exchange capacity associated to functional groups of type j	kmol/m^3
R	ideal gas constant	J/mol/K
T	temperature	K
x_i	ionic fraction of the ion $S_i^{Z_i+}$ in the solution	
X	% cross-linking degree	
y_i	overall ionic fraction of the ion $S_i^{Z_i+}$ in the resin	
$y_{j,i}$	ionic fraction of the ion $S_i^{Z_i+}$ in the resin functional groups of type j	
z_i	valence of the ion $S_i^{Z_i+}$	

Greek letters

γ_i	parameter defined by equation 10	
ε	average percentage error	
$\Delta G_{j,i}^o$	standard free-energy change associated to the reaction between counterion $S_i^{Z_i+}$ and reference counterion occurring on functional groups of type j	J/mol
$\overline{\Delta G_i^o}$	average standard free-energy change associated to the reaction between counterion $S_i^{Z_i+}$ and reference counterion	J/mol
σ_i	standard deviation of the distribution of the free-energy change associated to the reaction between counterion $S_i^{Z_i+}$ and reference counterion	J/mol

subscripts

r   resin phase
s   solution phase

References

1. D. Naden and M. Streat, "Ion Exchange Technology", Ellis Horwood Limited, Chirchester, 1984.
2. M. J. Slater, "Principles of Ion Exchange Technology", Butterworth-Heinemann, Oxford, 1991.
3. D. C. Shallcross, C. C. Herrmann and B. J. McCoy, *Chem. Eng. Sci.*, 1988, **43**, 279.
4. R. M. Allen, P. A. Addison and A. H. Dechapunya, *Chem. Eng. J.*, 1989, **40**, 151.
5. A. de Lucas Martinez, P. Cañizares and J. Zarca, *Chem. Eng. Tech.*, 1993, **16**, 35.
6. A. L. Myers and S. Byington, in "Ion Exchange: Science and Technology", A. E. Rodrigues, ed., NATO ASI Series E., No. 107, Martinus Nijhoff, Dordrecht, 1986, p. 119.
7. S. Melis, G. Cao and M. Morbidelli, *Ind. Eng. Chem. Res.*, 1995, **34**, 3916.
8. S. Melis, J. Markos, G. Cao and M. Morbidelli, *Ind. Eng. Chem. Res.*, 1995, submitted for publication.
9. R. M. Wheaton and W. C. Bauman, *Ind. Eng. Chem.*, 1951, **43**, 1088.
10. R. Khoroshko, V. P. Kol'nenkov, V. Soldatov, N. Sudarikova and N. G. Peryshkina, *Agrokhimiya*, 1974, **10**, 122.
11. J. S. Dranoff and L. Lapidus, *Ind. Eng. Chem.*, 1957, **49**, 1297.
12. O. D. Bonner, *J. Phys. Chem.*, 1955, **59**, 719.
13. F. Helfferich, "Ion Exchange", McGraw Hill, New York, 1962.
14. O. D. Bonner and L. L. Smith, *J. Phys. Chem.*, 1957, **61**, 326.
15. G. E. Boyd, F. Vaslow and S. Lindenbaum, *J. Phys. Chem.*, 1967, **71**, 2214.
16. C. C. Hsiao and W. Chen, in "Polymer Networks: Structure and Mechanical Properties", A. J. Chompff and S. Newman, eds., Plenum Press, New York, 1971, p. 395.
17. K. Jerabek and K. Setinek, *J. Polym. Sci.: Part A*, 1990, **28**, 1387.
18. L. Jones and G. Carta, *Ind. Eng. Chem. Res.*, 1993, **32**, 107.

MODELLING OF ION EXCHANGE IN RADIAL FLOW

Y. Tsaur and D. C. Shallcross

Department of Chemical Engineering
The University of Melbourne
Parkville, Victoria 3052
Australia

1 INTRODUCTION

Over the past decades the ion exchange process has found increased applications in a range of diverse fields. Most of these new applications however still use cylindrical ion exchange beds in which a solution is injected in one end and the effluent is collected at the other. The ion exchange process is approximated as a one-dimensional, linear operation. This paper describes a radial ion exchange process in which a solution is injected into a wedge-shaped ion exchange bed near its apex, and the effluent is collected from the bed's periphery.

The conventional one-dimensional, linear ion exchange column has been well investigated and many models have been developed to predict ion exchange performance in vertical flow.[1-3] However, most models assume no axial dispersion and either concentration independence of the rate constant or ideal behaviour of ion exchange equilibrium. These same assumptions are also made in the few models developed for radial flow in chromatography.[4,5] Although some radial models do consider radial dispersion, the dispersion coefficient is assumed to be constant, regardless of the fact that the radial dispersion is a function of the radial velocity which in turn varies with distance from the model axis.[6] For systems other than those dealing with dilute solutions and plug flow, the existing models lack accuracy in the prediction of break through which is fairly sensitive to non-idealities of flow and exchange behaviour.

Dispersion is however important; it reflects back mixing which occurs with flow and thus is a function of the fluid velocity. Also, depending on the system and solution concentration the exchanging ions can exhibit non-ideal behaviour. In this study, we have developed an ion exchange model which predicts ion exchange performance in radial flow through a packed bed of ion exchange resin. Our model takes account of both dispersion and the non-ideal behaviour of the exchanging ions in both the solution and solid (resin) phases and assumes infinitely fast kinetics. The radial ion exchange model predictions are tested by comparing them with the observations made during experiments in a radial, wedge-shaped bed.

In developing this model incorporating dispersion and non-ideal behaviour in both phases, a model was first developed for one-dimensional, linear flow. Once the model predictions had been validated by column experiments the linear model was then adapted to the radial geometry.

2 THEORETICAL MODEL DEVELOPMENT

The radial ion exchange model is based upon a simple material balance. It incorporates dispersion and uses an modification to the Mehablia equilibrium model[7] to predict ion exchange equilibrium. While the form of the model presented here considers binary cation exchange only, the fundamental concepts behind the model are not limited to such a simple system.

2.1 Radial Ion Exchange Model

An electrolyte solution containing cation species A is injected at a constant flow rate, L, at the centre of a uniformly packed thin cylindrical resin bed and moves radially outwards to the bed periphery. The effluent is produced at a rate uniform around the bed periphery. The resin bed has a constant thickness, h, and a constant voidage, e. The radius of the bed inlet is r_I and the radius of the bed outlet is r_O. The concentration of A cations in the injection solution is a constant, C_0.

Initially, the resin bed is in the B-form, i.e. all the exchange sites of resin are occupied by B cations; and the solution within the pores of the resin bed is free of A cations. At time $t = 0$, injection of the solution containing the A cations begins. For some time after injection begins, no A cations are present in the effluent. As the resin gradually takes up A cations, the concentration of A cations in the effluent rises from zero and eventually reaches the concentration of the injected solution. Plotting the concentration of A cations in the effluent against the volume of the effluent forms the sigmoidal-shape break through curve.

When the solution is in contact with resin bed, A cations in the solution exchange onto the resin and B cations on the resin exchange into the solution:

$$Z_B \ A^{Z_A} + Z_A \ R\text{-}B^{Z_B} \ \rightleftharpoons \ Z_B \ R\text{-}A^{Z_A} + Z_A \ B^{Z_B} \tag{1}$$

where, R- represents the resin; and Z_A and Z_B are the valencies of the exchanging cation A and B, respectively.

Now consider a thin but finite ring of the resin bed at position r, with a ring thickness of Δr. Through the elemental ring, $C_{m,A}$ is the solution concentration of the A cations, $q_{m,A}$ is the resin concentration of A cations, and $J_{m,A}$ is the dispersion flux. The material balance for A cations over the ring in cylindrical coordinates is:

$$e \ (V_{r+\Delta r} - V_r) \ \frac{\partial C_{m,A}}{\partial t} + (1 - e) \ (V_{r+\Delta r} - V_r) \ \frac{\partial q_{m,A}}{\partial t}$$

$$= \ e \ (A_r \ J_{m,A}|_r - A_{r+\Delta r} \ J_{m,A}|_{r+\Delta r}) + L \ (C_{m,A}|_r - C_{m,A}|_{r+\Delta r}) \tag{2}$$

where, at radial position r, the cross-section area is $A_r = 2\pi rh$ and the bed volume is $V_r = \pi \ (r^2 \text{-} r_I^2)h$; at the position $r+\Delta r$, the cross-section area is $A_{r+\Delta r} = 2\pi (r+\Delta r)h$ and the bed volume is $V_{r + \Delta r} = \pi[(r+\Delta r)^2 \text{-} r_I^2]h$; and neglecting the second order term of Δr, $(V_{r + \Delta r} - V_r) = A_r \Delta r$.

Dividing by ($A_r \, \Delta r$) and taking limits as $\Delta r \to 0$, $A_{r+\Delta r} \approx A_r$ equation (2) becomes:

$$\frac{\partial C_{m,A}}{\partial t} + \left(\frac{1-e}{e}\right) \frac{\partial q_{m,A}}{\partial t} = -\frac{\partial}{\partial r}(J_{m,A}) - \frac{L}{e \, A_r} \frac{\partial C_{m,A}}{\partial r} \tag{3}$$

Fick's Law is used to relate the dispersion flux to the concentration gradient:

$$J_{m,A} = -E_r \frac{\partial C_{m,A}}{\partial r} \tag{4}$$

where, the radial dispersion coefficient, E_r , is assumed to be independent of $C_{m,A}$ but is a function of the solution velocity through the porous resin bed. Since the solution is injected at constant flow rate and the resin bed has constant voidage, the radial dispersion coefficient is a function of the resin bed radius only: $E_r = f(r)$. In cylindrical coordinates, the derivative of the radial dispersion flux is:

$$\frac{\partial}{\partial r}(J_{m,A}) = -\frac{1}{r}\frac{\partial}{\partial r}\left(r \, E_r \frac{\partial C_{m,A}}{\partial r}\right)$$

$$= -\left[E_r \frac{\partial^2 C_{m,A}}{\partial r^2} + \frac{E_r}{r}\frac{\partial C_{m,A}}{\partial r} + \frac{\partial E_r}{\partial r}\frac{\partial C_{m,A}}{\partial r}\right] \tag{5}$$

Substitution of equation (5) into (3) gives:

$$\frac{\partial C_{m,A}}{\partial t} + \left(\frac{1-e}{e}\right)\frac{\partial q_{m,A}}{\partial t}$$

$$= E_r \frac{\partial^2 C_{m,A}}{\partial r^2} + \frac{E_r}{r}\frac{\partial C_{m,A}}{\partial r} + \frac{\partial E_r}{\partial r}\frac{\partial C_{m,A}}{\partial r} - \frac{L}{e \, A_r}\frac{\partial C_{m,A}}{\partial r} \tag{6}$$

The concentration of A cations in the resin phase, $q_{m,A}$, may be written in terms of the resin phase mole fraction, y_m: $q_{m,A} = y_m (Q\rho_R / Z_A)$, where, Q is the resin exchange capacity [meq/g], ρ_R is the resin density [kg/m^3]. Similarly, the concentration of A cations in the solution phase, $C_{m,A}$, may be written in terms of the solution phase mole fraction, x_m: $C_{m,A} = x_m (C_0 / Z_A)$, where C_0 is the total normality in the solution. C_0 does not change when ionic species of differing valencies exchange with one another. Re-writing equation (6) in terms of mole fraction and dividing by (C_0 / Z_A) yields:

$$\frac{\partial x_m}{\partial t} + \left(\frac{1-e}{e}\right)\frac{Q \, \rho_R}{C_0}\frac{\partial y_m}{\partial t}$$

$$= E_r \frac{\partial^2 x_m}{\partial r^2} + \frac{E_r}{r}\frac{\partial x_m}{\partial r} + \frac{\partial E_r}{\partial r}\frac{\partial x_m}{\partial r} - \frac{L}{e \, A_r}\frac{\partial x_m}{\partial r} \tag{7}$$

To express equation (7) in the dimensionless form, let $r_D = r / r_0$ and $t_D = tL / e2\pi hr_0^2$, where, r_D and t_D are dimensionless distance and time, respectively. The Péclet number for radial flow is defined as the ratio of bulk mass transfer to dispersive transfer in the radial direction:

$$Pe_r = \frac{U_{L,r}\ r}{E_r\ e} = \frac{L}{E_r\ e\ 2\ \pi\ h} \tag{8}$$

Multiplying through equation (7) by $(e2\pi hr_0^2 / L)$ yields:

$$\frac{\partial x_m}{\partial t_D} + \frac{(1-e)\ Q\ \rho_R}{e\ C_0}\ \frac{\partial y_m}{\partial t_D} = \frac{1}{Pe_r}\ \frac{\partial^2 x_m}{\partial r_D^2} + \left(\frac{1}{Pe_r} - 1\right)\frac{1}{r_D}\ \frac{\partial x_m}{\partial r_D}$$

$$+ \frac{1}{Pe_r}\ \frac{1}{E_r}\ r_0\ \frac{\partial E_r}{\partial r}\ \frac{\partial x_m}{\partial r_D} \tag{9}$$

The initial conditions at $t_D = 0$ are that the entire resin bed (i.e. $r_{D,I} \leq r_D \leq r_{D,O}$) is free of A cations in either phase; i.e. $y_m = 0$ and $x_m = 0$. The boundary conditions are that for $t_D > 0$, at the bed inlet (i.e. at $r_D = r_{D,I}$), the concentration of the injected solution remains constant C_0 (i.e. $x_m = 1$) and at the bed outlet (i.e. $r_D = r_{D,O}$) the gradient of the solution concentration is assumed to be constant (i.e. $\partial^2 x_m / \partial r^2 = 0$).

Equation (9) is the theoretical model equation for ion exchange performance in radial flow. It models ion exchange occurring with radial dispersion, both of which develop with time. The equation can be numerically solved by the finite difference technique. The numerical solutions give the prediction of either the concentration profile or concentration history of exchanging cations in both solution and resin phases.

There are several assumptions made in the model development: the resin bed is homogeneous and fixed; both fluid and resin beads compressibilities are negligible; fluid and resin bed are under isothermal conditions; molecular diffusion is negligible; dispersion occurs only in the radial direction; no resistances to the diffusion of exchanging ions exist in either the solution or the solid (resin) phase; and instantaneous pointwise ion exchange equilibrium is established throughout the resin bed.

2.2 Ion Exchange Equilibrium Model

The Mehablia model[7] is improved and then applied to predict ion exchange equilibrium. Non-idealities in both phases are considered by applying the Pitzer model[8] and the Wilson model[9] to calculate activity coefficients in the solution and solid resin phases, respectively and by accounting for the ion pair formation of electrolytes in the aqueous solution, i.e. only free ions are available for ion exchange. The thermodynamic equilibrium constant is calculated independently by the Argersinger approach[10].

For ion exchange as defined by equation (1), the thermodynamic equilibrium constants are defined as:

$$K_{AB} = \left(\frac{y_{m,A}^{Z_b}}{y_{m,B}^{Z_a}} \right) \left(\frac{C_{m,B}^{Z_a}}{C_{m,A}^{Z_b}} \right) \left(\frac{\gamma_B^{Z_a}}{\gamma_A^{Z_b}} \right) \left(\frac{\gamma_{RA}^{Z_b}}{\gamma_{RB}^{Z_a}} \right) \equiv K_{m,AB} \left(\frac{\gamma_{RA}^{Z_b}}{\gamma_{RB}^{Z_a}} \right) \qquad (10)$$

where $K_{m,AB}$ is the experimentally accessible equilibrium quotient. The solution phase activity coefficients, γ_i, are calculated by the Pitzer model using free ion concentrations. The solid (resin) phase activity coefficients, γ_{Ri}, are calculated using the Wilson model.

The relationship between the thermodynamic equilibrium constant and the equilibrium quotient is provided by Argersinger's approach:

$$\ln K_{AB} = \int_0^1 \ln K_{m,AB} \, \partial \, y_A \qquad (11)$$

where y_A is the ionic equivalent fraction of A cations in the solid resin phase.

By experiment, the equilibrium concentrations of both phases can be determined and thus the thermodynamic equilibrium constant and equilibrium quotient can be calculated. The known thermodynamic equilibrium constant allows the Wilson binary interaction parameters, Λ_{AB} and Λ_B, to be estimated. In this study, the ion exchange equilibrium experiments are conducted separately to obtain the equilibrium parameters for the systems of interest: K_{AB}, Λ_{AB} and Λ_B.

2.3 Radial Dispersion Model

Radial dispersion may significantly affect the ion exchange equilibrium conditions existing between the solution and solid (resin) phases. It reduces the driving force for mass transfer and eventually diminishes the ion exchange performance of the resin bed. The model for pure radial dispersion without any mass transfer can be obtained by simplifying equation (9) taking the packed bed to be chemical-inert:

$$\frac{\partial x_m}{\partial t_D} = \frac{1}{Pe_r} \frac{\partial^2 x_m}{\partial r_D^2} + \left(\frac{1}{Pe_r} - 1 \right) \frac{1}{r_D} \frac{\partial x_m}{\partial r_D} + \frac{1}{Pe_r} \frac{1}{E_r} r_0 \frac{\partial E_r}{\partial r} \frac{\partial x_m}{\partial r_D} \qquad (12)$$

The initial conditions at $t_D = 0$ are that the entire resin bed (i.e. $r_{D,I} \leq r_D \leq r_{D,O}$) is free of tracer; i.e. $x_m = 0$. The boundary conditions are that for $t_D > 0$, at the bed inlet (i.e. at $r_D = r_{D,I}$), the concentration of the injected tracer remains constant, C_0 (i.e. $x_m = 1$) and at the bed outlet (i.e. $r_D = r_{D,O}$) the gradient of the tracer concentration is assumed to be constant (i.e. $\partial^2 x_m / \partial r^2 = 0$).

Provided the form of radial dispersion coefficient function is known, equation (12) can be numerically solved. In this study, radial dispersion experiments are separately carried out in the resin bed. The radial dispersion within the bed is assumed to be directly proportional to the fluid velocity within the pores of the bed.

3 EXPERIMENTAL DESCRIPTION

The radial ion exchange apparatus consists of a wedge-shaped ion exchange resin bed. The wedge is 15 mm thick with 30° of arc, and has a 300 mm outer radius and a 30 mm inner radius. Curved sintered glass fitted at both ends serves to act as a solution distributor and a bed support to keep the resin in place. To ensure that the fluid is produced at a rate uniform around the bed's outer arc, an eleven channel peristaltic pump is connected to the outer end of the bed. Each of the pump's eleven inlets are equally spaced around the bed's outer arc. The solution is drawn into the bed through the bed's inner face.

The resin used is the commercially available 20-50 mesh cation exchange resin, DOWEX MSC-1, the macroporous polystyrene-divinylbenzene resin with active sulphonate groups. The electrolyte solutions are prepared by analytical-grade chemicals and distilled-deionized water. All the solutions contain chloride as the only anion species and are de-aerated before use. The effluent of the resin bed is collected at the bed outer face and the concentration of cations in the effluent is analysed by titration. The bed porosity varied between 31.0 % and 33.0 % depending upon the form of the resin.

The exchanger performance of the sodium-hydrogen binary system was studied experimentally and by using the model. Initially the bed was in the hydrogen form. A 0.5 M solution of NaCl was injected into the bed at a rate of 390 mL/hour.

Four experiments were also performed to study the dispersion within the radial bed. In the first two experiment the resin bed was initially completely in the hydrogen form. After being rinsed with de-ionized, distilled water a 0.1 M HCl solution was then used as a tracer, being injected at a rate of 408 mL/hour. At the outlet the effluent is collected and analysed by titration. This experiment was performed twice. The dispersion within the bed in the sodium form was performed in a similar manner in two experiments. A 0.1 M solution of NaOH was used as the tracer injected at a rate of 410 mL/hour. Again analysis was by titration.

4 DISCUSSION OF RESULTS

The results of the four tracer experiments show that dispersion within the radial model may be adequately represented by assuming it to be proportional to the fluid pore velocity within the bed. As the velocity decreases as the fluid moves away from the bed axis, so the level of dispersion varies within the bed. At the outer end of the bed the fluid is passing through the resin bed in laminar flow. Figures 1 and 2 show the variations in tracer concentrations in the effluent with volume produced. Figures 1 and 2 relate to the bed in the hydrogen and sodium forms respectively.

Using the radial dispersion model described in Section 2.3 the two breakthrough curves were fitted. For the resin in the hydrogen form the constant of proportionality between the dispersion coefficient and the velocity was found to be 0.008 m, while for the sodium form it was found to be 0.005 m. These differences arise due to resin swelling. The radial bed was packed with the same amount of resin for each of the four tracer experiments. As the resin swells when it is transformed from the sodium to the hydrogen form, the porosity of the bed changes. This in turn leads to a change in the level of the dispersion within the bed.

The proposed radial model is used to predict the ion exchange performance of the wedge bed. Initially the bed was in the hydrogen form. A 0.5 M solution of NaCl was

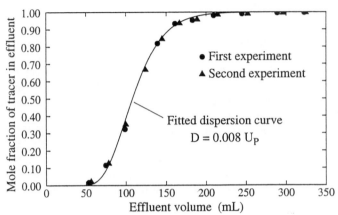

Figure 1 *Break through curve showing variation in effluent tracer concentration. Resin bed in hydrogen form*

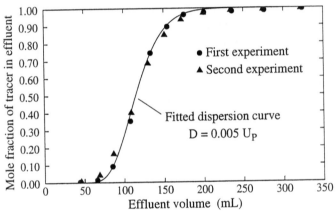

Figure 2 *Break through curve showing variation in effluent tracer concentration. Resin bed in sodium form*

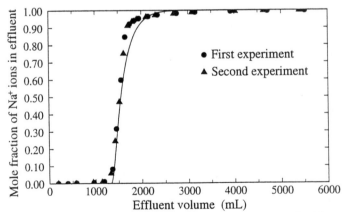

Figure 3 *Break through curve showing effluent Na^+ concentration for Na^+ - H^+ binary exchange in radial bed*

then injected through the inner face. Figure 3 shows the mole fraction of the sodium ions in the effluent following injection of the NaCl solution. Two identical experiments were performed and the observations clearly demonstrate the reproducibility of the results.

The model considers the change in the extent in swelling of the resin by allowing the porosity and the dispersion to vary with the resin form. Porosity varies between 31.0 % in the hydrogen form to 33.0 % in the sodium form. Dispersion is also assumed to vary with both velocity and the form of the resin.

The predicted break through curve made by the model is presented in Figure 3. As may be seen, agreement is good. It is believed that the disagreement between the values at about 1700 mL arises due to the dispersion model used. Further work is presently under way to study more closely the dispersive processes occurring within the model.

5 CONCLUDING REMARKS

A model has been developed to predict the performance of a radial ion exchange bed during binary cation exchange. Agreement between the experimentally-determined and predicted values is good, with dispersion determined independently.

References

1. F. Helfferich, "Ion Exchange", McGraw-Hill, New York, 1962, Chapter 9, p. 421.
2. K. Dorfner, "Ion Exchangers", Walter de Gruyter Berlin, New York, 1991, Chapter 2, p. 677.
3. M. J. Slater, "Principles of Ion Exchange Technology", Butterworth-Heinemann, Oxford, 1991, Chapter 7, p. 54.
4. V. V. Rachinskii, *J.Chromat.*, 1968, **33**, 234.
5. R. G. Rice, *Chem. Eng. Sci.*, 1982, **37**, 83.
6. R. G. Rice and B. K. Heft, *A. I. Ch. E. J.*, 1991, **37**, 629.
7. M. A. Mehablia, D. C. Shallcross and G. W. Stevens, *Chem. Eng. Sci.*, 1994, **49**, 2277.
8. K. S. Pitzer, "Activity Coefficients in Electrolyte Solutions", CRC Press, Florida, 1991, Chapter 3, p. 75; Chapter 6, p. 279; Chapter 7, p. 435.
9. G. M. Wilson, *J. Am. Chem. Soc.,* 1964, **86**, 127.
10. W. J. Argersinger, A. W. Davidson and O. D. Bonner, *Trans. Kansas. Acad. Sci.,* 1950, **53**, 404.

THERMODYNAMICS AND THEORETICAL CALCULATION OF ION EXCHANGE EQUILIBRIA IN ZEOLITES AND SOME ION EXCHANGE RESINS

A. M. Tolmachev and E. M. Kuznetsova

Department of Chemistry
The Moscow State University
Russia

1 INTRODUCTION

Methods for the thermodynamic description of ion exchange equilibria are widely considered in the literature. In the present report, this problem will be discussed within the framework of general theory of adsorption equilibria using full chemical potentials of ions by various choice of standard states of components in solution and ion exchanger[1-3].

In connection with the development of a new theory of strong electrolyte solutions[4,5], which can be (in contrast to the Debye-Huckel and other theories) the basis for calculating ion activity coefficients in individual and mixed electrolyte solutions in wide area of compositions and concentrations, it was interesting to consider the process of ion exchange as a distribution of ions between two solutions of various concentrations. Methods for the theoretical calculation of ion exchange equilibria in organic resins and zeolites developed on this basis will be submitted and discussed.

2 THERMODYNAMICS OF ION EXCHANGE

Processes of interphase (solution-exchanger) exchange of ions can be shown as:

$$v_A A^{z_A} + v_B \overline{B}^{z_B} = v_A \overline{A}^{z_A} + v_B B^{z_B} \tag{1}$$

where the exchange of ions neutralizing ionogenic groups of exchanger, is in general accompanied by adsorption of electrolytes and solvent and

$$v_i = f(z_A, z_B, c_A, c_B, P, T) \tag{2}$$

can be described on the basis of equality of full ion chemical potentials. This takes into account the work of ion transfer, connected with existence of electrostatic fields (ψ) in phases and overcoming the internal pressure forces of the exchanger matrix (σ).
At P, T = const:

$$\mu_i = \mu_i^0 + RT \ln m_i \gamma_i + z_i F(\psi - \psi_i^0) = \overline{\mu}_i =$$
$$= \overline{\mu}_i^0 + RT \ln \overline{m}_i \overline{\gamma}_i + z_i F(\overline{\psi} - \overline{\psi}_i^0) - \overline{v}_i(\sigma - \sigma_i^0) \tag{3}$$

Here and below

$$m_i = \frac{c_i}{(c_i\gamma_i)_{st.}}, \overline{m}_i = \frac{\overline{c}_i}{(\overline{c}_i\overline{\gamma}_i)_{st.}};$$

$c_i, c_{i,st.}, \overline{c}_i, \overline{c}_{i,st.}, \gamma_i, \gamma_{i,st.}, \overline{\gamma}_i, \overline{\gamma}_{i,st.}, \psi, \psi_i^0, \overline{\psi}, \overline{\psi}_i^0$ - molarities, activity coefficients, electrostatic potentials in solution and exchanger at equilibrium and in standard states for each component, $\overline{v}_i$ - partial molar volumes of hydrated ions in exchanger, σ, σ^0 - internal pressure (swelling pressure) of the exchanger at equilibrium and in standard states.

The expressions for the thermodynamic equilibrium constants ($\overline{K}$) of process (1) and corresponding thermodynamic functions can be obtained by combination of equations (3) for two ions. It is important to emphasize that for any type of function (2) all equations contain the terms z_i (valencies) and adsorption of electrolytes and solvent are taken into account in the values of concentrations:

$$-\Delta G^0 = RT\ln\overline{K} = \frac{1}{z_B}(\overline{\mu}_B^0 - \mu_B^0) - \frac{1}{z_A}(\overline{\mu}_A^0 - \mu_A^0) +$$

$$+ F(\overline{\psi}_A^0 - \overline{\psi}_B^0 + \psi_B^0 - \psi_A^0) + \frac{\overline{v}_B}{z_B}\sigma_B^0 - \frac{\overline{v}_A}{z_A}\sigma_A^0 = \qquad (4)$$

$$= RT\ln\frac{(\overline{m}_A\overline{\gamma}_A)^{1/z_A}(m_B\gamma_B)^{1/z_B}}{(m_A\gamma_A)^{1/z_A}(\overline{m}_B\overline{\gamma}_B)^{1/z_B}} + \sigma(\frac{\overline{v}_B}{z_B} - \frac{\overline{v}_A}{z_A})$$

Thus, there is minimum unknown information on interactions in equilibrium phases ($\gamma_i, \overline{\gamma}_i$), but the first member in the last of equalities (4) changes with the change of equilibrium compositions of phases due to change of σ. Only in the case of "hard" exchangers is transition of ions in the exchange process accompanied by the process of its partial dehydration, so that:

$$\frac{\overline{v}_B}{z_B} = \frac{\overline{v}_A}{z_A} = const \qquad (5)$$

It is expressed, in particular, in the impossibility of complete replacement of one ion with another in zeolites[6]. As the data necessary for calculation of the last member in (4) as a rule are unavailable, this unknown information is often included in the following value:

$$\overline{\gamma}_i^* = \overline{\gamma}_i\exp(-\overline{v}_i\sigma/RT) \qquad (6)$$

The values of $\overline{K}$, corresponding thermodynamic functions and $\overline{\gamma}_i$ can be obtained on the basis of experimental data by combination of (4) and Gibbs-Duhem equation for the exchanger phase. The simple solution can be obtained[7] only for "hard" exchangers, because if (5) is correct:

$$\overline{c}_Ad\ln\overline{\gamma}_A + \overline{c}_Bd\ln\overline{\gamma}_B + d(\overline{c}_A + \overline{c}_B) = 0 \qquad (7)$$

but in general, the information about concentration and activity coefficient of solvent in exchangers and values of $\bar{v}_i, \sigma$ are necessary[2,7]. The corresponding equations are well known[3,7-9]. We will briefly discuss three basic problems arising from practical usage of (4) and equations resulting from it.

Following the IUPAC recommendations for thermodynamic analysis of ion exchange equilibria it is necessary to use the following concentration scales: molarity, molality or full mole fraction of the component in phases. Usage of equivalent concentration, equivalent or non-full mole fraction is incorrect from the thermodynamic point of view. This circumstance has essentially lowered a value of detailed research[8] and many subsequent thermodynamic calculations based on it.

If the ion exchange process is accompanied by solvent and electrolyte adsorption, the determination of adsorption concentration can not be made precisely, as the measured change of the solution concentration allows to calculate only the "excess" values of adsorption. To calculate full (absolute) concentration in the exchanger used in all equations of a type (4), it is necessary to know $\bar{v}_i$ and to choose the interface location. The last problem can not be precisely solved by thermodynamical methods. Analysis of this problem for the processes of ion exchange is still inaccessible. The corresponding literature and appropriate discussion with reference to the processes of molecular adsorption are indicated in References 1 and 10.

The last problem is connected with the choice of component standard states in equilibrium phases. We would like to emphasize, that traditionally used standard states of components

$$(c_i \gamma_i)_{st.} = 1, \quad (\bar{c}_i \bar{\gamma}_i)_{st.} = 1 \tag{8}$$

are "hypothetical" and are not obtained in practice. The corresponding values of $\bar{K}$ and standard thermodynamic functions contain unknown standard parts of chemical potentials (see (4)), that makes their thermodynamic interpretation and development of methods of theoretical calculation of corresponding equilibria difficult. From our point of view it is more convenient to choose physically real and equilibrium standard states of components in phases, for example, saturated solutions of each electrolyte ($c_{i,s}$) and monoionic forms of exchangers equilibrium with them ($\bar{c}_i^*$):

$$(c_i \gamma_i)_{st.} = c_{i,s} \gamma_{i,s} \neq 1, \quad (\bar{c}_i \bar{\gamma}_i)_{st.} = (\bar{c}_i^* \bar{\gamma}_i^*) \neq 1 \tag{9}$$

In this case $\mu_A^0 = \bar{\mu}_A^0$, $\mu_B^0 = \bar{\mu}_B^0$ in (4) and the values of $\bar{K}$ and standard thermodynamic functions are determined by change of standard Gibbs free energy of exchanger (for transition from one of its monoionic form to another) and solvent (for transition from one saturated solution of electrolyte to another), that essentially simplifies their interpretation and development of equilibrium calculation methods[2,3]. The obvious advantages of such choice of standard states were demonstrated by example of adsorption equilibria calculations[1,12].

From (4) it follows that:

$$K = \frac{\overline{m}_A^{1/z_A} \overline{m}_B^{1/z_B}}{m_A^{1/z_A} m_B^{1/z_B}} = \overline{K} \frac{\gamma_A^{1/z_A} \gamma_B^{1/z_B}}{\overline{\gamma}_A^{1/z_A} \overline{\gamma}_B^{1/z_B}}$$ (10)

and the theoretical calculation of ion exchange equilibria (*i.e.* values of the equilibrium coefficient K and equilibrium concentrations of components in phases as functions of the ionic composition of the exchanger, *e.g.*, an equivalent fraction of i-th ion $\overline{N}_i$) at the strict thermodynamic approach is connected with the necessity of calculation of the thermodynamic constant and ion activity coefficients in equilibrium phases. Due to the fact that at the present time there are no satisfactory methods to calculate $\overline{K}$, this strict variant can not be used practically.

In this connection we will briefly consider a thermodynamic model, in which the ion exchange system is interpreted as two solutions of strong electrolytes of various concentration. It is obvious, that by strict consideration we will obtain result similar to (4, 10). We will, therefore, conduct an approximate analysis based on the following assumptions:

$$\mu_i^0 = \overline{\mu}_i^0, \psi_i^0 = \overline{\psi}_i^0, (c_i\gamma_i)_{st.} = (\overline{c}_i\overline{\gamma}_i)_{st.}, c_{i,st.} = \overline{c}_{i,st.} = 1$$ (11)

We will also consider, that at a combination of two equations (3):

$$\frac{\overline{v}_B}{z_B}(\sigma - \sigma_B^0) - \frac{\overline{v}_A}{z_A}(\sigma - \sigma_A^0) \approx 0$$ (12)

Assumptions (11, 12) mean that the standard characteristics of electrolytes in two solutions with different co-ions are accepted to be identical and that the influence of swelling pressure on ion exchange equilibria is negligible. It is important to emphasize, that according to this model the ion activity coefficients in two phases should be calculated on the basis of the theory of electrolyte solutions using $\gamma_{i,\pm}$, $\overline{\gamma}_{i,\pm}$ and reference states: $\gamma_i, \overline{\gamma}_i \to 1$ at $c_i, \overline{c}_i \to 0$. Since it is impossible to evaluate theoretically the possibility of using these assumptions, the equation obtained from (3) in this case (with $\overline{K} = 1$) is subject to experimental check up:

$$(K)_{\overline{N}_i} = \frac{\overline{c}_A^{1/z_A} c_B^{1/z_B}}{c_A^{1/z_A} \overline{c}_B^{1/z_B}} = \left(\frac{\overline{\gamma}_B^{1/z_B} \overline{\gamma}_R}{\overline{\gamma}_A^{1/z_A} \overline{\gamma}_R}\right)_{\overline{N}_i} \left(\frac{\gamma_A^{1/z_A}\gamma_X^{1/z_X}}{\gamma_B^{1/z_B}\gamma_X^{1/z_X}}\right)_{N_i} = \Delta_{\overline{N}_i}\delta_{N_i}$$ (13)

where: $(K)_{\overline{N}_i}$ - equilibrium coefficient at $\overline{N}_i$, $\Delta_{\overline{N}_i}, \delta_{N_i}$ - the corresponding ratios of activity coefficients in parentheses, N_i - equivalent fraction of i-th ion in solution.

Until recently it was impossible to check (13) up due to the lack of satisfactory ways of calculation of electrolyte activity coefficients in a wide range of concentrations. Such opportunity appeared only after the new theory of strong electrolyte solutions had been developed by Kuznetsova[4,5].

3 THEORETICAL CALCULATION OF ION EXCHANGE EQUILIBRIA

The new theory of strong electrolyte solutions has been advanced with the help of assumptions accepted by Debye and Huckel, but with the usage of modern quasi-crystalline model of an electrolyte solution. The calculation of ion interactions was carried out with the usage of radial distribution function and pair potential, which was calculated taking into account ion polarization and using crystallographic ion radii. Equations obtained have permitted calculation of the electrolyte activity coefficients to be made for individual and mixed solutions in various solvents in a wide range of concentrations (up to saturated solutions). That was proved in a series of publications (see References in 4, 5).

It has been of doubtless interest to check up (13) using the new theory and considering the exchanger as a concentrated binary solution.

After substitution of corresponding equations[4,13] in (13) and algebraic transformations we obtain, *e.g.*, for exchange of 1,1-electrolytes at 298K ($\delta_{N_i} = 1$):

$$(\ln K)_{\overline{N_A}} = \frac{0.7497 E\beta}{(\overline{x}_A + \beta \overline{x}_B)^2 \overline{\varepsilon}} \left[(d_{0,BR}^2 - d_{0,AR}^2)(1 + \frac{\theta}{2d^3}) - \frac{\theta_{BR} - \theta_{AR}}{d} \right] +$$

$$+ \frac{0.0798 E^{4/3} \beta}{(\overline{x}_A + \beta \overline{x}_B)^2 \overline{\varepsilon}} (\theta_{BR} - \theta_{AR}) \tag{14}$$

where: $E, \overline{\varepsilon}, \overline{x}_i, r_i, r_R, \alpha_i$ – ion exchange capacity (meq.mL^{-1}) and dielectric permeability of exchanger, mole fractions and radii of exchanged ions, radius of co-ion and polarizabilities of ions,

$$\beta = \left(\frac{r_A + r_R}{r_B + r_R} \right)^3, \quad d^2 = \sum d_{0,iR}^2 y_i, \quad y_A = \frac{x_A}{x_A + \beta x_B}, \quad y_B = \frac{\beta x_B}{x_A + \beta x_B},$$

$$d_{0,iR} = (r_i + r_R)(1 + \frac{0.013546 r_R \overline{\varepsilon}}{r_i} \ln \frac{1.327}{r_i}), \quad \theta = \sum \theta_i y_i, \quad \theta_i = \frac{\alpha_i z_R}{z_i} + \frac{\alpha_R z_i}{z_R} \tag{15}$$

The transition to 2, 2- and other symmetric electrolytes will only change numerical constants in (14). In case of asymmetric electrolytes the equation (14) will be more complicated[15] because of the difference between z_A, z_B; $\overline{\gamma}_{A\pm}, \overline{\gamma}_{B\pm}$; $\overline{N}_i, \overline{x}_i$.

If $\delta_{N_i} = const \neq 1$, it can be calculated using (14) and corresponding parameters for solution. If δ_{N_i} depends on N_i, it also can be calculated using the method of consecutive approximations.

As it follows from (14) and (15), all parameters included in (14) can be calculated from tabular data. There are only some problems when choosing values of E and $\overline{\varepsilon}$. Even in the case of "hard" exchanger, when the adsorption of electrolyte can be neglected, data on the specific weight of the exchanger is necessary. In the case of swelling exchangers there is a problem of calculating the change of exchanger volume with the change of ion composition. In literature there are only estimated values for $\overline{\varepsilon}$, the change of which due to change of an exchanger composition should also be taken into account.

The calculations which have been carried out showed, however, that for "hard" (zeolites) as well as for swelling exchangers the changes of E and $\bar{\varepsilon}$ probably compensate one another and they need not be taken into account. Thus, after choosing E the value of $\bar{\varepsilon}$ is specified using experimental data for one pair of ions on a given type of exchanger and then the calculations for all other electrolytes are carried out without using any additional experimental information, *i.e.* completely *a priori*. Since on the basis of literary data the empirical equation for dependence of $\bar{\varepsilon}$ on the contents of divinylbenzene (DVB, %) in organic resins was found[14], the value of $\bar{\varepsilon}$ can be specified only for one exchanger of such type.

The Tables given below contain our results (1, 4) as well as already published[13-15] ones (2,3). In all Tables, values of calculated (K_t) and experimental[6,14-21] (K_{ex}) equilibrium coefficients are compared taking into account δ for solution and concentration scales used in experimental work. Additional information is given in each table.

Table 1

Systems	Zeolites A, X,	$\bar{N}_i = 0.5$	Systems	KU-2, 7% DVB, $\bar{N}_i = 0.5$	
Ref 6	K_t	K_{ex}	Ref 14	K_t	K_{ex}
$Na^+ + KA$	0.72	0.74	$Na^+ + NH_4R$	0.66	0.69
$Li^+ + NaA$	0.11	0.10	$K^+ + NH_4R$	0.95	1.10
$Cs^+ + RbA$	0.34	0.34	$Na^+ + KR$	0.69	0.63
$K^+ + LiA$	13.5	12.7	$K^+ + LiR$	2.44	2.55
$Rb^+ + CsX$	0.54	0.55	$Rb^+ + KR$	1.06	1.04
$Br^- + FR$	Dowex 2, DVB (%), Reference 16						
$\bar{N}_i$	0.09	0.27	0.46	0.60	0.68	0.74	0.98	$\to 1$
K_t (4%)	11.0	12.3	14.6	17.2	18.9	20.6	30.1	31.2
K_{ex} (4%)	9.5	11.7	14.9	18.0	18.9	21.8	30.7	31.7
$\bar{N}_i$	0.05	0.09	0.17	0.25	0.57	0.75	0.95	$\to 1$
K_t (16%)	34.4	35.7	38.9	42.9	70.8	102	164	188
K_{ex} (16%)	26.4	28.4	32.0	40.8	74.4	108	157	180

Table 2

Systems	Dowex 50, (DVB %), Reference 17							
$K^+ + NaR$	4		8		12		16	
$\bar{N}_i$	K_t	K_{ex}	K_t	K_{ex}	K_t	K_{ex}	K_t	K_{ex}
$\to 0$	1.22	1.29	1.35	1.37	1.44	1.47	1.47	1.51
0.5	1.30	1.38	1.48	1.54	1.61	1.72	1.67	1.82
$\to 1$	1.44	1.49	1.72	1.71	1.92	1.95	2.02	2.12
$Cs^+ + NaR$	K_t	K_{ex}	K_t	K_{ex}	K_t	K_{ex}	K_t	K_{ex}
$\to 0$	1.22	1.27	1.35	1.16	1.43	1.10	1.46	1.10
0.5	1.39	1.58	1.62	1.69	1.78	1.81	1.82	1.86
$\to 1$	1.86	1.95	2.49	2.45	2.97	2.95	3.17	3.17

Table 3

Dowex 50, (DVB %), $\overline{N}_i = 0.5$						
References 18-20	4		8		16	
Systems	K_t	K_{ex}	K_t	K_{ex}	K_t	K_{ex}
$Ca^{2+} + CuR$	1.24	1.26	1.38	1.40	1.53	1.71
$Sr^{2+} + CuR$	1.49	1.45	1.82	1.72	2.20	2.30
$Ba^{2+} + CuR$	1.97	2.24	2.81	2.95	4.02	4.60
$Ca^{2+} + 2LiR$	3.94	4.14	5.12	5.16	6.25	7.27
$Sr^{2+} + 2LiR$	4.85	4.70	6.98	6.51	9.40	10.1
$Cu^{2+} + 2LiR$	3.15	3.29	3.65	3.85	4.01	4.46
$Ce^{3+} + 3LiR$	7.5	7.6	10.7	10.7	14.5	17.0

Table 4

Dowex 50 (12 % DVB), system K^+ - NH_4^+ - HR								
Reference 21			NH_4^+ - H^+		K^+ - H^+		K^+ - NH_4^+	
$\overline{N}_{NH_4}$	$\overline{N}_{Na}$	$\overline{N}_H$	K_t	K_{ex}	K_t	K_{ex}	K_t	K_{ex}
0.123	0.142	0.735	3.0	2.9	1.7	1.6	1.7	1.7
0.096	0.483	0.421	2.6	2.5	1.7	1.6	1.6	1.6
0.165	0.461	0.374	2.2	2.2	1.6	1.5	1.5	1.4
0.224	0.062	0.714	2.7	2.7	1.7	1.5	1.6	1.8
0.304	0.331	0.365	2.3	2.1	1.6	1.5	1.5	1.4
0.558	0.048	0.394	2.1	2.2	1.5	1.3	1.4	1.4

As can be seen from Tables 1-4, the equation (14) allows the calculation of the equilibrium coefficients of ion exchange with satisfactory accuracy. It means that assumptions of thermodynamical model in combination with approximations connected with the considering of ion exchangers as concentrated (3-15 meq.mL^{-1}) electrolyte solutions are mutually compensated. Clearing up of reasons leading to these results is still a subject to be analysed. It should be mentioned that in the case of "hard" exchangers (zeolites in particular) the dependence of $\overline{\gamma}_{i,\pm}$ on ion composition of exchangers has not been described very satisfactorily. It shows that in cases like these it is not enough to take into account only the increase of concentration in ion exchangers and probably the corrections of equations for $\overline{\gamma}_{i,\pm}$ will be required. From the practical point of view, the method developed may be convenient because it allows reduction in the volume of experimental research used under elaboration of the corresponding ion-exchanging processes, but for wide application it must be specified and checked up for different types of ion exchange systems.

References

1. A. M. Tolmachev, *Langmuir*, 1994, **7**, 1400.
2. A. M. Tolmachev, I. V. Baurova, *Vestn. Mosk. Univ., Khim.*, 1986, **27,** N 5, 465.
3. A. M. Tolmachev, I. V. Baurova, *Vestn. Mosk. Univ., Khim.*, 1986, **27,** N 6, 546.
4. E. M. Kuznetsova, G. M. Dakar, *Zh. Fiz. Khim.*, 1984, **58**, 2221.
5. E. M. Kuznetsova, *Zh. Fiz. Khim.*, 1993, **67**, 1765.
6. A. M. Tolmachev, V. A. Nikashina, N. F. Chelischev, "Ion Exchange", Moscow, 1981, p. 45.
7. A. M. Tolmachev, W. I. Gorshkov, *Zh. Fiz. Khim.*, 1966, **40**, 1924.
8. G. L. Gaines, H. C. Thomas, *J. Chem. Phys.*, 1953, **21**, 714.
9. V. S. Soldatov, "Simple Ion Exchange Equilibria", Nauka i Technika, Minsk, 1978.
10. A. M. Tolmachev, *Uspekhi Khimii*, 1981, **50**, 769.
11. A. M. Tolmachev, *Vestn. Mosk. Univ. Khim.*, 1994, **35**, 40.
12. A. M. Tolmachev, *Adsorption Sci. Technology*, 1993, **10**, 155.
13. E. M. Kuznetsova, *Zh. Fiz. Khim.*, 1994, **68**, 1278.
14. E. M. Kuznetsova, *Zh. Fiz. Khim.*, 1992, **66**, 2688.
15. E. M. Kuznetsova, *Zh. Fiz. Khim.*, 1994, **68**, 1283.
16. B. Sildano, D. Chesnut, *J Amer. Chem. Soc.,* 1955, **77**, N 5, 1334.
17. G. Mayer, G. Boyd, *J. Phys. Chem.*, 1956, **59**, 520.
18. O. D. Bonner, L. L. Smith, *J. Phys. Chem.*, 1957, **61**, 327.
19. O. D. Bonner, F. L. Livingston, *J. Phys. Chem.*, 1956, **60**, 530.
20. O. D. Bonner, S. F. Jumber, *J. Phys. Chem.*, 1958, **62**, 250.
21. V. S. Soldatov, V. A. Bichkova, "Ion Exchange Equilibria in Multicomponent Systems", Nauka i Tekhnika, Minsk, 1978.

MULTICOMPONENT ION EXCHANGE EQUILIBRIA OF WEAK ELECTROLYTE BIOMOLECULES

L. A. M. van der Wielen, M. L. Jansen and K. Ch. A. M. Luyben

Department of Biochemical Engineering
Delft University of Technology
Julianalaan 67
2628 BC Delft
The Netherlands

1 INTRODUCTION

Ion exchange is one of the powerful working horses in the recovery and purification of biotechnological products, such as antibiotics, amino acids and proteins. Rational design of bioseparation processes involving ion exchange, requires the reliable description of governing equilibria. The conventional description of multicomponent ion exchange equilibria[1] is based on stoichiometric, constant selectivity models and complete exclusion of co-ions from the resin matrix. Also, uptake mechanisms other than ion exchange are usually not accounted for explicitly.

Hence, the conventional description may not cover all occurring phenomena. Experimental deviations from constant selectivity models are observed as a rule, not as an exception[2]. For instance, for ion exchange of amino acids[3,4] and peptides[2] on ion exchange resins, a selectivity decline or even a selectivity reversal is observed. This could be attributed to a combination of steric effects such as partial volume exclusion of bulky biomolecules and heterogeneity of the resin matrix[5].

Secondly, most bioproducts are weak (poly) electrolytes and may be present in multiple ionic forms (ions, zwitterions and neutral component) depending on pH, concentration levels of the solutes and ionic strength. The overall selectivity of the resin depends also on sorption mechanisms, other than ion exchange, which have to be considered for neutral and zwitterionic species[6]

Thirdly, the Donnan effect[7] is usually interpreted as complete exclusion of co-ions from the resin phase. This common assumption, however, leads to non-realistic calculated concentrations of H^+ and OH^- in the resin. Although usually not too relevant for strong electrolytes, this assumption would lead to erroneous predictions of overall resin selectivity for weak electrolytes and limits the application range of conventional models essentially to dilute concentrations. Recently, Kawakita and Matsuishi[8] have shown for an aqueous ion exchange system involving lysine and ammonia at high pH that incorporation of these effects resulted in an adequate prediction of the overall uptake for a broad range of conditions. This approach has been extended by Jansen et al.[9] for multicomponent equilibria of weak and strong electrolytes on anion exchange resins.

In this paper, we present a description of multicomponent ion exchange equilibria for weak electrolytes using a general thermodynamic framework based on the identification of the surface excess properties in the resin phase, and on an assumed distribution function of functional groups. The model is partially based on the works by Novosad and

Myers[10] and Myers and Byington[5]. In addition, osmotic pressure and Donnan effects are incorporated in the model, following Jansen *et al.*[9]. The multicomponent model requires binary exchange data only. We will demonstrate the usefulness of the model for some selected cases of multicomponent ion exchange of weak electrolytes.

2 MULTI-COMPONENT ION EXCHANGE OF WEAK ELECTROLYTES: MODEL

2.1 Thermodynamic Framework

Following Donnan[7], the equality of chemical potentials of a charged species i over a liquid phase L and an ion exchange phase R leads to the subsequent relation:

$$\ln \frac{a_{Ri}}{a_{Li}} = \frac{\Delta \mu_i^o}{RT} + \frac{v_i}{RT} \Delta P + \frac{z_i F}{RT} \Delta \phi \tag{1}$$

$\Delta\mu_i^o$, a_i, v_i, and z_i are the difference in standard chemical potential of liquid and resin phase, activity, partial molar volume and charge of the target species i respectively. The differences in (osmotic) pressure ΔP and the electric potential $\Delta\phi$ between the liquid and resin phase are of course identical for all species. According, to the works of Myers and Byington[5], the difference in standard state of the liquid and resin phase $\Delta\mu_i^o$ is a distributed property due to steric effects and heterogeneity of the matrix. Hence, $\Delta\mu_i^o$ can be described with a distribution function f^q for site type q as follows :

$$\Delta \mu_i^o = \Delta \overline{\mu}_i^o + f^q \sigma_i \tag{2}$$

where $\Delta\overline{\mu}_i^o$ and σ_i are the mean value and the standard deviation of the distribution function respectively. Following Myers and Byington, we have used a discrete binomial distribution function of $n+1$ site types. The fraction of site-type q is given by its probability p_q according to:

$$p_q = \binom{n}{q} p^q (1-p)^{n-1} \quad \text{and} \quad f^q = \frac{q-np}{\sqrt{np(1-p)}} \tag{3}$$

where p is the skewness of the distribution function. The solution procedure of this general framework comprises several steps: 1) elimination of the (unknown) terms for osmotic pressure term ΔP and electrostatic potential difference $\Delta\phi$, 2) substitution of the site distribution function and 3) closing the resulting set of non-linear equations.

In the following, we shortly outline the procedure from 1 through 3. The (osmotic) pressure effect is mainly due to the water uptake (swelling phenomena) of the resin (hydration of functional groups). For water (w) without charge effects, we find:

$$v_w \Delta P = \Delta \mu_w^o + RT \ln(a_{Lw}/a_{Rw}) \tag{4}$$

Eliminating the osmotic pressure from the equilibrium relation for species i with equation 2, the following explicit expression for the Donnan potential difference is obtained in terms of the activities and standard states of water and an arbitrary species i.

$$\Delta\phi = \frac{RT}{z_i F}\left(\ln\frac{a_{Ri}}{a_{Li}} - \frac{\Delta\mu_i^o}{RT} + \frac{v_i}{v_w}\frac{\Delta\mu_w^o}{RT} - \frac{v_i}{v_w}\ln\frac{a_{Rw}^{v_i/v_w}}{a_{Lw}^{v_i/v_w}}\right) \tag{5}$$

Elimination of the electrostatic potential by subtraction of equation 5 for species i and j results in the following '*stoichiometric*' exchange relation for site of type q:

$$\frac{a_{Ri}^{1/z_i}\,a_{Lj}^{1/z_j}}{a_{Rj}^{1/z_j}\,a_{Li}^{1/z_i}} = \overline{S}_{ij}\,W_{ij}^{fq}\,\pi_{ij} = S_{ij}^{(q)} \tag{6}$$

The average selectivity $\overline{S}_{i,j}$ is given by

$$\overline{S}_{ij} = \exp\left(\frac{\Delta\overline{\mu}_i^o - v_i/v_w\Delta\overline{\mu}_w^o}{z_i RT} - \frac{\Delta\overline{\mu}_j^o - v_j/v_w\Delta\overline{\mu}_w^o}{z_j RT}\right) \tag{7}$$

and parameters representing the variance of the distribution function W_{ij} and the osmotic effect π_{ij} are given by

$$W_{ij} = \exp\left(\frac{\sigma_i - v_i/v_w\sigma_w}{z_i RT} - \frac{\sigma_j - v_j/v_w\sigma_w}{z_j RT}\right) \text{ and } \pi_{ij} = \left(\frac{a_w^{(R)}}{a_w^{(L)}}\right)^{(v_i/z_i - v_j/z_j)/v_w} \tag{8}$$

These relations are rigorous and valid for any pair of charged species. They could be interpreted as stoichiometric ion exchange for counterions of the matrix and as combined partitioning relations for co-ions of the resin. For neutral and zwitterionic species, the partitioning is described following similar lines and only the final result is given:

$$\frac{a_{Ri}^{(q)}}{a_{Li}^{(q)}} = \overline{K}_i\,W_i^{fq}\,\pi_i = K_i^{(q)} \tag{9}$$

with

$$\overline{K}_i = \exp\left(\frac{\Delta\overline{\mu}_i^o - v_i/v_w\Delta\overline{\mu}_w^o}{RT}\right); \; W_i = \exp\left(\frac{\sigma_i - v_i/v_w\sigma_w}{RT}\right); \; \pi_i = \left(\frac{a_{Rw}}{a_{Lw}}\right)^{v_i/v_w} \tag{10}$$

Note that the (binary) parameters are identical for all sites q and that the distribution of the resin's properties is only reflected via the distribution function f^q.

For a known liquid phase composition and an unknown composition of the resin phase in case of N_a anionic species, N_c cationic species and N_o neutral or zwitterionic species, this results in $(N_a+N_c+N_o-2)$ non-linear equations. Closure of the set of equations is possible via the electroneutrality relation for the resin phase and the dissociation equilibrium of water into H^+ and OH^- to relate cations and anions. In general, the set of equations has to be solved numerically but a closed form solution is possible for strictly monovalent species.

2.2 Elaboration for Monovalent Ions

For monovalent ions and thermodynamically ideal solutions, the equations are conveniently rewritten in terms of ionic fractions relative to the concentration of free charges for the liquid phase (x) and to the resin's capacity for the matrix (y). Then the mole fraction of species i at site q, relative to a reference component r is given by:

$$y_i^{(q)} = S_{ir} x_i / \sum_k^N S_{kr}^{(q)} x_k \tag{11}$$

When a two-site, binomial distribution function is assumed for a binary system, the following closed form expression is obtained:

$$S_{12} = \overline{S\pi} \frac{\overline{S\pi} W^{U+V} x_1 + (W^U (1-p) + W^V p) x_2}{\overline{S\pi} (W^V (1-p) + W^U p) x_1 + x_2} \tag{12}$$

where $\overline{S}_{i,j}$ and σ_j are the average value and the standard deviation of the distribution of the adsorption energies, and p is the skewness of the site distribution function. A closed-form expression for S_{ij} in multicomponent systems is given by Saunders *et al.*[3]. For anion exchange resins, the ionic fraction of OH^- in the resin phase composition is given by

$$y_{OH^-} = x_{OH^-} \frac{1 + \sqrt{1 + 4\gamma^2 AC}}{2A} \tag{13}$$

$$\text{with } \gamma = C_T / Q \ , \ A = \sum_a x_a S_{aOH^-} \ , \ C = \sum_c x_c S_{cH^+} \tag{14}$$

Q is the ion exchange capacity of the resin and C_T is the anion concentration in the liquid phase. The H^+-content of the resin is calculated through the water dissociation, and the content of all other anionic (a) and cationic (c) species are calculated through equation (6), relative to OH^- and H^+ respectively. Neutral and zwitterionic species are calculated from equations (9-10).

3 EXPERIMENTAL CASE STUDIES

In order to demonstrate the capability of the new model to describe ion exchange equilibria of multicomponent mixtures of strong and weak electrolytes over a wide range of operating conditions, some case studies are presented.

3.1 Ion Exchange of Carboxylic and Acetyl Amino Acids

Firstly, we seek to demonstrate the capability of the model to adequately account for the uptake at high concentrations and over a broad range of pH-values. Hence, we measured equilibria of relatively small strong (Cl^-, OH^-) and weak (acetate and acetylmethionine) electrolytes, for whom steric effects are be negligible. The strong-base anion exchange resin was the acrylate-based Macro-Prep Q resin, obtained from Bio-Rad (Hercules, CA, USA). The capacity of the resin was relatively low ($Q = 217$ mol/m^3), to allow large C_T/Q without departing too much from the infinite dilution standard state for the liquid phase. The resin was prepared as described by Jansen *et al*[9] and ion exchange experiments were performed using the batch method. The liquid phase was analyzed by HPLC for acetic acid and acetylmethionine and by Ion Chromatography for Cl^- as described by Jansen *et al*.[11]. The experimental results were correlated with the model using a non-linear parameter estimation technique and the resulting parameters are

$$\overline{S}_{Ac^-,OH^-} = 0.025, \qquad \overline{S}_{Na^+,H^+} = 28.7, \qquad \overline{K}_{HAc} = 0.88, \qquad W_{i(j)} = 1 \text{ and } p = 0.5.$$

Figure 1 gives the parity plot of experimental and calculated (overall) acetate content of the resin phase. The model could predict the ion exchange equilibria ranging from the dilute region to strongly exceeding the resin capacity with an average deviation of 14 %. For comparison, the acetate content of the resin phase as calculated from the conventional, constant stoichiometric model is given in Figure 1 as well. It can be observed that both models converge in the dilute region, as expected. However, the conventional model strongly underestimates the overall resin uptake at concentrations exceeding the resin's capacity (Q) and at pH values at which substantial amounts of the neutral acetic acid are present. This fraction decreases with increasing pH. The resin is practically saturated with acetate, except at high pH due to competition with OH^-. Increasing the overall acetate concentrations leads to an increased uptake, which exceeds the resin capacity by a factor 2.5. Similar results have been obtained for the Cl^- and acetylmethionine uptake.

The importance of these phenomena can be demonstrated by chromatographic experiments. A column was packed with the Macro-Prep resin and fed with a constant overall acetate concentration of 100 mol/m^3 at pH 4.7. During 4 liquid residence times, the pH was lowered to 3.0. Figure 2 shows the experimental data. The column responded with an uptake of the (neutral) acetic acid at the step-down in pH and with a release of acetic acid during the step-up of the pH. The column dynamics have been simulated with a local equilibrium model based on this work[11,12] and the calculated responses, using the above data, are shown in Figure 2 as well. It should be noted that the 'shoulder' in the pH-profile at the step-up, which does not occur at the step-down, corresponds to the unretained motion of the co-ion of acetate (Na^+). Similar phenomena were described by Helfferich and Bennett[13].

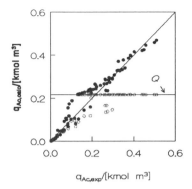

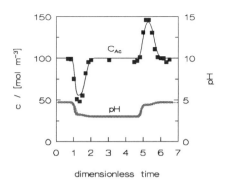

Figure 1 *Parity plot of experimental and calculated overall acetate uptake. Conventional model* (o) *and this work* (●)

Figure 2 *Experimental and calculated effluent composition of a Macro-Prep column. Feed described in text*

3.2 Anion Exchange of ß-lactam Antibiotics

Even relatively bulky biomolecules such as the ß-Lactam antibiotics penicillin G and 6-aminopenicillanic acid, or phenylacetic acid (PhAc) can be recovered by gel-type, strong anion exchange resins such as Amberlite IRA400[14]. Binary ion exchange equilibria of these components have been measured in batch experiments at 310 K relative to Cl^- (at pH 8[15]) and OH^-. The co-ion was Na^+ in all cases and the conditions were selected such that the uptake of anionic species dominated. Experimental detail can be found elsewhere[15]. Typical isotherms for penicillin G using Cl^- and OH^- as reference species are shown as Figure 3. The corresponding selectivities are shown in Figure 4. These results demonstrate the characteristic behavior of a high selectivity at a low ionic fraction and a substantial selectivity decline when increasing the ionic fraction of penicillin G. Similar data are obtained for 6-APA and PhAc. The experimental results were correlated with the model presented using a non-linear parameters estimation procedure. The binary parameters with respectively Cl^- and OH^- as reference species are given in Table 1. Calculated compositions and selectivities are shown as solid curves in Figures 3 and 4.

The selectivity decline at higher ionic fractions of the bulky solute is attributed to steric effects and resin heterogeneity. The distribution behavior at low ionic fractions can be related to the molecular interactions of solutes and resin's polymers as demonstrated by various other authors[2,4]. Hence, the infinite dilution selectivities can be correlated by a suitable measure for the solute-resin interaction. As the resin composition at infinite dilution of the solute (Pen G, 6-APA and PhAc) is practically constant within a series of constant reference ion, this interaction can be correlated with a single parameter, such as the hydrophobicity-based method outlined by Gude *et al.*[16].

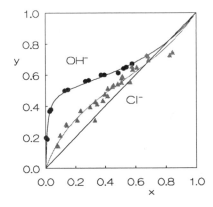

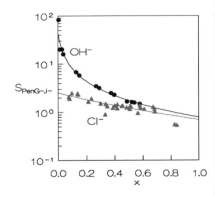

Figure 3 *Ion exchange isotherms for Penicillin G versus Cl⁻ (▲) and OH⁻ (●)*

Figure 4 *Selectivity of IRA400 for Penicillin G versus Cl⁻ (▲) and OH⁻ (●) as a function of liquid composition*

Table 1 *Equilibrium Parameters for the Exchange of X⁻/Cl⁻ and X⁻/OH⁻ on IRA400 at 310 K*

X⁻/Cl⁻	$\overline{S}$	W	p^1	S_o
Pen G	1.338	3.519	0.5	2.544
6-APA	0.307	1^1	0.5	0.307
PhAc	1.971	1^1	0.5	1.971
X⁻/OH⁻				
Pen G	5.843	14.738	0.5	43.3
6-APA	6.153	1^1	0.5	6.153
PhAc	12.41	1^1	0.5	12.41

¹   not fitted, value taken from Dye *et al.* (1990)

In this work, we have used the partitioning data in aqueous two-phase systems from Anderson *et al.*[17] as a hydrophobicity scale. Figure 5 shows the correlation between measures for the interaction in the solute-resin and solute-biphasic systems which are the logarithms of infinite dilution selectivity and partition coefficients respectively. This procedure allows for the estimation of binary parameters of unknown species once their relative hydrophobicities are established.

Another important feature of the model is the possibility to predict ion exchange equilibria of multicomponent mixtures. This is demonstrated in Figure 6 for the quaternary system PenG⁻, 6-APA⁻, PhAc⁻ and Cl⁻ in the form of a parity plot of calculated and experimental composition of the resin phase in batch experiments. The equivalent ionic fraction of the species was described with an average deviation of 30 % using binary data.

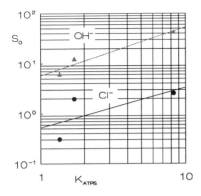

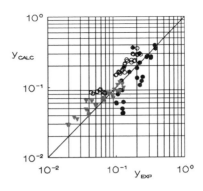

Figure 5 *Relation between infinite dilution selectivities for Cl⁻ (▲) and OH⁻ (●) as a function of hydrophobicity scale*

Figure 6 *Predicted and experimental resin phase composition for PenG (●), 6-APA (▲) and PhAc (o) in quaternary equilibria of Cl⁻-charged IRA400*

4 CONCLUSIONS

We have described a thermodynamic framework for multicomponent ion exchange systems with strong and weak electrolytes. The model account for a possible variation in exchange selectivities with system composition, which may occur due to steric effects and resin heterogeneity. The model is applied successfully at batch and column experiments of multicomponent systems containing relatively small strong and weak electrolytes. Stimulated by its success, the model will be used as a basis for more complex situations such as in the description of ion exchange chromatography of proteins.

Acknowledgements

Gist-brocades N.V. is acknowledged for the supply of penicillin G and DSM Research and the Netherlands Organization for Scientific Research (NWO) for the financial support. Technical support by B. L. Verlaan and M. J. A. Lankveld as well as analytical support by C. Ras and H. Corstjens is greatfully acknowledged.

Notation

a	activity
c	concentration in liquid phase
f	distribution function
K	partition coefficient
n	number of charges
P	pressure
q	resin phase concentration
Q	ion exchange capacity
R	gas constant
S	selectivity
T	temperature
U,V	defined in (13)
W	defined in eqns (8) and (10)
x,y	ionic fraction

Greek

ϕ	electrostatic potential
μ	chemical potential
v	(partial) molar volume
π	osmotic parameter
σ	variance

References

1. F. G. Helfferich, "Ion Exchange", McGraw-Hill, New York, 1962.
2. I. L. Jones and G. Carta, *Ind. Eng. Chem. Res.*, 1993, **32**, 107.
3. M. S. Saunders, J. B. Vierow and G. Carta, *A.I.Ch.E. J.*, 1989, **35(1)**, 53.
4. S. R. Dye, J. P. DeCarli II and G. Carta, *Ind. Eng. Chem. Res.*, 1990, **29**, 849.
5. A. L. Myers and S. Byington, "Ion Exchange: Science and Technology", NATO ASI Series Appl. Sci. E 107, 1986, p. 119.
6. S. Peterson and R. W. Jeffers, *J. Am. Chem. Soc.*, 1952, **74**, 1605.
7. F. G. Donnan, "The Theory of Membrane Equilibria", *Chem. Rev.*, 1925, **1**, 73.
8. T. Kawakita and T. Matsuishi, *Sep. Science Technol.*, 1991, **26(7)**, 991.
9. M. L. Jansen, A. J. J. Straathof, L. A. M. van der Wielen, K. Ch. A. M. Luyben and W. J. J. van den Tweel, *Submitted A. I. Ch. E. J.*, 1995a.
10. J. Novosad and A. L. Myers, *Can. J. Chem. Eng.*, 1982, **50**, 500.
11. M. L. Jansen, G. W. Hofland, J. Houwers, A. J. J. Straathof, L. A. M. van der Wielen, K. Ch. A. M. Luyben and W. J. J. van den Tweel, *Submitted A. I. Ch. E. J.*, 1995b.
12. G. Guiochon, S. Golshan-Shirazi and A. M. Katti, "Fundamentals of Preparative and Non-linear Chromatography", Academic Press, Boston, 1994.
13. F. G. Helfferich and B. J. Bennett, *Reactive Polymers,* 1984, **3**, 51.
14. L. A. M. van der Wielen, P. J. Diepen, A. J. J. Straathof and K. Ch. A. M. Luyben, *Annals N.Y. Acad. Sciences,* 1995a, **750**, 482.
15. L. A. M. van der Wielen, M. J. A. Lankveld and K. Ch. A. M. Luyben, *accepted J. Chem. Engng. Data*, 1995b.
16. M. T. Gude, L. A. M. van der Wielen and K. Ch. A. M. Luyben, *accepted Fluid Phase Equil.*, 1995.
17. E. Anderson, B. Mattiasson and B. Hahn-Hägerdahl, *Enz. Microb. Technol.*, 1984, **6**, 301.

REGENERATION OF ION EXCHANGE COMPLEXING RESINS IN A CSTR ADSORBER

L. M. Ferreira
Chemical Engineering Department, University of Coimbra
3000 Coimbra, Portugal

J. M. Loureiro and A. E. Rodrigues
Lab. of Separation and Reaction Engineering, School of Engineering, University of Porto
4099 Porto Codex, Portugal

1 INTRODUCTION

Ion exchange using complexing resins is now considered to be one of the most promising methods for the recovery of valuable metals. One of the best examples is the recovery of uranium from sea water.[1,2,3] Further examples, with economic interest, are the recovery of metals such as gold, platinum, silver, copper, zinc, chromium, etc., by treating electroplating wastes, hydrometallurgical liquors and waste mine waters.[4]

From the point of view of kinetics, in the majority of the systems particle diffusion is found to be the rate-controlling step. Kinetic models with intraparticle rate control have been summarized by Petruzzelli et al[5]. These models, based on Fick or Nernst-Planck equations, allow a good interpretation of rate data in simple systems, involving a purely electrostatic exchange of ions. In particular, systems where reactions between resin active sites and counter-ions occur, such as complex formation by using complexing resins, a reaction term in the material balance must be included and so, the interpretation of rate of ion exchange becomes more complex.

A theoretical analysis of various ion exchange processes involving ionic reactions has been given by Helfferich.[6] For systems, in which the counter-ion is consumed by an irreversible reaction, the conversion of the resin proceeds with an outer fully converted "shell" and an inner core still in its initial state separated by a front that moves from the particle surface to the center (shrinking core process or shell-progressive mechanism). Most of these systems have been described using the pseudo-steady-state diffusion assumption.[7,8,9] In this work, a front reaction model considering non-steady state diffusion was used to describe the regeneration process of a complexing resin.

The study of the regeneration step of the resin loaded with metal is essential when designing the overall process. In the sequence of previous works,[10,11] the main goal of this study was to better understand the mechanism of the regeneration process of a chelating resin Duolite ES 346 (with amidoxime functional groups) loaded with copper, zinc or lead, by using nitric acid. Here the system copper nitrate/Duolite ES 346 is used as an example.

Dynamic experiments in a CSTR adsorber were done in order to obtain effective intraparticle diffusivities. Results were well represented by the reaction front model which will be used later in the modelling of the ion exchange operation in a fixed bed.

2 EXPERIMENTAL

Continuous experiments were run in a perfectly mixed cylindrical basket adsorber (of Carberry type). In the loading step, after previous conditioning of the resin, a 2 g/L copper nitrate solution was fed until saturation of the resin was complete. In the regeneration step, nitric acid ($\approx$ 1 M) was used to remove and concentrate the metal. The metal (Cu^{++}) and hydrogen ion (H^+) concentrations were measured by atomic absorption spectrophotometry and potentiometric titration, respectively.

3 MODELLING

A reaction front model was developed to interpret the experimental results, based on the following assumptions:

1) Two equilibrium reactions between the species (metal and hydrogen ion) and the amidoxime group take place in a front which moves towards the center of the particle;

$$2\,H^+ + R_2M^= \rightleftharpoons M^{++} + 2\,RH^-$$

$$H^+ + RH^- \rightleftharpoons RH_2$$

2) Evolution of the reaction front is controlled by the hydrogen ion that releases the metal from the complex in the loaded resin;

3) Species concentrations in the reaction front ($r = r_c$) and in the unreacted-core ($r < r_c$) keep constant and equal to the initial concentrations;

4) Equilibrium reactions referred to previously are instantaneous and irreversible so that loaded resin (M^{++} form) is completely converted to clean resin (H^+ form) in the reaction front;

5) Mass transport of ion species beyond the front is negligible;

6) Film mass transfer is negligible.

The model equations are:

Mass balances inside particles

metal

$$\frac{1}{r^2}\frac{\partial}{\partial r}\,[r^2\,e_P\,D_{PM}\,\frac{\partial\,C_{PM}(r,\,t)}{\partial r}] = e_P\,\frac{\partial\,C_{PM}(r,\,t)}{\partial t} + \rho_{ap}\,\frac{\partial\,q_M(r,\,t)}{\partial t} \tag{1}$$

$$r_c < r \le r_o$$

hydrogen ion

$$\frac{1}{r^2}\frac{\partial}{\partial r}\,[r^2\,e_P\,D_{PH}\,\frac{\partial\,C_{PH}(r,\,t)}{\partial r}] = e_P\,\frac{\partial\,C_{PH}(r,\,t)}{\partial t} + \rho_{ap}\,(\frac{\partial\,q_{H^+}(r,\,t)}{\partial t} - \frac{\partial\,q_M(r,\,t)}{\partial t}) \tag{2}$$

$$r_c < r \le r_o$$

where $q_{H^+} = [RH_2]$ and $q_M = [R_2M^=]$. Considering the condition for the concentration of total fixed groups ($Q = q_{H^+} + q_H + q_M$, where $q_H = [RH^-]$) and assumption 4, we have

$$\frac{\partial q_{H^+}(r, t)}{\partial t} = \frac{\partial q_M(r, t)}{\partial t} = 0.$$

reaction front

$$-e_p D_{PH} \frac{\partial C_{PH}(r, t)}{\partial r}\Big|_{r = r_c} = [\rho_{ap}(2q_{M_o} + q_{H_o}) + e_p(C_{PH} - C_{PHo})] \frac{\partial r_c}{\partial t} \tag{3}$$

Species mass balances over the CSTR

metal

$$0 = L C_M + e V \frac{\partial C_M}{\partial t} + (1 - e) V e_P \frac{\partial \overline{C}_{PM}}{\partial t} + \frac{3V(1 - e) r_c^2 \, \rho_{ap} \, q_{Mo}}{r_o^3} \frac{\partial r_c}{\partial t} \tag{4}$$

hydrogen ion

$$LC_{HE} = L C_H + eV \frac{\partial C_H}{\partial t} + (1 - e)Ve_P \frac{\partial \overline{C}_{PH}}{\partial t} - \frac{3V(1 - e) r_c^2 \, \rho_{ap} \, (q_{Ho} + 2q_{Mo})}{r_o^3} \frac{\partial r_c}{\partial t} \tag{5}$$

in which $\overline{C}_{PM}$ and $\overline{C}_{PH}$ are intraparticle average concentrations for the metal and hydrogen ion, respectively.

Initial conditions

t=0

$$C_{PM}(r, 0) = C_{PMo} \tag{6}$$
$$C_{PH}(r, 0) = C_{PHo} \tag{7}$$
$$r_c = r_o \tag{8}$$

Boundary conditions

$r = r_o$

$$C_{PM}(r_o, 0) = C_M(t) \tag{9}$$
$$C_{PH}(r_o, 0) = C_H(t) \tag{10}$$

$r = r_c$

$$C_{PM}(r_c, t) = C_{PMo} = C_{Mo} \tag{11}$$
$$C_{PH}(r_c, t) = C_{PHo} = C_{Ho} \tag{12}$$

It should be noted that the derivatives of the average pore concentrations, $\dfrac{\partial \overline{C}_P}{\partial t}$, in equations (4) and (5) can be evaluated using equations (1) and (2), respectively, to get

$$\frac{\partial \overline{C}_P}{\partial t} = -\frac{3}{r_o^3}\ r_c^2 C_P\big|_{r=r_c}\ \frac{\partial r_c}{\partial t} + \frac{3}{r_o^3}\ [\ r_o^2 D_P\ \frac{\partial C_P(r,\,t)}{\partial r}\big|_{r=r_o} - r_c^2\ D_P\ \frac{\partial C_P(r,\,t)}{\partial r}\big|_{r=r_c}\ ]\ (13)$$

For the numerical integration of the model equations presented above, we used orthogonal collocation to discretize the radial particle coordinate by dividing it into N interior points; we get $2N + 3$ differential equations which were integrated using the package LSODE[12].

4 RESULTS AND DISCUSSION

The regeneration kinetics of a chelating amidoxime resin, pre-saturated with copper nitrate, are well described by the front reaction model developed in this work. Figures 1 and 2 show the experimental (points) and simulated (line) histories of concentration for copper and hydrogen ions. Experimental conditions used in regeneration are shown in Table 1.

The effective diffusivities, $\overline{D}_E$, calculated by fitting the model to the experimental results were for copper and hydrogen ion, 0.49×10^{-10} m^2/s and 6.4×10^{-10} m^2/s, respectively.

We can therefore conclude that the mathematical description of the regeneration of an amidoxime resin loaded with copper nitrate, treated in the present paper as a non-steady state diffusion process accompanied by two fast chemical reactions in a moving boundary, allows good prediction of the ionic concentrations evolution in solution. This methodology gives effective ionic diffusivities. The model seems suitable for describing the behavior of fixed bed ion exchange columns.

Table 1 *Experimental Conditions used in Regeneration with Nitric Acid of Duolite ES346 Saturated with Copper Nitrate*

Species concentrations (eq/m^3)	Resin properties	Adsorber and flow characteristics	Transport parameters (m^2/s)
$C_{Mo} = 50.11$	ep = 0.55	V = 1.095 x 10^{-3} m^3	$D_{m(Cu)} = 0.71 \times 10^{-9}$
$C_{Ho} = 1.259$	$r_o = 0.255$ mm	e = 0.90	$D_{m(H)} = 9.31 \times 10^{-9}$
$C_{HE} = 1.04$	Q = 4.8 eq/kg dry resin	L= 1.513x10^{-6} m^3/s	
	$\rho_{ap} = 517$ kg/m^3		

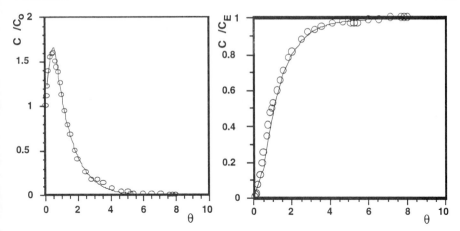

Figure 1 *History of concentration for copper ion, during regeneration of the adsorber*

Figure 2 *History of concentration for hydrogen ion, during regeneration of the adsorber*

Nomenclature

C	ionic concentration in solution
C_p	ionic concentration inside the pores
D_m	ionic diffusivity coefficient
D_p	diffusivity in resin pores
e	adsorber porosity
e_p	intraparticle porosity
L	volumetric flowrate
q	concentration in resin phase
Q	total resin capacity
r	radial position inside particles
r_c	radius of resin particle unreacted core
r_o	radius of resin particle
t	time
V	adsorber volume
[ ]	concentration

Greek symbols

ρ_{ap}	apparent density of the resin
θ	dimensionless time variable

Subscripts

ap	apparent
E	entrance
H	hydrogen ion
M	metal
o	initial

References

1. L. Astheimer, H. J. Schenk, E. G. White and K. Schwochau, *Sep. Sci. Tech.*, 1983, **18**, 307.
2. F. Vernon, and T. Shah, *Reactive Polymers*, 1983, **1**, 308.
3. T. Hirotsu, S. Katoh and K. Sugasaka, *Ind. Eng. Chem. Res.*, 1987, **26**, 1970.
4. R. W. Grimshaw and C. E. Harland, "Ion-exchange: Introduction to Theory and Practice", The Chemical Society, London, 1975.
5. D. Petruzzelli, F. G. Helfferich, L. Liberti, J. R. Millar and R. Passino, *Reactive Polymers,* 1987, **7**, 1.
6. F. Helfferich, *J. Chem. Phys.*, 1965, **4**, 1178.
7. G. Schmuckler, *Reactive Polymers*, 1984, **2**, 103.
8. M. Streat, *Reactive Polymers,* 1984, **2**, 79.
9. A. Fernandez, M. Diaz and A. E. Rodrigues, *The Chem. Eng. J.*, 1995, **57**, 17.
10. J. M. Loureiro, C. A. Costa and A. E. Rodrigues, in "Sep. Tech., Proceedings of the Engineering Foundation Conference", Schloss Elmau, 1987, p. 390.
11. J. M. Loureiro, C. A. Costa and A. E. Rodrigues, *Chem, Eng. Science*, 1988, **43**, 1115.
12. LSODE and LSODI, *ACM Signum News Letter*, 1980, **15**, 10.

OPERATING CONDITIONS OF CENTRIFUGAL COUNTERCURRENT ION EXCHANGE (CentrIX)

M. A. T. Bisschops, L. A. M. van der Wielen and K. Ch. A. M. Luyben

Department of Biochemical Engineering
Delft University of Technology
Julianalaan 67
2628 BC Delft
The Netherlands

1 INTRODUCTION

Ion exchange is a well established technique for removing low concentrated pollutants from large aqueous waste streams and for recovering diluted products from aqueous media such as fermentation broths. In conventional ion exchange equipment, the resin particle diameter is restricted due to hydrodynamic phenomena such as pressure drop or flooding under gravity. Typically, particle diameters are in the range of millimeters. As a consequence, conventional ion exchange equipment is relatively voluminous.

Application of very small particles, with diameters in the range of micrometers, is attractive because they have a high interfacial area and a small resistance to diffusion. This results in a high mass transfer efficiency and thus to compact separation equipment. Major drawback of micrometer particles (microadsobents) are the low settling rates under gravity. Therefore, an additional force is needed to obtain countercurrent flow.

Countercurrent Centrifugal Ion Exchange Technology (CentrIX) is a technique in which a centrifugal force is used to establish countercurrent flow between liquid and microadsorbents. The column or channel is mounted on a rotor, perpendicular to the axis of rotation. The centrifugal force is thus directed from the inner end of the channel, close to the axis of rotation, towards the outer end of the channel, at the periphery of the rotor. A schematic representation of the concept is shown in Figure 1.

Liquid, containing the sorbate, is fed at the outer end of the channel (at R_o) and is pumped towards the axis of rotation. The resin particles with a density higher than the liquid are fed at the inner end of the channel (at R_i) and are forced towards the outer end by the centrifugal field. In the channel, the two phases flow in continuous countercurrent direction.

Although CentrIX obviously involves complex equipment and difficult solids handling, its potential advantages are manifold. Most benefits are a result of the high sorption capacity in small equipment volumes. Major advantages are:

- Low space requirements, allowing application on sites with restricted space.
- Increased flexibility in type of resin and in capacity.
- Short contact times, allowing the processing of sensitive products under relatively harsh conditions.
- Low adsorbent inventory, allowing the use of selective and hence expensive ion exchange resins.

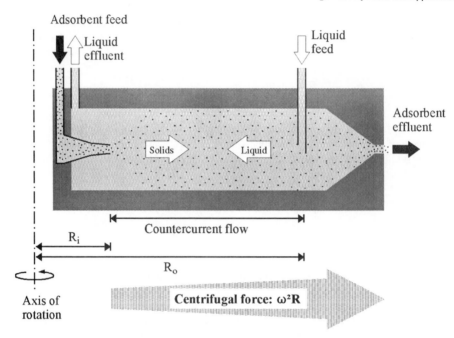

Figure 1 *Schematic representation of radial countercurrent flow in the CentrIX*

These features allow application in many fields, such as the downstream processing of biotechnological products from fermentation broths, as well as in waste water treatment plants for the removal of diluted contaminants.

A key factor in the design of CentrIX equipment is the hydrodynamics. It determines the capacity of the equipment related to the speed of rotation. Moreover it strongly influences the mass transfer efficiency, because the interfacial area per volume of equipment is proportional on the solids holdup.

2 THEORY

Two phase flow in centrifugal fields is substantially developed for solids recovery and extraction. We have extended the known theory to describe the hydrodynamics of countercurrent flow in the radial countercurrent ion exchange centrifuge.

2.1 Flow of Particles in a Centrifugal Field

The terminal settling velocity is defined as the velocity at which a single particle settles in a stagnant infinite medium. It can be derived from the balance between the net driving force (centrifugal - buoyancy) and the drag force. Coriolis force is negligible for very small particles and is thus not taken into account. For a particle in a centrifugal field, the force balance reads:

$$\underbrace{\frac{\pi}{6}d_p^3\left(\rho_s - \rho_L\right)\omega^2 R}_{\text{centrifugal-buoyancy}} = \underbrace{C_D\frac{\pi}{4}d_p^2\frac{v_\infty^2\rho_f}{2}}_{\text{friction}} \qquad (1)$$

In the laminar region the drag coefficient (C_D) obeys Stokes' law ($C_D = 24/Re$ for $Re < 1$) and in the turbulent region it is described by Newton's law ($C_D = 0.43$ for $Re > 10^3$). Several expressions have been published describing the drag coefficient for the intermediate region. Dallavalle[1] developed an expression that covers the entire range of Reynolds numbers, including the laminar and turbulent regime. Based on this relation, an explicit relation for the terminal settling velocity as function of the radius has been obtained.

Particles settling in a suspension have a sedimentation velocity which is lower than the terminal settling velocity of a single particle. To account for the influence of the particle concentration, the well known Richardson-Zaki[2] relation is used.

2.2 Countercurrent Flow in a Centrifugal Field

The slip velocity is the linear velocity of the resin particles relative to the water. It depends on the terminal settling velocity of the resin particles according to the Richardson-Zaki relation:

$$v_{slip} = v_S - v_L \Rightarrow$$

$$v_\infty (R) \varepsilon^{n-1} = \frac{u_S}{1 - \varepsilon} - \frac{u_L}{\varepsilon} \tag{2}$$

It should be emphasized that the terminal settling velocity is evaluated at the appropriate (local) position in the centrifugal field.

Wallis[3] and Rietema[4] outlined a graphical procedure for solving the slip velocity equation. The equation for the slip velocity can be rewritten as follows:

$$\varepsilon^n (1 - \varepsilon) = \frac{u_S}{v_\infty (R)} \varepsilon - \frac{u_L}{v_\infty (R)} (1 - \varepsilon) \tag{3}$$

In Figure 2 the left-hand side (LHS) and right-hand side (RHS) of this equation are plotted as a function of the void fraction (ε). The curve corresponding to the left-hand side of equation 3 only depends on the Richardson-Zaki coefficient (n), which is a weak function of the Reynolds number. The right-hand side of equation 3 represents a straight line, which crosses the left y-axis at $-u_L/v_\infty$ and the right y-axis at $+u_S/v_\infty$. The intersections of these two curves indicate the working points for the given set of conditions.

Equation 3 may give two solutions for the void fraction. Rietema[4] indicated these situations with the terms 'fluidized state' (for the low void fraction solution) and 'free settling state' (for the high void fraction solution). Mertes and Rhodes[5] used the terms 'N-phase' and 'P-phase' respectively. If the flow rates are increased, the left-hand side of equation 3 hardly changes because it only depends on the Richardson-Zaki exponent. The curve which corresponds to the right-hand side moves upwards. The maximum throughput is reached where the two curves are tangent and only have one single point in common. This point is called the 'maximum throughput point' or 'flooding point'. A further increase of the flow rates gives the situation without any solution to equation 3 and countercurrent flow is not possible.

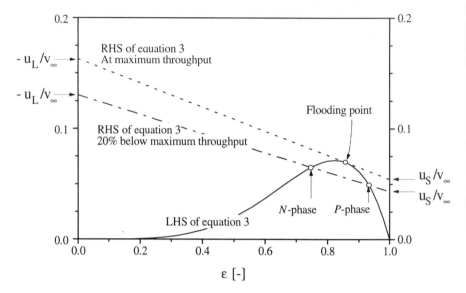

Figure 2 *Graphical solution of the slip velocity relation under laminar flow conditions (n = 4.65). The lines that correspond to the right-hand side (RHS) of equation 3 are drawn for constant flow ratios ($\phi_L/\phi_S = 3$)*

2.3 Flooding Criteria

Elgin and Foust[6] described flooding as the situation where the dispersed phase is rejected at the system entrance. In CentrIX equipment, the solids are fed to the channel at the inner radius (R_i), where the centrifugal force is at its minimum. The flooding criterion should therefore be chosen at the inner radius, which corresponds to the definition given by Elgin and Foust. Therefore, the operational envelope is bounded by flooding at the inner radius.

The maximum throughput can be found at the point where left-hand side and right-hand side of equation 3 are tangent, with the terminal settling velocity evaluated at the inner radius. Mathematically this is formulated as:

$$v_{\infty}(R_i)(1-\varepsilon)\varepsilon^{\,n} = u_S(1-\varepsilon) - u_L\varepsilon \tag{4}$$

and:

$$\frac{\partial\left[v_{\infty}(R_i)(1-\varepsilon)\varepsilon^{\,n}\right]}{\partial\varepsilon} = \frac{\partial\left[u_S(1-\varepsilon) - u_L\varepsilon\right]}{\partial\varepsilon} \tag{5}$$

From this criterion, the superficial velocities of both phases at flooding conditions can be derived. If the superficial velocities are written in dimensionless form, the solid throughput number (u_S/v_{∞}) and the liquid throughput number (u_L/v_{∞}) appear as a function of the void fraction:

$$\frac{u_S}{v_{\infty}(R_i)} = (1-\varepsilon)^2 n\varepsilon^{\,n-1} \tag{6}$$

and:

$$\frac{u_L}{v_\infty (R_i)} = ((1-\varepsilon)n-1)\varepsilon^n \tag{7}$$

The shape of the (dimensionless) flooding line only depends on the Richardson-Zaki coefficient (n). For laminar flow conditions ($n = 4.65$) the operational envelope is shown in Figure 3. The curve shows the maximum flow rates of the solid phase at a given liquid phase flow under given conditions and vice versa. Above this curve, countercurrent flow can not be obtained and solids will be rejected at the inner radius of the centrifuge.

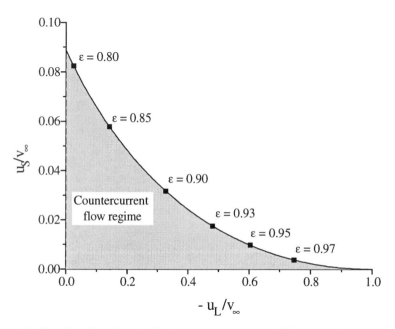

Figure 3 *Flooding line for the laminar region ($n = 4.65$) in its dimensionless form, calculated from equations 6 and 7*

3 DESIGN PROCEDURE

The design procedure for CentrIX equipment is analogous to that for conventional countercurrent equipment. It involves the calculation of the equilibrium flows and the capacity. Usually the mass transfer kinetics are evaluated in order to calculate the length of the channel. We will also discuss some other phenomena related to this quantity.

The flow ratio (ϕ_L/ϕ_S) is mainly determined by the ion exchange equilibrium. If the solids flow is chosen too low, the efficiency of the process will be very low. However, if the solids flow is too high, the ion exchange resin will be far from saturation. In general countercurrent processes should be designed with a separation factor ($S = K\,\phi_S/\phi_L$) somewhat above unity. As a rule of thumb separation factors in the range of $S = 1.10 - 1.25$ give reasonable separation efficiencies[7,8].

For a given distribution coefficient, the required flow ratio is set and the capacity only depends on the terminal settling velocity at the inner radius, where the solids are fed. This velocity varies with the relative centrifugal acceleration, which depends on equipment parameters, being the inner radius (R_i) and the speed of rotation (ω).

The mechanical stress on the periphery of the rotor is determined by the speed of rotation and the outer radius (R_o), and thus by the length of the channel. This length has a large influence on the number of transfer units and thus on the separation efficiency. It should therefore not be chosen too small. The stress varies linearly with the outer radius and quadratic with the speed of rotation. Therefore, for a given centrifugal force at the inner radius, the minimal stress at the periphery of the rotor is obtained at low speeds of rotation and large inner and outer radii. However, placing the channel at a large distance from the axis of rotation would result in a very high rotor diameter, which is not in agreement with the requirement of compact equipment design. The optimisation of the inner and outer radii — and consequently the speed of rotation — is a compromise between the mechanical strength of the equipment materials and the capacity and compactness of the equipment.

At high centrifugal forces, the mechanical balance of the rotor is an important aspect in the equipment design. From this point of view, it is best to mount at least two channels on the rotor and to position them in such a way that the rotor is balanced. Another option is to establish countercurrent contact in the entire circumference of the rotor, as is done in Podbielniak extractors[8]. This would guarantee optimal balance in stable operation, but it would lead to difficulties in extracting the solids evenly from the rotor. Therefore, separated straight channels are used.

CentrIX equipment should not be operated exactly at the flooding point, because any disturbance could cause maloperation. Mertes and Rhodes[5] reported that stable operation of countercurrent equipment under gravity conditions was obtained as high as 5-10% below flooding conditions.

Table 1   *Calculated Capacities of High Speed Industrial Scale CentrIX Equipment with 60 liter Equipment Volume. The Operation is taken at 10% below Maximum Throughput (flooding). ($\Delta\rho = 300 \; kg/m^3$, $\mu_L = 10^{-3} \; Pa.s$)*

d_P [μm]	$\phi_L/\phi_S = 20$		$\phi_L/\phi_S = 50$	
	$\omega^2 R_i = 3000 \; g$ $\phi_L \; [m^3/h]$	$\omega^2 R_i = 5000 \; g$ $\phi_L \; [m^3/h]$	$\omega^2 R_i = 3000 \; g$ $\phi_L \; [m^3/h]$	$\omega^2 R_i = 5000 \; g$ $\phi_L \; [m^3/h]$
1	0.28	0.47	0.38	0.63
2	1.1	1.9	1.5	2.5
5	6.7	11	8.9	15
10	25	39	32	52
20	81	126	106	164
50	302	439	385	557

Capacity calculations have been performed for an industrial scale centrifuge with 60 liters bowl volume. Industrial scale sedimentation centrifuges with a bowl volume up to 125 liters are not exceptional[9]. The equipment is operated at 10% below maximum throughput. The calculations have been performed for different speeds of rotation, characterised by the centrifugal acceleration at the inner radius ($\omega^2 R_i$). For instance, a centrifugal force of 3000 g corresponds to an inner radius of $R_i = 10$ cm and a speed of rotation of $\omega = 5200$ RPM. The distribution coefficient in ion exchange processes can vary over a wide range. As an example, the calculations have been performed for two flow ratios (ϕ_L/ϕ_S) which are of the same order of magnitude as found in the recovery of pharmaceuticals[10]. The results of the capacity calculations are shown in Table 1.

4   OBSERVATIONS

An experimental setup has been developed and is currently in operation. The hydrodynamic phenomena described in this paper will be validated in this apparatus. It consists of two parallel connected channels, with inner diameter of 25 mm. The solids feed location (R_i) can be varied from 15 mm to 240 mm. The liquid inlet is located at a distance of $R_o = 240$ mm from the axis of rotation. The speed of rotation can be varied up to 265 RPM, which corresponds to a relative centrifugal acceleration of 1.5 g at the (minimal) inner radius to 20 times gravity force at the periphery of the rotor. In order to study countercurrent flow in these low centrifugal force fields, model systems with a high density difference have been chosen, such as glass beads in water.

Apart from studying the hydrodynamics, the laboratory scale equipment was also used to identify and solve practical problems. The following difficulties have been encountered:

- A special rotary seal has been designed allowing the adsorbent and liquid feed at high rates. The seal is applicable in pilot scale and production scale equipment. Because the solids can not be fed as dry matter it is pumped in as a dense slurry.
- The solids have to be discharged at the periphery of the rotor, where centrifugal force is at its maximum. This should be done with a minimum of water leakage. A special device has been developed to control the discharge of the loaded adsorbent at the outer diameter.

The entire rotor is transparent to allow visual observations. Furthermore, pressure meters for measuring the variation in the solids holdup over the channel are being installed along the length of the channel.

Preliminary observations indicate that countercurrent flow is possible in a centrifugal field. Furthermore it was observed that the 'fluidised state' or 'N-phase' is most commonly obtained in the channel. A pilot scale centrifuge is being developed for experiments at centrifugal forces up to a few thousands g.

5 CONCLUSIONS

The sorption capacity of CentrIX equipment can be evaluated by hydrodynamic considerations. With the framework outlined in this paper the capacity can be related to process variables, such as the speed of rotation and flow ratio.

Preliminary observations indicate that stable countercurrent flow can be achieved in a centrifugal field and that the 'fluidised state' can be obtained. The solids holdup in the fluidised state is higher than in the free settling state, which is a big advantage with respect to mass transfer efficiency.

Capacity calculations show the enormous potential of CentrIX. Liquid flows in the order of magnitude of several hundreds m³/h per m³ of equipment volume can be processed in CentrIX equipment, with resin particles of 10 μm in diameter.

The capacity can be easily adjusted to the requirements of specific applications, by selecting the appropriate particle diameter and controlling the speed of rotation. This flexibility allows the application of CentrIX in many fields.

Acknowledgements

The research project is supported financially by the Dutch department of Economic Affairs via its IOP Environmental Technology (Prevention). Furthermore the technicians at the Mechanical Workshop of the Kluyverlaboratory for Biotechnology are gratefully acknowledged for their advise and assistance.

Nomenclature

C_D	Drag coefficient	[-]	v_L	Linear velocity of liquid	[m/s]
D_C	Channel diameter	[m]	v_S	Linear velocity of solids	[m/s]
d_P	Resin particle diameter	[m]	v_{slip}	Slip velocity	[m/s]
K	Equilibrium constant	[-]	v_∞	Terminal velocity	[m/s]
n	Richardson-Zaki exponent	[-]	ε	Voidage fraction	[-]
R	Radius in the channel	[m]	ϕ_L	Liquid flowrate	[m³/s]
R_i	Inner radius	[m]	ϕ_S	Solids flowrate	[m³/s]
R_o	Outer radius	[m]	μ_L	Liquid phase viscosity	[Pa.s]
S	Separation factor ($K\phi_S/\phi_L$)	[-]	ρ_L	Liquid phase density	[kg/m³]
u_L	Superficial velocity of liquid	[m/s]	ρ_S	Solid phase density	[kg/m³]
u_S	Superficial velocity of solids	[m/s]	ω	Speed of rotation	[rad/s]

References

1. J. M. Dallavalle, "Micromeritics, the Technology of Fine Particles", 2nd ed., Pitman & Sons, London, 1948.
2. J. F. Richardson, W. N. Zaki, *Trans. Inst. Chem. Engrs.*, 1954, **32**, 35.
3. G. B. Wallis, "One Dimensional Two-phase Flow", McGraw-Hill, 1969.
4. K. Rietema, *Chem. Eng. Sci.*, 1982, **37**, 1125.
5. T. S. Mertes, H. B. Rhodes, *Chem. Eng. Progr.*, 1955, **51**, 429 and 517.
6. J. C. Elgin, H. C. Foust, *Ind. Eng. Chem.*, 1950, **42**, 1127.
7. C. J. King, "Separation Processes", McGraw-Hill, 1971.
8. T. C. Lo, M. H. I. Baird, C. Hanson, "Handbook of Solvent Extraction", Krieger publ. comp., 1991.
9. H. Hemfort, "Separatoren, Zentrifugen für Klärung, Trennung, Extraktion", Technisch-wissenschaftliche Dokumentation Nr.1, Westfalia Separator AG, Oelde, 1983.
10. J. P. van der Wiel, J. A. Wesselingh, in: "Adsorption: Science and Technology", A. E. Rodrigues et al. eds., Kluwer Academic Publishers, 1989, 427.

ON THE APPLICATION OF GRANULAR INORGANIC ION EXCHANGERS IN THE DOMESTIC ADSORPTIVE FILTERS FOR DRINKING WATER PURIFICATION FROM HEAVY METAL IONS AND RADIONUCLIDES

A. I. Bortun, S. A. Khainakov, V. V. Strelko and I. A. Farbun

Institute for Sorption & Problems of the Endoecology of the National Academy of Sciences
32/34 Palladina Prosp.
252142 Kiev
Ukraine

1 INTRODUCTION

Considerable interest is being shown now in many countries for the development of effective portable filters for conditioning and purification of water used for drinking and cooking. This is now a special case for the Ukraine, because of its unfavorable ecological situation, that is characterized by heavy pollution of soil and water with heavy metals, radionuclides, pesticides and other toxic substances.

The analysis of the scientific literature indicates that among the variety of different kinds of known domestic filters, the most popular are those which use active charcoals and/or ion-exchange resins. Their efficiency greatly depends on the selectivity and capacity of the sorbents used.

At the same time it is well known that the selectivity of some inorganic ion exchangers for heavy metal ions and radionuclides is much higher than that of traditionally used organic resins.[1,2] Taking this into consideration we developed several new types of adsorptive filters containing as active elements both charcoals and granular inorganic adsorbents, in order to increase their protective ability on toxic heavy metal ions and radionuclides.

In this communication the results of our study of inorganic adsorbents and adsorptive filters in model systems and in real conditions are presented.

2 EXPERIMENTAL

2.1 Synthesis

Titanium and zirconium hydroxophosphates were synthesized by the gel method described in Reference 3. According to this method a 1M solution of $TiCl_4$ ($ZrOCl_2$) containing a certain amount of H_2O_2 (from 0.05M to 0.7M) was mixed under vigorous stirring with a 5M solution of phosphoric acid (molar ratio $M^{IV}:P=1:1.5$). After that, the mixture was dispersed into a vertical column filled with liquid aliphatic hydrocarbons, where the formation of spheres of hydrogel took place. The obtained granules were washed with water, dried at ambient temperature and then converted into the sodium form by treatment with a 1% NaCl solution.

The sorbents obtained by this method are amorphous and are characterized by the formula $M^{IV}O_2.0.5\ P_2O_5.3.5\text{-}4.0\ H_2O$. The granule's diameter varies from 0.25 to 2 mm. THP and ZHP specific surface areas are 250 and 50 m²/g, pore volumes (V_s) - 0.25 and 0.15 cm³/g, respectively, ion-exchange capacity 4-4.5 meqv/g.

2.2 Adsorption Experiments

Adsorption experiments were carried out in dynamic conditions at ambient temperature. For this purpose two types of polyethylene columns with inner diameter 10 and 45 mm were used. The volumes of the sorbent's bed were 3 and 100 mL, respectively. The water flow rate (W) was regulated from 30 to 150 bed volumes per hour with a peristaltic pump.

Based on the experimental data, the breakthrough curves were plotted in coordinates: C_i/C_o (C_o - initial concentration of ion, M; C_i - concentration of cation after passing the layer of adsorbent, M) against amounts of passed volumes of solution N (N = W/V, where W - volume of solution passed through adsorbent (mL); V - volume of adsorbent bed (mL)).

Heavy metal ion concentrations in the water before and after passing through the adsorbent bed were determined by AAS methods (SP9-800, Pye Unicam).

^{137}Cs and ^{90}Sr content in model solutions was measured on a β-spectrometer (RUB 01P with BDGB-06 1P detector, USSR). The efficiency of purification was characterized by purification coefficient (K_p) according to the formula:

$$K_p = (C_o - C_i)/C_i$$

where C_o, C_i - concentration (activity) of solution before and after passing through the sorbent bed, respectively.

2.3 Preparation of Model Solutions

Tap water, containing 48-55 mg Ca/L, 9-13 mg Mg/L, pH=6.2-6.8, was used for the preparation of model solutions by labeling with traces of heavy metal ions or ^{137}Cs and ^{90}Sr radionuclides.

3 RESULTS AND DISCUSSION

Our recent research has shown that the affinity of THP and ZHP exchangers for heavy metal ions is considerably higher than that for alkali or alkaline-earths cations:

for THP: $Fe^{3+} > Pb^{2+} >> Cd^{2+} > Cu^{2+} > Co^{2+} > Zn^{2+} > Ca^{2+} > Na^+$ [4]
for ZHP: $Pb^{2+} >> Cd^{2+} = Cu^{2+} > Fe^{3+} > Co^{2+} > Zn^{2+} > Ca^{2+} > Na^+$ [5]

This result suggests that such materials could be promising for drinking water treatment.

In order to evaluate the suitability of granular hydroxophosphate based adsorbents for column application, the influence of water flow rate, nature and concentration of heavy metal ion, etc., on the efficiency of simulant solutions purification was studied.

3.1 Water Flow Rate

For this experiment small size column, granular ZHP (0.3-0.5 mm) and simulant solution doped with copper (1.3 mg/L) were used. The breakthrough curves obtained are shown in Figure 1. Analysis of the data shows that the increased flow rate from 30 to 50 B.V./h has little effect on the copper adsorption: $C_i < 0.1$-0.14 mg/L. This indicates the possibility of decreasing the copper concentration in water far below the maximum permissible concentration ($MPC_{Cu}=1$ mg/L, Ukraine) at least in 6000 bed volumes of drinking water. We didn't study the possibility of purification of a larger quantity of water because the traditional adsorption filters are not designed for such a long-run service life. Starting with 100 B.V./h flow rate the extent of water purification by ZHP diminishes ($C_i > 0.2$-0.3), which is attributed to the relatively poor kinetic parameters of the inorganic sorbents.[6]

3.2 Concentration of Heavy Metal Ions in Model Solution

The breakthrough curves showing the efficiency of the treatment of model solutions, containing 0.8-4.5 mg Cu/L, are presented in Figure 2. The increase of Cu^{2+} concentration results in the decrease of the copper purification rate and leads to a reduction of the amount of water purified.

3.3 Nature of Heavy Metal Ions and Radionuclides

The data presented in Figure 3 show that both THP and ZHP adsorbents exhibit extremely high affinity for heavy metal cations. The highest K_p values were found for Pb^{2+} and Cd^{2+} ions adsorption. The tested inorganic ion exchangers are able to reduce Pb^{2+} and Cd^{2+} ions concentration from 10-50 MPC to the permissible level in 5000-10000 B.V. Such a high protective ability of the inorganic ion exchangers makes them indispensable in emergency cases (ingress of toxic substances in water supply system in the case of accidents on industrial plants, floods, etc.).

^{137}Cs and ^{90}Sr breakthrough curves are shown in Figure 4. THP and ZHP sorbents exhibit relatively high affinity for cesium and very poor affinity for strontium. Inorganic adsorbents are able to decrease the cesium content in the first several hundred bed volumes of solution in 30-60 times and only in 10-15 times after passing 1500-2000 B.V. This is connected with the facts that the amount of adsorption centers selective to cesium in THP and ZHP is limited and that the activity of the water simulant is high enough (1800 Bk/L). It is possible that in the case of solutions less contaminated with ^{137}Cs the volume of water purified with high K_p will increase.

The other way of improving THP's and ZHP's affinity for cesium is by modification of their surface with copper ferrocyanide (TPFC, 5% wt.). This results in a hundred-fold increase of the K_p values ($K_p=1000$-3000) as well as a considerable increase of the purified volume (up to 5000-10000 B.V.).

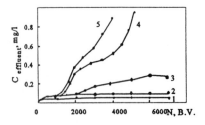

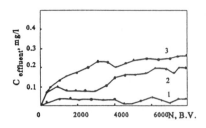

Figure 1 *The influence of the flow rate (in B.V./h): 20 (1), 50 (2), 75 (3), 100 (4) and 150 (5), on copper uptake by THP. $Cu^{2+}init = 1.3$ mg/L*

Figure 2 *The influence of the initial copper concentration in model solution (in mg/L): 0.8 (1), 1.9 (2) and 4.5 (3), on the efficiency of its uptake by ZHP. Flow rate is 30 B.V./h*

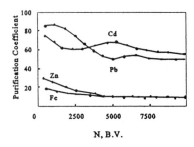

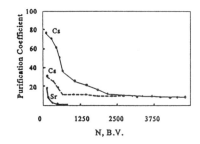

Figure 3 *Heavy metal ions uptake by ZHP. Flow rate is 30 B.V./h. Initial concentration of elements (in mg/L): $Fe^{3+}=0.35$, $Zn^{2+}=2.0$, $Pb^{2+}=0.30$ and $Cd^{2+}=0.10$*

Figure 4 *Adsorption of ^{137}Cs and ^{90}Sr by THP (—) and TPFC (- - -). Flow rate is 30 B.V./h. Initial activity of ^{137}Cs is 1800 Bq/L, of ^{90}Sr is 770 Bq/L. The K_p value for TPFC must be multiply by 100*

3.4 Granule's Size

As can be seen from Figure 5 the size of adsorbent's granules influences greatly the efficiency of d-metal ions uptake; it drops drastically with the size increase from D=0.3-0.5 mm to D=1.0-2.0 mm. This is connected with the poor kinetic characteristics of inorganic exchangers as a whole. In practice such a disadvantage of granular materials can be reduced to some extent by using smaller granules of exchanger or by operating with lower flow rates. At the same time knowing that THP and ZHP based ion

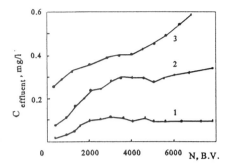

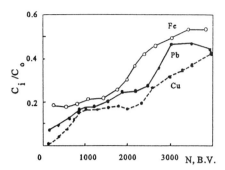

Figure 5 *The influence of the ZHP granule's size on the copper $(Cu^{2+}_{init} = 1.3 \ mg/L)$ adsorption. Flow rate is 30 B.V./h. Diameter of granules (in mm): 0.3-0.5 (1), 0.5-1.0 (2) and 1-2 (3)*

Figure 6 *Heavy metal ions adsorption by thin-film SiO_2-THP exchanger. Flow rate is 100 B.V./h. Initial concentration of cations (in $\mu g/L$): Fe^{3+} = 700, Pb^{2+} = 4, Cu^{2+} = 10*

exchangers have a high adsorption capacity for d-metal ions, another way to overcome this problem can be the use of adsorbents in the form of thin-films. An example of this is the results (Figure 6) obtained with the use of silica, impregnated with THP (or ZHP). Such a thin film exchanger as SiO_2-THP, containing 10% wt. of THP efficiently removes Cu^{2+} and Pb^{2+} ions from 2000 B.V. of water simulant.

Summing up the experimental data it is possible to make conclusions that titanium and zirconium hydroxophosphates possess high adsorption capacities, exhibit distinct affinity for heavy metal ions and radiocesium and are able to remove them efficiently in column processes.

3.5 Enlarged Tests of Inorganic Adsorbents

For the enlarged tests cartridges (diameter 40 mm, volume 100 cm³) containing inorganic sorbents (0.5-1.0 mm fraction) were used. 20 L of tap water, labeled with heavy metal cations, were passed downwards through the sorbent bed every day at a flow rate of 6-8 L/h. The breakthrough curves obtained from the use of THP and ZHP exchangers are shown in Figures 7, 8. Additionally, changes in water hardness, pH and release of phosphate ions (no traces of titanium or zirconium were detected in the effluent) from the tested exchangers were recorded. As can be seen from the breakthrough curves inorganic adsorbents decrease the heavy metal ion content in water up to 100-300 times. The better THP performance is attributed to its greater porosity, because the ion-exchange capacities of both titanium and zirconium hydroxophosphates are practically the same (4.1 and 4.4 meqv/g).

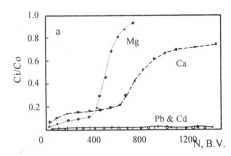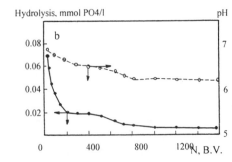

Figure 7 *a) Adsorption of Mg²⁺, Ca²⁺, Pb²⁺ and Cd²⁺ ions by THP. Initial concentrations (in mg/L): Mg²⁺ = 10.2, Ca²⁺ = 48.0, Pb²⁺ = 0.20, Cd²⁺ = 0.10. b) Phosphorus release from THP and the values of effluent pH*

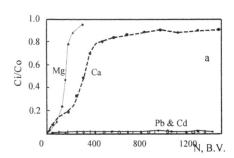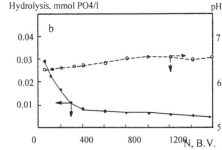

Figure 8 *a) Adsorption of Mg²⁺, Ca²⁺, Pb²⁺ and Cd²⁺ ions by ZHP. Initial concentrations (in mg/L): Mg²⁺ = 10.2, Ca²⁺ = 48.0, Pb²⁺ = 0.20, Cd²⁺ = 0.10. b) Phosphorus release from ZHP and the values of effluent pH*

It is worth noting that titanium hydroxophosphate not only remove practically all heavy metal ions but also considerably decreases the hardness of water. The total uptake of Ca and Mg ions by THP is comparable to that of organic carboxylic resins. But in contrast to resins THP does not completely adsorb all the alkaline earth metal ions from the first portions of passing water and as a result THP is able to soften larger volume of water (more than 1000 B.V.). ZHP's ability to reduce the water hardness is considerably less than that of THP.

Because the service life of inorganic adsorbents is much longer in the case of heavy metal ions and radionuclides recovery than in the case of reducing water hardness, additional experiments have been done studying the possibility of their regeneration and repeated use. It was found that treatment of the worked out THP with a 5% solution of NaCl or NaHCO₃ enables it to desorb 90-95% of Ca²⁺ and Mg²⁺ cations, without affecting

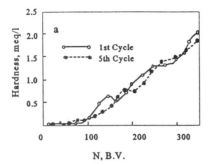 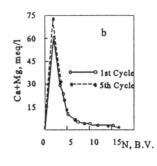

Figure 9 *a) Hardness of water passed through the THP on the first and on the fifth adsorption cycle. Initial water hardness is 3.9 meq/L. Flow rate is 100 B.V./h. b) Ca+Mg release during used THP regeneration (5% NaCl). 1st and 5th cycles of regeneration*

the heavy metal ions (which are trapped extremely firmly in the matrix), and restore practically all the initial adsorption ability of the material. This is illustrated by the data shown in Figure 9. Such periodic operation increases the efficiency and extends the service life of domestic filters several times.

Analysis of the experimental curves characterizing the phosphate release into the water (Figures 7,b and 8,b) shows that the hydrolytic stability of ZHP is much greater than that of THP. For the zirconium hydroxophosphate, the phosphorus concentration in tested samples of water doesn't exceed the MPC level whereas in the case of THP, in first 200-300 B.V., it exceeds the MPC in several times. This is in a good agreement with the chemical stability of phosphorus containing inorganic ion exchangers.[7] As a result of this THP must be used in more narrow pH ranges (pH=2-6.5) than ZHP (pH=1-9). One of the ways to overcome the problem of phosphate release from THP is the use of domestic filters containing a small amount of zirconium hydroxide which can effectively adsorb all the phosphate ions.

3.6 Field Tests of Adsorption Filters, containing Inorganic Ion Exchangers

For the illustration of the efficiency of inorganic ion exchangers as active elements in adsorption filters the experimental data on the content of some admixtures accumulated in ZHP layers of domestic filters used for a 2-3 month period (10,000 B.V. of water purified) in different regions of the Ukraine are presented in Tables 1, 2. It is seen that on the first place among the adsorbed admixtures are Ca^{2+} and Mg^{2+} ions (3.5-35 g), on the second place Fe and Zn (100-400 mg) and then other heavy metals, the total amount of which is not very high (from 0.7-2.3 mg for Cd to 50 mg for Cu). Analogous effects of concentration have been found in water purification of radionuclides (initial radioactivity less than 0.01 Bq/L). In the sorbent phase has been detected the following radionuclides: ^{134}Cs and ^{137}Cs, ^{106}Ru, ^{226}Ra, ^{232}Th, ^{125}Sb. In all cases, the main admixture was radioactive cesium. Considering the low initial content of radionuclides in the water, the obtained values of activity for the ZHP layer indicate a rather high efficiency of purification.

Table 1   *The Amounts of Cations fixed by ZHP Layer during Field Tests (in mg/kg of ZHP)*

Region	Ca	Mg	Fe	Zn	Cu	Pb	Ni	Cd
Kiev	32000	3500	393	124	50	18	6	1.4
Zaporozh'e	36500	3500	313	400	30	23	6	2.3
Dnepropetrovsk	37000	4300	160	440	8	2.5	6	0.7

Titanium and zirconium hydroxophosphates (and some of their modifications) in combination with charcoals and other materials have passed all the necessary medical, biological and toxicological tests, and we now have the permission of the Ukrainian Ministry of Health for their use in domestic filters for purification of drinking water. Large scale production of adsorptive filters, containing selective inorganic ion exchangers and charcoals, for the population of Ukraine and Belarus, that has suffered from the Chernobyl accident, is planned for several Ukrainian and Belorussian plants.

Table 2   *The Amounts of Radionuclides fixed by ZHP Layer during Field Tests (in Bq/kg of ZHP)*

Region	^{137}Cs	^{134}Cs	^{106}Ru	^{226}Ra	^{232}Th	^{125}Sb
Kiev	5.1	1.0	-	0.9	0.3	-
Zaporozh'e	6.1	1.0	-	1.4	0.4	1.1
Dnepropetrovsk	8.0	2.7	6.3	-	-	-

References

1.   C. B. Amphlett, "Inorganic Ion Exchangers", Elsevier, New York, 1964.
2.   A. Clearfield, ed., "Inorganic Ion Exchange Materials", CRC Press, Boca Raton, FL, 1982.
3.   V. V. Strelko, "Chemistry Role in the Environmental Protection", Naukova Dumka, Kiev, 1982.
4.   V. V. Strelko, A. I. Bortun, A. P. Kvashenko and V. N. Khryaschevsky, *Vysokochistye veschestva*, 1990, **N4**, 93.
5.   A. I. Bortun, V. N. Khryaschevsky and A. P. Kvashenko, *Soviet Progress in Chemistry*, 1991, **57**, 22.
6.   S. J. Harvie and G. H. Nancollas, *J. Inorg. and Nucl. Chem.*, 1968, **30**, 273.
7.   A. I. Bortun, T. A. Budovitskaya, V. V. Strelko, R. Garcia and J. Rodriguez, *Mater. Research Bull.*, in press.

SODIUM MICAS AS CESIUM ION SELECTIVE ADSORBENTS

L. N. Bortun, A. I. Bortun and A. Clearfield

Department of Chemistry
Texas A&M University
College Station
TX 77843, USA

1 INTRODUCTION

Micas are chemically and thermally stable layered aluminosilicates. They consist of negatively charged 2:1 layers that are compensated and bonded together by large positively charged interlayered cations. All the known natural micas exist in the potassium form. This is connected with the peculiarities of their crystal structure, namely, with tetrahedral SiO_4 crown-ether type formations of oxygen atoms that practically ideally correspond to the parameters of the potassium ion. According to theoretical calculations such adsorption centers should fit even better for cesium, which indicates that natural micas could be abundant, cheap and efficient adsorbents for radioactive cesium removal from different types of nuclear waste streams. The only problem in using the natural micas for such purpose is that potassium ion can not be exchanged directly for Cs^+ because of steric hindrance (K^+-mica's interlayer distance is about 10.0 Å, which is too small for cesium to access). This problem could be remedied by the preliminary conversion of potassium mica into the sodium form with a greater interlayer distance (12.2 Å). This is done by the application of the "sodium tetraphenylborate (STB) method", which allows the substitution of practically all the potassium by sodium by treating mica with STB at elevated temperature for a long period of time. Unfortunately, this method can not be used industrially because of the high cost of the STB reagent.

In this communication an alternative method for the conversion of potassium micas into their sodium form under mild hydrothermal conditions and the results of testing the materials selectivity towards cesium are presented.

2 EXPERIMENTAL

2.1 Reagents

All reagents were of analytical grade (Aldrich). Natural micas used were phlogopite and biotite (Ward's Scientific).

2.2 Analytical Procedures

The diffractometer used was a Scintag PAD 5 model with CuKα radiation. Thermal analysis was performed by a TA 4000 (under nitrogen, rate of heating 10°C/min).

2.3 Amine Intercalation

n-Alkylamine intercalation compounds were obtained by treatment of micas in the sodium form with a 2 M solution of the corresponding n-alkylamine hydrochloride for 1 day (V:m ratio 100:1; ambient temperature).

2.4 Ion Exchange Study

Adsorption of cesium ion on mica based exchangers was studied with different types of model solutions at V:m = 200:1 (mL:g) and room temperature. Contact time was 4 days. Distribution coefficients K_dCs were calculated according to the formula:

$$K_dCs = (C_o - C_i)/C_i * V/m \text{ (mL/g)},$$

where C_o and C_i are the initial and final cesium concentrations; V= volume of solution; m= weight of the sample.

The pH of model solutions after equilibration with the adsorbent were measured using an Orion SA-720 pH meter. The residual concentration of alkali metal ions in solutions were measured using a Varian SpectrAA-300 atomic absorption spectrometer.

3 RESULTS AND DISCUSSION

Natural powdered phlogopite and biotite (fractions: 0.15-0.5 mm and < 0.15 mm) were treated consecutively up to 5 times with 0.1-2 M NaCl solution, additionally containing in some cases 0.04-0.2 M NaOH, at V:m ratio 200:1 and hydrothermal conditions (teflon linen vessel, 190°C) for 24 h each time. After every step of hydrothermal treatment the mica samples were separated from solution by filtration and then thoroughly washed with deionized water. The influence of all studied factors on the conversion process are shown in Figures 1-5 (where *Conversion Rate* is $I_{12.6 Å}/(I_{10.0 Å} + I_{12.6 Å})$; $I_{10.0 Å}$ and $I_{12.6 Å}$ are the intensities of the 001 reflections of K- and Na-micas, respectively) and several typical XRD patterns illustrating the conversion route of phlogopite into the sodium form are presented in Figure 6.

3.1 Influence of the Reagent Concentration

It was found that the increase of NaCl concentration favors substitution in mica. The best results obtained by a one step treatment were in the case of using 2 M NaCl (Figure 1).

3.2 Alkalinity of Solution

The presence in the working solution of a small amount of NaOH (up to 0.02-0.04 M) enhances the conversion, but at higher NaOH content the process reverses (Figure 2).

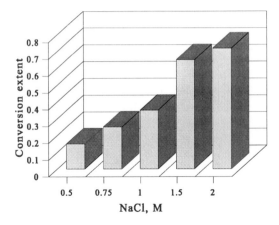

Figure 1 *Influence of NaCl concentration upon conversion rate of natural biotite into the sodium form. T=190°C, contact time 1 d*

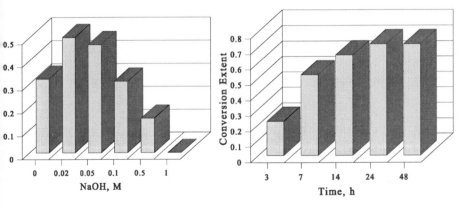

Figure 2 *Influence of the alkalinity of the model 1 M NaCl solution upon the conversion of natural biotite into the sodium form. T=190°C; 1 day treatment*

Figure 3 *Kinetics of natural biotite conversion into the sodium form. 1 M NaCl solution, T=190°C*

3.3 Duration of the Treatment

The substitution of potassium by Na^+ in biotite takes place even after 5-7 hours of treatment, but the best result in the one-step process was obtained after a 24 h treatment (Figure 3).

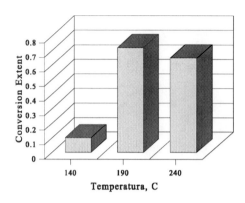

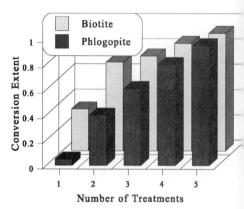

Figure 4 *Influence of the temperature of the hydrothermal treatment on the conversion rate of biotite into the sodium form. 1 M NaCl solution. 1 day treatment*

Figure 5 *Influence of the number of consecutive treatments on biotite and phlogopite conversion into the sodium form. 1 M NaCl solution; 190°C*

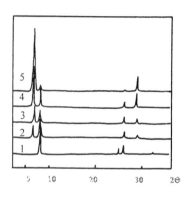

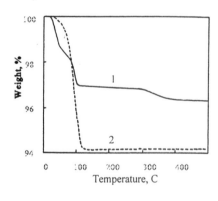

Figure 6 *The XRD patterns of materials obtained by successive treatment of phlogopite with 1 M NaCl solution at mild hydrothermal conditions (190°C). Number of treatments is shown on the graph*

Figure 7 *TGA curves of sodium biotite (1) and phlogopite (2)*

3.4 Temperature Factor

The conversion rate increases with temperature increase and stays constant after reaching 170-180°C (Figure 4).

3.5 Successive Treatment

Analysis of these data shows that the conversion of potassium phlogopite into the sodium form has a latent period, making the extent of transformation in the first stage of treatment negligibly small in comparison with biotite. Only after the second treatment does the active substitution of K^+ by Na^+ ions begin, which is indicated by the occurrence and increase of intensity of the 12.6 Å sodium mica reflection in the XRD patterns (Figures 5, 6). It was found that the 5 step treatment enables us to obtain practically pure sodium phlogopite and biotite without deteriorating its crystalline structure.

3.6 Thermoanalytical Study

Both sodium biotite and phlogopite are remarkably thermally stable materials. The only detected weight loss is in the 50-120°C interval and it is connected with the physically adsorbed water release. No other weight losses up to 1000°C were found (Figure 7).

3.7 Amine Pillaring

By the treatment of sodium micas with 2 M solutions of corresponding amine hydrochlorides dimethylamine-, diethylamine-, propylamine- and butylamine pillared phlogopite and biotite were successfully prepared.

3.8 Cesium Adsorption on Micas in Sodium Form

Cesium adsorption from the different model solutions for the potassium and sodium forms of biotite and phlogopite were studied by the batch technique. Some of these data are presented in the Table 1. Analysis of these data shows that both natural K^+- micas do not adsorb cesium under the examined conditions, while their sodium forms exhibit extremely high affinity for cesium in 0.001M CsCl solution (1st column). K_d values exceed 100,000. At the same time the corresponding K_d values drop drastically (to only several hundreds and lower) if the model solution additionally contains more than 1 M of background sodium nitrate (sodium hydroxide) electrolyte. It is also worth noting here a somewhat higher affinity of sodium biotite for cesium in comparison with phlogopite mica.

3.9 Cesium Adsorption on Amine Pillared Micas

Distribution coefficients for Cs^+ obtained on amine pillared micas are presented in Table 2. Analysis of these data shows that similar to the case of the initial micas, amine intercalated biotite possesses a higher affinity for cesium in comparison with amine pillared phlogopite. All the tested materials do adsorb cesium efficiently from water, but as a rule they show worse performance in the presence of sodium electrolytes. This is

Table 1 *Distribution Coefficient Values for Cesium Adsorption from Model Solutions,*
Containing 0.001 M Cs⁺, on Mica Samples

Mica	H_2O	0.1 M NaNO₃	1 M NaNO₃	5 M NaNO₃	1 M NaOH	5 M NaNO₃+ 1 M NaOH
Biotite-K	15	< 5	< 5	< 5	< 5	< 5
Biotite-Na	> 400,000	> 100,000	14300	750	-	450
Phlogopite-K	10	< 5	< 5	< 5	< 5	< 5
Phlogopite-Na	> 400,000	> 100,000	6250	500	500	120

connected with the fact that the pillaring of the layered materials (micas, phosphates, oxides, etc.) with amines, resulting in the expansion of their layers, can improve only the kinetics of adsorption and increase in some cases the effective adsorption capacity, but not the selectivity. In general, it is difficult to expect a distinct selectivity for layered ion-exchangers due to the lack of rigidity of their structure. As a result the layered adsorbents can accommodate ions of different size more or less easily. On the other hand, such a property as a lability of structure, could be used to the benefit of the adsorbent by modifying or intercalating it with different types of chelating or specific ion complexing reagents.

Taking this into consideration we have modified sodium biotite, phlogopite and synthetic fluoromica with a specific reagent for cesium - sodium tetraphenylborate (STB). This has been done at the stage of mica's preliminary conversion into the amine form.

3.10 Cesium Adsorption on the Micas Impregnated with Sodium Tertaphenylborate

Three micas impregnated with STB have been prepared for the investigation of ion exchange properties. In the first step of modification the sodium phlogopite, biotite and synthetic sodium fluoromica has been converted into the propylamine form by its treatment with 2 M PrNH₂·HCl solution (4 h, T=60-70°C) and then thoroughly washed with water (V:m=200-300) to remove excess amine. Then they were swollen in water for one day and treated with a 3% STB solution (24 h). The coagulation of material was observed as the result of this operation. Before drying in air at ambient temperature, the modified samples were washed with a small amount of water (V:m=3-5). The XRD patterns of the initial synthetic sodium fluoromica and the material impregnated with STB are presented in Figure 8. The treatment of the amine pillared mica with STB leads to a shift of the first XRD reflection (in the wet samples) from 14.1 Å to 38.4 Å, a result of sodium tetraphenylborate penetration into the interlayer space with the formation of the correspondent amine salt. The drying of the modified mica (without preliminary washing) leads to substantial shrinkage of the layers (d=13.05 Å) and to a decrease of the XRD peaks intensities of intercalated STB. At the same time practically no changes were detected in the XRD patterns of biotite and phlogopite as a result of intercalation of the STB reagent.

The efficiency of cesium uptake on modified biotite has been tested from model 5 M NaNO₃ + 1 M NaOH solution containing 10^{-3} M CsNO₃ and 10^{-3} M KCl (V:m ratio is 100:1). The distribution coefficients found in all cases were >>100,000 for both Cs and K ion uptake, which is hundreds of times higher than the best known ion exchangers show.

Table 2 *Distribution Coefficients Values for Cs⁺ Sorption on Amine Pillared Phlogopite and Biotite Samples. Initial Cs⁺ Concentration 0.001 M*

Mica	Amine	H_2O	5 M $NaNO_3$	5 M $NaNO_3$ + 1 M NaOH
Phlogopite	$(CH_3)_2NH$	1000	180	120
Phlogopite	$(C_2H_5)_2NH$	> 100,000	240	80
Phlogopite	$C_3H_7NH_2$	> 100,000	250	110
Phlogopite	$C_4H_9NH_2$	> 100,000	460	60
Biotite	$C_3H_7NH_2$	> 100,000	700	590

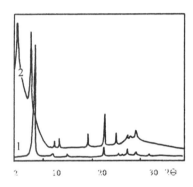

Figure 8 *The XRD patterns of sodium fluoromica pillared with propylamine (1) and of the product of STB intercalation (2)*

This indicates an extremely high affinity of STB modified micas to heavy alkali metal ions as well as on their high adsorption capacity.

In order to clarify the reason for such a performance of STB containing micas, the STB modified biotite was contacted for 1 day with 5 M $NaNO_3$ + 1 M NaOH solution containing 0.01 M of $CsNO_3$, filtered, washed with water and dried at ambient temperature. According to the X-ray data the first reflection in the XRD pattern was at 10.0 Å. This indicates the desorption of all intercalated organic molecules and the complete conversion of the exchanger into the cesium form. Based on these data we suggest that the cesium uptake by STB modified micas is a complex process including in the first step the formation of low soluble Cs-STB complex between the layers or on the surface of the clay (precipitation stage), followed by the solid-solid exchange reaction resulting in cesium trapping between the mica layers (ion exchange stage).

In order to estimate the efficiency of materials operating by such a precipitation-adsorption mechanism the cesium uptake by sodium and STB modified biotites and by crystalline sodium titanium silicate $Na_2Ti_2O_3SiO_4$ with channel structure (for comparison) were studied from model 5 M $NaNO_3$ + 1 M NaOH solutions containing different amounts (0.001 - 0.01 M) of $CsNO_3$. The data obtained are presented in Figure 9 as the adsorption capacity values versus Log[Cs⁺] at equilibrium point. It is seen that by the adsorption capacity for cesium the studied materials can be placed in the order: STB-biotite > sodium biotite >> sodium titanium silicate. Biotite modified with STB reagent shows high IEC_{Cs} values (0.8-0.85 meq/g) in an extremely wide range of Cs⁺ concentrations, which is very important for solving different practical tasks.

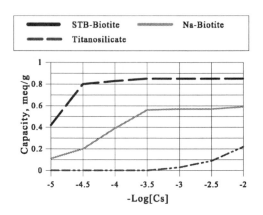

Figure 9 *Cesium uptake by STB-biotite, Na-biotite and sodium titanium silicate as a function of equilibrium Cs⁺ concentration in model 5 M NaNO₃+1 M NaOH solution*

In considering a sorbent for nuclear waste remediation there are three important factors to consider, high selectivity for the ion of interest, rapid kinetics of exchange and a reasonable capacity. The sodium titanium silicate exhibits a high selectivity for Cs⁺ because this ion fits snugly into the tunnel at the midpoint of 8 equidistant oxygen atoms. However, once into this position there is a large barrier to further diffusion into the interior of the tunnels. This difficulty is greatly affected by the presence of sodium ions which can diffuse more rapidly to occupy the exchange sites. Thus, the overall capacity for cesium is very low in alkaline nuclear waste solutions. In contrast the STB treated biotite exhibits high selectivities, rapid kinetics and a moderately high capacity. In fact at a Cs⁺ concentration of 10^{-3} M the STB mica sorbent has ~20 times the capacity of the sodium titanium silicate. An additional advantage could result from reslurring the Cs-TB-mica complex in water whereupon the sodium ion in the mica diffuses out to form STB and the Cs⁺ ion is trapped between the mica layers. This product could be heated to seal the Cs⁺ permanently within the mica or could be fed directly into the glass vitrifier without loss of Cs⁺ and without generating benzene as a by-product. The only disadvantage is the high affinity of the STB-mica complex for K⁺. While the K_d's are lower for K⁺, they are still too high for solutions of high potassium content. We are continuing studies to improve the selectivity of Cs⁺ relative to K⁺.

Acknowledgment

The authors acknowledge with thanks financial support of this study by the Department of Energy and Battelle Memorial Institute, PNL on contract 198567-A-F1.

UPTAKE OF ACTINIDES AND OTHER IONS BY DIPHOSIL, A NEW SILICA-BASED CHELATING ION EXCHANGE RESIN

R. Chiarizia, E. P. Horwitz and K. A. D'Arcy
Chemistry Division, Argonne National Laboratory
Argonne, Illinois 60439, USA

S. D. Alexandratos and A. W. Trochimczuk
Department of Chemistry, University of Tennessee
Knoxville, TN 37996, USA

1 INTRODUCTION

A new chelating ion exchange resin, Diphonix, containing geminally substituted diphosphonic acid ligands bonded to a styrene-based polymeric matrix, was recently developed and characterized.[1-8] Its high affinity for actinides even in very acidic solutions has been attributed to the capacity of the resin to chelate actinides through either ionized or neutral diphosphonic acid ligands, leading to metal complexes of high stability.[1] Based on this property, the Diphonix resin has found application in mixed waste treatment and in analytical procedures for separation and determination of actinides in soil and bioassay samples.[9,10]

The only effective way of stripping actinides from the resin is to use as stripping agents compounds belonging to the family of aqueous soluble diphosphonic acids, which contain the same ligand group as the resin.[1] After stripping, the aqueous diphosphonic acid has to be thermally degraded for further processing of the actinides, leading to solutions containing high concentrations of phosphoric acid. An alternative procedure is to destroy, through wet oxidation, the whole resin bed used to sorb the actinides. This procedure is time and reagent consuming. Therefore, it may be desirable in most cases to leave the actinides on the resin which becomes a solid waste. In this case it would be preferable that some or all of the polymeric resin matrix be replaced by an inorganic material, to avoid possible generation of gases due to slow radiolytical degradation of the organic polymer. Therefore, a new version of the Diphonix resin has been prepared, where the chelating diphosphonic acid groups are grafted to a silica support. The new material is called Diphosil, for Diphonix on silica.

In this paper, some results on the equilibrium and the kinetics of uptake of a number of actinide species and of other metal ions of nuclear, environmental or hydrometallurgical interest by the new Diphosil material are reported. A more complete account of the investigation on the metal uptake properties of the Diphosil resin will be published elsewhere.[11]

2 EXPERIMENTAL

2.1 Resins

The Diphosil resin was prepared as follows; 10 g of porous silica (Davisil™, 60-100 mesh size, Aldrich) were reacted at reflux for 12 hours with a mixture of trichloroethylsilane (TCES) and trichlorovinylsilane (TCVS) in 30 mL toluene. After toluene and acetone washes, the material was dried and grafted with vinylbenzyl chloride (VBC). The grafting procedure was performed by immersing the silica-TCVS/TCES material in a mixture of VBC and azobis(isobutyro)nitrile (AIBN) initiator in toluene, heating at 80 °C for 8 hours under continuous stirring. The grafted material was then washed with toluene and extracted with this solvent for 10 hours. In a separate flask, the tetraisopropyl ester of methylenediphosphonic acid was dissolved in toluene and reacted with sodium metal. After the reaction was completed, the VBC grafted silica-TCVS/TCES material was added and the mixture heated for 24 hours at reflux. The material was then washed with acetone, acetone/water and water, hydrolysed with 3 $\underline{M}$ HCl for 3 hours and finally extensively washed with water.

The Diphonix resin in the H^+ form and 50-100 mesh size was obtained from Eichrom Industries, Inc., Darien, Illinois. Figure 1 reports representative structures of the two resins. Table 1 reports their capacity and density determined following literature procedures.[12] Phosphorus elemental analysis was carried out by digesting a resin sample with perchloric/nitric acid (2:1 v/v) followed by reaction with amidol and ammonium molybdate.[13] The capacity values reported in Table 1 are consistent with the resin formulae given in Figure 1.

2.2 Uptake Measurements

The sorption of ^{241}Am, ^{239}Pu, ^{59}Fe, ^{45}Ca, and freshly purified ^{230}Th and ^{233}U at tracer level concentration from aqueous solutions of different compositions was measured at room temperature (24 ± 1 °C) following the same procedure reported previously.[1,8] Dry weight distribution ratios, D, were calculated as

$$D = \left(\frac{A_o - A_f}{A_f} \right) \left(\frac{V}{w} \right)$$

(1)

where A_o and A_f are the aqueous phase activity (proportional to concentration) before and after equilibration, respectively, w is the weight of dry resin (g) and V is the volume of the aqueous phase (mL). In the experiments on the uptake of Al(III), analyses of the metal were performed by spectrometry. In each experiment, the amounts of metal and of resin were chosen in such a way that the resin capacity was always in large excess over the metal. Duplicate experiments showed that the reproducibility of the D measurements was generally within 10 %, although the uncertainty interval was substantially higher for the highest D values ($D \geq 10^3$).

The experiments on uptake kinetics were performed following the same discontinuous technique described previously.[4] The primary kinetic data, D vs. time, were normalized by converting them into fractional attainment of equilibrium vs. time data, using eq. 1. The kinetic data were then plotted according to the previously derived

$$\ln(1 - F) = -kt \tag{2}$$

where k is the experimentally observed overall rate constant and the fractional attainment of equilibrium, F, is defined as the amount of metal in the resin phase at time t divided by the amount of metal in the resin phase at equilibrium.

Equation 2 has been reported to hold for film-diffusion control with infinite solution volume for the limiting case of isotopic exchange on ion exchange resins.[14,15] Based on the considerations discussed in our previous study of the kinetics of metal uptake by the Diphonix resin,[4] eq. 2 should apply to the kinetic data obtained in this work.

DIPHONIX RESIN

DIPHOSIL RESIN

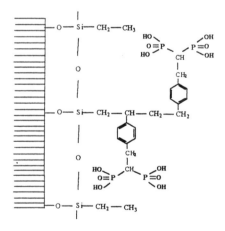

Figure 1 *Chemical structures of the Diphonix and Diphosil resins*

Table 1 *Capacity and Density of the Diphonix and Diphosil Resins*

	Diphonix	*Diphosil*
P capacity (mmol/dry g)	1.64	0.68
Total H capacity (mmol/dry g)	7.31	1.36
Sulfonic H capacity	2.42	/
Wet density (g/mL)	1.16	1.50
Bed density (g dry resin / mL bed)	0.30	0.39

3 RESULTS AND DISCUSSION

3.1 Uptake of Metal Ions

In comparing the tracer concentration level metal uptake data obtained with the Diphonix and the Diphosil resins, a number of factors must be considered.

The first factor is the different nature of the two resins. Although the Diphonix resin has a higher P capacity (see Table 1) than the Diphosil resin, in the former material the diphosphonic acid groups are homogeneously dispersed through the whole volume of the bead, while in the Diphosil resin they are concentrated in a surface layer where the ligand groups are closely packed. The higher concentration of ligand groups on the surface of the Diphosil resin may facilitate the interaction of incoming metal ions with more than one ligand at the same time, thus favoring complexation of metal ions with a higher charge and more complex coordination geometry, such as the tetravalent actinides.

Another factor to be considered is that the benzene ring separated by a CH_2 group from the diphosphonic acid group of the Diphosil material will somewhat reduce the basicity of the P=O group, reducing the tendency of the ligand for metal complexation.

Finally, the presence of other functional groups on the Diphonix resin, such as the sulfonic acid group, may play a role in the binding of metals. The sulfonic acid groups may take part in metal uptake by providing negatively charged sites which can contribute to the metal charge neutralization.

All these factors, that is, concentration and packing of the ligand, P=O basicity, coordination number and geometry of the metal species, charge neutralization requirements, surface reaction (Diphosil) vs. involvement of bulk material (Diphonix), make a complete description of the interaction between the two resins and metal ions quite difficult and the relative behavior of each single metal species difficult to predict.

This is clearly illustrated in Figures 2 and 3. Figure 2 shows a comparison of the uptake of actinide species by the two resins measured as function of the aqueous acidity. With both resins, the almost complete lack of acid dependency for hexavalent and tetravalent actinides indicates coordination of neutral metal salts by the P=O groups of the ligand without displacement of hydrogen ions. These results have been discussed at length in some of our previous studies.[1,6] Comparing the behavior of the two resins, for Am(III) and U(VI) the data show little change in going from the Diphonix to the Diphosil resin, with the former exhibiting slightly higher D values. However, the previously anticipated higher uptake of tetravalent actinides is clearly shown by the Pu(IV) data. The Th(IV) data show that the uptake by the Diphosil material is much less sensitive to the acidity than with the Diphonix resin. The slow decline of the D values of tetravalent actinides in the range of high aqueous nitric acid concentrations has been explained as mostly due to the competition of nitric acid with the metal for the phosphoryl groups of the resin.[1] This competition should be less effective if the basicity of the phosphoryl groups is reduced, in agreement with the trend indicated by the Th(IV)-Diphosil data.

Figure 3 reports similar data for Al(III), Fe(III) and Ca(II). The comparison of the uptake data of Al(III) and Fe(III) shows that the Diphosil resin exhibits a much higher affinity for trivalent iron and a much lower affinity for aluminum relative to the Diphonix resin. It is known that aqueous soluble diphosphonic acids form stronger complexes with iron (III) than with aluminum. For example, with methylenediphosphonic acid (MDPA), 1:1 and 1:2 complexes are reported with log values equal to 19.9 and 26.6 for Fe(III) and to 14.1 and 23.0 for Al(III).[16] Thus, a better complexation of Fe(III) over Al(III) by the two resins is expected and indeed verified. However, the Diphosil resin seems to magnify the difference in complexation of the two metal species by the diphosphonic acid group. This, in turn, may be an indication that in the Diphonix resin the sulfonic acid groups also play a role in the binding of Al(III), a role which would obviously be absent in the case of the Diphosil resin.

As already discussed in the case of Th(IV), the Fe(III)-Diphosil D values in the high acidity range do not drop with increasing acidity. Again, this can be explained by the reduced competition of the acid with the metal for the less basic P=O groups of the Diphosil resin.

Ca(II) exhibits substantially lower D values with the Diphosil resin. It is difficult to attribute this behavior to a single reason. The most obvious explanation would be the involvement of the sulfonic groups of the Diphonix resin in the uptake of these metals. Yet, at least in the case of the alkali earth cations, it has been demonstrated previously, based on the identical relative affinity measured with the two resins for the cations of the series Ca, Sr, Ba, and Ra, that in the Diphonix resin the cations are bound mainly to the diphosphonic acid groups.[7]

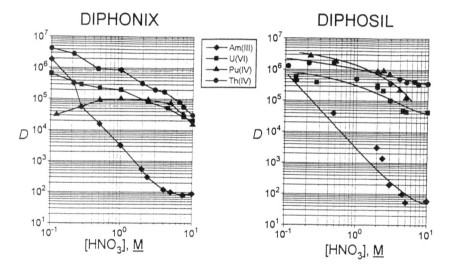

Figure 2 *Comparison of the uptake of Am(III), U(VI), Pu(IV) and Th(IV) by the Diphonix and the Diphosil resins from nitric acid solutions*

3.2 Kinetics of Metal Uptake

Figures 4 reports the rate of uptake of Am(III) by the Diphonix and the Diphosil resins. The data are plotted according to eq. 2, that is, in the form of fractional attainment of equilibrium vs time plots. The data show that in most cases about 99% of the equilibrium distribution is reached within 5 - 10 minutes of contact, that is, with both resins, and especially with the Diphosil one, the uptake of metal species at tracer concentration level is very rapid. From a more general point of view, the data show, as expected based on the very low metal concentration, that the metal uptake is film-diffusion controlled for at least 90 % of the process.

The data of Figures 4 show that the metal uptake is faster at the lower concentrations of aqueous acid, that is when the equilibrium distribution ratio is higher. Similar data, not

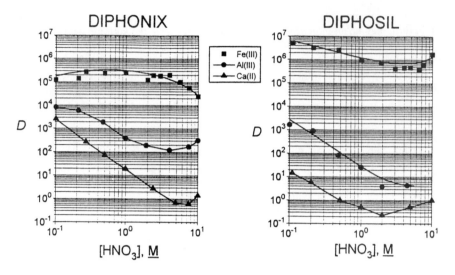

Figure 3 *Comparison of the uptake of Al(III), Fe(III) and Ca(II) by the Diphonix and the Diphosil resins from nitric acid solutions*

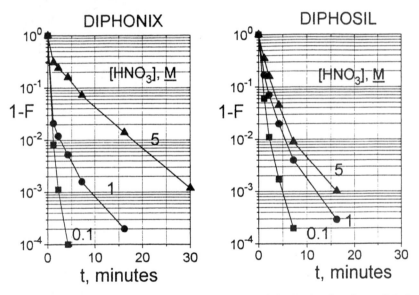

Figure 4 *Kinetics of Am(III) uptake by the Diphonix and the Diphosil resins at 0.1, 1 and 5 M HNO₃*

reported here, have been obtained for U(VI), Th(IV) and Fe(III). A complete discussion of the effect of selectivity on the ion exchange rate can be found in the literature.[15] Similar results obtained with the Diphonix resin have been previously reported and discussed in detail.[4] Here it is worth mentioning that the effect is larger with the Diphonix than with the Diphosil resin. The reason for this difference can be understood by considering that with the Diphosil resin the metal ions react only with the ligands contained in a surface layer, and the resin bead is not penetrated by the external solution. The different kinetic behavior at the different acidities is then the true selectivity effect on the rate of uptake. With the Diphonix resin, on the other hand, the metal uptake is accompanied by the resin swelling, which is larger for more diluted external solutions.

4 CONCLUSIONS

The Diphosil resin is a new material containing chelating diphosphonic acid groups grafted to a silica support. The equilibrium and kinetics of the uptake of actinide ions by the new resin have been measured in a variety of conditions and compared with those of the previously investigated Diphonix resin, where the same ligand groups are bonded to a styrene-based polymeric matrix.

The measurements reported in this paper have allowed us to conclude that :

1. In the uptake of actinide and other ions at trace concentration level the two resins behave quite similarly although with some minor differences.
2. The Diphosil resin exhibits even higher affinity for tetravalent actinides than the Diphonix resin. The uptake of tri- and hexa-valent actinides is essentially the same for the two resins.
3. Ca(II) is substantially less sorbed by the Diphosil resin. Among the trivalent cations, Al(III) is less sorbed and Fe(III) is more sorbed by the Diphosil as compared to the Diphonix resin.
4. The Diphosil resin sorbs the actinide ions at trace concentration level very rapidly even at high acidities, in spite of the lack of swelling.
5. Based on the above results, the Diphosil resin appears particularly well suited for those cases where the recovery and further processing of the sorbed actinides is not required and the loaded resin can be considered as a solid waste. Because about 90 % of the Diphosil weight is silica, the problem of possible generation of gaseous compounds due to the slow radiolytical degradation of the organic components of the resin is minimized.

Acknowledgement

Work performed under the auspices of the Office of Basic Energy Sciences, Division of Chemical Sciences, U.S. Department of Energy, under contract number W-31-109-ENG-38.

References

1. E. P. Horwitz, R. Chiarizia, H. Diamond, R. C. Gatrone, S. D. Alexandratos, A.W. Trochimzuk and D. W. Creek, *Solvent Extr. Ion Exch.*, 1993, **11**(5), 943.
2. R. Chiarizia, E. P. Horwitz, R. C. Gatrone, S. D. Alexandratos, A. W. Trochimzuk and D. W. Creek, *Solvent Extr. Ion Exch.*, 1993, **11**(5), 967.
3. K. L. Nash, P. G. Rickert, J. V. Muntean and S. D. Alexandratos, *Solvent Extr. Ion Exch.*, 1994, **12**(1), 193.
4. R. Chiarizia, E. P. Horwitz and S. D. Alexandratos, *Solvent Extr. Ion Exch.*, 1994, **12**(1), 211.

5. E. P. Horwitz, R. Chiarizia and S. D. Alexandratos, *Solvent Extr. Ion Exch.*, 1994, **12**(4), 831.

6. R. Chiarizia and E. P. Horwitz, *Solvent Extr. Ion Exch.*, 1994, **12**(4), 847.

7. R. Chiarizia, J. R. Ferraro, K. A. D'Arcy and E. P. Horwitz, *Solvent Extr. Ion Exch.*, 1995, **13**(6), in press.

8. R. Chiarizia, K. A. D'Arcy, E. P. Horwitz, S. D. Alexandratos and A. W. Trochimczuk, *Solvent Extr. Ion Exch.*, 1995, submitted.

9. J. J. Hines, H. Diamond, J. E. Young, W. Mulac, R. Chiarizia and E. P. Horwitz, *Separ. Sci. Technol.*, 1995, **30**(7-9), 1373.

10. L. L. Smith, J. S. Crain, J. S. Yaeger, E. P. Horwitz, H. Diamond and R. Chiarizia, *J. Radioanal. Nucl. Chem.*, 1995, submitted.

11. R. Chiarizia, E. P. Horwitz, K. A. D'Arcy, S. D. Alexandratos and A. W. Trochimczuk, *Solvent Extr. Ion Exch.*, 1995, in preparation.

12. J. Korkish, "Handbook of Ion Exchange Resins : Their Application to Inorganic Analytical Chemistry", CRC Press, Boca Raton, Florida, 1989.

13. S. D. Alexandratos, A. W. Trochimczuk, D. W. Crick, E. P. Horwitz, R. A. Gatrone and R. Chiarizia, *Macromolecules*, in press.

14. F. Helfferich, in "Ion-Exchange", J. A. Marinsky, ed., Marcel Dekker, New York, 1966, Vol.1, Chap. 2.

15. F. Helfferich, "Ion Exchange", McGraw-Hill, New York, 1962.

16. E. N. Rizkalla, *Reviews in Inorg. Chem.*, 1983, **5**(3), 223.

THE REMOVAL OF CHLORIDE ION USING A NEW INORGANIC ANION EXCHANGER

H. Kodama and A. Watanabe

National Institute for Research in Inorganic Materials
Namiki 1-1
Tsukuba, Ibaraki
305 Japan

1 INTRODUCTION

Although many studies have been carried out on inorganic ion exchangers, they are mainly for cation removal and the study of removing anions is very limited. One of the reasons is that we cannot find so many compounds which contain OH or H_2O in their composition. These kind of compounds have the possibility to react with anions but, generally speaking, the compounds which contains OH or H_2O are not stable in strong acidic or strong alkaline solution. This also makes it difficult to find out an useful inorganic anion exchanger.

We have been studying the synthesis of new compounds belonging to Bi-O-NO_3 system and their ion exchange properties.[1-5] They do not contain OH or H_2O in their composition but contain NO_3, which can be expected to exchange with other anions. For example, new compounds such as $Bi_5O_7NO_3$ and $BiPbO_2NO_3$ were found out and their ion exchange properties were examined.

The present study reports a method to remove chloride ions from solution by use of a new compound belonging to Bi-O-NO_3 system. Its synthesis, structure, ion exchange property and ion exchange mechanism are studied. The present results are expected to be useful for removal and solidification of radioactive chloride ions.

2 SYNTHESIS AND STRUCTURE OF A NEW COMPOUND

2.1 Experimental

A method and procedure for synthesizing a new material were almost the same as those used in the previous study for synthesizing $Bi_5O_7NO_3$.[2] That is, the preparation was carried out by thermal decomposition of basic bismuth nitrate, $4Bi(NO_3)(OH)_2.BiO(OH)$ in air, the decomposition of which was controlled by heating temperature and heating time.

At first, a temperature range necessary for preparing the new material was obtained from the previous data, which was evaluated by means of thermal analysis-mass spectrometry.[2] The previous experimental results on the thermal decomposition of basic bismuth nitrate shows that the thermal decomposition above 350°C is suitable for preparing a compound belonging to Bi-O-NO$_3$ system, since the decomposed products do not contain H$_2$O or OH.

Moreover, in the previous study, the thermal decomposition of basic bismuth nitrate was studied over the temperature range 400°C to 515°C and Bi$_5$O$_7$NO$_3$ was obtained. In the present experiment, the thermal decomposition of basic bismuth nitrate was studied below the temperature 400°C, because in the previous study, we found an unknown compound produced with Bi$_5$O$_7$NO$_3$ by insufficient decomposition of the starting material. Hence, the present decomposition is studied over the temperature range 350°C to 400°C.

Within the evaluated temperature range, the heating temperature and the heating time for preparing the compound were determined as follows: basic bismuth nitrate (about 2 g) was charged in a platinum crucible and heated in air at a constant temperature for a constant time. After heating, decomposed products were quenched and characterized by X-ray diffraction analysis. The NO$_3$ content was determined by means of mass spectrometry and thermogravimetric analysis. The thermal and mass data were obtained at the same time on a computer-interfaced TG-DTA/MS system of MAC Science Co. using a heating rate of 10°C/min.

Thus, by changing the heating temperature and heating time at some intervals, the best conditions for synthesizing a pure compound were established.

2.2 Results

The experimental results are shown in Table 1. A new compound was obtained by heating basic bismuth nitrate at the temperature and the time shown in Table 1. Its composition is represented by a formula, Bi$_{10}$O$_{13+X}$(NO$_3$)$_{4-2X}$ (-0.18 $\leq$ X $\leq$ 0.29). As shown in the formula, its composition is nonstoichiometric. The composition and its nonstochiometric range was determined experimentally by means of mass spectrometry and thermogravimetric analysis and by using X-ray diffraction (XRD) patterns.

Table 1 *The Successful Experimental Conditions of the Synthesis of*
Bi$_{10}$O$_{13+X}$(NO$_3$)$_{4-2X}$ (-0.18 $\leq$ X $\leq$ 0.29)

Heating Temperature / °C	Heating Time / h
350	16.0 ~ 70
360	5.5 ~ 40
375	2.0 ~ 10

The XRD data of Bi$_{10}$O$_{13+X}$(NO$_3$)$_{4-2X}$ (-0.18 $\leq$ X $\leq$ 0.29) are given in Table 2. All the diffraction peaks were indexed on the tetragonal cell of a = 7.965 Å and c = 20.417 Å. Each peak of the XRD pattern is broad and this suggests the product is not so well crystallized. The reaction product is white and fluffy powder.

Table 2   *X-ray Powder Diffraction Data for $Bi_{10}O_{13+X}(NO_3)_{4-2X}$ (- 0.18 ≤ X ≤ 0.29)*

(h k l)	$d_{cal}(Å)$	$d_{obs}(Å)$	$I_{obs}(\%)$
0 0 2	9.928	10.209	13
2 0 2	3.729	3.710	7
2 1 3	3.153	3.156	100
2 2 0	2.811	2.816	39
2 0 6	2.586	2.587	10
3 2 1	2.191	2.196	6
4 0 1	1.986	1.982	12
3 1 7	1.903	1.906	12
4 1 6	1.680	1.680	15
4 2 6	1.578	1.578	7

3 ION EXCHANGE REACTION

3.1 Experimental

$Bi_{10}O_{13}(NO_3)_4$ was equilibrated in solution. The reaction was carried out with shaking in a plastic tube stopped tightly with a lid and placed in a thermostatic container. The test tube was confirmed to be airtight by measuring its mass both before and after the reaction.

After the reaction, solid was separated from solution and identified by X-ray powder diffraction patterns. The anion concentration was determined by means of ion chromatography, using a DIONEX 4500 i instrument.

3.2 Extent of Ion Exchange Reaction

The existent of the ion exchange reaction of $Bi_{10}O_{13}(NO_3)_4$ with aqueous halogenide ions was examined as a function of time at 50°C. The experiments were carried out in solution without adjusting pH. The reaction was examined in NaX (X = F, Cl, Br, I) solutions. The experimental conditions were as follows: mass of $Bi_{10}O_{13}(NO_3)_4$, 244 mg; concentration of NaX solution, 0.1 mol dm^{-3}; volume of NaX solution; 0.1 mL. The result is given in Figure 1, where only chloride ion is removed very well. For example, after the reaction of 48 hours, chloride ion more than 99 % was removed, but in the case of the other ions, more than 40 % of them remained in solution.

The extent of the ion exchange reaction of $Bi_{10}O_{13}(NO_3)_4$ with chloride ion was examined as a function of time, at 25°C and 75°C. The experimental conditions were the same as in the experiment at 50°C. The curves 1, 2 and 3 in Figure 2 correspond to the results at 25°C, 50°C and 75°C. They show that the extent of the reaction increased with an increase in temperature. For example, the reaction time necessary for removing more than 99 % of chloride ions is 72 hours at 25°C, 48 hours at 50°C and only 3 hours at 75°C.

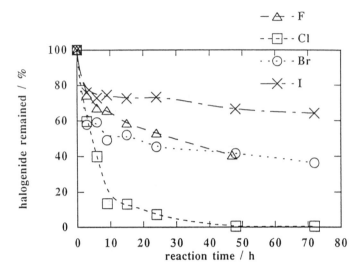

Figure 1 *Extent of reaction of various halogenide at 50°C,*
halogenide remained vs. reaction time

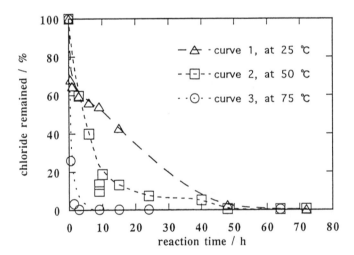

Figure 2 *Extent of reaction of chloride at 25°C, 50°C and 75°C,*
chloride remained vs. reaction time

3.3   Ion Exchange Reaction

The ion exchange reaction of $Bi_{10}O_{13}(NO_3)_4$ with chloride ion was studied in 0.05, 0.005 and 0.0005 mol dm^{-3} NaCl solutions, pH of which were not adjusted.   Powder $Bi_{10}O_{13}(NO_3)_4$ (2.00 g) was placed in 10 mL of solution at 50°C for 72 hours.

Table 3 gives the experimental results.   Chloride ion was removed very well from all the solutions.

Table 3   *The Results of the Ion Exchange Reaction*

The Concentration of Chloride Ion / mol dm^{-3}	
Before Reaction	After Reaction
5×10^{-2}	1.408×10^{-4}
(1773 ppm)	(4.99 ppm)
5×10^{-3}	0.457×10^{-5}
(177.3 ppm)	(0.16 ppm)
5×10^{-4}	0.502×10^{-6}
(17.7 ppm)	(0.01 ppm)

3.4   Ion Exchange Capacity

Accurately weighted $Bi_{10}O_{13}(NO_3)_4$ (100.0 ~ 102.0 mg) was equilibrated with 0.2 mol dm^{-3} NaCl solutions (1 mL) for 48 hours at 25°C, 50°C and 75°C. The ion exchange capacity was measured with the solution pH adjusted to various values from 1 to 13. Experimental results are shown in Figure 3, where the curve 1 corresponds to the experiment at 50°C. The measurements at 25°C and 75°C were carried out in only three solutions of pH = 1, 6.7 and 13, and the results are given by points.

Figure 3 shows that the ion exchange capacity of $Bi_{10}O_{13}(NO_3)_4$ has the largest value in the solution adjusted pH to 1 and the smallest value in the solution adjusted pH to 13. The ion exchange capacity measured at 50°C and 75°C were higher than the one measured at 25°C without the measurement in the solution of pH = 13.

If this ion exchange reaction proceeds only by the exchange of $(NO_3)^-$ with Cl$^-$, the calculated maximum value of the ion exchange capacity is 1.57 meqe / g, which is shown by a dot line in Figure 3. The ion exchange capacity observed in the solution of pH = 1 is bigger than the maximum value and the ion exchange capacity observed in the pH range from 2 to 11 is almost the same as the calculated maximum value. Exchange mechanism of chloride ion with $(NO_3)^-$ was very complicated.

4 MECHANISM OF ION EXCHANGE

Mechanism of ion exchange reaction was studied using $Bi_{10}O_{13}(NO_3)_4$. If all of the $(NO_3)^-$ in $Bi_{10}O_{13}(NO_3)_4$ are exchanged with Cl$^-$ in keeping with it's chemical formula, the ion exchange reaction should be represented as follows;

$$Bi_{10}O_{13}(NO_3)_4 + 4Cl^- \rightarrow Bi_{10}O_{13}Cl_4 + 4(NO_3)^- \qquad (1)$$

When we discuss this reaction, it is important whether the compound, $Bi_{10}O_{13}Cl_4$ can actually exist or not. Compounds belonging to Bi-O-Cl system, five kinds of compounds, $Bi_{12}O_{17}Cl_2$, Bi_3O_4Cl, $Bi_{24}O_{31}Cl_{10}$, $Bi_4O_5Cl_2$[6,7] and BiOCl[8] are known but $Bi_{10}O_{13}Cl_4$ has not been reported. Therefore, the possibility that the ion exchange reaction proceeds by the formula (1) is quite low.

The XRD patterns and TG-DTA/MS data were used to examine the reaction mechanism. Solid separated after reaction from solution was identified by XRD patterns and thermal decomposition was observed by means of TG-DTA/MS system to get information on the composition of the reaction product. After the decomposition, the XRD patterns of the products were observed again.

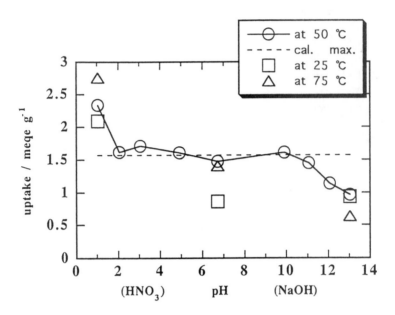

Figure 3 *Ion exchange capacity in solution adjusted pH to various values*

4.1 Reaction in Solution of pH = 1

The XRD pattern suggests that the solid is a mixture of the starting material, $Bi_{10}O_{13}(NO_3)_4$ and a reaction product, BiOCl. All of the peaks in the later profile are broad and some of them are slightly shifted to the high angle side.

The TG-DTA and MS data are shown in Figures 4 and 5. They show that the solid is decomposed especially intensely near the temperatures 240°C and 420°C. After the TG-DTA/MS experiments, the decomposed products were identified by XRD patterns. The pattern suggested it was a mixture of BiOCl and $Bi_{24}O_{31}Cl_{10}$. On basis of these data, we propose that the first decomposition near 240°C corresponds to the decomposition of BiOCl and the second decomposition near 420°C corresponds to the decomposition of $Bi_{10}O_{13}(NO_3)_4$. This means that the reaction product, BiOCl contains OH⁻ at a part of Cl⁻ site. The presence of $BiOCl_{1-x}(OH)_x$ is already reported by previous researchers.[9] We have also confirmed by other experiments that $BiOCl_{1-x}(OH)_x$ decomposes near 240°C with release of OH and that $Bi_{10}O_{13}(NO_3)_4$ decomposes near 400°C with release of NO. Pure BiOCl is stable up to about 600°C when heated in a flow of Ar.

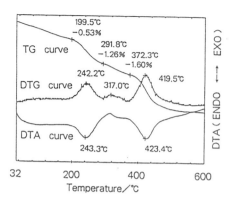

Figure 4 *TG, DTG and DTA curves of the solid produced in solution of pH = 1*

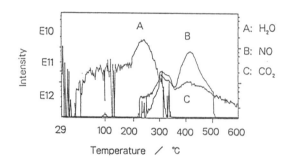

Figure 5 *The profile of ion intensity (logarithmic display) against temperature for the thermal decomposition of the solid produced in solution of pH = 1*

4.2 Reaction in Solution of pH = 13

The observed XRD pattern suggests that the reaction product is a mixture of Bi_3O_4Cl and $Bi_{12}O_{17}Cl_2$. The obtained TG-DTA/MS data are shown in Figures 6 and 7. The decomposed products in the experiments were a mixture of Bi_3O_4Cl and $Bi_{12}O_{17}Cl_2$ and these compounds are the same as the one before heating but the ratio of the amount of solid was changed. That is, by judging from their XRD patterns, Bi_3O_4Cl decreased and $Bi_{12}O_{17}Cl_2$ increased after the TG-DTA/MS experiment. As shown in Figure 7, the reaction product releases OH^- near 70°C. In this case, the OH^- may be not included as one of the composition but is adsorbed on the material surfaces, because the releasing temperature is very low and near the releasing temperature, any DTA peak was not observed. In Figure 7, release of fragment of CO_2 is observed at the temperature near 440°C. These species may come from the decomposition of bismutite, because it is formed easily and in large quantity in alkaline solution. The formation of bismutite disturbs the reaction of $Bi_{10}O_{13}(NO_3)_4$ with chloride ion.

Moreover, reaction products in solution pH = 5 was examined. It could not be identified by the observed XRPD pattern. The TG-DTA/MS data were almost same as Figures 4 and 5. The decomposed product in the experiment was $Bi_{24}O_{31}Cl_{10}$.

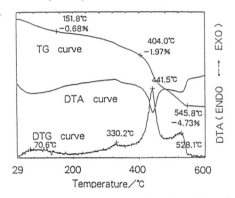

Figure 6 *TG, DTG and DTA curves of the solid produced in solution of pH = 13*

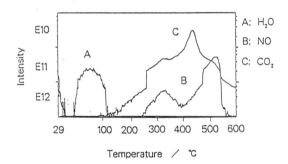

Figure 7 *The profile of ion intensity (logarithmic display) against temperature for the thermal decomposition of the solid produced in solution of pH = 13*

5 THE EFFECT OF OTHER ANIONS

The ion exchange reaction of $Bi_{10}O_{13}(NO_3)_4$ with aqueous chloride was studied under coexistence with F^-, Br^-, I^-, NO_3^- and HCO_3^- in neutral solution. The experiments were carried out by the following procedure:

Powered $Bi_{10}O_{13}(NO_3)_4$ (490 mg) was equilibrated at 50°C for 48 hours with a mixed solution of 0.1 or 0.01 mol dm^{-1} NaX (X = above anions). The total volume of solution is 1 mL.

The experimental results for Br^- and I^- are shown in Tables 4 and 5, which suggest that the coexistence of these ions disturbs significantly the reaction of $Bi_{10}O_{13}(NO_3)_4$ with aqueous chloride. In the case of other ions, their influence was too small to be measurable.

Table 4 *Effect of Coexisting Br⁻*

Before Reaction		After Reaction
$NaCl$ / $mol\ dm^{-3}$	$NaBr$ / $mol\ dm^{-3}$	Cl^- remained / %
0.1	0.01	0.3
0.1	0.1	92.0
0.01	0.1	6.6

Table 5 *Effect of Coexisting I⁻*

Before Reaction		After Reaction
$NaCl$ / $mol\ dm^{-3}$	NaI / $mol\ dm^{-3}$	Cl^- remained / %
0.1	0.01	30.4
0.1	0.1	100
0.01	0.1	6.6

References

1. H. Kodama, *The Proceedings of the Ion-Ex'93 Conference*, North East Wales Institute, Wrexham, UK, 1993, p. 55.
2. H. Kodama *J. Solid State Chem.*, 1994, **112**, 27.
3. H. Kodama, *Bull. Chem. Soc. Jpn.*, 1994, **67**, 1788.
4. H. Kodama, *The Proceedings of the Ion-Ex'95 Conference*, North East Wales Institute, Wrexham, UK, 1995, to be published.
5. H. Kodama, *The Proceeding of the ICIE'95*, Takamatsu, Japan, 1995, p. 285.
6. B. Z. Nugaliev, T. F. Vasekina, A. E. Baron, B. A. Popovkin, and A. V. Novoselova, *Russ. J. Inorg. Chem.*, 1983, **28**, 415.
7. B. Z. Urgaliev, B. A. Popovkin, and S. Yu. Stefanovich, *Russ J. Inorg. Chem.*, 1983, **28**, 1252.
8. Swanson et al., NBS Circular 539 vol. **IV**, 1953, p. 54.
9. "Gmerins Handbuch der Anorganichen Chem.", Gmerin-Institute, Wismut (system nummer 19), p. 692.

SYNTHESIS AND INVESTIGATION OF ION EXCHANGE PROPERTIES OF THE SODIUM TITANIUM SILICATE $Na_2Ti_2O_3(SiO_4) \cdot 2H_2O$

A. Clearfield, A. I. Bortun and L. N. Bortun

Department of Chemistry
Texas A&M University
College Station
TX 77843, USA

1 INTRODUCTION

Among the inorganic adsorbents zeolites are the most widely used now in industry, agriculture, for environment protection, etc.[1,2] This is connected with their valuable properties: high thermal and radiation stability, high adsorption capacity, marked selectivity, as well as with the possibility of their preparation in a granular form. Unfortunately aluminosilicate based zeolites have one essential drawback, a low chemical stability in acid and alkaline media, which imposes certain restrictions on their application.

An active search for new types of zeolite-like materials undertaken in the last decades has shown that other types of inorganic compounds are also able to form zeolitic type structures. Special attention among such new materials is given to titanosilicates, possessing high chemical stability, and which are regarded now as promising materials for heterogeneous shape selective catalysis.[3,4] At the same time the presence of ion exchange functional groups and the existence in their structure of well defined channels and cavities suggests that crystalline titanosilicates will be able to exhibit unique ion exchange properties, namely, high selectivity to certain ions. Confirmation of this statement is derived from the data[5,6] that the crystalline sodium titanium silicate (STS) $Na_2Ti_2O_3SiO_4 \cdot 2H_2O$ shows a high affinity to cesium, which makes it now one of the most promising materials for nuclear waste remediation. On studying the STS structure Clearfield and co-authors[7] have found that the possible reason for such behavior is connected with the correspondence of the geometrical parameters of its channels to the size of the selectively adsorbed Cs^+ ion.

There are known now several methods of $Na_2Ti_2O_3SiO_4 \cdot 2H_2O$ synthesis and all of them include the direct reaction of organic or inorganic Ti- and Si-containing compounds in alkaline media at mild hydrothermal conditions.[6,8] The molar Ti:Si ratio, duration of synthesis and some other nuances of preparation greatly affect the crystallinity and hence the selectivity of materials to cesium. In this communication a new method of titanium silicate preparation of composition $Na_2Ti_2O_3SiO_4 \cdot 2H_2O$ by the hydrothermal hydrolytic decomposition of crystalline or amorphous titanosilicophosphates in alkaline media is presented and the systematic study of STS ion exchange properties are discussed.

2 EXPERIMENTAL

2.1 Reagents

All reagents were of analytical grade (Aldrich).

2.2 Synthesis of Sodium Titanosilicate

The essence of the proposed method is the use as the initial reagent compounds of formula $TiO_2 \cdot xSiO_2 \cdot P_2O_5 \cdot H_2O$ (x = 1.0-2.0) instead of the individual titanium or silicon-containing reagents. It is well known that the phosphates of the Group IV Elements are hydrolytically unstable in alkaline media.[9,10] They easily release phosphorus at pH higher than 8-10 with the formation of a highly reactive corresponding hydrous oxide in the salt form. Using this property the crystalline sodium titanium silicate $Na_2Ti_2O_3SiO_4 \cdot 2H_2O$ was obtained by a one-step hydrolytic decomposition of titanosilicophosphates (TiSP) in 1M NaOH solution at mild hydrothermal conditions (180°C) for 24-48 hours, according to the overall reaction (1):

$$2TiO_2 \cdot xSiO_2 \cdot P_2O_5 \cdot H_2O + [2(2x-1)+14]NaOH =$$
$$Na_2Ti_2O_3SiO_4 \cdot 2H_2O + (2x-1)Na_2SiO_3 + 4Na_3PO_4 + [(2x-1)+8]H_2O \qquad (1)$$

2.3 Analytical Procedures

The diffractometer used was a Seifert-Scintag PAD5 with CuK_α radiation. Thermal analysis was performed by a TA 4000 (under nitrogen, rate of heating 10°C/min). The content of titanium and silicon in the solids were determined by using a SpectraSpec Spectrometer DCP-AEC.

2.4 Ion Exchange Study

Adsorption experiments in batch conditions have been carried out from diffrent types of model solutions at V:m = (100-200) : 1 (mL:g) and ambient temperature. Contact time was 4 days (periodic shaking). The pH of model solutions after equilibration with adsorbent were measured using an Orion SA-720 pH meter. Residual concentrations of ions of interest in solutions were measured using Varian SpectrAA-250 plus atomic absorption spectrometer.

The affinity of crystalline sodium titanium silicates to cesium, strontium and some other elements was expressed through the distribution coefficient (K_d, mL/g) values that were found according to the formula (2)

$$K_d = (C_o - C_i / C_i) \cdot V/m \qquad (2)$$

where C_o, C_i - are the ion concentrations in the initial solution and in the solution after equilibration with adsorbent, respectively; V/m - is the volume to mass ratio.

3 RESULTS AND DISCUSSION

Analysis of the experimental data presented in Table 1 and in Figure 1 shows that in all the cases studied only the one type of material $Na_2Ti_2O_3SiO_4 \cdot H_2O$, characterized by the first XRD peak at 7.9 Å, is formed. It is also worth noting the high reproducibility of the material's crystallinity, which suggests the uniformity of their ion exchange properties.

In order to follow the course of reaction (1), where initial layered crystalline or amorphous material is transformed into the zeolite-like one, a series of additional experiments were carried out using step by step hydrolytic treatment of titanosilicophosphates with a small amount of sodium hydroxide in each stage. Certainly, such experiments can not represent absolutely adequately all the processes taking place in (1) due to interruption of the reaction, washing the intermediate products resulting in removal of some soluble ingredients, etc. But at the same time it gave us the possibility to isolate the main types of intermediate compounds formed in the course of the TiSP hydrolytic decomposition. Some of their XRD patterns are given in Figure 2.

Table 1 *Experimental Conditions of STS Synthesis (T=180°C)*

STS sample	Initial titanosilicophosphate	Amount of 1 M NaOH/ 1 g TiSP, mL	Time, days
STS1	$TiO_2 1.0SiO_2 P_2O_5 H_2O$	28	2
STS2	$TiO_2 1.5SiO_2 P_2O_5 H_2O$	25	2
STS3	$TiO_2 2.0SiO_2 P_2O_5 H_2O$	25	2

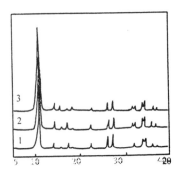

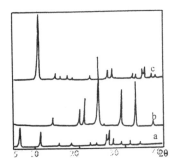

Figure 1 *The XRD patterns of crystalline sodium titanosilicates STS1 (1), STS2 (2) and STS3 (3)*

Figure 2 *The XRD patterns of the intermediate products that are formed during hydrolytic decomposition of TiSP: sodium titanosilicophosphate (a), $NaTi_2(PO_4)_3$ (b) and STS1 (c)*

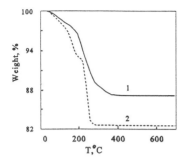

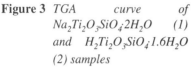

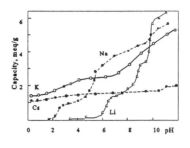

Figure 3 *TGA curve of $Na_2Ti_2O_3SiO_4·2H_2O$ (1) and $H_2Ti_2O_3SiO_4·1.6H_2O$ (2) samples*

Figure 4 *Potentiometric titration curves of $H_2Ti_2O_3SiO_4·1.6H_2O$ with 0.05 M LiOH, NaOH, KOH and CsOH*

Analysis of these data show that in the initial stages the formation of a mixture of the sodium or partially hydrolyzed sodium titanosilicophosphate phase[11] takes place (Figure 2,a). Further hydrothermal treatment in the presence of NaOH leads to the formation of only one crystalline phase $NaTi_2(PO_4)_3$ (Figure 2,b). This indicates that the TiSP begins to lose phosphorus and undergoes a cardinal structural change from layered to non-layered compound, which is accompanied by the release and amorphization of silicon-containing component. As a result of the continuation of alkaline hydrolysis a total loss of phosphorus takes place with the formation of titanium silicate $Na_2Ti_2O_3SiO_4·2H_2O$ (Figure 2,c) in the final stage.

A thermogravimetric analysis (TGA) study of the synthesized material shows that crystalline sodium titanosilicate and its hydrogen form are thermally stable compounds. As it is seen from Figure 3 the weight loss for sample STS1 occurs in several steps in the temperature range 50-400°C and is connected with the release of physically bound water and water localized in zeolite-like cavities.

Performance evaluation of the ion exchange behavior of the synthesized titanosilicates with channel structure towards alkali and alkaline-earth metal cations has been made by means of the potentiometric titration method. Analysis of the experimental data (Figure 4) shows that the ion exchange behavior is strongly dependent on the type of exchanged cation. Li^+ and Na^+ ions uptake on $H_2Ti_2O_3SiO_4·1.6H_2O$ occurs in three steps. Li^+ adsorption starts at pH > 3.5 and the first step ($pK_{a1} = 6.2$) is accomplished at pH = 7.5 (1.7 meq/g). The second step takes place in the pH range 7.5-9.1 ($pK_{a2} = 8.75$). $IEC_2 = 3.4$ meq/g. The third step of Li^+ uptake is observed in pH range 9.1-11.5 ($pK_{a3} = 9.6$). Maximum IEC_{Li} is 6.5 meq/g. In the case of Na^+ exchange only 1 meq of Na^+/g is adsorbed in the first step (pH < 3.5, $pK_{a1} = 2.5$). In the second stage (pH = 3.5-7.5, $pK_{a2} = 5.6$) an additional 3 meq Na^+/g is taken up. The maximum IEC_{Na} is 6.0 meq/g. Only one non distinct step is observed in the case of K^+ ion adsorption. Its uptake starts in acid solution with extremely high IEC_K value (2 meq/g, pH = 1). The increase of K^+ uptake starts at pH > 7 and the maximum IEC_K value is 5.3 meq/g at pH = 12. Adsorption of Cs^+ starts at pH < 0.5 (1.1 meq Cs^+/g, pH = 0.5) and the IEC_{Cs} values are practically independent on the pH of model solution.

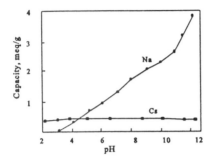

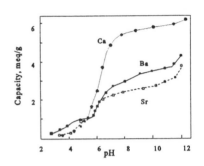

Figure 5 *Cesium   and   sodium*
potentiometric titration curves
of H₂Ti₂O₃SiO₄·1.6H₂O with
0.05M CsOH-NaOH solution

Figure 6 *Potentiometric titration curves*
of H₂Ti₂O₃SiO₄·1.6H₂O with
0.025 M Ca(OH)₂, Sr(OH)₂
and Ba(OH)₂

The maximum IEC_{Cs} is 2.05 meq/g (pH = 11.6) which does not differ significantly from the IEC values in low acid and neutral solutions (1.3-1.5 meq/g). It follows from here, that $H_2Ti_2O_3SiO_4 \cdot 1.6H_2O$ exhibits distinct ion sieve properties. In acid solution it preferably adsorbs K^+ and not Cs^+ ion, while in alkaline media a completely reversed apparent selectivity sequence is observed with $Li^+ > Na^+ > K^+ >> Cs^+$.

Certainly, adsorption from individual solutions does not represent adequately the selectivity of studied material. Taking this into consideration, additional experiments have been done from several complex model solutions. As it is seen from the potentiometric titration curves given in Figure 5 the presence in the system simultaneously of sodium and cesium ions results in decrease of the total sodium uptake and cesium ion uptake is reduced fourfold. It also suppresses drastically the acidity of the Ti-OH functional groups, responsible for the Na^+ ion exchange: adsorption of sodium ions starts only at pH higher than 3, instead of pH > 1.7, as was the case for the pure NaCl-NaOH solution. This indicates an antagonistic behavior of ions: the sodium ion poisons the cesium ion uptake and vise versa. Such a behavior can be explained by the peculiarities of the crystalline structure of $H_2Ti_2O_3SiO_4 \cdot 1.6H_2O$. It is known that the geometrical parameters of the STS channels are similar to the radius of the Cs^+ ion.[8] According to this concept the adsorption of one cesium ion per unit cell should prevent, by steric reason, the uptake of 3 sodium ions and, on the other hand, the adsorption of even one sodium ion should prevent (practically completely) the uptake of cesium ions.

Because nuclear waste solutions usually contain large amounts of sodium ion (and in some cases K^+ ion) cesium adsorption on STS was also studied in the presence of Na^+ and K^+ under a variety of conditions. The distribution coefficient (K_d) values characterizing the selectivity of materials for cesium are presented in Table 2. It is seen from Table 2 that sodium titanosilicates exhibit high preference to cesium over sodium in acid and neutral solutions but in basic media the K_d values are drastically decreased (to 20 in 5M Na^+ + 1M OH^- solution).

Table 2 K_d *Values for Cesium Sorption from Model Solutions, Containing 10^{-3} M CsCl, on Sodium Titanosilicates*

Solution	STS1	STS2	STS3
0.1 M KCl	7200	6800	7000
0.1 M KCl + 0.1 M HCl	13400	12800	12600
1 M NaNO$_3$ + 2 M HNO$_3$	1100	1200	1200
1 M NaNO$_3$ + 0.1 M HNO$_3$	> 30000	27000	24000
1 M NaNO$_3$	800	900	400
1 M NaOH	50	50	37
5 M NaNO$_3$ + 1 M NaOH	< 20	< 20	< 20

It is interesting to note that STS exchangers preferably adsorb cesium even in the presence of a hundred-fold excess of potassium which, at first glance, is in contradiction to the potentiometric titration data (Figure 4). In reality the potentiometric titration represents the macro effects in sorption (high filling rates) when both selective and non-selective adsorption centers are in action. When adsorption of micro amounts takes place, only selective ion exchange sites (amounts of which are negligibly small in comparison to the total IEC of exchanger < 2-3%) are responsible for cesium uptake. Analysis of the data suggests that STS adsorbents are very attractive materials for a selective cesium recovery from acid nuclear wastes, but they are not feasible for alkaline nuclear wastes remediation.

Potentiometric titration curves of sample STS1 with alkaline earth metal hydroxides are presented in Figure 6. The alkaline earth metal cations uptake by $H_2Ti_2O_3SiO_4 \cdot 1.6H_2O$ starts at pH > 2.5. The normal selectivity sequence was found in acid media: $Ba^{2+} > Sr^{2+} > Ca^{2+}$, although the absolute values of cation uptake are relatively low (less than 1 meq/g or less than 12% of theoretical IEC). The first step in the potentiometric titration curves is observed in the pH range 5-7 for Sr^{2+} and Ba^{2+} ions. In all cases the IEC values are about 2.5 meq/g, which is fairly close to 33% of theoretical IEC. Sr^{2+} and Ba^{2+} ions uptake in the pH range 7-11.5 increases monotonically from 2.5 meq/g to 3.0 meq/g for Sr^{2+} and 4.0 meq/g for Ba^{2+} ions. The second step on their potentiometric curves is observed only at pH > 11.5-12.0. In the case of calcium sorption a drastic increase in uptake is observed in the pH range 5.5-6.5. IEC_{Ca} is 6.2 meq/g (pH = 11), which is fairly close to those found for alkali metal cations.

The main peculiarity of alkaline earth metal cations adsorption on STS is that the IEC values in acid solutions are considerably lower than that for cesium and potassium ions. This allows us to suggest that alkaline earth metal cation will not be competitive for selective sorption of cesium from complex solutions.

The preliminary estimation of the affinity of sodium titanosilicate exchangers to strontium has been done from model 10^{-3}M $SrCl_2$-0.01M $CaCl_2$ solution (V:m=200:1). It was found that in neutral and low alkaline solutions $Na_2Ti_2O_3SiO_4$ exhibits an unexpectedly high affinity for Sr^{2+}: Sr/Ca separation coefficients are in the range 50-100. The same high preference to Sr^{2+} exhibited by STS in the presence of large amounts of sodium, potassium and magnesium cations. As an illustration there can be practically complete Sr^{2+} recovery (> 99.9%) from more than 1000 bed volumes of sea water simulating solution (Sr_{init}=2.7 mg/L) in column regime, which can not be done with the use of any other known organic or inorganic adsorbents.

Table 3 *Adsorption of two and three charged cations on STS*

Cation	K_d	pH
Cr^{3+}	> 100,000	7.2
Hg^{2+}	310	8.2
Pb^{2+}	13,100	8.5
Cd^{2+}	2,150	7.9
Cu^{2+}	16,600	7.2
Co^{2+}	8,300	7.7

Adsorption of some di- and trivalent cations (in trace amounts - 10^{-3} M) on $Na_2Ti_2O_3SiO_4$ has been studied from 1M NaCl ($NaNO_3$) model solutions. Analysis of the experimental data presented in Table 3 shows that STS exhibits extremely high affinity to Cr^{3+} and Cu^{2+} ions. The selectivity sequence found is the following: $Cr^{3+} >> Cu^{2+} > Pb^{2+} > Co^{2+} > Cd^{2+} >> Hg^{2+}$. It is therefore reasonable to expect that STS adsorbents can be used for the selective removal of heavy metal cations in the processing of high quality reagents.

4 CONCLUSIONS

A new method of preparation of sodium titanosilicate, $Na_2Ti_2O_3SiO_4$, with a porous channel structure has been developed utilizing the hydrothermal, hydrolytic decomposition of crystalline or amorphous titanosilicophosphates in alkaline media. The obtained materials were characterized with the use of X-ray and TGA methods. The ion exchange properties towards alkali, alkaline earth and some transition metal cations in model individual and complex solutions have been examined in both batch and column conditions. It was found that sodium titanium silicate $Na_2Ti_2O_3SiO_4$ exhibits high affinity to heavy alkali (K^+, Cs^+) and alkaline earth (Ba^{2+}, Sr^{2+}) metal cations in a wide pH range, which makes this adsorbent a promising material for acid and neutral nuclear waste remediation. Taking also into account that sodium titanosilicate is easily prepared and non toxic it is possible to expect its application for radioanalytical purposes and as an enterosorbent ($^{134+137}$Cs and ^{90}Sr antidotes) in medicine.

Acknowledgment

The authors acknowledge with thanks financial support of this study by the Department of Energy and Battelle memoriale Institute, PNL on contract 198567-A-F1.

References

1. C. B. Amphlett, "Inorganic Ion Exchangers", Elsevier, New York, 1964.
2. D. W. Breck, "Zeolite Molecular Sieves", Wiley and Sons, New York, 1973.
3. S. M. Kuznicki, K. A. Trush, F. M. Allen, S. M. Levine, M. M. Hamil, D. T. Hayhurst and M. Mansour, "Synthesis of Microporous Materials", 1992, p. 427.
4. D. M. Chapman and A. L. Roe, *Zeolites*, 1990, **10**, 730.
5. Chem. & Eng. News, 1992, July 13, 26.
6. R. G. Anthony, C. V. Philip and R. G. Dosch, *Waste Management*, 1993, **13**, 503.
7. D. M. Poojary, R. A. Cahill and A. Clearfield, *Chem. Mater.,* 1994, **6**, 2364.
8. A. Clearfield, A. Bortun and L. Bortun, *Solvent Extraction and Ion Exchange*, in press.
9. A. Clearfield, "Inorganic Ion Exchange Materials", CRC Press, Boca Raton, FL, 1982.
10. I. V. Tananaev, "Phosphates of the IV Group Elements", Nauka, Moscow, 1972.
11. J. R. Garcia, R. Llavona, M. Suarez and J. Rodriguez, *Trends Inorg. Chem.*, 1993, **3**, 209.

REMOVAL OF RADIUM AND THORIUM FROM WATER AND SALINE SOLUTIONS USING INORGANIC-ORGANIC COMPOSITE ABSORBERS WITH POLYACRYLONITRILE BINDING MATRIX

F. Šebesta, J. John and A. Motl
Department of Nuclear Chemistry, Czech Technical University in Prague
Brehová 7, 115 19 Prague 1, Czech Republic

E. W. Hooper
AEA Technology, Harwell Laboratory
Didcot, Oxfordshire OX11 0RA, UK

1 INTRODUCTION

Composite inorganic-organic absorbers represent a group of inorganic ion-exchangers modified by using binding organic material for preparation of larger size particles having higher granular strength. Such modification of originally powdered or microcrystalline inorganic ion-exchangers makes their application in packed beds possible.

Modified polyacrylonitrile (PAN) was proposed[1,2] as a universal binding polymer for practically any inorganic ion-exchanger (active component). The method developed enables preparation of granular absorbers even from such inorganic ion-exchangers that are impossible, or very difficult, to synthesise in granular form. The principle scheme of the preparation of composite absorbers containing PAN has been presented by Šebesta et al.[3]

The use of PAN-based organic binding polymer has a number of advantages due to the relatively easy modification of its physico-chemical properties (hydrophilicity, porosity, mechanical strength). The kinetics of ion exchange and sorption capacity of such composite absorbers is not influenced by the binding polymer mentioned above. The content of active component in the composite absorber can be varied over a very broad range (5 - 95 % of the dry weight of the composite absorber).

The proposed method can be applied to most of the inorganic ion-exchangers known. Up to now, some 24 types of composite absorbers have been prepared at the Czech Technical University in Prague. A full list of the composite absorbers has recently been published.[3-6] The procedure developed also enables preparation of composite absorbers containing mixtures of various active components. Properties of such a composite absorber can thus be exactly "tailored" to the specific features of the foreseen application.

These composite absorbers were proposed and tested for separation and concentration of various contaminants from aqueous solutions. Their high selectivity and sorption efficiency are advantageous for treatment of various radioactive and/or industrial waste waters, removal of natural and/or artificial radionuclides and heavy or toxic metals from underground water, determination of radionuclides (^{137}Cs, ^{60}Co, U, Ra) in the environment, etc. Examples of some of these applications were reviewed in several recent publications.[3-8]

This research was conducted in collaboration with the University of Oxford, Particle & Nuclear Physics Department, Nuclear Physics Laboratory with prospect of application of the composite absorbers at the Sudbury Neutrino Observatory. SNO, which is in the final stage of its construction in the Creighton mine will need to purify H_2O, D_2O and a chloride salt such as NaCl, used in the preparation of 0.25 % solution in D_2O for the neutrino detector, down to 10^{-15} g/g of uranium, thorium, and their daughter radionuclides.[9] Composite absorbers containing the following active components have been chosen for testing - barium sulphate activated by calcium, manganese dioxide, titanium dioxide, and sodium titanate. The results of this study may be also applied for the removal of radium and thorium from various surface or waste waters, e.g. from decommissioned uranium mines and mills.

2 EXPERIMENTAL

2.1 Chemicals and Instrumentation

A water suspension of hydrated titanium dioxide produced as industrial intermediate from the sulphate process of production of titanium pigments (Prerov Chemical Factory) was used for the preparation of titanium dioxide. All other chemicals were commercial preparations. A stock solution of ^{226}Ra (EB5) was supplied by the Czech Metrological Institute and contained 0.99 x 10^{-5} g ^{226}RaBr$_2$ per 1 g of solution containing of 1 g/L of BaCl$_2$.

A TESLA single-chamber analyser, with a scintillation tube with bare photomultiplier, was used for ^{226}Ra measurements. A Canberra multichannel analyser, with coaxial germanium detector (Princeton-Gamma Technologies), was used for the gamma-spectrometric determination of ^{234}Th. Beta-counting of ^{234}Th was carried out on a TESLA low-background counter with anticoincidence shielded gas-flow GM detector.

For the column experiments BIO-RAD Poly-Prep columns with 2 mL of absorbers were used. The flow rate of feed solution was controlled with peristaltic miniflow pump ELMED. In addition standard laboratory equipment (fraction collector, shaker, pH-meter, mixer, centrifuge, drying kiln) was used.

2.2 Methods

2.2.1 Preparation of Active Components. Barium sulphate activated by calcium (Ba[Ca]SO$_4$) was precipitated following the procedure described by Berák[10] by addition of one part of mixed solution of 1M BaCl$_2$ + 1M CaCl$_2$ (7 : 1) to two parts of 1M Na$_2$SO$_4$. *Manganese dioxide (MnO)* was prepared according to the equation

$$2 \text{ KMnO}_4 + 3 \text{ MnSO}_4 + 2 \text{ H}_2\text{O} \rightarrow 5 \text{ MnO}_2 + \text{K}_2\text{SO}_4 + 2 \text{ H}_2\text{SO}_4$$

by slow addition of a stoichiometric amount of a solution of potassium permanganate ($C_m = 0.22$) to a solution of manganese sulphate ($C_m = 0.5$) at 80 - 90°C with stirring. Titanium dioxide (TiO) was prepared by filtering and carefully washing the received suspension of industrial intermediate. Sodium titanate (NaTiO) was prepared according to Heinonen[11] by heating a suspension of titanium dioxide in a mixture of C_2H_5OH and NaOH solution. All the filtered products were dried in a drying kiln at 45 - 50°C (except for NaTiO that was dried at 105°C). The dried preparations were powdered and sieved.

2.2.2 Preparation of Composite Absorbers. All the composite absorbers were prepared following the general method of their preparation[2,3] from the respective powdered (< 0.1 mm) active components. The content of active component was 85 %

w/w dry weight for all the composite absorbers except for NaTiO-PAN, which was 92 % w/w dry weight. Composite absorbers with grain size 0.18 - 0.6 mm were used for the experiments with MnO-PAN, TiO-PAN, and NaTiO-PAN. In experiments with Ba[Ca]SO$_4$-PAN, absorber with grain size < 0.6 mm (fines removed by sedimentation) was used.

All absorbers were conditioned to the pH of the solutions to be tested (column and kinetics experiments) or to pH ~ 7 (K$_D$ determination). Successive additions of NaOH or HNO$_3$ solutions followed by contacting with the respective solutions were used.

2.2.3 Preparation of Labelled Solutions. Labelled solutions of [226]Ra were prepared by 10^5 times diluting the stock solution, they contained ~ 10^{-7} g/L of [226]Ra (3.7 kBq/L) and ~ 4.8 x 10^{-8} mol/L of Ba. Labelled solutions of [234]Th were prepared from carrier-free [234]Th separated from uranyl nitrate following to Alian et al.[12] and purified using modified procedure of Berman.[13] The labelled solutions were prepared 1 - 2 days before experiments.

2.2.4 Determination of [226]Ra and [234]Th. For the [226]Ra determination counting of gross alpha activity of [226]Ra and its radioactive decay daughter products was used. The procedure is based on the coprecipitation of radium with BaSO$_4$ in the presence of surplus sulphate anions at pH ~ 1.[14] Ba[Ra]SO$_4$ precipitate was mixed with ZnS[Ag] scintillator and filtered using a Whatman GF/C glass microfibre filter. Samples prepared from feed solutions and solutions contacted with absorbers were always measured at the same time. Gamma-spectrometric determination of [234]Th (E$_\gamma$ = 92.4 keV) was used for the evaluation of column experiments. Gross beta counting of [234]Th and its radioactive daughters ([234m]Pa, [234]Pa) was used for the evaluation of kinetics of sorption. 1 mL aliquots of solutions for analysis soaked into a paper disk were counted the day after preparation.

2.2.5 Column Experiments. A suspension of the absorber in the solution used for conditioning was transferred into a Poly-Prep column. The bed of the absorber (V$_{BED}$ = 2 mL) was briefly washed. Feeds were pumped through the columns in downward direction at a flow rate of 5.5 - 7.0 V$_{BED}$/hour. Fractions were collected in 1 hour intervals. In some fractions pH was measured and [226]Ra or [234]Th concentrations were determined. The breakthrough of [226]Ra or [234]Th (%) was calculated from their activities in the feeds and the single fractions.

2.2.6 Kinetics of Sorption. 1 mL of preconditioned absorber was transferred into a glass vial with 50 mL of solution containing [226]Ra or [234]Th. The contents of the vial were intensively mixed using a glass propeller. Samples of the suspension of absorber (2 mL) were taken at certain time intervals. The aqueous phase was separated by centrifugation for 2 - 3 minutes at 3000 rpm. Sorption yield (%) was calculated relative to the feed activity. Distribution coefficients after contact time of 120 minutes - (K$_D$)$_{120}$ were calculated as described below.

2.2.7 Determination of Distribution Coefficients. 1 mL of preconditioned absorber was separated from the solution by filtration through GF/C filter. The absorber (with the filter) was transferred into a polyethylene bottle with 50 mL of solution. The bottle was shaken for 24 hours at (21±1)°C. Blank experiment was performed with the same solution without absorber but with GF/C filter. pH was measured at the end of experiment. Aqueous phase was separated by centrifugation for 1min at 3000 rpm. K$_D$ was calculated using equation:

$$K_D = \frac{A_o - A}{A} \times \frac{V_w}{V_a} \qquad [mL/mL]$$

where A_o and A are count rates of aliquots of feed and the aqueous phase, respectively, and V_w and V_a are volume of aqueous phase and volume of the absorber, respectively.

3 RESULTS AND DISCUSSION

3.1 Kinetics of Sorption

The aim of this study was to compare the rate of radium and thorium uptake by all the absorbers studied and to determine any effect of sodium chloride concentration. Comparison of the rate of uptake of radium by the absorbers studied was performed for a 3 % NaCl solution (Figure 1). For all the absorbers the uptake of radium is rather rapid. The rate of radium absorption decreases in the order MnO-PAN ≈ NaTiO-PAN >>>> Ba[Ca]SO_4-PAN ≈ TiO-PAN. The effect of NaCl concentration on the uptake kinetics of ^{226}Ra was studied for MnO-PAN absorber (Figure 2) and was found to be very small.

The rate of thorium uptake by single absorbers was studied from water without sodium chloride (Figure 3). Thorium uptake by Ba[Ca]SO_4-PAN and TiO-PAN absorbers is slower than the uptake of radium. Uptake of thorium by MnO-PAN and NaTiO-PAN is very low, no conclusions on sorption rate can be easily made. The effect of concentration of NaCl on the uptake kinetics of ^{234}Th was studied for the TiO-PAN absorber (Figure 4). The rate of thorium uptake from 3 % NaCl solution was high, further increasing the NaCl concentration (10 % solution) resulted in a partial decrease. Pronounced increase of rate of thorium uptake after the addition of NaCl indicates that slow rate of thorium sorption from water without NaCl is probably caused by hydrolysis of thorium and/or presence of other non-ionic forms of thorium and not the properties of the absorbers.

3.2 Distribution Coefficients Measurement

Distribution coefficients of radium found for the absorbers studied are given in Table 1 together with equilibrium pH values. All the data are average values of duplicate analysis of two parallel experiments. Blank experiments revealed that in the case of 0 % NaCl, a large proportion of ^{226}Ra was sorbed onto the GF/C glass microfibre filter. The sorption of ^{226}Ra onto the PE - bottle and/or GF/C filter decreases with increasing concentration of NaCl.

From the comparison of measured data it follows that, for water without sodium chloride, distribution coefficients for the absorbers studied decrease in the order TiO-PAN >> Ba[Ca]SO_4-PAN > MnO-PAN >> NaTiO-PAN. The lower K_D values found in the case of MnO-PAN absorber may possibly be caused by partial peptisation and even washing of the active component out of the absorber.

For a 3 % solution of NaCl the order of distribution coefficients is MnO-PAN >> NaTiO-PAN > Ba[Ca]SO_4-PAN ≈ TiO-PAN. In the case of 10 % NaCl solution the K_D values decrease in the order MnO-PAN ≈ NaTiO-PAN > TiO-PAN >> Ba[Ca]SO_4-PAN.

For ^{234}Th, "equilibrium" distribution coefficients were not determined. The values of distribution coefficients after 120 minutes of contact $(K_D)_{120}$ (calculated from kinetics measurements data) are summarised in Table 2.

From the data obtained it follows that for water without NaCl distribution coefficients $(K_D)_{120}$ for the absorbers studied decrease in order Ba[Ca]SO$_4$-PAN >> TiO-PAN >> MnO-PAN > NaTiO-PAN. The influence of NaCl concentration on $(K_D)_{120}$ values for TiO-PAN is not very pronounced. It should be noted that not all the systems were close to equilibrium after 120 minutes of contact (see Figures 3 and 4).

3.3 Column Experiments

Breakthrough curves for ^{226}Ra and ^{234}Th were measured, for all the composite absorbers prepared, up to ~ 220 V_{BED} of 0 %, 3 % and 10 % NaCl solutions. Examples of the measured breakthrough curves are shown in Figures 5 - 10.

Sorption of ^{226}Ra on Ba[Ca]SO$_4$-PAN absorber (Figure 5) from water without NaCl did not depend on the volume of treated feed solution. The average breakthrough was found to be < 0.7 % (i.e. decontamination factor D_f > 140). In sodium chloride solutions the sorption of radium decreases with increasing concentration of sodium chloride. For 3 % NaCl the average breakthrough was found to be < 0.4 % (i.e. D_f > 250). ^{226}Ra uptake is thus more efficient from 3 % NaCl solution than from distilled water. Application of Ba[Ca]SO$_4$-PAN absorber for treatment of 10 % NaCl is not very advantageous (D_f ~ 20).

Sorption of ^{226}Ra on MnO-PAN absorber (Figure 6) from water without NaCl is similar to that on Ba[Ca]SO$_4$-PAN absorber. However, MnO-PAN absorber is much more effective for treatment of solutions containing NaCl. The average breakthrough of ^{226}Ra for the treatment of 200 V_{BED} of feed solution was found to be ~ 0.1 % (3 % NaCl) and ~ 0.2 % (10% NaCl) which corresponds to decontamination factors D_f ~ 1000 and D_f ~ 500, respectively. From the shape of the breakthrough curves it can be deduced that much larger volumes of sodium chloride solutions could be treated with similar efficiency. The lower efficiency of MnO-PAN absorber for the sorption of ^{226}Ra from water without sodium chloride might be caused by partial peptisation of the active component of the absorber.

TiO-PAN absorber is the most effective for separation of ^{226}Ra from water without sodium chloride. The average breakthrough was found to be 0.2 - 0.3 % (i.e. D_f ~ 500 - 300) when treating 200 V_{BED} of water. For 10 % NaCl solution the breakthrough increases rapidly from 0.1 % (at 50 V_{BED}) to ~ 6 % (at 200 V_{BED}).

The reason for testing NaTiO-PAN absorber was namely comparison of this absorber with TiO-PAN. From the results obtained it follows that NaTiO-PAN absorber is less effective for the separation of ^{226}Ra from water without NaCl. For 10 % NaCl NaTiO-PAN is more effective than TiO-PAN. It can be thus concluded, that there is no significant difference in sorption properties of both the absorbers towards ^{226}Ra.

In Figure 7, comparison of breakthrough curves of ^{226}Ra from 10% NaCl on all the absorbers studied is shown. These curves are a good illustration of correspondence between the distribution coefficients found (see Table 1) and breakthrough curves. The order of the curves in Figure 7 exactly matches the order of the distribution coefficients.

Column experiments with ^{234}Th have shown that separation of thorium from the studied solutions is much more complicated. Especially difficult is separation of ^{234}Th

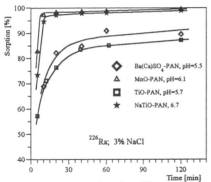

Figure 1 *Kinetics of uptake of ^{226}Ra from 3 % NaCl solution by the composite absorbers studied*

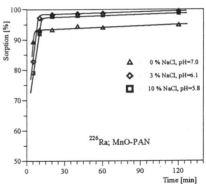

Figure 2 *Influence of NaCl concentration on the kinetics of uptake of ^{226}Ra by MnO-PAN composite absorber*

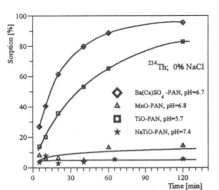

Figure 3 *Kinetics of uptake of ^{234}Th from distilled water without NaCl by the composite absorbers studied*

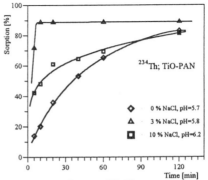

Figure 4 *Influence of NaCl concentration on the kinetics of uptake of ^{234}Th by TiO-PAN composite absorber*

Table 1 K_D *[mL/mL] Values for* ^{226}Ra

Absorber	0 % NaCl		3 % NaCl		10 % NaCl	
	pH	K_D	pH	K_D	pH	K_D
Ba[Ca]SO$_4$-PAN	6.9	2550	6.9	8250	7.2	1180
MnO-PAN	7.1	1430	6.8	13750	7.1	11790
TiO-PAN	6.7	3520	6.8	7200	7.0	8810
NaTiO-PAN	7.6	290	6.7	11630	6.7	12402

Table 2 $(K_D)_{120}$ *[mL/mL] Values for* ^{234}Th

Absorber	0 % NaCl		3 % NaCl		10 % NaCl	
	pH	$(K_D)_{120}$	pH	$(K_D)_{120}$	pH	$(K_D)_{120}$
Ba[Ca]SO$_4$-PAN	6.7	1100	-	-	-	-
MnO-PAN	6.8	8.5	-	-	-	-
TiO-PAN	5.7	240	5.8	370	6.2	220
NaTiO-PAN	7.4	3.0	-	-	-	-

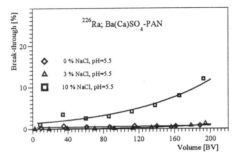

Figure 5   *Influence of NaCl concentration on the break-through curves of ^{226}Ra on Ba(Ca)SO$_4$-PAN composite absorber*

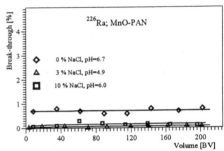

Figure 6   *Influence of NaCl concentration on the break-through curves of ^{226}Ra on MnO-PAN composite absorber*

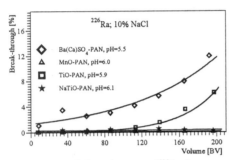

Figure 7   *Break-through curves of ^{226}Ra from 10 % NaCl solution for the composite absorbers studied*

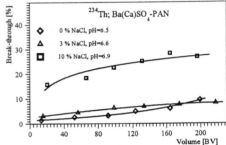

Figure 8   *Influence of NaCl concentration on the break-through curves of ^{234}Th on Ba(Ca)SO$_4$-PAN composite absorber*

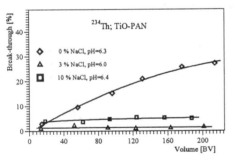

Figure 9   *Influence of NaCl concentration on the break-through curves of ^{234}Th on TiO-PAN composite absorber*

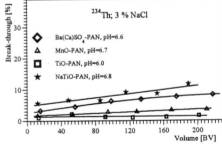

Figure 10   *Break-through curves of ^{234}Th from 3 % NaCl solution for the composite absorbers studied*

from water without NaCl. All the absorbers except for Ba[Ca]SO$_4$-PAN are for this purpose unsuitable. For this absorber, average breakthrough for treatment of 200 V$_{BED}$ of water was ~ 5% which corresponds to D$_f$ ~ 20 (Figure 8). Low sorption of thorium from water without NaCl at pH ~ 6 - 7 might be caused by hydrolysis of thorium and/or slow kinetics of sorption. This conclusion follows e.g. from relating high breakthrough of ^{234}Th through a column of TiO-PAN absorber (Figure 9) to $(K_D)_{120}$ distribution coefficient values (see Table 2) and data on kinetics of ^{234}Th from water without NaCl (Figure 4).

Separation of ^{234}Th on Ba[Ca]SO$_4$-PAN absorber from 3 % NaCl solution is slightly worse than from water without NaCl (average D$_f$ ~ 16). For all the other absorbers sorption of ^{234}Th is significantly higher from 3 % NaCl solution than from water without NaCl (for example see Figure 9). The average decontamination factors are D$_f$ ~ 35 (MnO-PAN), D$_f$ ~ 70 (TiO-PAN), D$_f$ ~ 12 (NaTiO-PAN); for the breakthrough curves see Figure 10.

The efficiency of separation of ^{234}Th from 10 % NaCl solution is lower when compared to that from 3 % NaCl. Best results were achieved with TiO-PAN absorber, the average decontamination factor for treatment of 200 V$_{BED}$ of solution was found to be D$_f$ ~ 20. The MnO-PAN absorber was only slightly less effective (D$_f$ ~ 18). From the comparison of TiO-PAN and NaTiO-PAN absorbers it follows that the efficiency of separation of ^{234}Th is, for all the solutions, worse for NaTiO-PAN absorber than for the TiO-PAN one.

4 CONCLUSIONS

For separation of radium from water without NaCl, TiO-PAN absorber was found to be most suitable (D$_f$ ~ 300 - 500). From the results of column experiments with sodium chloride solutions MnO-PAN absorber seems to show the best performance (D$_f$ ~ 500 - 1000).

For separation of thorium from water without sodium chloride only Ba[Ca]SO$_4$-PAN absorber may be used (D$_f$ ~ 20). TiO-PAN absorber is the most suitable for separation of thorium from sodium chloride solutions (D$_f$ ~ 70 for 3 % NaCl and D$_f$ ~ 20 for 10 % NaCl). Generally, it can concluded that the separation of thorium from the studied solutions is much more complicated than separation of radium.

For some prospective applications of composite absorbers it might be advantageous if both radium and thorium could be separated in one step using just one absorber. For purification of water without sodium chloride only Ba[Ca]SO$_4$-PAN might be used because of low sorption of thorium by all the other absorbers studied. TiO-PAN absorber is the most effective for separation of both radium and thorium from sodium chloride solutions.

Acknowledgement

Financial support for this research was provided by the University of Oxford, award No. DL 38685, and is gratefully acknowledged.

References

1. V. Stoy, F. Šebesta, V. Jirásek and A. Stoy, Czech Patent A. O. 181605, 1980.
2. F. Šebesta, Czech Patent A. O. 273369, 1992.
3. F. Šebesta, A. Motl, J. John, M. Prazský and J. Binka, *Proc. 1993 Int. Conf. Nucl. Wastes Management and Environmental Remediation*, Prague, Czech Republic, ASME, New York, 1993, Vol. 3, p. 871.
4. F. Šebesta, J. John and A. Motl, *Proc. SPECTRUM '94*, Atlanta, USA, ANS, LaGrange Park, IL, 1994, Vol. 2, p. 856.
5. F. Šebesta, J. John and A. Motl, *Proc. BUDAPEST '94, Second International Symposium and Exhibition of Environemntal Contamination in Central and Eastern Europe*, Budapest, Hungary, 1994, 20-23 September, p. 1052.
6. F. Šebesta and J. John, "An Overview of the Development, Testing and Application of Composite Absorbers", LA-12875MS, February, 1995.
7. F. Šebesta, *Proc. Radionuclides and Ionising Radiation in Water Management*, Harrachov, Czech Republic, Dum techniky CSVTS, Ústí nad Labem, 1990, p. 51.
8. F. Šebesta, J. John and A. Motl, *Progress Reports of the Research Contracts No. 7381/RB and 7381/R1/RB*, Faculty of Nuclear Sciences and Physical Engineering, Czech Technical University in Prague, January 1994 and February 1995, respectively.
9. G. T. Ewan, *Nucl. Instr. Meth. Phys. Res.*, 1992, **A314**, 373.
10. L. Berák, *Coll. Czech. Chem. Comm.*, 1965, **30**, 1490.
11. O. J. Heinonen, J. Lehto and J. Miettinen, *Radiochim. Acta.*, 1981, **28**, 93.
12. J. Alian et al., *J. Radioanal. Nucl. Chem. Letters*, 1986, **108**, 317.
13. S. S. Berman et al., *Talanta*, 1960, **4**, 153.
14. J. Sedlácek, F. Šebesta and E. Raimová, *Proc. Symp. The Mining Príbram in Science and Technology, Sect. Water Econ.*, CSVTS, Príbram, Czech Republic 1979, p. 211.

KINETICS AND THERMODYNAMICS OF ION EXCHANGE ON AMORPHOUS IRON (III) ANTIMONATE IN MIXED SOLVENTS

H. F. Aly, E. S. Zakaria, N. Zakaria and I. M. El-Naggar

Atomic Energy Authority, Hot Lab. Centre
Atomic Energy Post Code 13759
Cairo
Egypt

1 INTRODUCTION

Acid salts of multivalent metals exhibit a high selectivity to some elements and their exchange properties are often connected with a fairly good stability towards high temperature and ionizing radiations. The factors underlying the behaviour of these materials as ion exchangers are not clearly understood and contradictory data are often obtained[1,2].

A research programme on ion exchange of the acidic salts of multivalent metals in aqueous and mixed media has been initiated with the aim of elucidating their ion exchange behaviour and selectivity phenomena in general. To achieve this goal, the kinetic and thermodynamic cation exchange behaviour of cerium(IV) and tin(IV) antimonates in aqueous solutions have been previously examined[3-5]. Comparison of the results obtained with other results for other metals antimonates, particularly iron (III) antimonate, has led to some useful conclusions regarding the selectivity behaviour and the ion exchange kinetics of the acidic salts of multivalent metals and has also emphasized the major role of the acidity of the exchange sites in determining the selectivity. However, thermodynamics is a powerful tool to study conditions at equilibrium but kinetics have an important role in the study of the mechanism of ion exchange. In addition, the study of the kinetics of ion exchange processes is important since it provides information on the mechanism of rate controlling process and on the reactions accompanying the ion exchange as well as on the internal physical structure of the exchanger and the extent of hydration of the exchanging ions.

The present work reports on the kinetics and thermodynamics of exchange on iron(III) antimonate in aqueous solution and in the presence of up to 90% (w/w) acetone, methanol and isopropanol. The results obtained are compared with those on other inorganic exchangers and organic resins.

2 EXPERIMENTAL

Preparation of Iron (III) Antimonate (FeSb)

Iron(III) antimonate was prepared as previously mentioned[6] by the dropwise addition of 0.5M solution of antimony (V) chloride to 0.5M solution of ferric chloride, at a Fe/Sb molar ratio of 1.3. The precipitate was kept in the mother solution overnight, filtered and washed with distilled water and then dried at room temperature. The product was ground and sieved to the desired particle sizes.

3 PROCEDURE

Kinetic measurements and ion exchange isotherms were carried out as reported earlier[3-5].

4 RESULTS AND DISCUSSION

The experimental conditions were set for particle diffusion mechanism only[7]. The equation developed by Boyd et al[8] and improved by Reichenberg[9] was used for evaluating the kinetic parameters.

Figure 1 shows that the rate of exchange for Cs^+ on FeSb is directly proportional to the temperature (where F given in Figure 1 is the fractional attainment of equilibrium, Bt values are taken from Reichenberg's Table[10]). The linearity of Bt vs. t plots for Cs^+ which pass through the origin indicates that the rate is controlled by particle diffusion

where $B = \dfrac{\Pi^2 D_i}{r^2}$, D_i being the effective diffusion coefficient of the exchanging ion

inside the particle and r the radius of the particle. Similar findings were found for Na^+ on the same matrix. It was found that the rate is dependent on the radius of the particle as can be expected for particle diffusion-controlled kinetics. The plot of log D_i against $1/T$ is linear, enabling the energy of activation (Ea) and the pre-exponential constant (D_o) to be estimated from the Arrhenius equation $D = D_o \exp^{(-Ea/RT)}$. The entropy of activation ΔS^* is calculated from the equation $D_o = 2.72 \ (KTd^2/n) \ \exp^{\Delta S^*/R}$, where R is the gas constant, K the Boltzmann constant, T is taken as $273°K$, d is the average distance between two successive positions in the process of diffusion. The results of D_i, Ea and ΔS^* under different conditions are given in Tables 1-4. The values of D_i given in Tables 1-4, are appreciably higher than those for zeolites, strongly cationic resins or even weakly cationic resins and semicrystalline niobium phosphate[10] zirconium[11] and tin(IV) antimonates[4]. Thus a comparison of the diffusion coefficient reported here to those of organic resins and other inorganic exchangers indicates that the rates of exchange on iron (III) antimonate are appreciably lower than those of strongly cationic resins (D_i for Na^+/H^+ exchange is reported to be $1.1-7.3 \times 10^{-6}$ cm^2 sec^{-1} for strong catonic sulfonated polystyrene resin)[9] but are nearly equal to some inorganic exchangers (the D_i values of Na^+/H^+ and Cs^+/H^+ exchanges on hydrous titania are 4.51×10^{-8} and 2.13×10^{-8} cm^2 sec^{-1}, respectively)[12].

The relatively small activation energy values calculated under different conditions given in Tables 1-4 for both Na^+ and Cs^+ on iron (III) antimonate, suggest that the rate of exchange is particle diffusion. Moreover, the negative value for the entropy of activation, where the entropy changes is small in the solid phase, suggesting that no significant structural change occurred in iron(III) antimonate. The effect of solvent concentration on the rate of exchange of Na^+ and Cs^+ on FeSb at different reaction temperatures (25, 40 and 60°C) are investigated for methanol, acetone and isopropanol.

The calculated values of D_i are given in Tables 2-4. These Tables show that the D_i of Na^+ and Cs^+ generally but not always decrease with the increase of solvent concentration in solution. This is in agreement with the data obtained by others[13,14].

The values of the activation energies of cation diffusion processes in mixed solvents are generally higher as compared to those in aqueous solution (Tables 2-4). This may be due to the higher basicity of the alcoholic solutions which increases the acidity of the exchange sites of the matrices. The increase of acidity leads to generally stronger electrostatic interaction of the cations, reducing their mobility in the exchanger.

The isotherms for replacement of H^+ in iron (III) antimonate by Cs^+ and Na^+ in aqueous solution and mixed solvent media are not given here for sake of brevity. The selectivity coefficient (K_c) with loading ($\overline{X}_M$) for H^+/Cs^+ exchanges on FeSb are given in Figures 2 and 3. The thermodynamic equilibrium constants (K_a) were obtained from the relation[15].

$$\ln K_a = \int_0^1 \ln K_c \, d\overline{X}_M$$

The equation of Gains and Thomas[15] can be used for both aqueous and mixed media provided that the total molality is ≤ 0.1 M (which is the case in the present paper) and the solvent composition is constant. The solvent composition in the case of the mixed solvents can be considered constant based on the rigid structure of FeSb with almost no swelling together with the high V/m ratio used in the present work.

The integrals were calculated from the areas under the curves. The calculated values of K_a are given in Table 5. The values of the differences in the thermodynamic quantities at 25°C are calculated as previously reported[5].

The various thermodynamic data of H^+ / Na^+ and H^+ / Cs^+ exchanges at different medium conditions are given in Table 5. The selectivity series increase in the order

Aqueous < methanol< acetone   for Na^+
and   Aqueous < methanol < acetone <isopropanol   for Cs^+

Table 5 shows that FeSb prefers hydrogen ion than Na^+ because of the entropy term. This is similar for the H^+-Na^+ and H^+-Cs^+ exchanges on cerium (IV) antimonate, whereas for the H^+- Cs^+ exchange isotherms on tin (IV) antimonate the reverse is true, i.e. the tin (IV) antimonate has a preference for Cs^+ ion[5]. The enthalpy change is the reason for the selectivity to cesium over hydrogen ion, Cs^+ is also preferred by FeSb other than Na^+.

From Table 5, it can be generally said that ΔH^o and ΔS^o decrease (become more negative) on increasing organic solvent concentrations. In the presence of the organic solvents, the Na^+ and Cs^+ are expected to be more solvated and therefore more dehydrated than in the purely aqueous solution. Therefore, less heat energy will be consumed in the dehydration of Na^+ and Cs^+ ions and in the removal of the H^+ ions than in the aqueous solution. This is expected to lead to a lower (negative) heat of exchange in the presence of the organic solvents.

The expected major effects of addition of organic solvents on the solid phase are probably connected with a chemisorption of these solvents and a dehydration of the ingoing alkali ion with its subsequent stronger interaction with the exchange sites. Chemisorption of these solvent may be expected to lead to negative enthalpy and entropy changes, while dehydration of Na^+ and Cs^+ are expected to lead to positive enthalpy and entropy changes. Thus the obtained enthalpy and entropy changes may be the resultants of the changes accompanying these processes. On going from 30 to 90% methanol, acetone and isopropanol, a relative change is observed which is expected to play some role in the interactions due to dehydration, may be expected to be more important.

Table 1   *Diffusion Coefficients and Other Thermodynamic Parameters Calculated for the Exchange Of Na^+ and Cs^+ on FeSb at Various Drying Temperatures (at Particle Diameter 0.25mm)*

exchange system	drying temp. °C	reaction temp. °K	D_i cm² sec⁻¹	D_o cm² sec⁻¹	Ea kJ mol⁻¹	ΔS^* J mol⁻¹
		298	1.19×10^{-8}	1.22×10^{-5}		-67.0
	50°C	313	1.85×10^{-8}	1.36×10^{-5}	17.1	-66.1
		333	2.67×10^{-8}	1.32×10^{-5}		-66.3
		298	1.11×10^{-8}	2.54×10^{-5}		-60.9
Na^+/H^+	200°C	313	1.85×10^{-8}	1.36×10^{-5}	19.1	-66.0
		333	2.32×10^{-8}	2.25×10^{-5}		-61.9
		298	8.57×10^{-9}	4.55×10^{-5}		-56.0
	400°C	313	1.53×10^{-9}	5.48×10^{-5}	21.2	-54.5
		333	1.97×10^{-8}	4.32×10^{-5}		-56.5
		298	1.25×10^{-8}	1.26×10^{-5}		-66.7
	50°C	313	1.81×10^{-8}	1.31×10^{-5}	17.1	-66.4
		333	3.64×10^{-8}	1.60×10^{-5}		-64.7
		298	8.39×10^{-9}	1.93×10^{-5}		-63.2
Cs^+ / H^+	200°C	313	1.39×10^{-8}	2.21×10^{-5}	19.1	-62.0
		333	2.50×10^{-8}	2.52×10^{-5}		-60.9
		298	7.34×10^{-9}	3.78×10^{-4}		-38.4
	400°C	313	1.48×10^{-8}	3.49×10^{-4}	26.8	-37.0
		333	2.04×10^{-8}	3.34×10^{-4}		-52.4

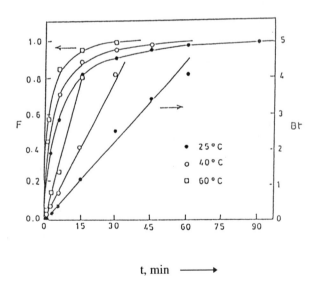

Figure 1   *Plots of F and Bt against time for exchange of Cs^+ on FeSb dried at 50°C from aqueous medium at different temperatures 25, 40 and 60°C*

Table 2 *Diffusion Coefficients and Other Thermodynamic Parameters Calculated for the Exchange of Na+ and Cs+ on FeSb from Methanol Solution at Different Temperatures (at Particle Diameter 0.25mm)*

exchange system	medium	reaction temp. $°K$	D_i $cm^2 sec^{-1}$	D_o $cm^2 sec^{-1}$	Ea $kJ\ mol^{-1}$	ΔS^* $J\ mol^{-1}$
	H_2O aqueous	298	1.19×10^{-8}	1.22×10^{-5}		-67.0
		313	1.85×10^{-8}	1.36×10^{-5}	17.1	-66.1
		333	2.67×10^{-8}	1.32×10^{-5}		-66.3
Na+ / H+	30% methanol	298	6.45×10^{-9}	1.48×10^{-4}		-46.3
		313	1.12×10^{-9}	1.59×10^{-4}	24.8	-45-6
		333	1.73×10^{-8}	1.39×10^{-4}		-46.7
	60% methanol	298	7.67×10^{-9}	1.75×10^{-5}		-46.0
		313	1.41×10^{-8}	2.20×10^{-5}	19.1	-62.1
		333	1.58×10^{-8}	1.66×10^{-5}		-64.4
	H_2O aqueous	298	1.25×10^{-8}	1.26×10^{-5}		-66.7
		313	1.81×10^{-8}	1.31×10^{-5}	17.1	-66.4
		333	3.64×10^{-8}	1.60×10^{-5}		-64.7
Cs+ / H+	30% methanol	298	2.79×10^{-9}	8.39×10^{-5}		-50.6
		313	5.27×10^{-9}	1.03×10^{-4}	25.5	-49.3
		333	8.83×10^{-9}	8.99×10^{-5}		-50.4
	60% methanol	298	5.57×10^{-9}	1.84×10^{-5}		-63.6
		313	8.04×10^{-9}	1.89×10^{-5}	20.0	-63.3
		333	1.15×10^{-8}	1.64×10^{-5}		-64.5

Table 3 *Diffusion Coefficients and Other Thermodynamic Parameters Calculated for the Exchange of Na+ and Cs+ on FeSb from Acetone Solution at Different Temperatures (at Particle Diameter 0.25mm)*

exchange system	medium	reaction temp. $°K$	D_i $cm^2 sec^{-1}$	D_o $cm^2 sec^{-1}$	Ea $kJ\ mol^{-1}$	ΔS^* $J\ mol^{-1}$
	H_2O aqueous	298	1.19×10^{-8}	1.22×10^{-5}		-67.0
		313	1.85×10^{-8}	1.36×10^{-5}	17.1	-66.1
		333	2.67×10^{-8}	1.32×10^{-5}		-66.3
Na+ / H+	30% acetone	298	1.11×10^{-8}	1.64×10^{-4}		-45.3
		313	1.76×10^{-8}	1.65×10^{-4}	33.7	-45.3
		333	2.78×10^{-8}	1.51×10^{-4}		-46.1
	60% acetone	298	6.50×10^{-9}	1.52×10^{-5}		-65.1
		313	9.29×10^{-9}	1.32×10^{-5}	19.1	-66.3
		333	1.39×10^{-9}	1.44×10^{-5}		-65.6
	H_2O aqueous	298	1.25×10^{-8}	1.26×10^{-5}		-66.7
		313	1.81×10^{-8}	1.31×10^{-5}	17.1	-66.4
		333	3.64×10^{-8}	1.60×10^{-5}		-64.7
Cs+ / H+	30% acetone	298	6.19×10^{-9}	7.89×10^{-7}		-89.8
		313	7.74×10^{-9}	7.70×10^{-7}	11.9	-90.0
		333	9.26×10^{-9}	7.02×10^{-7}		-90.7
	60% acetone	298	6.49×10^{-9}	3.57×10^{-6}		-77.2
		313	9.76×10^{-9}	3.90×10^{-6}	15.5	-76.5
		333	1.48×10^{-8}	4.22×10^{-6}		-75.8

Table 4 *Diffusion Coefficients and Other Thermodynamic Parameters Calculated for the Exchange of Na^+ and Cs^+ on FeSb from Isopropanol Solution at Different Temperatures (at Particle Diameter 0.25mm)*

exchange system	medium	reaction temp. °K	D_i $cm^2\ sec^{-1}$	D_o $cm^2\ sec^{-1}$	Ea $kJ\ mol^{-1}$	ΔS^* $J\ mol^{-1}$
	H_2O	298	1.19×10^{-8}	1.22×10^{-5}		-67.0
	aqueous	313	1.85×10^{-8}	1.36×10^{-5}	17.1	-66.1
		333	2.67×10^{-8}	1.32×10^{-5}		-66.3
		298	3.94×10^{-9}	8.07×10^{-6}		-70.4
Na^+ / H^+	30%	313	5.91×10^{-9}	8.42×10^{-6}	18.5	-70.1
	isopropanol	333	8.45×10^{-9}	7.76×10^{-6}		-70.7
		298	9.29×10^{-9}	7.72×10^{-6}		-70.8
	60%	313	1.03×10^{-9}	7.90×10^{-6}	16.6	-70.6
	isopropanol	333	1.90×10^{-8}	7.78×10^{-6}		-70.7
	H_2O	298	1.25×10^{-8}	1.26×10^{-5}		-66.7
	aqueous	313	1.81×10^{-8}	1.31×10^{-5}	17.1	-66.4
		333	3.46×10^{-8}	1.60×10^{-5}		-64.7
		298	7.89×10^{-9}	9.38×10^{-6}		-69.2
Cs^+ / H^+	30%	313	1.05×10^{-8}	9.02×10^{-6}	17.5	-69.5
	isopropanol	333	1.50×10^{-8}	9.31×10^{-6}		-70.2
		298	6.19×10^{-9}	1.30×10^{-5}		-66.4
	60%	313	9.28×10^{-9}	1.33×10^{-5}	18.9	-66.2
	isopropanol	333	1.36×10^{-8}	1.27×10^{-5}		-66.6

Table 5 *Thermodynamic Data for H^+ / Na^+ and H^+ / Cs^+ Exchange on FeSb in Aqueous and Mixed Solvent Solution at 25°C*

Exchange System	Medium	K_a	ΔG^o $KJ\ mol^{-1}$	ΔH^o $kJ\ mol^{-1}$	ΔS^o $J\ mol^{-1}\ {}^oK^{-1}$
	aqueous	1.25	0.5	-12.7	-41
	30%	1.06	-0.5	-10.9	-31
H^+ / Na^+	60% methanol	1.11	-0.2	-14.3	-47
	90%	1.79	-0.5	-14.8	-51
	aqueous	0.08	-4.4	-21.1	-56
	30%	1.20	-0.4	-9.5	-30
	60% methanol	3.21	-2.8	-25.4	-75
	90%	2.65	-2.4	-38.2	-120
	30% Iso-	1.61	-1.1	-9.5	-28
H^+ / Cs^+	60% propanol	3.72	-3.2	-25.5	-74
	90%	4.42	-3.6	-25.5	-73
	30%	2.09	-1.8	-20.6	-63
	60% acetone	2.43	-2.2	-20.1	-6-
	90%	3.32	-2.9	-20.6	-70

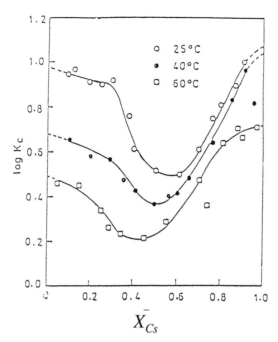

Figure 2 *Variation of log selectivity coefficients with Cs⁺ - loading for H⁺/Cs⁺ exchange on FeSb in aqueous medium*

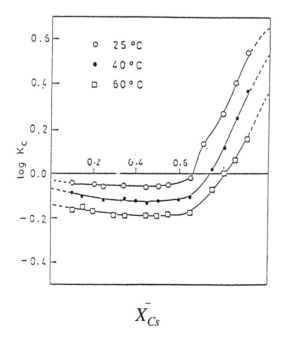

Figure 3 *Variation of log selectivity coefficients with Cs⁺- loading for H⁺/Cs⁺ exchange on FeSb in 30% methanol*

5 CONCLUSION

1. The effective diffusion coefficients (D_i) of Na^+ and Cs^+ decrease with the increase of solvent concentrations in solution and D_i is lower than in the aqueous solutions.
2. The values of activation energies of cation diffusion process in mixed solvent are higher compared with the cation diffusion in aqueous solutions. This may be due to the higher basicity of the alcoholic solutions that increases the acidity of the exchange sites of the matrices.
3. The selectivity series for H^+/Na^+ and H^+/Cs^+ exchange on FeSb increases in the order,   Aqueous < methanol for Na^+ and
            Aqueous < methanol < acetone < isopropanol for Cs^+

References

1. V. Vesely, and V. Pekarek, *Talanta*, 1972, **12**, 219.
2. A. Clearfield, "Inorganic Ion Exchange Materials", CRC Press, Boca Raton, Florida, 1982.
3. I. M. El-Naggar, M. R. El-Absy, H. N. Salma, M. A. F. Fattah and N. Souka. *Radiochimica Acta*, 1993, **62**, 91.
4. I. M. El-Naggar, M. A. El-Absy and S. I. Aly, *Solid State Ionics*, 1992, **50**, 241.
5. I. M. El-Naggar, M. A. El-Absy and H. F. Aly, *Colloids and Surfaces*, 1992, **66**, 281.
6. I. M. El-Naggar, E. S. Zakaria and H. F. Aly, *Reactive Polymers*, accepted (1995).
7. F. Helfferich, "Ion Exchange", McGraw-Hill, New York, 1962.
8. G. E. Boyd, A. W. Adamson and L. S. Myers, *J. Am. Chem. Soc.*, 1947, **69**, 2836.
9. D. Reichenberg, *J. Am. Chem. Soc.*, 1953, **75**, 589.
10. M. Qureshi and A. Ahmed, *Solv Ext. Ion Exch.*, 1986, **4**, 823.
11. V. I. Garashkov, G. M. Pnchenkov and T. V. Ivanova, *Z. H. Hiz. Khim.*, 1962, **36**, 19690.
12. I. M. El-Naggar and M. A. El-Absy, *J. Radio. Anal. Nucl. Chem. Articles*, 1992, **157**, 313.
13. D. Nadan and A. R. Gupta, *J. Phys. Chem.*, 1975, **79**, 180.
14. E. S. Zakaria, Thesis, Ain Shams University, Cairo, Egypt, 1994.
15. G. L. Gains and H. C. Thomas, *J. Chem. Phys.*, 1953, **21**, 714.

SOLVENT IMPREGNATED RESINS CONTAINING KELEX® 100 : AQUEOUS SOLUBILITY OF KELEX® 100 AND DISTRIBUTION EQUILIBRIA OF GERMANIUM(IV)

G. Cote, D. Bauer and S. Esteban

Laboratoire de Chimie Analytique (Unité associée au CNRS No 437)
E.S.P.C.I., 10, rue Vauquelin
75005 Paris
France

1 INTRODUCTION

Solvent Impregnated Resins (SIR) have been postulated as a promising technological alternative for problems associated with metal separation and recovery.[1-4] It was initially developed out of a need for ion-specific resins because of the deficiency of suitable methods for chemical functionalization of polymeric supports. The impregnated resins bridge the gap between solvent extraction and ion exchange resins because they combine the properties of the organic phase in solvent extraction with the equipment and operation facilities of the ion exchange technique. The development and application of these systems in metal extraction processes have been intensively investigated, especially during the past few years.[5-21]

In the present work, various impregnated resins have been prepared by introducing variable quantities of Kelex® 100 (7-(4-ethyl-1-methyloctyl)-8-quinolinol also denoted hereafter HL) into Amberlite® XAD-7 which is a porous polymethacrylate-based polymer having a high internal surface area (450 m²/g). Impregnated resins are typically heterogeneous materials and therefore a fundamental question arises concerning the description of the chemical activity of the impregnated reagents, especially in the absence of organic diluent. For instance, in the present case, should impregnated Kelex® 100 be considered as a film of pure Kelex® 100 having a constant activity, or alternatively, as a compound "diluted" in the polymeric matrix and exhibiting a variable activity depending on its concentration in the impregnated phase? For examining such a question, the aqueous solubility of impregnated Kelex® 100 and the extraction equilibria of germanium(IV) with impregnated Kelex® 100 have been investigated as reported below. We point out that germanium(IV) is present in the electrolytic solutions of zinc sulphate. The removal of germanium(IV) from such baths is necessary as it acts as an inhibitor of electrolysis, but its recovery is also economically interesting as germanium is an element having a high added value.

2 EXPERIMENTAL

2.1 Reagents

Kelex® 100 was kindly supplied by Schering (Germany) and used as delivered. It contained 7-(4-ethyl-1-methyloctyl)-8-hydroxyquinoline as the main component and small amounts of two inert furoquinoline type impurities, namely 3-ethylfuro[2,3-*h*]quinoline and 2-(2-ethylhexyl)-3-methylfuro[2,3-*h*]quinoline.[22] The Amberlite® XAD-7 polymer (Rohm and Haas) of practical-grade quality (size 0.3 - 1.2 mm) was carefully washed with ethanol and water, then dried and finally impregnated with Kelex® 100 according to the conventional dry impregnation method.[23,24] Germanium(IV) solutions were prepared from germanium dioxide (Riedel-de Haën). All the other reagents were of analytical grade.

2.2 Methods

The aqueous solubility of impregnated Kelex® 100 was investigated by equilibrating a given mass of impregnated resin with a given volume of aqueous solution. The aqueous concentration of Kelex® 100 was then determined by UV-visible spectrometry after re-extraction in heptane.[25] The losses of Kelex® 100 during long-term experiments in columns have also been investigated. In such a purpose, a 0.6 mol/kg $ZnSO_4$ solution containing 0.5 mol/kg H_2SO_4 (pH = 0.3) was used as the percolating aqueous phase. The typical experimental conditions were as follows : 1 g of impregnated resin, bed height = 3 cm, column diameter = 1.1 cm, flow rate = 15 bed volume per hour. The Kelex® 100 dissolved in the percolating aqueous phase was not directly determined by UV-visible spectrometry as indicated above, but collected on a adsorbent (unloaded Amberlite® XAD-7) and periodically eluted with ethanol. The concentration of Kelex® 100 in ethanol was then determined by UV-visible spectrometry. Alternatively, the ethanolic solution was pumped dry under vacuum and the amount of Kelex® 100 was determined gravimetrically. The extraction of germanium(IV) by impregnated Kelex® 100 has been studied in batch. Germanium(IV) in aqueous solutions was determined by acidimetry in the presence of mannitol.[26,27] The pH measurements were carried out with a combined glass electrode. All the experiments have been performed at 22 ± 2 °C.

3 RESULTS AND DISCUSSION

3.1 Preliminary Remarks

3.1.1 Swelling Phenomena. It has been noticed that the penetration of Kelex® 100 and/or water into the Amberlite XAD-7 phase causes an important swelling of its polymeric matrix (Figure 1). Throughout this work, the resin was used both impregnated and wet, thus no further change in apparent volume was observed.

3.1.2 Distribution of Kelex® 100 inside the Amberlite XAD-7 Support. Amberlite® XAD-7 is a porous polymethacrylate-based polymer in which three categories of pores can be found, namely the micropores (ϕ_d < 4 nm), the mesopores (4 nm < ϕ_d < 60 nm) and the macropores (ϕ_d > 60 nm). The characteristics (specific volume and specific area) of these three types of pores and the manner according to which Kelex® 100 distributes among them were studied by porometry with nitrogen at the temperature of liquid nitrogen (77.4 K).[28] The main results obtained are summarized in Table 1. Examination of this table shows that the microporous space is totally occupied by Kelex® 100 as soon as [Kelex® 100]$_{resin}$ = 0.60 mol/kg. It can also be noticed that between 0.60 and 1.34 mol/kg, Kelex® 100 mainly penetrates inside the mesoporous space, but that above

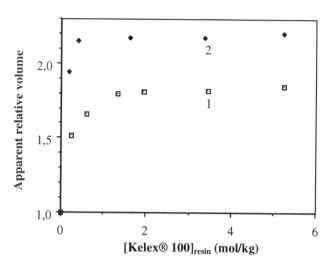

Figure 1 *Apparent relative volume of Amberlite® XAD-7 impregnated with Kelex® 100. (1) dry impregnated resin, (2) wet impregnated resin*

Table 1 *Distribution of Kelex® 100 inside the Porous Matrix of Amberlite® XAD-7*

$[Kelex® 100]_{resin}$ (mol/kg)	0	0.23	0.60	1.34	1.94	3.44
$[Kelex® 100]_{micropores}$ (mol/kg)	0	0.23	0.36	0.39	0.39	0.39
$[Kelex® 100]_{mesopores}$ (mol/kg)	0	0	0.23	0.91	1.41	2.22
$[Kelex® 100]_{macropores}$ (mol/kg)	0	0	0.01	0.04	0.14	0.83

1.34 mol/kg, Kelex® 100 simultaneously penetrates into the mesoporous and macroporous regions, in spite of the very different specific area still available for adsorption in these two types of pores (at $[Kelex® 100]_{resin} = 1.34$ mol/kg, $S_{mesopores} = 1.5 \times 10^5$ m²/kg and $S_{macropores} = 10^4$ m²/kg. These observations, and especially the last one, suggest that the interaction between Kelex® 100 and Amberlite® XAD-7 is rather weak and that a multilayer-type distribution takes place rather than a monolayer-type one. This is in agreement with the conclusions of a previous study.[29]

3.2 Aqueous Solubility of Impregnated Kelex® 100

In Table 2, the aqueous concentration of Kelex® 100 is given versus pH for two concentrations of Kelex® 100 in Amberlite® XAD-7. Examination of this table shows that the aqueous solubility of impregnated Kelex® 100 reaches its lowest values between about pH = 4 and pH = 11. Below pH = 4 and above pH = 11, the aqueous solubility of

Table 2   *Decimal Logarithm of the Aqueous Solubility of Impregnated Kelex® 100 as a Function of pH*

pH	0	1	3	4	5	7	9	11	12	14
[Kelex® 100]$_{resin}$ = 0.21 mol/kg	-5.72	-6.25	-7.54	-7.90	-8.05	-8.15	-8.15	-8.00	-7.72	-6.75
[Kelex® 100]$_{resin}$ = 1.44 mol/kg	-3.92	-4.50	-5.90	-6.22	-6.35	-6.40	-6.40	-6.36	-6.24	-5.70

impregnated Kelex® 100 increases as the pH is decreased or increased, as a result of the formation of H_2L^+ and L^- in the bulk of the aqueous phase, respectively. The same phenomena were reported in liquid-liquid systems.[25]

Figure 2 represents the logarithmic plot of the aqueous concentration of Kelex® 100 versus its concentration in Amberlite® XAD-7, for pH = 1 and pH = 7. Examination of Figure 2 shows that the aqueous concentration of Kelex® 100 increases as its concentration in the impregnated phase is increased. Such an observation unambiguously indicates that impregnated Kelex® 100 does not behave as a film of pure Kelex® 100, at least when Amberlite® XAD-7 is used as an adsorbent. Indeed, a film of pure Kelex® 100 would impose a constant value for its aqueous concentration at a given pH, regardless its total quantity in the polymeric phase. Further examination reveals that the two logarithmic plots represented in Figure 2 can roughly be simulated by two straight lines the slopes of which are equal to 1.94 (pH = 1, $R^2 = 0.954$) and 1.84 (pH = 7, $R^2 = 0.978$), respectively. This indicates that the aqueous solubility of Kelex® 100 is not proportional to its concentration in the polymeric matrix, but roughly varies as the latter at the power two (i.e., $C_{aq} = k [C_{resin}]^2$). Thus, Kelex® 100 does not behave in this impregnated system as an ideal solute in an ideal diluent, as it does in liquid-liquid systems where its aqueous and organic concentrations have been found to satisfy the mass action law (i.e., $C_{aq} = C_{org}/K_d$ with K_d = constant) within a large range of concentrations.[25] For simulating the distribution of Kelex® 100 between the aqueous phase and the impregnated polymer, the Brunauer, Emmett and Teller (B.E.T.) equation can be considered.[30,31] In the simplest case where the presence of the protonated (H_2L^+) and anionic (L^-) forms of Kelex® 100 is negligible, i.e., between about pH = 5 and pH = 10, the B.E.T. relationship can be written as Eq. (1) :

$$\frac{\dfrac{(C_{HL})_{aq}}{(C_{HL})^\infty_{aq}}}{x\left[1 - \dfrac{(C_{HL})_{aq}}{(C_{HL})^\infty_{aq}}\right]} = \frac{1}{x_m C'} + \frac{(C'-1)\left[\dfrac{(C_{HL})_{aq}}{(C_{HL})^\infty_{aq}}\right]}{x_m C'} \tag{1}$$

where x denotes the concentration of Kelex® 100 in the polymer (i.e., x = [Kelex® 100]$_{resin}$), x_m is the theoretical quantity of Kelex® 100 needed to totally recover the internal surface of the polymer with a monolayer, $(C_{HL})_{aq}$ represents the aqueous concentration of Kelex® in equilibrium with the impregnated resin, $(C_{HL})^\infty_{aq}$ is equal to 1.1×10^{-6} mol/kg at 22 °C and finally C' is a constant characteristic of the adsorption process. From Eq. (1), one can calculate the value of $\left(C_{HL}\right)_{aq}$ for any value of $x = [\text{Kelex® 100}]_{resin}$ and then plot $\log \left(C_{HL}\right)_{aq}$ versus $\log [\text{Kelex® 100}]_{resin}$. Comparison between open circles and black triangles in Figure 2 (curve 2), shows that the B.E.T. equation allows an acceptable simulation of the distribution of Kelex® 100 in impregnated system. The preceding simulation can be extended to any value of pH by taking into account the formation of the protonated (H_2L^+) and anionic (L^-) forms of Kelex® 100.

The preceding results show that the aqueous solubility of impregnated Kelex® 100 sharply decreases as its concentration in the polymer is decreased. From a physical point of view, it is of interest that the aqueous solubility of Kelex® 100 reaches its lowest values when the latter is present only in the microporous volume of Amberlite XAD-7 (i.e., $[\text{Kelex® 100}]_{resin} < 0.6$ mol/kg). Thus, for obtaining the lowest relative losses of Kelex® 100 by aqueous solubility, the loading of the polymeric matrix should be as low as permitted in the frame of the considered application. This tendency is illustrated in Figure 3 where the results of long-term experiments in column are reported. Indeed, in this case where an aqueous phase simulating an electrolytic bath was used, the relative losses of Kelex® 100 are significantly lower with the resin initially loaded at 1.0 mol/kg than with the one loaded at 2.5 mol/kg. Examination of Figure 3 also shows that the absolute losses of Kelex® 100 are important in long term experiments, but it should be pointed out that they are less important than in liquid-liquid systems under equivalent conditions, except for high loading (i.e., $[\text{Kelex® 100}]_{resin} > 2.5$ mol/kg).

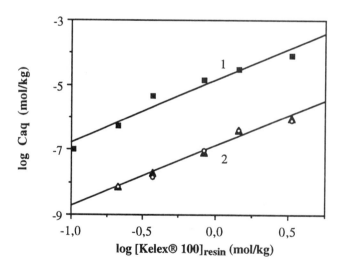

Figure 2 *Logarithmic plot of the aqueous concentration of Kelex® 100 versus its concentration in the Amberlite® XAD-7 support, for pH = 1 and pH = 7. Black squares and triangles represents the experimental points. Open circles (curve 2) correspond to the simulated points derived from Eq. (1) with C' = 45 and $x_m = 0.94$ mol/kg*

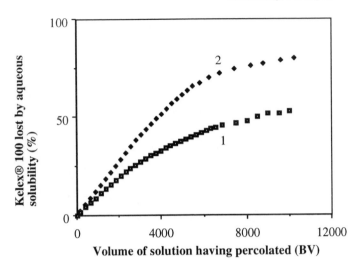

Figure 3 *Percentage of Kelex® 100 eluted as a function of the volume of aqueous solution (0.6 mol/kg ZnSO₄ + 0.5 mol/kg H₂SO₄ (pH = 0.3)) having percolated through the column. Flow rate = 15 ± 2 BVH. Initial concentration of Kelex® 100 in Amberlite® XAD-7 (mol/kg) : (1) 1.0 , (2) 2.5*

3.3 Equilibria of Germanium(IV) Extraction with Impregnated Kelex® 100

In previous works, it has been shown that germanium(IV) can be efficiently and selectively extracted from acidic zinc sulphate solutions with Kelex® 100.[26,27,32] The extraction is as a result of the formation of two species, $GeL_2(OH)_2$ and GeL_3^+,X^- (with $X^- = HSO_4^-$, Cl^-, etc.), the proportions of which depend on the acidity of the aqueous phase. $GeL_2(OH)_2$ forms above pH 2 whereas GeL_3^+,X^- appears below pH 3. The formation of these two species occurs according to the two following reactions:

$$Ge(OH)_{4\ aq} + 2\ HL_{org} \Leftrightarrow GeL_2(OH)_{2\ org} + 2\ H_2O \tag{2}$$

$$Ge(OH)_{4\ aq} + 3\ HL_{org} + X^-_{aq} + H^+_{aq} \Leftrightarrow (GeL_3^+, X^-)_{org} + 4\ H_2O \tag{3}$$

The subscripts "aq" and "org" refer to the species in the aqueous and organic phases, respectively. In liquid-liquid system, the distribution of germanium(IV) can satisfactorily be described by using the two following apparent constants which involve only the concentrations of the considered species (i.e., no activity coefficient is necessary) :

$$K_1 = \frac{[GeL_2(OH)_4]_{org}}{[Ge(OH)_4]_{aq}[HL]^2_{org}}$$

$$\tag{4}$$

$$K_2 = \frac{[GeL_3^+, X^-]_{org}}{[Ge(OH)_4]_{aq}[HL]_{org}^3[X^-]_{aq}[H^+]_{aq}} \tag{5}$$

Numerically, it has been found that $\log K_1$ and $\log K_2$ are equal to $(2.24 \pm 0.09)_{95\%}$ and $(6.44 \pm 0.35)_{95\%}$, respectively, when kerosene containing 10% (v/v) 1-octanol is used as a diluent.[26]

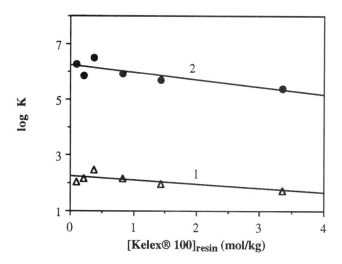

Figure 4 *Apparent constants K_1 (curve 1) and K_2 (curve 2) as a function of the concentration of Kelex® 100 in Amberlite® XAD-7*

Considering the complex properties of impregnated Kelex® 100 exhibited in § 3.2, it is of interest to examine how do Eqs. (4) and (5) allow a description of the partition of germanium(IV) in the present impregnated system. In such a purpose, the extraction of germanium(IV) has been investigated at different pH and for various concentrations of Kelex® 100 in the Amberlite® XAD-7 support. From the obtained results, the values of the apparent constant K_1 and K_2 have been determined (the subscript "org" refers this time to the species in the impregnated resin). These values are plotted in Figure 4 as a function of the initial concentration of Kelex® 100 in Amberlite® XAD-7. Examination of this figure shows that K_1 and K_2 keep roughly constant values, which indicates that the distribution equilibria of germanium(IV) can reasonably be simulated by using the mass action law without having to introduce activity coefficients for the involved species, including Kelex®100 and the extracted complexes. Thus, in spite of the complexity of the distribution law of Kelex® 100 and probably of that of its metal complexes (the partition of their individual molecules also likely follows a B.E.T.-type equation), the global extraction behaviours can merely be represented in a simple way. Such a result is highly interesting from a practical point of view.

Notations

HL	Kelex® 100 (7-(4-ethyl-1-methyloctyl)-8-quinolinol)
C_{HL}, C_{aq}, C_{org}	concentrations in solution expressed in molality (mol/kg)
C'	constant characteristic of the adsorption process
$[X]_{resin}$, C_{resin}	concentrations in the impregnated phase expressed in mol of X per kg of dry unloaded Amberlite XAD-7 (mol/kg)
K_1, K_2	extraction constants
K_d	partition constant
k	coefficient
x, x_m	concentration of Kelex® 100 in the polymer
ϕ_d	pore diameter

Acknowledgments

The authors acknowledge the French Ministry for Research and Technology (M.R.T.) for financial support (Grant No 91.R.0261) and Metaleurop Recherche for authorization of publication.

References

1.   A. Warshawsky, *Trans. Inst. Min. Metall.*, 1974, **83**, C101.
2.   A. Warshawsky, *Talanta*, 1974, **21**, 624.
3.   A. Warshawsky, *Talanta*, 1974, **21**, 962.
4.   D. S. Flett, *Chem. Ind. (London)*, 1977, 641.
5.   J. L. Cortina, N. Miralles, M. Aguilar and A. M. Sastre, In : "Solvent Extraction 1990", T. Sekine and S. Kusakabe, eds., Elsevier, Amsterdam (1992), Part A, p. 159.
6.   I. Villaescusa, M. Aguilar, J. de Pablo, M. Valiente and V. Salvado, In : "Solvent Extraction 1990", T. Sekine and S. Kusakabe, eds., Elsevier, Amsterdam (1992), Part A, p. 171.
7.   H. Matsunaga, Y. Wakui and T. M. Suzuki, In : "Solvent Extraction in the Process Industries", D. H. Logsdail and M. J. Slater, eds., Elsevier Applied Science, London, 1993, Volume 1, p. 615.
8.   I. Villaescusa, V. Salvado, J. de Pablo, M. Valiente and M. Aguilar, *Reactive Polymers*, 1992, **17**, 69.
9.   J. L. Cortina, N. Miralles, A. M. Sastre, M. Aguilar, A. Profumo and M. Pesavento, *Reactive Polymers,* 1992, **18**, 67.
10. E. P. Horwitz, R. Chiarizia and M. L. Dietz, *Solvent Extr. Ion Exch.*, 1992, **10**, 313.
11. E. P. Horwitz, M. L. Dietz, R. Chiarizia, H. Diamond, A. M. Essling and D. Graczyk, *Anal. Chim. Acta,* 1992, **266**, 25.
12. E. P. Horwitz, R. Chiarizia, M. L. Dietz, H. Diamond and D. M. Nelson, *Anal. Chim. Acta,* 1993, **281**, 361.
13. J. L. Cortina, N. Miralles, A. M. Sastre, M. Aguilar, A. Profumo and M. Pesavento, *Reactive Polymers,* 1993, **21**, 89.

14. J. L. Cortina, N. Miralles, A. M. Sastre, M. Aguilar, A. Profumo and M. Pesavento, *Reactive Polymers,* 1993, **21**, 103.
15. J. L. Cortina, N. Miralles, M. Aguilar and A. Warshawsky, In : "Solvent Extraction in the Process Industries", D. H. Logsdail and M. J. Slater, eds., Elsevier Applied Science, London, 1993, p. 962.
16. J. L. Cortina, N. Miralles, M. Aguilar and A. M. Sastre, *Hydrometallurgy,* 1994, **36**, 131.
17. J. L. Cortina, N. Miralles, M. Aguilar and A. M. Sastre, *Solvent Extr. Ion Exch.,* 1994, **12**, 349.
18. J. L. Cortina, N. Miralles, M. Aguilar and A. M. Sastre, *Solvent Extr. Ion Exch.,* 1994, **12**, 371.
19. J. L. Cortina, N. Miralles, A. M. Sastre and M. Aguilar, *Hydrometallurgy,* 1995, **37**, 301.
20. S. Amer, J. M. Figueiredo and A. Luis, *Hydrometallurgy,* 1995, **37**, 323.
21. G. Zuo and M. Muhammed, *Solvent Extr. Ion Exch.,* 1995, **13**, 879.
22. E. Dziwinski, G. Cote, D. Bauer and J. Szymanowski, *Hydrometallurgy,* 1995, **37**, 243.
23. A. Warshawsky and A. Patchornik, *Isr. J. Chem.,* 1978, **17**, 307.
24. F. Vernon, *Sep. Sci. Technol.,* 1978, **13**, 587.
25. G. Cote and D. Bauer, *J. Inorg. Nucl. Chem.,* 1981, **43**, 1023.
26. B. Marchon, G. Cote and D. Bauer, *J. Inorg. Nucl. Chem.,* 1979, **41**, 1353.
27. G. Cote and D. Bauer, *Hydrometallurgy,* 1980, **5**, 149.
28. S. Esteban, G. Cote, J.-L. Bonardet and D. Bauer, to be published.
29. L. Bokobza and G. Cote, *Polyhedron,* 1985, **4**, 1499.
30. S. Brunauer, P. H. Emmett and E. Teller, *J. Am. Chem. Soc.,* 1938, **60**, 309.
31. H. Hommel and A. P. Legrand, *Reactive Polymers,* 1983, **1**, 267.
32. D. Rouillard-Bauer, G. Cote, P. Fossi and B. Marchon, US Pat. 4,389,379 (1983) and 4,568,526 (1986).

DEVELOPMENTS IN IMPREGNATED AND ION EXCHANGE RESINS FOR GOLD CYANIDE EXTRACTION

E. Meinhardt, V. Marti, A. Sastre, M. Aguilar and J. L. Cortina

Chemical Engineering Department (ETSEIB)
Universitat Politècnica de Catalunya
Barcelona, 08028
Spain

1 INTRODUCTION

In recent years, zinc dust precipitation has been replaced to a large extent by carbon adsorption processes for recovery of gold from dilute cyanide leach solutions. The success of carbon adsorption processes has prompted interest in alternative processes, such as solvent extraction and ion exchange. Economic analysis of process alternatives for gold recovery from dilute alkaline cyanide solutions suggests that a Resin-in-Pulp (RIP) or Resin-in-Leach (RIL) process may offer some advantages over the other options[1-2]. Recent research has shown that adsorption from an alkaline solution can be achieved with neutral polymeric adsorbents[3] and using various ion exchange resins with high selectivity to extract gold and/or silver from cyanide solutions[4-5].

It is well known that strong-base resins have a high capacity for gold but low selectivity, as well as being difficult to strip or elute. Weak-base resins, on the other hand, are easy to strip, offer some selectivity but have a low capacity in alkaline solutions. More recently, Impregnated Resins have also been considered as a process alternative to conventional carbon adsorption and ion exchange processes in the gold industry to increase the concentration efficiency and selectivity vis-a-vis other metal cyanide complexes present in the solutions[4]. Although such an alternative process for the recovery of Au from dilute alkaline cyanide solution offers potential advantages, appropiate solvent systems have only been identified recently, and performance characteristics have not yet been established[6-7]. In this regard, Akser et al.[3] have used solvent-impregnated resins for Au recovery from alkaline cyanide solution using modified weak-base amines. The effectiveness of these modified amine-impregnated resins was evaluated in terms of loading/stripping characteristics, selectivity, and solvent losses. These studies on gold extraction from alkaline cyanide solution indicate that weak-base amines, e.g., ditridecylamine (ADOGEN 283), when modified with alkyl phosphorous esters such as tri-n-butylphosphate, tri-n-octylphosphate and di-n-butylphosphonate can efficiently extract gold from alkaline cyanide solution[6-7]. Furthermore, the selectivity of these alkyl phosphorous ester/modified-amine systems is remarkably high for the aurocyanide anion over a host of other possible cyanoanions. Organic losses caused by entrainment and volatility can be decreased or even eliminated if the organic phase is firmly adsorbed at an adequate substrate.

In this work a family of impregnated and ion exchange resins have been studied for their application in gold extraction from cyanide solutions. Solid-liquid extraction studies of gold cyanide by these resin families using batch experiments have been performed. The influence of both reagent and polymer functionality structure in their extraction ability and the influence of the aqueous composition (pH, cyanide and metalcyanide concentrations) of two different mine leaching liquors have been studied.

2 EXPERIMENTAL

2.1 Reagents

Alamine 336 (Henkel) was used without further purification. Samples of anionic resins were obtained from Rohm and Haas and Bayer AG and the resins are listed in Table 1, along with the total capacity (moles of active group per gram of dry or wet settled resin).

Table 1 *List of Anionic Resins and Chemical Properties*

Resin	*Millimoles of active group per gram of dry or wet settled resin*	
- Anionic resins		
MP 64 (Bayer AG)	tertiary	2.14
Amberlite IRA 94 S (Rohm and Haas)	tertiary	1.04
- Impregnated Resin		
Alamine 336 (Henkel)	tertiary	0.73

Stock solutions of $Au(CN)_2^-$, $Ag(CN)_2^-$, $Fe(CN)_6^{4-}$ and $Cu(CN)_3^{2-}$ ($1g.dm^{-3}$) were prepared by dissolving the corresponding salts (Johnson Mathey, A.R. grade) in sodium cyanide solution. Amberlite XAD-2 Resin (Rohm and Haas), size 0.3-0.9 mm, was used. Real cyanide leach liquors were obtained by leaching of a gold mineral ore from Brazil. The composition of this solution is shown in Table 2.

Table 2 *Composition of the Metal Cyanide Solutions*

	Mining Leach Solution 1	*Mining Leach Solution 2*	*Synthetic Solution 2*
Metal	Concentration (ppm)	Concentration (ppm)	Concentration (ppm)
Gold (Au)	3.50	10.10	10.11
Silver(Ag)	0.57	2.06	3.95
Cooper(Cu)	3.20	5.15	18.94
Iron (Fe)	6.86	10.10	19.31

2.2 Impregnation Process

XAD2-Alamine 336 resins were prepared according to a modified version of the dry impregnation method described previously[9]. The amount of Alamine impregnated was evaluated by washing a known amount of resin with anhydrous acetic acid, which completely elutes these extractants for subsequent determination by $HClO_4$ potentiometric titrations[10].

2.3   Experimental Procedure

The extraction of $Au(CN)_2^-$ and other metal cyanide complexes was carried out using batch experiments.   Samples between 0.05-0.2 g of Ion Exchange resins or Impregnated Resins, were mixed mechanically in special glass stoppered tubes with an aqueous solution (20-250 mL) until equilibrium was achieved.   The composition of the aqueous solutions varied depending on the nature of the experiment.   According to Figure 1, where the extraction percentage of $Au(CN)_2^-$ is plotted as a function of time, three hours were enough to reach equilibrium.   After phase separation with a high-speed centrifuge, the equilibrium pH was measured using a Methrom AG 9100 combined electrode connected to a CRISON digital pH-meter, model Digilab 517.   Metal content in both phases was determined by Atomic Absorption Spectrophotometry (Perkin-Elmer 2380 AAS with air-acetylene flame), Inductively Coupled Plasma Spectrophotometry (ICP) (SpectroFlame, Kleve, Germany) and Capillary Zone Electrophoresis (Isco, Lincoln, NE, USA) depending on solution composition.

Reagent losses were determined from analyses of aliquots of the aqueous phase.   The values found (less than 0.5ppm) were lower than those obtained for Alamine 336 as a pure component (5ppm).

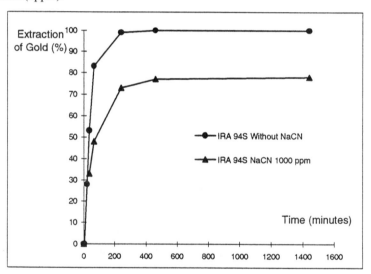

Figure 1 *Gold extraction percentage (%E) as a function of time (min) for Amberlite*
IRA94S. Phase ratio was 0.1g resin/250mL of aqueous solution
([Au(CN)$_2^-$]=10ppm)

3 RESULTS AND DISCUSSION

Since typical gold leach solutions contain gold as gold(I) dicyanide complexes, $Au(CN)_2^-$, and have a pH of about 10, it was reasoned that the objectives might be accomplished if a reagent, which would operate on the hydrogen cycle shown below, could be developed:

$$R_{res} + H^+ \Leftrightarrow RH^+_{res}$$

where the equilibrium lies far to the left at pH$\leq$10 and far to the right at pH$\geq$13. The reagent in the protonated form, (RH$^+$), at pH$\leq$10 would be an active anion extractant, the reagent in the neutral form, at pH $\geq$13, would not be an anion extractant.

Available extractants do not meet the above objectives. Weak base amines whether alone or modified with organophosphorous esters and/or oxides, do not extract gold well at pH>9. Quaternary amines have very good extraction properties for gold at pH 10; however the gold will not strip at 13.5-14.0. In fact, the gold is so difficult to strip that a solvent extraction process for gold from cyanide leach solutions where the loaded quaternary amine is incinerated in order to recover high-purity metallic gold recently has been proposed[11-12].

The experimental results suggest that the extraction of Au(CN)$_2^-$ and other metal cyanide complexes by Ion Exchange Resins and Impregnated Resins containing tertiary amine groups proceeds by means of the following extraction reaction:

$$R_3NH^+X^-_{res} + Au(CN)_2^- \Leftrightarrow R_3NH^+Au(CN)_2^-{}_{res} + X^-$$

i.e. MP64, IRA94S, XAD2-Alamine

Tertiary amine resins were evaluated in our study on goldcyanide extraction.

3.1 Extraction Kinetics

The extraction kinetics of these resins when operating in conditions of Resin in Leaching experiments, have been studied and are shown in Figure 2. In this Figure, the extraction kinetics of different metalcyanides using Amberlite IRA94S shows that times of 3 to 4 hours were enough to reach equilibrium values. Similar results were obtained with the other resins, MP64 and XAD2-Alamine 336.

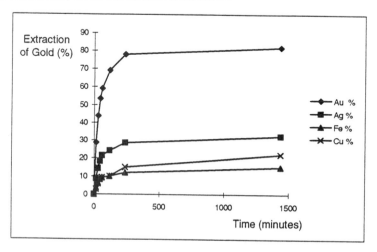

Figure 2 *Metal cyanide extraction percentage (%E) as a function of time (min) for Amberlite IRA94S. Phase ratio was 0.2g resin/20mL of solution with composition given in Table 2*

3.2 Metal Extraction Studies

The metal extraction studies on the efficiency of the resins used in this work were performed simulating the experimental conditions expected from the leach solutions obtained in the cyanidation step (pH, free cyanide concentration and presence of other metal cyanide interference's).

3.2.1 Cyanide Concentration Effect. The effect of total cyanide concentration on goldcyanide extraction was evaluated in the concentration range expected from the processing of mineral ores. Cyanide concentration changes through the mineral leaching step from values of 1000 ppm (starting point) to 100-200 ppm (closing point). Figure 3 shows equilibrium gold distribution isotherms for the different ion exchange and impregnated resins as a function of cyanide concentration in the aqueous phase. As can be seen from Figure 3, the extraction of $Au(CN)_2^-$ is unaffected by the presence of total cyanide content for MP64, Amberlite 94S and XAD2-Alamine 336 impregnated resins in concentrations lower than 500 ppm. For cyanide concentration higher than 500 ppm, simulating the intermediate and initial leach solutions, the extraction percentage decrease to 75% for Amberlite 94S and MP64 and is close to 30% for XAD2-Alamine 336 resins. This effect shows the cyanide competition of CN^- on $Au(CN)_2^-$ extraction. This effect should not affect the extraction process when using RIP or RIL where free cyanide concentrations expected should lower than 500 ppm.

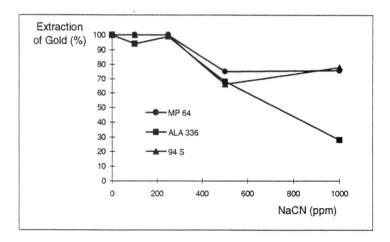

Figure 3 *Gold extraction percentage (%E) for the different ion exchange and impregnated resins as a function of cyanide concentration on the aqueous phase ([CN⁻]). Phase ratio was 0.2g resin/20mL of aqueous solution ([Au(CN)₂⁻]=10ppm)*

3.2.2 Extraction Efficiency and Selectivity. The extraction efficiency and selectivity of $Au(CN)_2^-$ and other metal cyanides complexes in mining leach solutions under different experimental conditions was evaluated. Figure 4a shows the selectivity factors of $Au(CN)_2^-$ and other metal cyanide complexes for Amberlite 94S, MP64 and XAD2-

Alamine 336 resins using a synthetic solution whose composition is given in Table 2. As a general trend, ion exchange resins Amberlite 94S and MP64 show a higher efficiency on $Au(CN)_2^-$, with extraction percentages close to 80%, and low selectivity on $Au(CN)_2^-$ extraction when compared with the other metal cyanide complexes (Fe, Ag, Cu). Impregnated resins with Alamine 336 show low extraction percentages but high selectivity factors over the other metalcyanide anions.

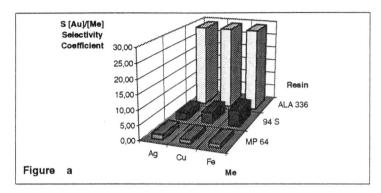

Figure   a

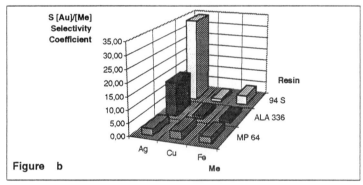

Figure   b

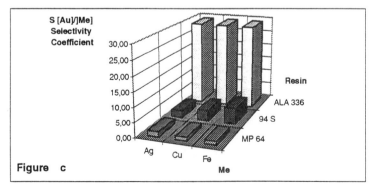

Figure   c

Figure 4 *Selectivity factors for the different metal cyanide complexes for MP64, Amberlite IRA94S and XAD2-Alamine 336 resins in (a) synthetic solution (b) mining leach solution 1 and (c) mining leach solution 2*

Selectivity factors in experiments using two mining leach solutions obtained from minerals whose composition is shown in Table 2 are presented in Figures 4b-4c. While tertiary type ion exchange resins (MP64 and Amberlite 94S) show the similar trend of high gold extraction percentages with a decrease in the selectivity patterns. Different behaviour was observed on tertiary amine type impregnated resins XAD2-Alamine 336 where the increase in gold cyanide, rising to 80%, was also followed by a decrease in selectivity patterns.

4 CONCLUSIONS

The kinetic behavior shown by the resins tested in this work indicate that residence times of 3 to 4 hours are needed to reach steady state. This period of time is similar to those used in Resin-in-Leach and Resin-in-Pulp processes in technological applications.

The extraction efficiency of Ion Exchange Resins (Amberlite IRA94S, MP64) and impregnated resins XAD2-Alamine 336 showed high extraction percentages for $Au(CN)_2^-$ nevertheless the selectivity against other metalcyanide ions varied depending on the metal cyanide concentration and free cyanide concentration. From the three resins studied XAD2-Alamine 336 showed the highest goldcyanide selectivity factors with moderate extraction efficiency. In this sense, modifications to the impregnated resins prepared by a) increased ligand loading in the resin phase and b) the combination of Alamine 336 mixtures with solvating extractants as (TBP, TOPO and TIPBS) are under development.

The selectivity of resins containing amine type groups for $Au(CN)_2^-$ and other metallocyanide complexes appears to be determined by a combination of factors. Among these, the following could be pointed out: (a) nature of the amine group (b) the spatial distribution of the active groups; and (3) the degree of hydrophobicity of the polymeric matrix. The high hydrophobic nature of styrene-divinylbenzene polymer (Amberlite XAD2) increased by tri-decyl and tri-octyl (C_8-C_{10}) chains of Alamine 336 could favor the loading of poorly hydrated ions such as $Au(CN)_2^-$ against multivalent hydrated ions, such as $Cu(CN)_4^{3-}$ and $Fe(CN)_6^{4-}$.

Acknowledgements

We wish to acknowledge to Prof. C. Hoffmann, and Rubens Kautzmann, Mineral Processing Laboratory (LAPROM), Department of Metallurgy, Federal University of Rio Grande Do Sul, Porto Alagre (Brazil) for their support in sample supply and mineralogical characterization, and to Rohm and Haas Hispania and Bayer AG for resin supply. Finally, this work was supported by CICYT Project **MAT 93-6212** (Ministerio de Educación y Ciencia de España).

References

1. M. B. Moiman, J. D. Miller, J. B. Hiskey, and A. R. Hendricksz, in "Au and Ag Heap and Dump Leaching Practice", J. B. Hiskey, ed., AIME, 1984, p. 93.
2. C. A. Fleming and G. Gromberge, *J. S. Afr. Inst. Min. Metall.*, 1984, **84**, 369.
3. M. Akser, R. Y. Wan and J. D. Miller, The Metallurgical Society (AIME), *The Reinhardt Schuhmann International Symposium on Innovative Technology and Reactor Design in Extraction Metallurgy*, Colorado Springs, CO, 1986, **87**, 72.
4. B. R. Green, *Mintek 50-Proceedings of the International Conference on Mineral Science and Technology*, L. F. Haughton, ed., Johannesburg, South Africa, 1985, Vol. 2, p. 627.
5. P. A. Riveros, and W. C. Cooper, *Solvent Extr. & Ion Exch.*, 1988, **6**, 479.
6. M. B. Moiman, and J. D. Miller, *Proceedings of ISEC 83*, A. I. Ch. E., 1983, 530.
7. J. D. Miller and M. B. Moiman, *Sep. Sci. & Tech.*, 1984, **19**, 895.
8. P. L. Sibrell, and J. D. Miller, *Proceedings of ISEC 86*, Munich, 1986.
9. J. L. Cortina, A. M. Sastre, N. Miralles, M. Aguilar, A. Profumo, and M. Pesavento, *React. Polym.*, 1992, **18**, 67.
10. St. J. H. Blakeley, and V. J. Zatka, *Analytica Chimica Acta*, 1975, **74**, 139.
11. P. A. Riveros, *Hydrometallurgy*, 1993, **33**, 43.
12. P. A. Riveros, *Hydrometallurgy*, 1990, **24**, 135.

THE USE OF ION EXCHANGE RESINS TO RECOVER RARE EARTHS FROM APATITE GYPSUM RESIDUE

A. M. A. Padayachee, M. W. Johns and B. R. Green

Mintek
Private Bag X3015
Randburg 2125
South Africa

1 INTRODUCTION

The phosphate type mineral, apatite, used in the production of phosphoric acid has a rare earth oxide (REO) content between 0.4 and 0.9 percent. A sulphuric acid attack is carried out on the ore, producing a dilute impure phosphoric acid stream and gypsum, the latter being dumped. Although some of the rare earth elements (REE) report to the acid stream, the bulk is precipitated with the gypsum. During concentration of the acid stream a sludge is produced which also contains REE[1]. Hence, a low tonnage/high grade REE stream (the sludge) and a high tonnage/low grade REE stream (the gypsum) are produced.

Recovery of REE from the sludge was investigated and a process involving an acid leach, solid/liquid separation and solvent extraction with phosphate type extractants was proposed. Subsequent investigations revealed that the use of ion exchange resins[2] to recover REE from the sludge in a resin-in-pulp process, resulted in an increase in the leaching efficiency of REE.

As a larger proportion of REE report to the gypsum, it was considered that recovery of REE from gypsum, using a resin in leach (RIL) process similar to that proposed for the sludge might be successful. The use of RIL instead of solvent extraction (SX) removes the need for a costly solid/liquid separation unit operation required when processing low grade/high tonnage material. This paper describes work to evaluate the technical feasibility of using resin to recover REE from the gypsum residue.

2 EXPERIMENTAL

2.1 Selection of the Resin

Using the speciation programme, MINTEQA2[3], and the available stability constant data[4] for Ce^{3+} and Nd^{3+} complexed to SO_4^{2-}, Ce and Nd were found to exist predominantly as the unassociated triply charged cationic species, at a pH value of 1.0. Thus C20MB (sulphonic acid), ES467 (amino phosphonic acid), as supplied by *Duolite*, and other resins with phosphorus acid groups should be suitable for REE extraction under the expected RIL conditions. Furthermore, since REE are known to form complexes with acetylacetone and citric acid it was decided to synthesise resins with these moieties chemically attached. Tri-n-butyl phosphate (TBP) is a known extractant for REE and as the closest convenient structure to TBP for a resin is a phosphonate ester resin it was decided to prepare this resin as well. Two additional resins, containing phosphonic acid groups (A and PA in Table 1), that were believed to be worthy of consideration were also prepared. In preparing solvent impregnated resins, conventional extractants used in solvent extraction of REE, e.g. TBP, di-2-ethyl hexylphosphoric acid (D2EHPA), were adsorbed onto resins or carbon. The "syntheses" of these resins were reported by Warshawsky in 1972[5], and more recently by Britz and Cloete[6]. The methods used to impregnate adsorbents, (see experimental details below) were based on previous work by Green[7].

2.2 Resin Synthesis and Characterisation

2.2.1 Functionalised Resins. A macroporous chloromethylated polystyrene divinylbenzene matrix (20 percent chlorine), was chosen as the base polymer. The resins were characterised by phosphorus and chlorine micro analysis, water retention capacity, wet settled density, exchange capacity and infra red spectroscopy; details of which are given in Table 1.

Table 1 *Characteristics of Resins Synthesised*

Resin **	P (%)	Cl (%)	Proton exchange capacity (mmol/g)	Infra red data (cm⁻¹)
(polystyrene)–CH₂–P(=O)(OEt)(OEt) **E[8]**	7 (12)*	1.2	-	1047 (C-O)
(polystyrene)–CH₂–P(=O)(OH)(OH) **A[8]**	12 (16)*	3.3	7.6	1175-1200 (P=O), 3400 (O-H)
(polystyrene)–CH₂–CH(C(CH₃)=O)(C(CH₃)=O) **AC[9]**	-	4.5	-	1718 (C=O)
(polystyrene)–CH₂O–(CH₂CH₂O)ₓ–CH₂CH₂OH **PEG[10]**	-	6.9	-	1100 (C-O), 3450 (O-H)
(polystyrene)–P(=O)(OH)(OH) **PA[11]**	13	2.0	5.2	1175-1200 (P=O), 3400 (O-H)

** Syntheses based on methods described in the literature
* Value expected for total reaction

2.2.2 Solvent Impregnated Resins/Adsorbents. Acetylacetone (ACAC), di-n-butyl phosphate (DBBP), polyethylene glycol 600 and 400 (PEG), TBP, D2EHPA, cupferron, citric, oxalic and acetic acids were chosen as extractants for adsorbent impregnation.

The functionalised resins, solvent impregnated resins and commercially available resins, viz. ES467 and C20MB, were subjected to resin-in-leach tests with gypsum.

2.3 Leach and Resin-in-Leach Tests

2.3.1 Unprocessed Gypsum: Resin-in-Leach Tests. A slurry of 25 g phosphogypsum (2000 ppm REE) and 100 mL of lixiviant (water, nitric or sulphuric acid) was added to 5 mL of resin and agitated at room temperature for 24 h in a rolling bottle. The resin, filtrate and gypsum were analysed for their REE content.

2.3.2 Unprocessed Gypsum: Leach Tests. A slurry of 1000 g phosphogypsum (2600 ppm REE) and 4000 mL of lixiviant (water and enough sulphuric acid to ensure pH values of 1.0, 1.5 and 2.0) was agitated at room temperature for 10 days in a rolling bottle.

2.3.3 Cycloned Gypsum: Leach Tests. A slurry of phosphogypsum fines (8900 ppm REE) and lixiviant (water and enough sulphuric acid to ensure pH value of 1.0) was agitated at room temperature in a rolling bottle. A matrix of experiments was set up where the slurry was agitated at various temperatures (22 or 40 °C), times (½, 1, 2, 5, 8 and 24 hours), fines: solution ratio (1:4, 1:20, 1:40) and in the presence of alum, alumina and NaF to ascertain if leaching could be increased. The filtrates in all tests were analysed for their REE concentration.

2.3.4 Cycloned Gypsum: Resin-in-Leach Tests. A slurry of phosphogypsum fines (9540 ppm REE) and lixiviant (water and enough sulphuric acid to ensure pH value of 1.0) was agitated at a solid:liquid:resin ratio of 5g:20mL:1mL at room temperature in a rolling bottle. Resins used were C20MB, ES467, A and PA (see Table 1 for details of A and PA). In addition, resin-in-leach tests in the presence of additives were undertaken.

2.4 Methods of Analyses

Infra red spectra were measured on a Fourier Transform Infra Red Spectrophotometer using diffuse reflectance techniques. Water retention capacity, proton exchange capacity and density were determined by standard methods. REE were determined by ICP-MS.

2.5 Hydrocycloning of Gypsum

The dump gypsum, (50 percent passing 80 µm), was hydrocycloned to yield a fines fraction (50 percent passing 17 µm) and a coarse fraction. The concentration of REE in the fines fraction was 8000 ppm in 8 percent of the mass of gypsum.

3 RESULTS AND DISCUSSION

3.1 Mechanism

The REE leached from apatite during phosphoric acid production are partitioned between the gypsum residue and the dilute acid. It is thought that the solubility of the

REE is limited by the presence of fluoride or phosphate which causes them to coprecipitate with the gypsum. Hence, the REE in the acid medium are in equilibrium with the REE flouride and phosphate in the gypsum as follows.

$$LF_3 \rightleftharpoons L^{3+} + 3F^-$$

where L = a Lanthanide.

The adsorption of REE by a resin can be described by the reaction

$$3R\text{-}H + L^{3+} \rightleftharpoons 3\,R\text{-}L + 3H^+$$

The use of an ion exchange resins to extract REE from solution would be expected to shift the equilibrium to the right and thus increase leaching efficiency. This is described by the overall reaction

$$LF_3 + 3R\text{-}H \rightleftharpoons 3\,R\text{-}L + 3HF$$

It might also be expected that the use of certain additives, e.g. alum and alumina, might help to shift the equilibrium reaction to the right by complexing the flouride or phosphate anions thereby releasing the REE into solution.

3.2 Screening of Resins

Resin-in-leach tests were carried out initially at neutral pH values. Recovery of REE was, at best, only 4.5 percent as shown in Table 2 (only the best results are given as the experiments were too numerous to be reported here). The adsorbents impregnated with D2EHPA and resin C20MB gave the best recoveries of REE.

Table 2 *Results of RIL Tests on Unprocessed Gypsum at Neutral pH*

Resin	REE loaded (mg/kg)	Ca loaded (mg/kg)	REE loaded (%) [#]
Solvent Impregnates			
XAD-1, D2EHPA	413	n.d	4.5
Carbon, D2EHPA	184	n.d	3.3
Carbon, TBP	33	n.d	0.5
XAD-1, TBP	23	n.d	0.3
Functionalised resin			
C20 MB	149	64100	3.6
E	69	1240	0.7
ES467	83	14400	0.6
AC	29	6530	0.3

[#] percent REE loaded is the REE loaded from the gypsum onto the resin during a RIL test.

As these tests (reported in Table 2), were carried out at neutral pH and since it was observed in leach tests (Section 3.3.1) that the optimum pH at which leaching occurs is 1.0, the tests were repeated at this pH value using the best resins and the closest resin analogues of the best impregnates, i.e. resins A and PA. Results of these tests at the optimum pH of extraction, viz. 1.0, and at room temperature are given in Table 3.

Table 3   *Results of RIL Tests on Unprocessed Gypsum at a pH Value of 1.0*

Resin	REE loaded (g/kg)	Ca loaded (g/kg)	REE loaded (%)
ES467	7.8	8.9	19
PA	7.2	10.5	18
C20MB	4.7	74.4	15
A	5.1	5.7	14

The resins all recovered a similar percentage of REE, however, the phosphorus containing resins had higher affinities for REE than had C20MB. Although the leach tests (Section 3.3.1) indicated that optimum leaching occurred at a pH value of 1.0, further RIL tests were carried out at lower pH value of 0.65 to determine if higher recoveries of REE could be attained. Results of these tests are given in Table 4 below.

Table 4   *Results of RIL Tests on Unprocessed Gypsum at a pH Value of 0.65*

Resin	REE loaded (g/kg)	Ca loaded (g/kg)	REE loaded (%)
ES467	7.9	8.9	19
C20MB	5.4	47.7	17
A	1.9	6.5	8

The results for ES467 was unaffected. Resin C20MB seemed to load more REE at a lower pH value while the loading of REE on resin A was depressed. A probable reason for the latter result might be the increased competition of protons with REE for sites on the resin. Resin C20MB, being a stronger acid resin, should not be protonated under the RIL test conditions and loading of REE on this resin should not be depressed.

The screening of resins and impregnated adsorbents has suggested a number of possibilities for future investigation, however, since resins ES467 and C20MB are commercially available they were used in subsequent testwork.

3.3   Optimization of Process Parameters

As only a fraction of REE were recovered in the initial experiments, the leaching characteristics of gypsum were investigated.

3.3.1 Unprocessed Gypsum : Leach Tests. A leach test carried out at a pH value of 1.0 for 24 hours resulted in 22 percent of REE being leached (referred to as the standard test on unprocessed gypsum). If the standard test was carried out over longer time periods (72, 96 and 240 h) no significant increases in extraction of REE occurred. Leaching efficiencies of only 1 to 3 percent were achieved in leach tests at higher pH values of 1.5 and 2.0 over 24, 72, 96 and 240 hours. Thus optimum leaching occurred under the standard test conditions.

A leached residue was subjected to a number of successive leaches with fresh lixiviant. It was apparent from these tests that the extractability of REE is limited to a maximum of 27 percent of the REE in the gypsum.

Although the results suggest that the proposed mechanism might be true for some of the REE content of the gypsum, a substantial portion appears to be insoluble.

3.3.2 Cycloned Gypsum : Leach Tests. It was considered that if the leachable REE were present in smaller particles it might be worthwhile concentrating these REE so that a reduced mass could be processed. Hydrocycloning the gypsum resulted in an increase in REE concentration from about 2500 g/kg to about 9000 g/kg in a fines fraction of 10 percent of the mass of gypsum used.

In the standard leach test (pH value of 1.0, 24 hours, ambient temperature and a liquid:solid ratio of 4:1) only 10 percent of REE was leached. Further tests were carried out where each of the variables were changed to determine their effect on the leaching of REE. The effect of time indicated that leaching of the fines was rapid and that no more than an hour at room temperature was required to attain equilibrium. An increase in temperature seemed to result in a slight decrease in leaching of REE. At a liquid:solid (L:S) ratio of 40:1, 48 percent of REE were leached. When the leached fines were subjected to a second leach test with fresh lixiviant an additional 16 percent of REE were found to be leached. The increased extraction efficiency observed as the L:S ratio is increased or with repeated leaches support the contention that the equilibrium will be disturbed by the presence of resin or additives.

From the results of leach tests in the presence of additives (Table 5) it is seen that the effect of the additives was to increase the leaching of REE from 10 to 45 percent thereby further supporting the mechanism.

Table 5   *Effect of Additives on the Leaching of REE from the Fines*

Test details	REE in solution (mg/L)	REE leached (%)#
Standard test on fines (STD)	220	10
STD + 5 % alum	1148	45
STD + 2 % each of alum and alumina	956	38
STD + 2 % alumina	440	20

percent REE leached is the REE leached from the gypsum in the leach tests

3.3.3 Cycloned Gypsum : Resin-in-Leach Tests. Four resins were tested in resin-in-leach tests, with and without the addition of alum or alumina. Results of these tests (Table 6) indicated that at most about 30 percent of the REE in the fines could be recovered. Percentages of REE recovered are calculated from the REE concentration in the fines fraction used (viz. 9540 g/kg). The overall extraction of REE was higher in the RIL tests on cycloned gypsum when compared to the standard leach test on the fines. When alum or alumina was added to the resin-in-leach test the loading of REE increased for the resins ES467 and PA2, but decreased for C20MB. However, these effects were marginal. In leach tests on the fines the concentration of alum added was 5 percent, while in these tests the concentration was only 2 percent. Therefore a larger amount of alum might be required in a resin-in-leach test to further improve recoveries of REE. An appreciable amount of REE remained in solution. It is possible that with a multistage counter current resin-in-leach process, where the depleted solution is in contact with fresh resin, the leaching reaction could be enhanced and recoveries improved.

A comparison of results for resins-in-leach tests on unprocessed and cycloned gypsum (Tables 3 and 6) indicate that REE loadings on resins in the tests with cycloned gypsum were up to 5 times higher.

Table 6   *Results of RIL Tests on Hydrocycloned Gypsum*

Resin	REE remaining in solution (mg/L)	Ca on resin (g/kg)	REE on resin (g/kg)	REE loaded (%)	REE recovered (%)#
C20MB	79	50.7	34.0	25	29
ES467	114	2 .5	29.6	16	21
PA2	125	2.3	21.1	15	21
A4	215	2.7	6.8	4	14
C20 MB + 2% alum	84	47.4	32.8	23	27
ES 467 + 2% alum	271	0.6	33.0	18	29
PA2 + 2% alum	264	1.2	25.0	17	28
PA2 + 2% alumina	160	1.5	25.4	17	26

percent REE recovered is the sum of the REE in the gypsum that were leached and loaded during a RIL test.

3.4 Proposed Process Flowsheet

Although a significant quantity of dump material already exists and current arisings are meaningful, the gypsum residue from the production of phosphoric acid contains REE at a low concentration. Therefore a desired process would have to be simple to be cost effective. As suggested by the present work a degree of concentration would be preferable and furthermore a solid/liquid separation step would be undesirable. Hence, a concentration step involving cycloning to concentrate on size, followed by a RIL process is suggested. The resins can be eluted with NaCl. Hence, the process envisaged is simple but the cost effectiveness of processing the dump needs to be assessed.

4 CONCLUSION

Processing of the mineral, apatite, for phosphoric acid produces large quantities of gypsum containing rare earth elements (REE) in low concentration. Recovery of the REE is technically feasible by a resin-in-leach (RIL) process.

It seems that there is a limit to the extractability of REE in gypsum. Hydrocycloning the gypsum to concentrate the leachable REE in a smaller and finer particle size mass fraction resulted in an increase in REE concentration from about 2500 g/kg to about 9000 g/kg.

The use of an ion exchange resin to extract REE from the cycloned gypsum shifted the equilibrium reaction, thereby allowing an increase in the leaching efficiency. The use of additives that complex certain anionic species also increased the leaching efficiency of REE.

Screening of resins and impregnates suggested a number of possible adsorbents to be investigated in the future. A comparison of results for RIL tests using the resin ES467 and C20MB on unprocessed and cycloned gypsum showed that REE loadings in the presence of cycloned gypsum was up to five times higher than that in unprocessed gypsum.

A process involving hydrocycloning of gypsum followed by resin-in-leach (with NaCl elution) is proposed.

Acknowledgements

The technical assistance of C. Tolken, M. Conway, T. Singh as well as useful discussions with Dr M. Bryson, Dr R. Paul, Dr P. Harris, Dr J. S. Preston and Ms M. Perry are gratefully acknowledged. This paper is published by permission of Mintek.

References

1. ZA 894878, Sentrachem.
2. ZA 893025, Sentrachem and Mintek.
3. MINTEQA2, *Equilibrium Metal Speciation Model,* Version 3.11, US Environmental Protection Agency, Athens, Georgia, 1991.
4. R. M Smith, A. E Martell, "Critical Stability Constants", 1976, Vol. 4, Inorganic Complexes, Plenum Press, New York.
5. A. Warshawsky, R. Kalir, H. Berkovitz, *Trans. Ins. Metall.,* 1979, March, C31.
6. A. J. Britz, F. L. D. Cloete, *Hydrometallurgy,* 1990, **25**, 213.
7. B. R. Green, R. D. Hancock, *Sep. Sci. Tech.,* 1980, **15(5)**, 1229.
8. S. D. Alexandratos, D. R. Quillen, M. E. Bates, *Macromolecules,* 1987, **20**, 1191.
9. R. S. Wright, M. Sc. Thesis, University of Witwatersrand, South Africa, 1989.
10. A Warshawsky *et al., J. Am. Chem,* Soc., 1979, **101**, 4249.
11. S. D. Alexandratos, *Macromolecules,* 1985, **18(5)**, 829.

KINETICS OF CHROMIC ACID REMOVAL BY ANION EXCHANGE

H. K. S. Tan

Department of Applied Chemistry
National Chiao Tung University
Hsinchu
Taiwan, ROC

1 INTRODUCTION

Ion exchange is widely used in the treatment of waste rinse water from electroplating processes. Among the chemicals present in the plating waste rinse water, the hexavalent chromium ion is one of the most toxic substances that is harmful to environmental health. The use of ion exchange for removing toxic chromium ions in rinse water not only meets EPA discharged standards but also makes possible the recovery of concentrated acid upon regeneration on the resin bed. Although the practice of treating chrome plating rinse water by ion exchange have been known for a long time[1-3], practically no studies have been made on the exchange kinetics between the resin and the dilute chromic acid. The fundamental understanding on the rate of chromium ion uptake in the anion resin is important in the design and operation for industrial applications.

The study on the kinetics of ion exchange of chromic acid is hampered by the fact that the exchange between chromate ions in solution and the OH^- ion in the resin is not an ordinary exchange process. The presence of multiple chromate species as pH varies also contributes to the complexity of the study. In order to simplify the task of interpreting and analyze experimental data quantitatively, several assumptions were made in order to use the simple kinetics model available.

2 METHOD OF STUDY

2.1 Theoretical

The composition of chromate species at a given value of pH or total Cr(VI) concentration can be calculated by ionic equilibrium relationships and by charge and mass balance. The equilibrium relationships for the major chromate species are

$$K_1 = [H^+][CrO_4^=] / [H CrO_4^=] = 3.2 \times 10^{-7}$$

$$K_2 = [H CrO_4^=]^2 / [Cr_2O_7^=] = 0.0302$$

These two equilibrium expressions together with a mass balance for total Cr(VI) and a charge balance determine the chromate species composition as a function of pH and

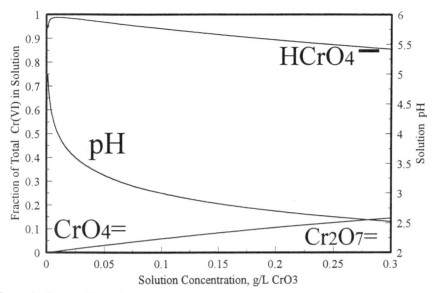

Figure 1 *Composition of chromates in chromic acid*

total Cr(VI) concentration. Figure 1 is a plot of the three major chromate species distribution in dilute chromic acid for Cr(VI) concentration up to 0.3 g/L CrO₃.

As can be seen from Figure 1, hydrogen chromate ($HCrO_4^-$) is the predominant species present. Even at a concentration of 0.3 g/L CrO_3 the fraction of $Cr_2O_7^=$ present is not more than 0.15. Thus as an approximation, the physical and chemical properties of $HCrO_4^-$ can be used to represent all the chromate species in solution. The rate equation is formulated for total Cr(VI) content since experimental determination of individual chromate species is difficult to carry out. The total Cr(VI) content in the resin at any time is related to that in solution by a mass balance. This mass balance is

$$V_r(q - q_o) = V(C_o - C)$$

where V_r is the volume of the resin, V is the volume of the solution, q and C denote, respectively, the concentration of total Cr(VI) in the resin and in the solution, subscript o stands for the initial condition. For resin initially containing no Cr(VI), $q_o = 0$, and in terms of dimensionless quantities, the mass balance becomes

$$x = (1 - R y)$$

where $x = C/C_o$, $y = q/Q$, with Q representing the resin capacity and R is a stoichiometric parameter equal to $(V_rQ)/(VC_o)$ or $(WQ')/(VC_o)$, where W is the weight of the resin and Q', the resin capacity in eq/kg.

The linear driving force rate equation for film diffusion control is formulated for total chromate species present in solution as

$$dy/dt = k_La_p (C_o/Q)(x - x^*)$$

where, k_L is the liquid film mass transfer coefficient and a_p, surface area per unit volume of resin particles and x^* the equilibrium value.

We should assume a two stage reaction model for the uptake of chromium ions. In the first stage of reaction, OH^- released from anion resin is neutralized by the free hydrogen ion in the solution. The resin then converts to monochromate form (R_2CrO_4). Since the neutralization is very rapid, the hydrogen concentration is determined by the dissociation of water and H^+ becomes insignificant and hence the concentration of all the chromate species at the interface is assumed to be zero. In the second stage of reaction, resin is completely depleted of OH^- ion, and exchange involves only $Cr_2O_7^=$, $CrO_4^=$ and $HCrO_4^-$ ions. In the second stage of exchange, it is assumed that x^* is small compared to x, thus a rate equation in the form similar to the first order irreversible chemical reaction is used for analyzing the experimental data in this work. Combining the simplified rate equation and the mass balance equation gives

$$dy/dt = k_L a_p (C_o/Q)(1 - R y)$$

Solving for time variation of y(t) and x(t)

$$y(t) = [1 - \exp(-k_L a_p V_r t /V)] /R \quad \text{and} \quad x(t) = \exp(-k_L a_p V_r t /V)$$

The value of $k_L a_p$ determined from fitting experimental curve of x(t) is then used to calculate the value of D/δ, with $a_p = 3/r_o$. Here D is the diffusion coefficient, δ the liquid film thickness and r_o radius of the resin beads. Following the approach of Helfferich[4], k_L is related to D/δ by $k_L = (C_o/Q)(D/\delta)$.

2.2 Experimental

Ten experimental runs were carried out in a batch system. A 6 liter capacity vessel equipped with a magnetic stirrer is used in the study. The solution concentration prepared for the experimental work contains 0.1 to 0.3 g/L CrO_3. Three different sizes of strong anion resin, Dowex 1 X8 (20-50 mesh, 50-100 mesh, 100-200 mesh) were employed for the study. The capacity of each type of resin was determined independently. In each run, a known amount of resin in OH^- form was introduced to the vessel together with the known volume of prepared CrO_3 solution. Small samples of solution were withdrawn at appropriate time intervals and its total Cr(VI) concentration analyzed and pH value recorded. Each experiment is terminated only when the Cr(VI) concentration in the solution practically reaches the final steady state value. A spectrophotometric method is used in the analysis of total Cr(VI) concentration in the solution.

3 RESULTS AND DISCUSSION

Table 1 lists the operating conditions of the ten experimental runs. The results are summarized and shown in Figures 2 to 6. Figure 2 shows the result of a typical run. The experimental x(t) values are obtained directly by measurement while the experimental y(t) values are determined indirectly by mass balance. The calculated values of x(t) and y(t) are obtained by using equations derived above. The fitted value of $k_L a_p$, as determined from experimental x(t) data, does not vary significantly for each data point. In fact, in all the runs, the standard deviation of $k_L a_p$ value is less than 20%. The experimental solution pH variation with time is also shown in Figure 2. Figures 3 to 6 show the results of the parametric effect on the rate of chromium uptake on the exchanger. For comparison purposes, the fractional approach to equilibrium is plotted against time.

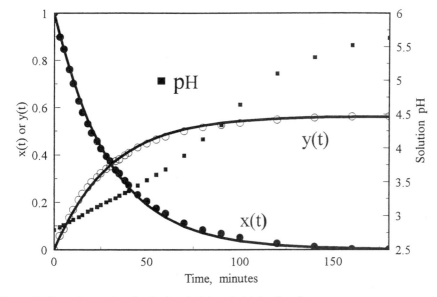

Figure 2 *Experimental and calculated x(t) and y(t) for Run 5*

The fractional approach to equilibrium is defined by $q(t)/q^*$. In Figure 3, results of the three different initial concentration runs show a slightly higher rate of exchange for larger initial concentration. The values of $k_L a_p$ determined for these three runs are very close. The higher rate for the higher concentration run must be due to the greater concentration gradient. Runs 4, 5 and 6 were performed using the same volume of solution and the same initial concentration. The amount of resin used was 3, 5 and 7 grams and the corresponding stoichiometric ratios are 1.03, 1.72 and 2.41. The rate of attainment of equilibrium for these three runs is shown in Figure 4.

Table 1 *Operating Conditions of Experimental Runs*

Run	W	V	Q'	C_o	Stirring	r_o
	g	mL	meq/g	g/L	rate, ppm	mesh no.
1	3.001	3000	3.466	.1031	1000	20-50
2	3.005	3000	3.446	.2010	1000	20-50
3	3.001	3000	3.446	.2935	1000	20-50
4	3.002	5000	3.446	.1981	1000	20-50
5	5.021	5000	3.446	.1946	1000	20-50
6	7.001	5000	3.446	.2037	1000	20-50
7	3.007	3000	3.446	.2002	600	20-50
8	3.001	3000	3.446	.2007	800	20-50
9	3.002	3000	3.607	.1998	1000	50-100
10	3.003	3000	3.756	.1965	1000	100-200

The value of $k_L a_p$ determined for the three different stoichiometric ratio (R) runs are very close. The higher rate observed for larger values of R is due to fact that the rate is expressed in terms of change of q/q^* with respect to time. In this case, with the same values of V and C_o, lower R values correspond to higher values of q^* from the mass balance relation. The effect of stirring rates, as illustrated in Figure 5, indicates that the rate of fractional attainment of equilibrium is almost identical for stirring speeds of 800 rpm and 1000 rpm. As pointed out by Helfferich[5], a limiting hydrodynamic efficiency is reached beyond which further increase in stirring speed will have no effect on the rate of exchange. The results, shown in Figure 6 , indicate definitely the higher rate that is being realized by using small sizes of resin. The ratio of values of $k_L a_p$ determined using 100-200 mesh resin to that of 50-100 mesh is about two. Although no independent measurement of average resin radius was performed, a factor of two, relating the average radius for the two different sizes of resin, appears not to be out of order. As indicated in many studies, the rate control step is sometimes determined by examining the rate data for various resin sizes used. For film diffusion control, the rate is inversely proportional to the average resin radius. Whereas for the particle diffusion control, the rate is inversely proportional to the square of resin radius. The value of $k_L a_p$ determined in this work supports the assumption that the rate of exchange is most likely film diffusion control.

The values of $k_L a_p$ determined by fitting experimental data are tabulated in Table 2. The corresponding calculated values of $(D/r_o)/\delta$ are also listed in the Table.

Table 2   *Calculated Values of $k_L a_p$ and $(D/r_o)/\delta$*

Run		1	2	3	4	5	6	7	8	9	10
$k_L a_p (V_r/V)$	L/min	.0428	.0434	.0387	.0221	.0332	0452	.141	.314	.0228	.0397
$k_L a_p$	L/s	642	.651	.581	.552	.498	.484	2.12	4.71	.342	.595
$(D/r_o)/\delta$	L/s	.214	.217	.193	.184	.166	.161	.706	1.57	.114	.198

Blickenstaff et al.[6] had studied the kinetics of exchange between strong acidic cation resin in H^+ form with dilute NaOH solution. Using the film diffusion control rate model and the theoretical equation derived by Helfferich[7], they obtained values of D_{NaOH}/δ for resin size of $r_o = 0.0179$ cm (40-50 mesh) and $r_o = 0.0388$ cm (20-25 mesh). The values reported are $(D/r_o)/\delta = 0.8436$/sec for $r_o = 0.0179$ cm and 0.4356/sec for $r_o = 0.0388$ cm. The values obtained for the chromate system are smaller than the NaOH system. It appears that the mobility of the chromate ions is lower than caustic ions. In any case the experimentally determined values of $(D/r_o)/\delta$ for close batch system can serve as a useful reference in the calculation of column performance. For fixed bed processes with film diffusion rate controlling, Glueckauf[8] proposed that the mass transfer rate parameter, $k_L a_p$ be calculated from the value of $(3D)/(2r_o\delta)$.

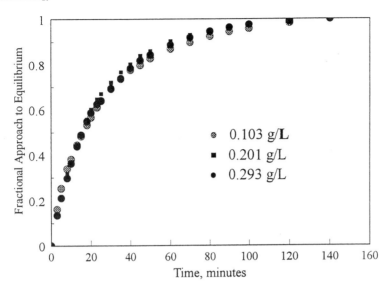

Figure 3 *Effect of initial solution concentration*

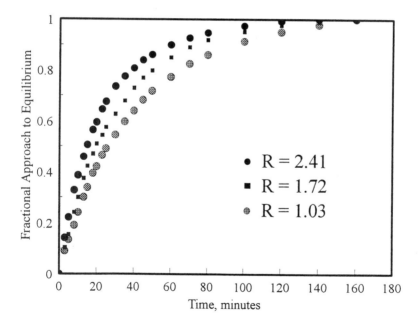

Figure 4 *Effect of stoichiometric ratio*

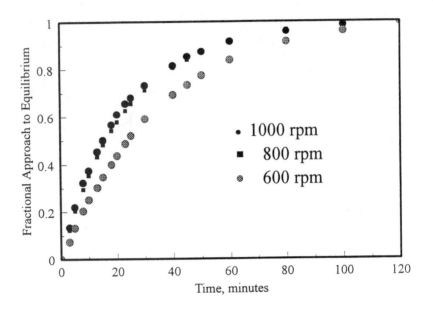

Figure 5 *Effect of stirring speed*

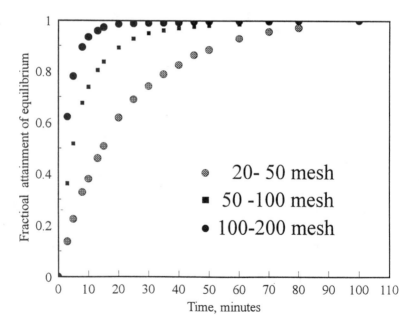

Figure 6 *Effect of resin size*

4 CONCLUSIONS

In this work, a study of the kinetics of removal of dilute chromic acid by anion exchanger in OH⁻ form was undertaken. The experimental work was conducted by employing a batch system with finite solution volume. Linear driving force rate equation based on total Cr(VI) species transfer from solution to resin was utilized for quantitatively analyzing the experimental data. The results obtained confirm the suggestion that the rate of exchange is film diffusion rate controlled. Values of D/δ obtained in this work serve as useful reference in the design of fixed bed processes for industrial applications.

Nomenclature

a_p	surface area per unit volume of particle	m^2/m^3
C	solution concentration of Cr(VI)	eq/m^3
C_o	initial solution concentration of Cr(VI)	eq/m^3
D	diffusivity in solution	m^2/s
k_L	liquid film mass transfer coefficient	m/s
q	resin Cr(VI) concentration	eq/m^3
Q	resin capacity	eq/m^3
Q'	resin capacity	eq/kg
r_o	radius of resin bead	m
R	stoichiometric ratio	
t	time	s
V	solution volume	m^3
V_r	resin volume	m^3
W	weight of resin	g
x	fraction of initial solution concentration of Cr(VI)	
y	fraction of resin capacity in Cr(VI) form	

Greek symbol

δ	liquid film thickness	m

References

1. S. Sussman, F. C. Nachold, and W. Wood, *Ind. Eng. Chem.*, 1945, **37,** 618.
2. H. Gold, "Ion Exchange in Pollution Control", C. Calmon, and H. Gold, eds., Boca Raton, Florida, 1979, 173.
3. T. Nadeau, and M. Dejak, *Plating and Surface Finishing*, 1986, **73 (4)**, 48.
4. F. Helfferich, "Ion Exchange", McGraw-Hill Co., New York, 1962, p. 262.
5. F. Helfferich, "Ion Exchange", McGraw-Hill Co., New York, 1962, p. 285.
6. R. A. Blickenstaff, J. D.Wagner, and J. S. Dranoff, *J. Phys. Chem.,* 1967, **71,** 1665.
7. F. Helfferich, *J. Phys. Chem.,* 1965, **69,** 1178.
8. E. Glueackauf, in "Ion Exchange and its Applications", Society of Chemical Industry, London, 1955.

KINETIC STUDIES ON HEAVY METAL IONS REMOVAL BY IMPREGNATED RESINS CONTAINING ORGANOPHOSPHORUS EXTRACTANTS

J. L. Cortina, N. Miralles and M. Aguilar

Chemical Engineering Department (ETSEIB)
Universitat Politècnica de Catalunya
Barcelona, 08028
Spain

1 INTRODUCTION

The increasing concern towards optimization of industrial processes dealing with metals, impose the need for the development of advanced separation techniques and, in particular, for liquid wastes. In this context, the preparation of selective adsorption systems by physical immobilization of metal extractants, on macroporous resins to give impregnated resins (IR) has been presented as a technological alternative to solvent extraction and ion exchange technologies[1,2]. The idea behind the development of impregnated resins is the ability to combine the selectivity and specificity of conventional extractants with the advantages of a discrete polymer support material. Since the pioneering work the development and application of these systems in metal extraction processes has been intensively investigated for hydrometallurgical[3-5] and analytical applications[6-7]. In the last years, our research group has been working on the development of Impregnated Resins (IR) for recovery and separation of metal ions from dilute solutions. The impregnated resins have been characterized physically and chemically and their behavior in the extraction of base metals has been studied[8-11]. However, the application of these systems in industrial scale equipment using fixed column or fluidized bed technology requires a knowledge of the operating hydraulic behavior, the equilibrium data and kinetics of metal extraction processes. The determination of kinetic parameters has two objectives: (i) to approach, as accurately as possible, the real physical-chemistry of the metal extraction process and (ii) to get empirical or semiempirical equations for the design of the equipment.

This paper describes a study of metal extraction kinetics with Impregnated Resins prepared by direct adsorption of mixtures of DEHPA(HL) and TOPO(S) into Amberlite XAD2. For this purpose, resin loadings as a function of the contact time were monitored.

2 EXPERIMENTAL

2.1 Reagents

Di(2-ethylhexyl)phosphoric acid (BDH) and Tri-n-octylphosphine oxide (Merck) were used without further purification. Stock solutions of Zn(II), Cu(II) and Cd(II)

(1g.dm^{-3}) were prepared by dissolving the corresponding salts (Merck, A.R. grade) in water. Amberlite XAD-2 Resin (Rohm and Haas), size 0.3-0.9 mm, was used.

2.2 Impregnation Process

XAD2-DEHPA-TOPO impregnated resins were prepared according to a modified version of the dry impregnation method described previously[8]. The amount of DEHPA ($[HL]_r$) and TOPO ($[S]_r$) impregnated was evaluated by washing a known amount of resin with ethanol. This completely elutes both extractants for subsequent determination. After elution of both extractants with ethanol, the DEHPA content was determined by NaOH potentiometric titrations from an aliquot. The TOPO content was determined by spectrophotometry measurements at 230 nm[12].

2.3 Metal Extraction Kinetic Measurements

The testwork was performed using the shallow bed technique on a micro scale. According to this technique[13], an aqueous metal solution is passed at high flow rate through a thin layer of resin beads in a column. The objective of this procedure is to avoid the formation of a concentration gradient along the resin bed. Thus, the composition of the external solution remains practically constant throughout the experiment. The flow of the solution is periodically stopped and the resin is washed and analyzed to provide data on resin composition as a function of time. A typical resin bed contained 40 beads (2-3 mm depth) and the flow rate employed was 200 mL/h.

The resin composition was determined by passing volumes of 25 or 50 mL of 0.2 M hydrochloric acid solution at low flow rate 50 mL/h through the thin layer of resin beads in the column. After appropriate dilution, the metals were analyzed by atomic absorption spectrophotometry. The results are expressed as mmol of metal per gram of dry impregnated resin XAD2-DT-S. The fraction of largest beads, having a radius between 0.5 to 0.7 mm, was selected for the test work.

3 RESULTS AND DISCUSSION

3.1 Metal Extraction Kinetic Studies in Impregnated Resins. Kinetic Models

Metal extraction reactions in Impregnated Resins, as in other heterogeneous reactions between solids and fluids, are explained through a number of sequential processes that determine the rate of reaction: a) Diffusion of ions through the liquid film surrounding the particle, b) Diffusion of ions through the polymeric matrix, c) Chemical reaction with the functional groups attached to the matrix. One of the steps usually offers much greater resistance than the others, so it can be considered as the rate limiting step of the process.

Accordingly, the extraction mechanism will involve counter diffusion of M^{2+} ions from the aqueous solution and H^+ ions from the resin phase through a number of possible resistances. The species originally in the solution phase must diffuse across the liquid film surrounding the impregnated-resin particle, transfer across the solution/particle interface, diffuse into the bulk of the impregnated-resin particle and possibly interact with a impregnated extractant molecule adsorbed in the surface of the macroporous support.

The species within the impregnated-resin simultaneously experience these same sequences reverse order.

The exchange of M^{2+}/H^+ can be rigorously described by the Nernst Planck equation. This applies to counter diffusion of two species in a quasi-homogeneous media. Resin phase controlled diffusion of ions from an infinite volume of solution into a spherical ion exchange particle can be described by the following equation[14]

$$-\ln\left(1 - X^2\right) = 2kt \quad \text{where} \quad k = \frac{D_r \pi^2}{r_o^2} \tag{1}$$

If liquid film diffusion controls the rate of exchange, the following analogous expression can be used:

$$-\ln\left(1 - X\right) = K_{li} \quad \text{where} \quad K_{li} = \frac{3DC}{r_o C_r} \tag{2}$$

When the porosity of the polymer is small and thus practically impervious to the fluid reactant, the reaction may be explained by the "Shell Progressive" approach. The kinetic concept of a "Shell Progressive" mechanism can be described in terms of the concentration profile of a liquid reactant containing a counterion A advancing into a spherical bead of a partially substituted ion exchanger. As the reaction progresses in the bead, the material balance of counterion A follows Fick's diffusion equation with spherical coordinates. In this case, the relationship between reaction time and degree of conversion gives the following expressions [15-16]:

a) When the fluid film is in control:

$$t = \frac{a r_o C_{so}}{3 C_{Ao} K_{mA}} X \tag{3}$$

b) When the diffusion though the reacted layer control

$$t = \frac{a r_o^2 C_{so}}{6 D_{e,r} C_{Ao}} \left[3 - 3\left(1 - X\right)^{2/3} - 2X\right] \tag{4}$$

c) When the chemical reaction control

$$t = \frac{r_o}{k_s C_{Ao}} \left[1 - \left(1 - X\right)^{1/3}\right] \tag{5}$$

In the present work the metal/proton extraction process in the impregnated resins prepared will be quantitatively discussed. The computation methods used in this case may serve as a general model for treating all the processes in which the kinetics are controlled by layer diffusion, film diffusion or chemical reaction.

The use of DEHPA-TOPO mixtures physically immobilized into Amberlite XAD2 supports has shown a high efficiency in the extraction of Zn(II), Cu(II) and Cd(II). The extraction of these metal ions involves the formation of mixed species in the resin phase with a general composition $ML_2(HL)_{q,r}$ and $ML_{(2-t)}(NO_3)_t(HL)_{q,r}$ where q, t values take different values depending on the metal[12].

3.2 Metal Extraction Kinetic Studies in Impregnated Resins. Kinetic Results

Kinetics measurements in the extraction of Zn(II), Cu and Cd(II) with XAD2-DP-S at two different levels of metal concentrations ($1x10^{-2}$ M and $1x10^{-4}$ M) were made. In all cases the rate of extraction was measured under conditions simulating those in a counter current extraction process, i.e., conditions in which the concentration of metal in solution remains approximately constant as the metal loading on the resin increases and approaches its equilibrium value. These conditions were obtained using the shallow bed technique described above.

3.2.1 High-Solution Concentration Kinetics. Figures 1a, b show the results of the extraction kinetics of Zn(II) from single element solutions in the form of different kinetic models defined by equations eq. 1 and eq. 4 as a function of contact time. This series of experiments was run at a relatively high concentration ($1x10^{-2}$mol.L^{-1}) with the purpose of promoting particle diffusion control. Only the first hour of the process was studied in this experiment.

Chemical reaction can be discarded as the controlling step since the fit did not give a straight line. Both Homogeneous Particle Diffusion (based on Fick's Law) and the Shell Progressive model seemed to fit the data satisfactorily, though the homogeneous particle diffusion model fits over the extended range of the process. A relatively good straight line could also be obtained over a smaller range for the film diffusion model. Similar results were obtained for Cu(II) and Cd(II).

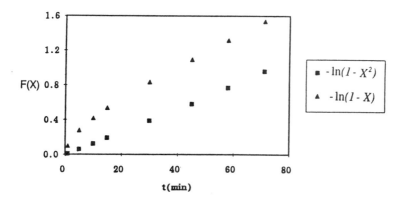

Figure 1a *Kinetics of Zn(II) extraction and test of mathematical models proposed by homogenous particle diffusion for XAD2-DEHPA-TOPO resins. (Experiments at high Zn(II) concentrations [Zn(II)]= 1x10^{-2}M))*

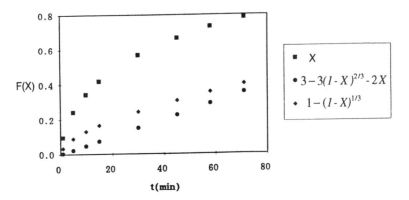

Figure 1b   *Kinetics of Zn(II) extraction and test of mathematical models proposed by shell progressive model. (Experiments at high Zn(II) concentrations [Zn(II)]= 1x10⁻²M))*

The results of the linear regression analysis for both functions ($-\ln(1-X^2)$ and ($3-3(1-X)^{2/3}-2X$) for the three metal ions are summarized on Tables 1a, b. The straight lines that are obtained in all cases do not pass, as they should, through the origin because of a slight deviation from linearity near the origin. This result can be explained by the fact that at the beginning of the reaction the thickness of the reacted layer is still very small and thus comparable with that of the liquid film adhering to the particle. The film's resistance to diffusion of the reactant can therefore become comparable with the resistance provided by the resin outer shell.

Table 1a *Linear Regression Analysis of Functions $-\ln(1-X^2)$ versus Time (t) at High Metal Concentrations ($1x10^{-2}M$). (Homogeneous Diffusion Model)*

Metal	Intercept	Slope $(h^{-1})^{(1)}$	Linear Correl.	$D_r(m^2s^{-1})$
Zn(II)	-0.0099	0.8091	0.9996	$4.09x10^{-8}$
Cu(II)	-0.0023	0.2924	0.9991	$1.48x10^{-8}$
Cd(II)	0.0016	0.2354	0.9992	$1.19x10^{-8}$

(1) slope 2K where $K=D_r\pi^2/r_o^2$

Table 1b *Linear Regression Analysis of Functions $(1-3(1-X)^{2/3}-2X)$ versus Time (t) at High Metal Concentrations ($1x10^{-2}M$). (Shell Progressive Model)*

Metal	Intercept	Slope $(h^{-1})^{(2)}$	Linear Correl.	$D_{e,r}(m^2s^{-1})$
Zn(II)	-0.0026	0.3021	0.9999	$0.51x10^{-6}$
Cu(II)	-0.0001	0.1125	0.9995	$0.12x10^{-6}$
Cd(II)	0.0022	0.6009	0.9999	$2.77x10^{-8}$

(2) slope $6D_{e,r}C_{Ad}/aC_{so}r_o^2$

The linear correlations indicate a good fit for both models. The slope values can be used to calculate effective diffusion coefficients for each cation using equations eq. 1 and eq. 4. The diffusion coefficient calculated from the value of the Diffusion Coefficients is in fact a measure of the mean interdiffusion coefficient of the various species involved in the ion exchange process. A comparison of the diffusion coefficients proposed for both kinetic models shows that there are significant differences in diffusivity for the three metal ions. The relative order of diffusivity in the resin matrix is: Zn(II) > Cu(II) and Cd(II).

3.2.2 Low-Solution Concentration Kinetics. Figure 2 shows both the fractional approach to equilibrium (X) and the function -ln(1-X) as a function of contact time for the last series of experiments at low metal concentration levels for Zn(II) extraction. The linearity of the graphs suggests a regime of film diffusion control since the Nernst diffusion layer and the external solution concentration are constant. Similar results for Cu(II) and Cd(II) were obtained.

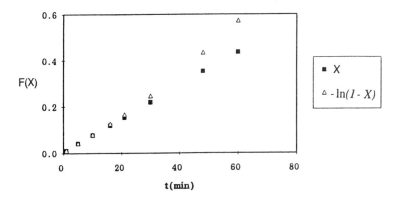

Figure 2 *Kinetics of Zn(II) extraction and test of mathematical models proposed by shell progressive model for XAD2-DEHPA-TOPO resins. (Experiments at low Zn(II) concentrations [Zn(II)]= 1x10⁻⁴M))*

Tables 2a, b contain the regression analysis of both functions (X and -ln(1-X)). Both the intercept close to zero, indicating negligible contribution to the rate from the resin diffusion, and linear correlations indicate good fit to a linear relation. Both Shell Progressive and Homogenous Diffusion Models predict that in the case of film diffusion control, the fractional approach to equilibrium (X) and -ln(1-X) function bear a linear relationship with time as shown by equations eq 2 and eq 3. Both equations are valid for a constant external solution and spherical resin beads of uniform and constant size. The slope values from Table 2a, b were used to calculate relative values of both K_m and K_{li}. A comparison of the mass transfer coefficients proposed for both kinetic models shows that there are significant differences in diffusivity for the three metal ions through the liquid film. The relative order diffusivity in the liquid film is: Zn(II) > Cu(II) and Cd(II).

Table 2.a   *Linear Regression Analysis of Functions -ln(1-X) versus Time (t) at High*
Metal Concentrations (1x10⁻⁴M). (Homogeneous Diffusion Model)

Metal	Intercept	Slope $(h^{-1})^{(3)}$	Linear Correl.	$K_{li}(s^{-1})$
Zn(II)	0.0017	0.4804	0.9988	1.33×10^{-4}
Cu(II)	-0.0094	0.4219	0.9984	1.17×10^{-4}
Cd(II)	0.0020	0.1127	0.9993	3.13×10^{-5}

$^{(3)}$ slope K_{li}

Table 2.b   *Linear Regression Analysis of Functions X versus Time (t) at High Metal*
Concentrations (1x10⁻⁴M). (Shell Progressive Model)

Metal	Intercept	Slope $(h^{-1})^{(4)}$	Linear Correl.	$K_{mA}(m.s^{-1})$
Zn(II)	-0.0015	0.5672	0.9936	3.21×10^{-3}
Cu(II)	-0.0014	0.3606	0.9996	1.27×10^{-3}
Cd(II)	0.0029	0.1064	0.9990	3.76×10^{-3}

$^{(4)}$ slope $3C_{Ao}K_{li}/ar_oC_{so}$

4 CONCLUSIONS

Due to the lack of precise information on the metal exchange reactions in impregnated resins, it is unlikely that a rigorous kinetic model can be developed. We have adapted the Fick's Law approach and the Shell Progressive Models to fit our experimental data and, in a first approach both models can be used in the study of the metal extraction processes in Impregnated Resins. Results obtained in this work indicated: (a) Working at high metal ion concentrations, the rate-determining step of the metal extraction is the resin-phase diffusion, (b) Working at low metal ion concentrations, the rate-determining step of the metal extraction rate is the liquid-film diffusion and (c) Both models proposed a mean or average interdiffusion coefficient which provides an insight into the diffusion mechanism and a parameter for subsequent design calculations.

It is felt that Fick's Law approach and the Shell Progressive mechanisms represent good general approaches to the kinetics of metal extraction reactions on polymeric macroporous supports containing impregnated reagents.

Acknowledgments

This work was supported by CICYT Project **MAT 93-6212** (Ministerio de Educación y Ciencia, España).

Nomenclature

a	stoichiometric coefficient
C	Total concentration of both exchanging species, M
C_r	Total concentration of both exchanging species in the ion exchanger, M
C_{Ao}	Concentration of species A in bulk solution, M
C_{so}	Concentration of solid reactant at the bead's unreacted core, M
t	time, s
D	Diffusion coefficient in solution phase, m^2s^{-1}
D_r	Diffusion coefficient in solid phase, m^2s^{-1}
K_{li}	Rate constant for film diffusion (infinite solution volume condition)
X	Fractional attainment of equilibrium or extent of resin conversion
$D_{e,r}$	Diffusion coefficient in solid phase, m^2s^{-1}
K_{mA}	Mass transfer coefficient of species A through the liquid film, $m.s^{-1}$
K_s	reaction constant based on surface, $m.s^{-1}$

References

1. A. Warshawsky, "Ion Exchange and Solvent Extraction", J. A. Marinsky and Y. Marcus, eds., Marcel Dekker, New York, 1981, Vol. 8, p. 229.
2. L. L. Tavlarides, J. H. Bae, and C. K. Lee, *Sep. Sci. Technol.*, 1987, **22**, 581.
3. S. Gonzalez-Luque and M. Streat, *Hydrometallurgy*, 1983, **11**, 207.
4. A. C. Muscatello and J. D. Navratil, *J. Radional. Nucl. Chem. Letters*, 1988, **128**, 463.
5. K. Yoshizuka, Y. Sakamoto, Y. Baba, and K. Ionue, *Hydrometallurgy*, 1990, **23**, 309.
6. Y. Wakui, H. Matsunaga and T. Suzuki, *Anal. Sci.*, 1989, **5**, 189.
7. E. Horwitz, M. Dietz, D. Nelson, J. La Rosa, and W. Fairman, *Anal. Chim. Acta*, 1990, **238**, 263.
8. O. Abollino, E. Mentasti, V. Porta, and C. Sazarnani, *Anal. Chem.*, 1990, **62**, 21.
9. J. L. Cortina, A. M. Sastre, N. Miralles, M. Aguilar, A. Profumo, and M. Pesavento, *React. Polym.*, 1992, **18**, 67.
10. J. L. Cortina, A. Sastre, N. Miralles, and M. Aguilar, *Hydrometallurgy*, 1994, **36**, 131.
11. J. L. Cortina, N. Miralles, M. Aguilar, and A. Sastre, *Solv. Extr. Ion Exch.*, 1994, **12**, 371.
12. J. L. Cortina, N. Miralles, M. Aguilar, and A. Warshawsky, *React. Polym.*, 1995, **27**, 61.
13. J. L. Cortina, N. Miralles, M. Aguilar, and A. M. Sastre, *Hydrometallurgy*, 1995, **37**, 301.
14. F. Helfferich, in "Ion Exchange", Mc Graw-Hill, New York, USA, 1962.
15. L. Liberti, and R. Passino, "Ion Exchange and Solvent Extraction", J. A. Marinsky, and Y. Marcus, eds., Marcel Dekker Inc., New York, 1977, Vol. 7, Chapter 3.
16. G. Schmuckler, and S. Golstein "Ion Exchange and Solvent Extraction", J. A. Marinsky and Y. Marcus, eds., Marcel Dekker Inc., New York, 1977, Vol. 7, Chapter 1.

ELIMINATION OF HEAVY METALS FROM WATER BY MEANS OF WEAKLY BASIC ANION EXCHANGE RESINS

W. H. Höll

Karlsruhe Research Center
Institute for Technical Chemistry
Section "Water Technology and Geotechnology"
P.O. Box 3640
D-76021 Karlsruhe

1 INTRODUCTION

The elimination of heavy metals from water by means of ion exchange often meets the problem that cations of the background ionic composition are present at much higher concentrations. As a consequence, a sufficient selectivity for heavy metals is required. Because standard cation exchangers did not exhibit a marked preference, specifically selective resins have been developed[1-3]. The particular properties of such chelating resins is due to electron donor atoms in the functional groups which act as LEWIS-bases. By this means, the sorption of metal species is caused both by ionic and coordinative forces. A great variety of such resins has been synthesized, sometimes selective only for one kind of counterion[1-3]. A common functional group is based on iminodiacetic acid. To increase the selectivity mainly for copper, resins with two or even three nitrogen atoms per functional group have been developed[4,5]. Their application has widely been studied and described in literature[5-7].

Iminodiacetic resins are amphoteric exchangers with mixed weakly basic and weakly acidic functionalties. At pH < 2 the nitrogen atom is protonated and, therefore, the resins act as an anion exchanger. Between pH-values 2 and 4 only chelation bonds with heavy metal species are formed, at pH > 6, however, the carboxylic groups are dissociated and allow also an uptake of alkaline earth ions. As a consequence, the effective capacity for heavy metals becomes smaller.

Standard weakly basic exchangers exhibit a similar performance with the exception that there is no dissociation of acidic groups at elevated pH. As long as the nitrogen atoms are protonated, they act as anion exchangers. At pH values greater than the resins' pK, the LEWIS base properties lead to the sorption of heavy metals. Since alkaline earth ions are excluded, there is an extreme selectivity. The sorption of heavy metals ions has first been demonstrated by Saldadze and coworkers with modified standard anion exchangers and specially synthesized resins[8-12]. Further modified standard products were also investigated by other authors[13-15]. Based upon experiences with selective polymers[16-18] chelating resins basically being weakly basic resins have been developed on a commercial basis. Some equilibrium data have been presented by Inoue et al[19].

The application of commercially available weakly basic (and of chelating) resins for the treatment of heavy metals-bearing solutions has first been reported by Hubicki[20]. However, neither equilibrium data nor breakthrough curves were presented.

2 SORPTION/DESORPTION MECHANISM

When a weakly basic exchange resin in the free base form is contacted e.g. with a $CuSO_4$ bearing solution, the resin will adsorb Cu^{2+} ions due to coordinate bonds. Corresponding to the negligible degree of protonation of the nitrogen atoms, an exchange of SO_4^{2-} for OH^- ions is negligible. However, the pure uptake of copper species would violate the condition of electroneutrality in both the aqueous and the resin phases. As a consequence, a simultaneous sorption of both, Cu^{2+} and SO_4^{2-} species has to occur:

$$\overline{M-NR'R''} + Cu^{2+} + SO_4^{2-} \Leftrightarrow \overline{M-NR'R''(Cu^{2+},SO_4^{2-})}$$

(R' and R" are arbitrary alkyl chains of the resin structure, overbarred symbols denote the resin phase.) Corresponding to individual resin properties, different structures of the complexes between resin and heavy metal ion can occur[8]. Protons are bound much stronger than heavy metal species, therefore, the resins can be regenerated by means of strong acid followed by conversion to the free base form by means of NaOH:

$$\overline{M-NR'R''(Cu^{2+},SO_4^{2-})} + H_2SO_4 \Leftrightarrow \overline{M-N^+R'R''HSO_4^-} + CuSO_4$$

$$\overline{M-N^+R'R''HSO_4^-} + 2\,NaOH \Leftrightarrow \overline{M-NR'R''} + 2\,H_2O$$

3 EXPERIMENTAL

Most of the experiments were carried out using the commercially available resins DUOLITE A7 (secondary amine, phenol-DVB) and AMBERLITE IRA 35 (tertiary amine, acrylic-DVB). For investigations of the sorption kinetics also AMBERLITE IRA 93 (tertiary amine, styrene-DVB) and DUOLITE A 561 (tertiary amine, phenol-DVB) were studied. To remove monomers and other contaminations due to manufacturing, the resins were pretreated by several cycles with 1 molar HCl and NaOH solutions.

To obtain sorption isotherms several samples of ≈ 1 g of wet centrifugated resin material, in the free base form, were stored in small shallow filter columns and equilibrated with 2 - 5 liters of solution slowly flowing across the resin material. Equilibrium was assumed to be achieved when the feed and effluent pH values were approximately equal. In some cases, batch experiments with automatic pH control had to be applied. After adjustment of equilibria, the resin samples were eluted by means of 1 molar HCl or H_2SO_4.

Sorption kinetics were investigated in batch experiments by means of a centrifugal stirrer apparatus as used in earlier studies of ion exchange kinetics[21].

Filter experiments were carried out at the laboratory scale using columns of 1 or 2 cm diameter and bed heights between 3 and 15 cm. The feed consisted either of pure solutions of $CuSO_4$, $CuCl_2$, $NiSO_4$, $ZnSO_4$, $CdSO_4$, and $Pb(NO_3)_2$ in distilled water or of mixtures of $CuSO_4$ and $NiSO_4$ or of heavy metal-bearing natural water. The concentrations of the pure solutions amounted to 1 - 5 mmol/L, those of the natural water were ≤ 0.1 mmol/L. Throughput varied between 2.5 and 10 BV/h. For regeneration 0.01 - 1 molar HCl or H_2SO_4 solutions were applied at different throughput conditions.

Heavy metals and anion concentrations were determined from effluent samples by means of atomic absorption or inductive coupled plasma and by ion chromatography measurements, respectively. In the filter experiments, the pH value was recorded automatically.

4 SORPTION EQUILIBRIA

4.1 Sorption Capacities

The maximum possible loading is obtained if each nitrogen atom forms a coordinative bond with one metal ion. Therefore, this loading should correspond to the resin capacity for the uptake of e.g. HCl. For DUOLITE A7 and AMBERLITE IRA 35, the maximum loadings with Cu^{2+}, Ni^{2+}, and Zn^{2+} ions obtained by means of 25 mmol/L solutions and the capacities for uptake of HCl are listed in Table 1. It can be expected that even greater loadings can be found by applying still more concentrated solutions.

Table 1 *Sorption Capacities of DUOLITE A7 and AMBERLITE IRA 35 in mol/g at Experimental pH[22]*

	DUOLITE A7	AMBERLITE IRA 35
Cl^-	1.66	1.50
Cu^{2+}	1.07 (5.0)	0.86 (5.0)
Ni^{2+}	0.51 (6.5)	0.75 (7.0)
Zn^{2+}	0.48 (5.0)	0.73 (6.7)

For copper, the results indicate at least part of the ions have to be fixed to only one nitrogen atom. For nickel and zinc, the loadings obtained are too small for similar conclusions.

4.2 Complexation Constants

For the metal-resin complexes, stability constants can be defined in analogy to those in aqueous phases[22]:

$$K = \frac{\bar{c}(Me^{2+})}{c(Me^{2+}) \cdot \left[\bar{c}(Me^{2+})_{max} - \bar{c}(Me^{2+})\right]^n}$$

The numerator gives the concentration of resin-metal complexes whereas the expression in brackets in the denominator corresponds to the concentration of free resin ligands.

It can be assumed that the sequence of stability roughly corresponds to that of ordinary amine complexes with ligands of the types

$$R_1 - \underline{N} - (CH_2)_2 - \underline{N} - R_2$$

being similar to the structure of DUOLITE A7 and

$$R_1 - \underline{N} - (CH_2)_3 - \underline{N}(CH_3)_2$$

which approximately corresponds to that of AMBERLITE IRA 35. The same sequence could only be found for n = 1. As a consequence, one metal ion forms coordinate bonds only with one single chain. Due to the length of R_1 and R_2 in the resins phases, however, the numerical values are smaller that those of ordinary amine complexes[23]. The results are summarized in Table 2.

Table 2 *Formation Constants for Resin Complexes with Copper, Nickel, and Zinc Ions. Liquid Phase Concentration: 10 mmol/L MeSO$_4$* [22]

	DUOLITE A7	AMBERLITE IRA 35
log K (R-Cu)	3.8	4.7
log K (R-Ni)	2.5	3.2
log K (R-Zn)	2.4	2.6

4.3 Sorption Isotherms

Isotherms from the sorption of Cu^{2+} ions from CuSO$_4$ solutions at pH values of 5, 4, and 2.8, respectively are presented in Figure 1. The developments of the isotherms depicts that the loading generally increases with increasing concentration in the liquid phase. Furthermore, the maximum loadings decrease with decreasing pH as can be concluded from the competing sorption of protons and metal ions. For the secondary amine type resin DUOLITE A7, rather high loadings are reached within the concentration range applied in the experiments. Comparison of the isotherms at pH = 2.8 demonstrates that with DUOLITE A7 still relatively high copper loadings are found whereas AMBERLITE IRA 35 is practically in the regenerated state. This has to be contributed to the pK values of the two resins reported to be 4.3 for DUOLITE A7 and 7.7 for AMBERLITE IRA 35[24].

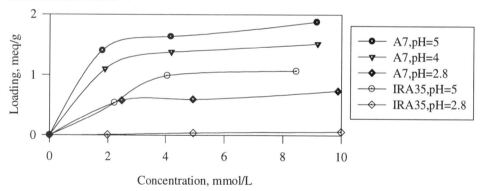

Figure 1 *Isotherms of the sorption of Cu^{2+}, Ni^{2+}, and Zn^{2+} ions onto DUOLITE A7; pH = 5* [24]

5 SORPTION KINETICS

Figure 2 shows the development of the concentration of Cu^{2+} for three similar experiments with three different resins. Throughout the experiments, pH was adjusted to

5±0.1. The data demonstrate that the rate of sorption is small and that the equilibrium has not been achieved even after 4 hours. This is in agreement with results from the sorption of acids by weakly basic exchange resins[25]. Due to the pK values of the resins, only a poor degree of sorption is observed for AMBERLITE IRA 35. For the other resins, the decrease of Cu^{2+} concentration is approximately equal. Despite the poorer uptake of $CuSO_4$ by AMBERLITE IRA 35, however, the sorption seems to develop faster than with the other resins.

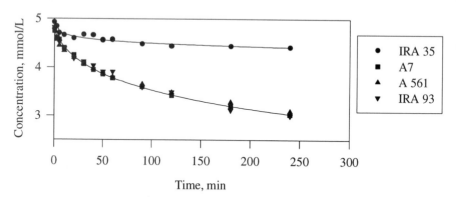

Figure 2 *Development of the sorption of Cu^{2+} species onto DUOLITE A7, DUOLITE A 561, AMBERLITE IRA 35, and AMBERLITE IRA 93. Liquid volume: 2 L, resin quantity: 6 g, diameter of resin beads: 0.9 - 1 mm, pH ≈ 5*

6 COLUMN EXPERIMENTS

In the first series with pure solutions, the filter columns with an inner diameter of 2 cm and bed heights of 10 or 15 cm were used. Results from experiments with DUOLITE A7 and AMBERLITE IRA 35 with $NiSO_4$ feed solutions of 5 mmol/L are given in Figure 3.

For DUOLITE A7, the breakthrough curves for both Ni^{2+} and SO_4^{2-} ions are identical indicating the simultaneous elimination of equivalent amounts. For AMBERLITE IRA 35, a slight discrepancy between the breakthrough curves of nickel and sulphate species occurs at the beginning of the run. For the major part of the experiments, however, both developments are identical, too. For DUOLITE A7, the breakthrough occurs earlier than for AMBERLITE IRA 35. This has to be credited to the slower kinetics of sorption (see Figure 2).

For DUOLITE A7 and at feed concentrations of 5 mmol/L, the following series of selectivity could be deduced:

$$Cu^{2+} > Pb^{2+} > Cd^{2+} > Zn^{2+} > Ni^{2+}$$

Results from experiments with mixed $CuSO_4/NiSO_4$ solutions, both resins are plotted in Figure 4. The development depicts the preferred sorption of Cu^{2+} species and indicates the efficient elimination of copper species. Furthermore, it seems that in such mixtures, the breakthrough of Cu^{2+} occurs later than with pure solutions.

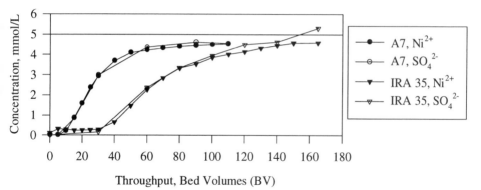

Figure 3 *Breakthrough curves of Ni²⁺ and SO₄²⁻. Feed concentration: 5 mmolL, throughput: 5 BV/h* [26]

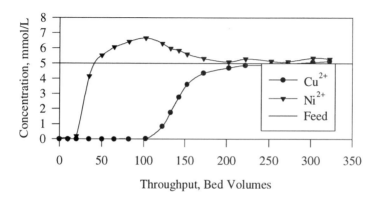

Figure 4 *Copper and nickel effluent concentrations during an experiment with a mixture of CuSO₄ and NiSO₄. Feed concentrations: 5 mmolL, throughput: 2.5 BV/h, resin: DUOLITE A7* [26]

Due to the expected extreme preference of heavy metals over alkaline earth ions, weakly basic exchange resins should offer a favourable possibility to eliminate traces of heavy metals e.g. from potable water. Therefore, in a third series of experiments, the feed consisted either of tap water spiked with traces of copper or nickel or of heavy metal-bearing natural water. For the experiments, the smaller column with only 1 cm inner diameter and 3 cm bed height were used to reduce the time and feed volumes. Figure 5 shows results from an experiment with tap water spiked with $CuSO_4$. In the preparation of the feed solution, the precipitation of copper-bearing salts in the storage vessel was observed. Therefore, the concentration at the column entry was only about 0.016 mmol/L ($\approx$1 mg/L) instead of 0.1 mmol/L. Feed pH was 6.52. In the effluent, a pH value of about 8.25 was detected indicating some small exchange of anions for OH⁻ ions. The relatively high initial effluent concentration was caused by channelling effects. The experiment had to be interrupted after a duration of 160 hours (about 7 days). No regeneration was carried out.

In a similar experiment, zinc was eliminated from a mine drainage water which is to be used for the production of drinking water by the local water supply authority. The feed concentration amounted about 2.5 mg/L. Up to a total throughput of 1850 bed volumes, the concentration of Zn^{2+} was practically at the limit of detection by ICP^{24}. Again, no regeneration was carried out.

Regeneration of the heavy metal-loaded resins was studied after several of the loading experiments with pure solutions. Figure 6 shows the development of Cu^{2+} concentration during an experiment in which the resin had been preloaded using a pure $CuCl_2$ solution. Regeneration was carried out by means of 0.01 molar H_2SO_4 at a throughput of 10 BV/h. The concentration history of copper in the effluent demonstrates that rather high concentrations can be reached which should allow the recovery of copper and possibly also of other metals.

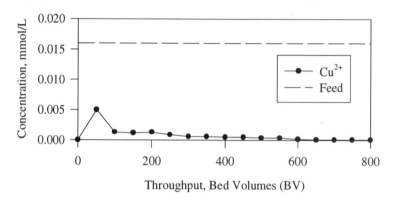

Figure 5 *Breakthrough curve from the elimination of Cu^{2+} from tap water. Throughput: 5 BV/h[24]*

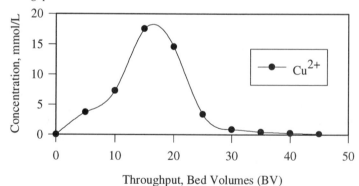

Figure 6 *Development of regeneration of DUOLITE A7[26]*

7 CONCLUSIONS

The results of the investigations have demonstrated that standard weakly basic anion exchange resins in their free base forms can be applied for the elimination of heavy metal salts from aqueous solutions. The sorption is favoured by small pK values. Since any competitive uptake of alkaline earth ions is excluded, trace amounts of heavy metals can easily be removed. The sorption step develops relatively slowly. As a consequence, the rate of filtration in column operations has to be small. Due to individual affinities for different heavy metal ions, mixtures can be separated. Regeneration is achieved by consecutively treating the exhausted resins with concentrated acid and with NaOH.

References

1. S. K. Sahni, J. Reedijk, *Coord. Chem. Rev.*, 1977, **59**, 1.
2. A. Warshawsky, in: "Ion Exchange and Sorption Processes in Hydrometallurgy", M. Streat, D. Naden, eds., Society of Chemical Industry, J. Wiley & Sons, Chichester, 1987, Vol. 19, p. 166.
3. M. J. Hudson, in: "Ion Exchange: Science and Technology", NATO ASI Series, A. E. Rodrigues, ed., Martinus Nijhoff, Boston, 1986, p. 35.
4. B. R. Green, R. D. Hancock, *J. South African Inst. Mining and Metallurgy*, 1982, **82**, 303.
5. R. R. Grinstead, in: "Ion Exchange Technology", D. Naden, M. Streat, eds., Ellis Horwood, Chichester, 1984, p. 509.
6. R. R. Grinstead, *Hydrometallurgy*, 1984, **12**, 387.
7. Y. Zhu, E. Millan, A. S. Sengupta, *Reactive Polymers*, 1990, **13**, 241.
8. K. M. Saldadze, V. D. Kopylova, *Russ. J. Anal. Chem.*, 1972, **27**, 956.
9. K. M. Saldadze, et al., *Russ. J. Anal. Chem.*, 1970, **25**, 1462.
10. V. D. Kopylova, K. M. Saldadze, G. D. Asambadze, *Russ. J. Anal. Chem.*, 1970, **25**, 1069.
11. V. D. Kopylova, K. M. Saldadze, G. D. Asambadze, *Russ. J. Anal. Chem.*, 1970, **25**, 1287.
12. V. D. Kopylova, K. M. Saldadze, G. D. Asambadze, *Russ. J Inorg. Chem.,* 1970, **15**, 332.
13. Z. S. Liu, G. L.Rempel, *Reactive Polymers*, 1991, **14**, 229.
14. Z. Mateijka, Z. Zitkova, R. Weber, K. Novotna, in: "New Developments in Ion Exchange", M. Abe, T. Kataoka, T. Suzuki, eds, Kodansha, Tokyo, 1991, p. 591.
15. D. Krauß, DE 4128837 A1, April 4, 1993.
16. G. Nickless, G. R. Marshall, *Chromatography Reviews*, 1964, **6**, 154.
17. A. Warshawsky, *Angew. Makromol. Chem.*, 1982, **109/110**, 171.
18. M. J. Hudson, in: "Fundmentals and Applications of Ion Exchange", L. Liberti, J. Miller, eds., NATO ASI Series, Ser. E, Appl. Sci. No. 198, Martinus Nijhoff Publishers, Doordrecht, Boston, Lancaster, 1985, p. 7.
19. K. Inoue, K. Yoshizuka, Y. Baba, *Solvent Extraction and Ion Exchange*, 1990, **8**, 309.
20. Z. Hubicki, *Hydrometallurgy*, 1986, **16**, 361.
21. W. H. Höll, R. Kirch, *Desalination*, 1978, **26**, 153.
22. A. Ebensperger, Diploma Thesis, University of Karlsruhe, 1994.
23. R. M. Smith, A. E. Martell, "Critical Stability Constants", Vol. 2: Amines, Plenum Press, New York, London, 1977.
24. G. Hölscher, Diploma Thesis, University of Karlsruhe, 1995.
25. F. G. Helfferich, *J. Phys. Chem.*, 1965, **69**, 1178.
26. C. Christoforou, Diploma Thesis, University of Karlsruhe, 1995.

SYNTHESIS OF A NEW ION EXCHANGE RESIN FOR THE RECOVERY OF NOBLE METALS

E. J. De Oliveira and G. O. Chierice

Grupo de Química Analítica e Tecnologia de Polímeros
Instituto de Química de São Carlos
Universidade de São Paulo, Caixa Postal 780
CEP 13560-970 - São Carlos - SP- Brasil

1 INTRODUCTION

The precious metals - platinum metals, gold and silver - are recovered from a wide variety of sources that present different analytical as well as metallurgical problems. The recovery of gold from various raw materials has traditionally been carried out by dissolution with aqua regia or cyanide, and in most cases these solutions contain base metals beside gold. Alternative processes for gold recovery have been reviewed, and of these processes, resin ion-exchange has received the most attention. Ion exchangers should ideally have high capacities and selectivities for particular metals and be capable of being regenerated; unfortunately resins that are specific for metals, or at least groups of metals tend to be expensive. The results of these new processes, using ion-exchange resins have been compared to conventional technology[1]. The advantages of polyurethane foam compared with activated carbon are the same for ion exchange resins: elution is possible at room temperatures; polyurethane foams are superior to activated carbon with respect to the rate and equilibrium loading of gold complex; they do not require periodic thermal reactivation for the removal of adsorbed organic materials, a step that is necessary with activated carbon; the problem of poisoning by organic species, which can severely inhibit the loading of gold onto activated carbon is not a problem.

2 HISTORICAL BACKGROUND

The use of polyurethane foam as an extractant for organic and inorganic compounds has been developed since the publication of Bowen[2], who was the first to discover the sorption properties of polyurethane foams toward species in aqueous solution. Since the appearance of this work considerable progress has been made in recent years in the use of polyurethane foams for separation and preconcentration purposes. Several reviews[3-9] dealing with the extraction of species by the foam have appeared. Considerable work has been done on the extraction of species by foam, e.g. Braun in Hungary and Chow and co-workers at the University of Manitoba), but only a few papers have discussed the possibility of using polyurethane foam as a sorbent for precious metals from aqueous mineral wastes. In acidic chloride media it is possible to preconcentrate gold[10-12] using polyurethane foam. The possibility of using untreated polyurethane foams for the collection and recovery of gold from an alkaline cyanide solution has also been examined[13], but this work recommended that gold could be recovered in two ways: by burning the foam and dissolving the residue in aqua regia, or by dissolving the foam itself in aqua regia. Although both procedures gave satisfactory results, the foam destruction is not desirable for industrial operations. In the present work, the possibility of using this cheap sorbent for the recovery of gold from synthetic leaching solutions and from clarified leaching solutions of ores is examined. The present method does not require the destruction of the foam, the gold can be recovered by elution of the foam with a solvent, and the foam can be recycled. This procedure can be valuable in the recovery of precious metals from industrial liquors.

3 EXPERIMENTAL

Gold cyanide and chloride solutions of appropriate gold concentration, pH and ionic strength were prepared using reagent-grade chemicals. For selectivity tests, some metal cyanides were prepared from their appropriate salts and NaCN.

3.1 Synthetic Leaching Solutions

The hydrometallurgy of gold involves the leaching of gold ore in oxygenated cyanide solutions. This treatment results in the formation and stabilization of Au^+ as the dicyanoaurate(I) anion $Au(CN)_2^-$. As the cyanide leach system is not particularly selective, a wide range of metal cyanocomplexes also appears in the hydrometallurgical circuit. In order to simulate the leaching solutions from a hydrometallurgical circuit, the following synthetic solutions prepared from stock standard solutions, as suggested by Haddad[14]: Al^{3+} 0.4; As^{3+} 1.2; Ca^{2+} 480.0; Co^{3+} 1.1; Cu^+ 12.1; Fe^{2+} 6.2; K^+ 140.0; Mg^{2+} 1420.0; Mn^{2+} 2.4; Na^+ 1.0; S^{2-} 1050.0; Zn^{2+} 0.6; Au^+ 5.0 mg/L. All water used was distilled and deionized. All experiments were performed at 25.0±0.1°C and atmospheric pressure, unless stated otherwise.

3.2 Foam Synthesis

The castor oil polyurethane foam is synthesised by reacting castor oil (with three reactive hydroxyl groups per molecule) with an isocyanate; in this study 2,4-toluene diisocyanate (TDI) was used.

3.3 Preparation of Foam Cubes

Foam cubes of approximately 0.2 g (10 mm³) each were washed with 1.5 M hydrochloric acid for 12 hours to remove any possible inorganic contaminant (e.g. metallic compounds used as catalysts in foam synthesis), washed with water until acid free and with acetone to remove any organic contaminant. Washed cubes were then air-dried at room temperature and stored in a dark coloured bottle.

3.4 Materials and Equipment

A Perkin-Elmer 305 flame atomic spectrophotometer was used to quantify the gold sorbed by foam (fuel-lean air/acetylene flame). Interference effects were avoided by suitable matrix-matching. For temperature control a thermostatted water bath was used. The pH meter was calibrated each day with appropriate buffer solutions.

3.5 Batch Studies

To investigate the extraction of gold by polyurethane foam, an aliquot (20 mL) of aqueous solution containing gold in cyanide media was placed in a glass cell containing the desired amount of foam. The foam was squeezed by means of a Teflon plunger in order to bring fresh solution in contact with it. The plunger was operated by an automatic squeezer consisting of an eccentric cam turned by a motor, the frequency of squeezing used was 94 rotations per minute. The degree of gold extraction, E%, and the distribution coefficient, D, were calculated as shown by equations 1 and 2 :

$$E\% = \frac{(C_{Au_o} - C_{Au}^*)}{C_{Au_o}} \cdot 100 \qquad (1)$$

$$D = \frac{E\% . V}{m_f . (100 - E\%)} \qquad (2)$$

where C_{Au_o} is the initial gold concentration (in mg / L) before extraction, C_{Au}^* is the gold concentration after the extraction period, V is the volume of leaching solution in mL, and m_f is the mass of the foam sample in grams. Gold loading on the foam was determined from solution depletion as measured by atomic absorption spectrophotometry, and reported per unit weight of dry foam. Sorption behavior was determined as a function of temperature, ionic strength, pH, time of contact and initial gold concentration. In the case of elution experiments, previously loaded foams were rinsed with water several times prior to stripping.

4 RESULTS AND DISCUSSION

4.1 The Effect of Time Extraction

To study the influence of the agitation time on gold extraction by polyurethane foam, several extractions were carried out using equal volumes of gold cyanide solution with the desired gold concentration. The agitation time varied from 2 min to 2 hours. The results show that, under these experimental conditions, the extraction is not dependent upon the agitation time and the degree of gold extraction, E%, is constant after 120 min of contact (Figure 1).

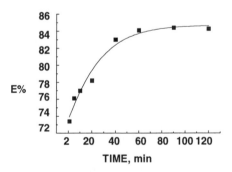

Figure 1 *The effect of time extraction (five replicates)*

4.2 The Effect of Temperature

The loading of gold by the foam was examined at various temperatures and the results are presented in Figure 2. The leaching solution and the foam were placed in a glass cell attached to a thermostatted water bath, samples were taken and analyzed for gold. As expected, gold sorption decreases as temperature increases. The temperature dependence of the sorption reaction can be attributed to the nature of ion-pair formation. The tendency for ion-pair formation decreases with an increase of temperature and is predicted by the Bjerrum equation[15].

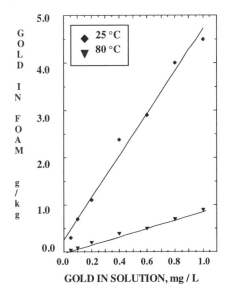

Figure 2 *Effect of temperature on gold loading for polyurethane foam*

4.3 The Effect of pH

The pH value of a solution containing 10 mg/L of gold as gold chloride was varied from 1 to 12. Adjustments to the pH value of the solution were made with hydrochloric acid or sodium hydroxide. The results are shown in Figure 3 and indicate that the pH value of solution has a marked effect on degree of gold extraction, E%, especially at low values. This result agrees with McDougall et al[16] and is the basis of the foam elution procedure.

4.4 The Effect of Ionic Strength

The rate of extraction and equilibrium loading were determined from solutions of variable ionic strength containing 10 mg/L of gold. Adjustments to the ionic strength value of the solution were made with sodium chloride. The results, Table 1, indicate that the degree of gold extraction, E%, and the distribution coefficient, D, increase with increasing ionic strength.

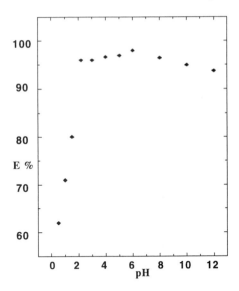

Figure 3 *Effect of pH on gold loading for polyurethane foam*

Table 1 *Effect of Ionic Strength on Degree of Gold Extraction (E%) and Distribution Coefficient*

Ionic strength, M	E%	D (L/kg)
~ 0	90.0	900
0.005	91.5	1076
0.010	93.0	1328
0.050	97.4	1786
0.500	95.1	1940
0.600	96.6	2841
0.700	97.1	3348

Conditions: mass of foam ~0.2 g, gold in solution 10.0 mg/L

4.5 The Effect of Competing Ions

Foreign ions (e.g. Al^{3+}, Cu^{2+}, Zn^{2+}, Fe^{3+}, Si^{4+}) are always associated with gold, a synthetic solution is described by Haddad[14], in cyanide leach liquors. Therefore, the study of the effects of impurities during the sorption of gold by polyurethane foam is of prime importance. The effect of these ions were studied by using the synthetic leaching solution described above, and their behavior was studied by comparing their concentrations in the cyanide solution before and after sorption. The sorption and elution of gold by the castor oil foam in presence of competing ions is shown in the Table 2. There was no evidence of any adverse effect on the gold-extraction efficiency when the foam was recycled to the sorption process (each foam cube was recycled 20 times with approximately the same degree of gold recovery).

Table 2 *Sorption and Recovery of Gold by the Foam in Presence of Competing Ions*

Sample	Au+ feed solution mg	Au eluted from the foam, mg	Au recovery %
1	0.100	0.098	98.4
2	0.200	0.196	97.8
3	0.300	0.303	101.1
4	0.402	0.386	96.1
5	0.500	0.504	100.8
6	0.599	0.587	98.0

Conditions: V = 20 mL; time of extraction = 30 min; mass of foam = 0.2 g; pH = 10

4.6 Comparison between Castor Oil Foam and Activated Carbon

A comparison between the gold extraction by two samples of castor oil foam and that of activated carbon is shown in the Table 3. The foam has much lower value when compared with activated carbon, but the recovery of gold is easier in the foam.

Table 3 *Loading of Gold Complex onto Activated Coconut-shell Carbon, a Synthetic Polymeric Sorbent Duolite S-761 and Castor Oil Foam*

Sorbent	Sample	Loading mg/g
Castor oil foam	1	2
Duolite S-761	2	3
Coconut shell carbon	--	45

In each experiment, 0.25 g of sorbent was contacted for 2 hours with 25 mL of solution containing the gold complex at 250 mg/L (initial concentration). The mode of sorption of inorganic compounds onto activated carbon is by a reduction mechanism[17], this conclusion is supported by the fact that Duolite S-671, which is able to duplicate the adsorptive behaviour of carbon with respect to certain organic molecules, e.g. phenol, is unable to load the gold complex to a significant extent[18]. The sorption in castor oil foam is by a cation chelation mechanism[19] this can explain the massive difference in gold loading.

4.7 Gold Recovery Tested on Cyanide Pulp Prepared from Ore Sample

In view of the success obtained with synthetic solutions simulating gold pregnant liquors, an attempt was made to recover gold from a solution prepared by the cyanidation of gold ore sample carried out using 0.1% sodium cyanide, 0.05% calcium oxide and in presence of atmospheric oxygen. The pulp was placed in a glass cell containing the desired amount of foam, the foam was squeezed using the same procedure as described above (batch studies). The pregnant pulp had the composition shown in Table 4.

Table 4   *Composition of Pregnant Pulp, Barren Pulp and Elements Recovered from the Foam*

Element	Pregnant Pulp mg / L	Barren Pulp mg / L	Recovery from Foam (eluate) mg / L
Fe	10	8.22	1.76
Si	10	nd	nd
Al	2	1.50	0.47
K	i	-	-
Ca	0.9	0.80	0.10
Na	1.8	1.78	1.77
Ni	0.01	-	-
Cu	0.2	0.02	0.17
Mg	0.1	-	-
Zn	0.03	-	-
Pb	0.01	-	-
Au	0.8	0.01	0.78
Ag	1.0	0.01	0.97

Note :   i - interference; nd - not determined

Conditions: V = 20 mL (pulp); time of extraction = 30 min; mass of foam = 0.2 g; pH = 10; V = 20 mL (eluate); eluate: 90% ethanol - 10 %HCl

4.8 Column Studies

In order to consider the application of the foam derivated from castor oil, an experiment was carried out to determine the stability of the foam to repeated treatment with the eluent (ethanol-HCl). A series of cycles (four) were carried out, each consisting of an adsorption and elution stage using a column together with a peristaltic pump. The column size used was, 1160 mm X 5 mm (80 g dry foam, 2285 cm^3 per bed volume). Each cycle consisted of four stages: 1) the foam was loaded by passing through 2 liters containing 94.5 mg/L of gold as aurocyanide in 0.15 % sodium cyanide at a rate of 100 mL per hour; 2) the column was washed with 100 mL which was combined with the effluent from step 1; 3) the column was eluted with 250 mL of 90% ethanol - 10% HCl at a rate of 50 mL per hour; 4) the column was washed with 100 mL of water which was combined with the eluate from step 3. The extraction profiles from studies using columns and the cycles of sorption and desorption of gold is shown in Table 5. Castor oil foam shows a good affinity for Au^+. After four cycles the recovery capacity of column had fallen by 7%. The loss in capacity is probably due to the accumulation of traces of other metals. It is conceivable that loaded metals may be recovered from the foam through selective elution, but in this study the elution profile was restricted to gold. The eluent used was the mixture ethanol-HCl. Organic solvents containing mineral acids are useful as eluting agents for aurocyanide based upon the assumption that the presence of mineral acid would bring about a partial ionization of the gold-foam complex.

Table 5 *Cycles of Sorption and Desorption of Gold in Column*

	Amount of Gold		Amount Eluted	% Recovery Column Capacity
	mg		mg	
cycles	in effluent	in foam	in eluate	
1	1.7	187.3	179.8	96.0
2	6.6	192.3	181.9	94.6
3	10.4	188.5	175.7	93.2
4	16.8	179.7	159.9	89.0

5 CONCLUSIONS

Since the loaded gold can be eluted from the foam by using an ethanol-hydrochloric acid eluent the gold can be recovered without it being necessary to destroy the foam which can be regenerated and recycled. This study has shown that Au^+ as $Au(CN)_2^-$ can be effectively concentrated and separated with castor oil foam sorbent from a sample matrix containing metal cyano complexes and other species. A possible use for castor oil foam may therefore be the preconcentration of gold to analysis or to separate gold from base metals. The recovery from the foam reached 97% (on average) so the high efficiency of gold sorption has been demonstrated. The castor oil foam has a basic nitrogen and a double bond which are versatile enough to undergo further modifications by introduction of functional groups, thereby opening the route to a new family of resins useful in hydrometallurgy and chemical analysis. Another possibility is to investigate the recovery other metals such as copper, nickel, cobalt and gallium, directly from pregnant pulp solutions. Pilot plant operations will be necessary in enlarging the scale of the process, before any conclusion of the economics of the process could be made.

References

1. M. B. Mooiman, J. D. Miller, J. B. Hiskey, A. R. Hendriksz, in "Gold and Silver Heap and Dump Leaching Practice", J. B. Hiskey, ed., AIME, 1984, p. 93.
2. H. J. M. Bowen, *J. Chem. Soc. (A)*, 1970, 1082.
3. T. Braun, J. D. Navratil, A. B. Farag, CRC Press, Boca Raton, 1985.
4. H. D. Gesser, E. Bock, W. G. Badwin, A. Chow, D. W. McBride, W. Lipinsky, *Sep. Sci.*, 1976, **11**, 315.
5. S. Al-Bazi. A. Chow, *Anal. Chem.*, 1981, **53**, 1073.
6. J. J. Oren, K. M. Geugh, H. D. Gesser, *Can. J. Chem.*, 1979, **57**, 2032.
7. V. S. K. Lo, A. Chow, *Talanta*, 1981, **28**, 157.
8. G. J. Moody, J. D. R. Thomas, *Analyst*, 1979, **104**, 1.
9. T. Braun, *Z. Anal. Chem.*, 1983, **314**, 652.
10. P. Schiller, G. B. Cook, *Anal. Chim. Acta.*, 1971, **54**, 364.
11. S. Sukiman, *Radiochem. Radioanal. Lett.*, 1974, **18**, 129.
12. T. Braun, A. B. Farag, *Anal. Chim. Acta.*, 1973, **66**, 419.
13. T. Braun, A. B. Farag, *Anal. Chim. Acta.*, 1983, **153**, 319.

14. P. R. Haddad, N. E. Rochester, *J. Chromatography*, 1988, **439**, 23.

15. J. O. M. Bockris, A. K. N. Reddy, "Modern Electrochemistry I", Plenum, New York, 1977.

16. G. J. McDougall, R. D. Hancock, M. J. Nicol, O. L. Wellington, R. G. Copperwaite, *J. S. Afr. Inst. Min. Metall.*, 1980, **80(11)**, 344.

17. M. Streat and D. Naden, "Ion Exchange and Sorption Processes in Hydrometallurgy", Wiley and Sons, New York, 1987.

18. G. J. McDougall, "Adsorption on Activated Carbon", ChemSA, 1982, 24.

19. R. F. Hamon, A. S. Khan, A. Chow, *Talanta*, 1982, **29**, 13.

A NEW METHOD OF PRODUCING METAL HYDROXIDES BY ELECTRODIALYSIS

R. Lumbroso

15 rue Ribera
F. 75016 Paris
France

1 INTRODUCTION

The future of reloading batteries is connected to the nickel/cadmium couple of electrodes.

From now on, we can see electric motor cars, for which we estimated that their development was in connection with Zinc/Air or Aluminium/Air couple of electrodes. But up to now, even if research is promising we never meet cars equipped with that kind of batteries, but with Ni/Cd or acid-lead accumulators.

The problem had to be attacked by an amelioration the qualities of Ni/Cd electrode supports and of chemicals constituting electrodes.

Speaking about electrodes supports, we can estimate that the problem is now solved by producing metallic foams and felts. These metallic shapes have a better electric d.c. conduction and especially a notable lightening of the manufactured battery.

Everybody has understood that the electric generators lightening is primordial for electric vehicles. The introduction of foams and felts in batteries equipment gives a notable lightening and consequently gives electric properties, bounded to the weight, largely higher to those classic batteries with plain and chemically handled electrodes.

Another parameter of lightening can be the production of nickel and/or cadmium hydroxides having tap densities higher than those we actually find on the market.

These hydroxides will be the active chemicals permitting to form, mechanically, electrodes, by foams and felts empasting and subsequent calendering.

This process will facilitate the production of that type of electrodes.

2 ORIENTATION OF THE STUDY

2.1 Classical Method

Metallic hydroxides are generally made by action of alkaline solutions on metal soluble salts.

Those hydroxides have a thin divided aspect, offently gelatinous. Its formula is generally done as $M-(OH)_n$.

The obtained gelatinous shape doesn't facilitate washing of the hydroxide prepared in an alkaline medium, and its separation, by filtration, from washing waters. It is very hard to obtain it in powdered form and with a suitable degree of purity.

2.2 Target of the Study

Researching an increase of tap density of the nickel hydroxide, some authors[1] have tried to use chemical processes, acting by diffusion of reactive solutions, working at low temperatures (always under 10°C).

Known manufacturers as TANAKA in Japan and STARK in Holland supply metallic hydroxides in spheric form, having a tap density approaching 2.

Our basic study[2] is made on the nickel hydroxide. The results of the technique are extended to mixtures of Ni/Cd and Ni/Co/Cd hydroxides as to acid effluents proceeding from installations treating uranium salts, as we describe it, below, in the section 5.

The technique is also suitable, as we can see it in the section 6, to the precipitation of metal amphoteric hydroxides like aluminium in very high basic medium as obtained in electrochemical metals forming factories.

2.3 Considered Method

Having had a chemical education, studying ion exchange materials, we try, in a first step, to realise, in the middle compartment of an electrodialysis cell, the controlled meeting of Ni^{++} and OH^- ions (see Figure 1).

You can observe that the cell configuration is not the classic one, this allows:

- the migration of OH^- ions across MA,

- the meeting of Ni^{++} and OH^- ions in the middle compartment
- the migration of H^+ ions across MC to maintain a constant pH

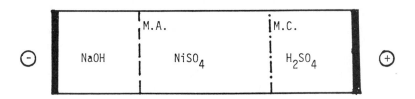

Figure 1 *Three compartment cell*

The result obtained was not the expected one, that is to say, we didn't have the controlled precipitation of nickel hydroxide, $Ni(OH)_2$, in the total volume of the central compartment, but a green almond precipitate close against the anionic membrane MA. The deposit had a crystalline form, easily detachable from the anionic membrane, easily washable until a correct degree of purity. We have attributed to this hydroxide, after analysis, the formula:

OH - Ni - O - Ni - OH

3 SELECTED METHOD

3.1 Essential Target Definition

In our industrial consideration of the study, we had to sacrifice the determination of some parameters, which may have been helpful if this work had been an academic one.

Our essential preoccupation was to produce an hydroxide with tap density around 2 and presenting, in fact, a better massic electric capacity. That necessity lead us to prepare a nickel hydroxide, having constant physical, chemical and electric properties. We compared them to products actually on the market.

3.2 Electrodialysis Cell

After ascertaining that it was useless to feed, in the central compartment, H^+ ions to neutralize hydroxyl ions expected in our first hypothesis, we choose to work on a two compartment cell as seen below (Figure 2).

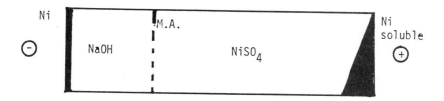

Figure 2 *Two compartment cell*

- catholyte: sodium hydroxide solution with nickel cathode
- anolyte: nickel sulphate solution with soluble nickel anode
- anionic membrane as active separator between the two compartments.

3.3 Parameters Determined during the Study

- Voltage	V (volts)
- Current density	I (amperes)
- pH of the anolyte	
- Anolyte temperature	T°C
- Time of d.c. distribution	t (minutes)
- Theoretical weight hydroxide to obtain	Q' (grammes)
- Weight of hydroxide obtained	Q (grammes)
- Faradic yield	Q/Q' %
- Tap density	d (g/liter)
- Electric energy	E (Ah/kg Ni)
- Swelling	G %

Tap density is measured on a dry and screened product placed in a preweighed graduated glass tube (Po). The final volume (v) of the product is obtained after taps until stabilisation. The graduated glass tube is reweighed (p) and a rapid calculation gives the tap density (d).

Electric energy is measured on an electrode prepared in a standardized mode. This electrode is submitted to a standard number cycles of charges and discharges. The result shows electrode ability to receive and give back electric energy. The theoretical electric energy for nickel is 290 Ah/kg of Nickel.

Swelling is an important property, its evolution, during electric energy cycles measurements, gives the stability of the electrode/elctrolyte arrangement and will reveal risks of deformation and, at the extend, possible risks of explosion.

3.4 Electrodialysis Pilot Cell

We proceed to all pilot trials using electrodialysis cell similar to the one represented Figure 3.

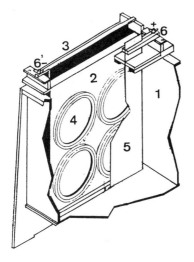

In the anolyte compartment (1) we place the catholyte casket (2) equipped with a nickel cathode. This casket is pierced by portholes in which the anionic membranes are inserted. The area of each piece of membrane is 2.5 dm². Face to the portholes (about five centimeters) is placed the nickel soluble anode (5). The direct current is carried by the contacts (6) and (6').

A circulation is foreseen in both anionic and cationic compartments.

The anionic flow runs across a refrigerated intermediate tank where the temperature is maintained around 35°C.

Figure 3 *Electrodialysis pilot cell*

4 RESULTS

4.1 Parameters Maintained Constant during all the Study

- *Anolyte* : It is, during all the study, a nickel sulphate solution, 4N - Nickel concentration. The problem of concentration constancy will be discussed later, in connection with the pH values observed.

- *Anolyte Flow Rate* : It is maintained at 25 L/h by feeding with a controlled leaking.

- *Temperature* : It is maintained constant by crysocopic equipment, regulated, to stay between 30°C to 35°C. The equipment is placed in the intermediate tank, on the anolyte flow.

- *Membrane* : We have tried many samples of anionic exchanging membranes. We selected IONAC MA.3475, having moderate electric qualities, but high robustness, supporting, without damage, relatively high d.c. specific densities (25 to 30 A/dm²).

- *Catholyte and Cathode* : In every trial, cathode is nickel made and catholyte is a 2N - sodium hydroxide solution. There is no consumption of sodium hydroxide.

4.2   Parameter to Change during the Study

- *Specific Current Intensity* : We choose to use two specific current densities, for a voltage comprise between 13 and 15 volts, of 10 A/dm^2 and 20 A/dm^2, for a total membrane area of 10 dm^2.

- *Anode* : We used, in a first step, lead anode, but when we have had to regulate the pH, we used nickel soluble anode (plain plates or beads placed in titanium basket).

- *pH* : It is an important parameter for the managing of the operation. At the beginning, the pH in the anolyte was around 1.8. With the use of a lead anode, the pH decreases continuously to reach, after two or three hours of work, a stable value around 0.5. The Ni^{++} concentration decreases the same way. With a soluble nickel anode, the starting pH is also the same, namely 1.8, the pH decreases more slowly till 1.5 - 1.6 after the same time of working. The pH will be stable at 1.8 if a permeable bag containing nickel carbonate is placed in the intermediate tank on the anolyte flow. This carbonate delivers, in relation with the pH, Ni^{++} ions, and maintains it at its initial value.

4.3   Conditions to Reach the Desired Results

- *Operative Parameters of Fabrication*   We can summarize those parameters as below:

$$V \quad = 13 \text{ to } 15 \text{ volts}$$
$$I \quad = 10 \text{ to } 25 \text{ A/dm}^2$$
$$t \quad = 120 - 180 - 420 \text{ minutes}$$
$$T \quad = 30 \text{ to } 35°C \text{ in the anolyte compartment}$$
$$pH \quad = 0.5 \text{ to } 1.8$$

According those values we obtain faradic yields of 50% to 80%.

- *Electric Qualities Parameters* : They are directly connected to the nickel electrodes preparation, as described in paragraph 3.3, by empasting nickel hydroxide on a substrate, foams or felts, of metallic nickel.

$$d \quad = 1.75 \text{ to } 1.85$$
$$E \quad = 80\% \text{ to } 100\% \text{ (theoretical 290 Ah/kg nickel)}$$
$$G \quad = 25 \text{ to } 35\%$$

5 DEVELOPMENT OF THE STUDY

5.1   Direct Production of Mixed Hydroxide

In the perspective developed in our introduction, in other words, the preparation of nickel electrode to equip reloading Ni/Cd batteries, it was necessary to solve the problem of direct production of a nickel hydroxide "doped" by cadmium, in the proportion of 3% cadmium in regard of nickel metal.

The work made previously allows us, starting from a mixture solution of nickel and cadmium sulphate, in the proportion of 3% Cd, to prepare directly a mixed hydroxide, having the physical property of density around those expected and presenting electric capacities of 80% to 100% and swelling (G) never under 30%.

We can, using the same way, produce Ni/Zn, Ni/Co, Ni/Cd/Co hydroxides.

5.2 Treatment of Rejects from Plants Working on Uranium Compounds

This method allows also to consider rejects treatment containing uranium salts.
- *Destructive Method*

It consists in the simple precipitation, from a basic solution of uranyl carbonate-bicarbonate, by an excess of strong basic solution, of an uranyl hydroxide and dispose it in drums after filtration and desiccation.
- *Electrodialytic Method[3]*

It was made in an electrodialysis cell with two compartments separated by a cation exchange membrane. The solution to treat is placed in the cathodic compartment, the sodium hydroxide was in the anodic compartment.

The d.c. current flow permits Na^+ ions to cross the cationic membrane, in a first step, and to reach the pH value of 12 in the catholyte.

The precipitation occurs, later, in a second step, with, during both steps, a continuous consumption of sodium hydroxide.

This technique was not possible to use for the treatment of acidified solutions, going from extraction of potassium diuranate, having a pH value below 1.5.
- *New Electrodialytic Method*

In this case, where the electrodialysis cell is also with two compartments, the separative membrane is an anionic one; the solution to treat is placed in the anionic compartment, a OH^- ions generating solution is placed in the cationic compartment, in fact, in an inverse configuration than that described just above.

The d.c. current flow gives immediately an uranyl hydroxide precipitate close to the anionic membrane, as described in the paragraph 2.3.

We made studies on raw rejects containing 4 g/L uranium under a pH value of 10, or on rejects preacidified till pH is 1.3 by hydrochloric acid.

Under a voltage of 15 volts and a specific current density of 15 A/dm^2, we have, in both case, a thin and filterable precipitation of uranyl hydroxide, and in the depolluted solution, a residual concentration of uranium below 1 mg/L.

During the operation we note no sodium hydroxide consumption, but only a cost of electricity around 225 kWh/kg uranium metal.

6 EXTENSION OF THE METHOD

In an extension of the method, we can, eventually, speak of aluminium hydroxide formation in the anodic compartment according to the sequence (Figure 4) described below.

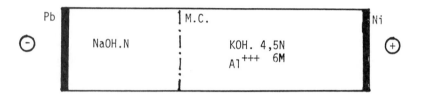

Figure 4 *Configuration of a two compartment cell for potash solution regeneration*

The basic aluminate solution, going from a solution of aluminium electrochemical forming, or from a circulating solution into an Al/Air battery, is charged by aluminium and looses its efficiency when the concentrations reach:

$$\text{KOH} \quad 4.5 \text{ N} \qquad \text{Al}^{+++} \quad 6 \text{ M}$$

Our target is the continuous regeneration of a circulating potash solution and to realize a reloading battery able to fit out a mobile or stable station generating electricity.

The space allowed is too short to describe correctly, the extension of this method applicated to other cells and batteries types.

Nomenclature

$\oplus$	Anode		
A/dm^2	Specific d.c. density	Ampere by square decim.	
C	Electric capacity	mAh/cm^2	
°C	Temperature	° Celsius	
$\ominus$	Cathode		
D	Anolyte flow rate	Liter/hour	L/h
d	Tap density	gramme/liter	g/L
d.c.	Direct current		
G	Swelling	%	
I	Current intensity	Ampere	
MA	Anionic membrane		
MC	Cationic membrane		
Q, Q'	Weight	grammes	
t	d.c. distribution time	minutes	
V	Voltage	volts	

Acknowledgements

I have to give my thanks to Dr Denis Doniat for his hospitality in his laboratories in Fontenay sous Bois (Val de Marne) in France, during all my work.

References

1. Kansai Shokubai Kagaku K. K., Japan Patent 4 - 68249, 30 October 1992.
2. R. Lumbroso, French Patent 2,688,235, 23 June 1995; U. S. Patent 5,384,017, 24 January 1995; Patents filed in Europe, Israel, Japan and Canada.
3. R. Kunin, U. S. Patent 2,832,728, 29 April 1958.

STUDIES ON THE DESORPTION OF ION EXCHANGE RESIN LOADED WITH HUMIC ACID-GOLD

J. Peng and Z. Fan

Institute of Chemical Metallurgy
Academia Sinica
Beijing 100080
China

1 INTRODUCTION

Gold leaching with humic acid is a cheap and poisonless process[1-3]. The modified humic acid can extract gold from several kinds of natural ore as well as sodium cyanide performs. Many methods such as activated carbon adsorption, solvent extraction, precipitation etc. have been applied to recover gold from humic acid leaching solutions. Ion exchange with strongly basic anion resin (type 717, a Chinese resin) is considered to be the most successful method for gold recovery. A load capacity of 2.2 mg gold/g resin has been reached even from a very low concentration feedstock (0.7 mg/L).

Removing the adsorbed gold from resin is very difficult. Several chemical solutions have been used to desorb humic acid-gold. Thiourea solution is found to be the most effective one, but the ion exchange resin loaded with humic acid-gold must be pretreated with dilute HCl solution before desorption. The influence on the rate of gold desorption of a number of parameters such as time, temperature, the concentration of thiourea solution, the pretreatment process and the solution to resin mass ratio have been examined. The desorption rate increases as temperature, thiourea concentration or the solution to resin mass ratio increases. The influence of temperature on the desorption process is very strong. At 40°C, the recovery is 52% while at 77°C the recovery reaches 95%. For gold loaded resin containing 2.2 mg gold/g resin, the recovery can reach 95% or more.

2 EXPERIMENTAL

2.1 Chemicals and Apparatus

High purity distilled water was used throughout, other chemicals were of analytical grade. Gold content of the resin was analyzed by an atomic absorption spectrometer GGX-5. The pH value of the desorption solution was measured with a pH meter (type 5985-75). Experiments were carried out in stoppered glass tubes immersed in a constant temperature bath controlled to ± 0.1°C. The tubes were held in the bath on a magnetic stirring apparatus.

2.2 Procedure

Accurately weighed amounts of dry resin loaded with gold were placed in dry glass tubes. Measured volumes of desorption chemical solutions were then added. The tubes were immersed in the bath which was already at the selected temperature, and the bath was placed on the magnetic stirring apparatus. The desorbed resin was separated from the desorption solution and gold content was analyzed.

3 RESULTS AND DISCUSSION

Several chemical solutions have been used to desorb gold from the resin. The extent of desorption varies from 0.8% to 95.4%. Table 1 shows the experimental results of 9 chemical solutions.

After pretreatment with 2 wt% HCl aqueous solution for 2 hours, 95.4% of the gold can be desorbed from the resin with 2 wt% thiourea solution, so the thiourea solution was adopted as the desorption solution.

Table 1 *Desorption Results of Several Chemical Solutions* *

	Chemical Solutions		*Time (h)*	*Temperature (°C)*	*Weight of Resin (g)*	*Volume of Solution (mL)*	*Extent of Desorption (%)*
1	$(NH_2)_2CS$ HCl	2.5 wt% 1.0 wt%	3	40	2.5	50	30.0
2	HCl	5.0 wt%	2	40	4.0	14	37.6
3	NaOH	7.0 wt%	2	25	2.5	20	8.8
4	NH_4HCO_3	22 wt%	8	25	2.5	20	0.8
5	NaCl NaOH NaClO	15 wt% 5 wt% 0.75 wt%	8	25	2.5	20	16.4
6	$(NH_2)_2CS$ H_2SO_4	8.0 wt% 3.0 wt%	2	25	1.5	15	58.1
7	$CuSO_4$ $Na_2S_2O_3$	1.5 wt% 22 wt%	1.5	25	0.5	10	8.4
8	$Na_2S_2O_3$	5 wt%	1.5	25	0.5	10	17.3
9	$(NH_2)_2CS$	2 wt%	2	80	0.5	10	95.4

* initial gold content of resin 527 mg/L

Table 2   *Effect of Pretreatment Solution Concentration on Desorption Process*

HCl Concentration in Pretreat. Solution (wt%)	Gold Content of Resin (mg/kg)		Extent of Desorption (%)
	before desorption	after desorption	
0	527.1	527.1	0.0
2	527.1	13.7	97.3
4	527.1	17.7	96.6

Table 3   *Effect of Pretreatment Solution Volume on Desorption Process*

Pretreat. Solution Volume (mL)	Gold Content of Resin (mg/kg)		Extent of Desorption (%)
	before desorption	after desorption	
15	527.1	24.2	95.4
10	527.1	21.9	95.8
5	527.1	28.5	94.6

3.1 Effect of Pretreatment

The resin loaded with humic acid-gold must be pretreated with dilute HCl solution before desorption, otherwise gold can not be desorbed from the resin with thiourea solution even if the temperature is as high as 80°C. The effect of pretreatment process on desorption is shown in Tables 2 and 3. Both the concentration and the volume of the pretreatment solution (dilute HCl solution) have little influence on the desorption process, but the pretreatment process is necessary for the desorption process.

3.2 Effect of Desorption Time

Figure 1 shows the relation between the effect of gold desorption with time at 80°C when the thiourea solution desorbs gold from the resin. The desorption increases rapidly in the first hour then there is little desorption after 2 hours. The extent of desorption reaches 88.3% for 1 hour and 95.4% for 2 hours.

3.3 Effect of Desorption Temperature

The effect of varying the contact temperature from 40°C to 80°C is shown in Figure 2. The higher the temperature the greater the desorption ratio. When the temperature is over 70°C, the desorption increases slightly with the increasing temperature. This is typical for the desorption process of an ion exchange resin.

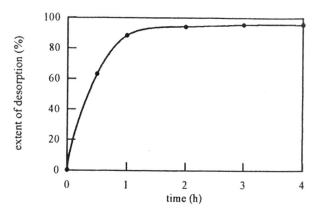

Figure 1 *Variation of desorption with time*

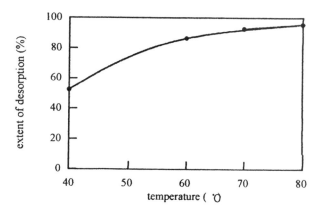

Figure 2 *Variation of desorption with temperature*

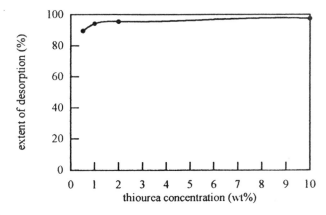

Figure 3 *Variation of desorption with thiourea concentration*

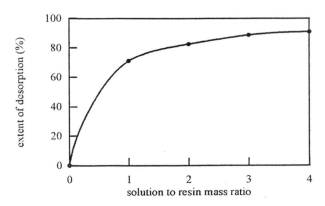

Figure 4 *Variation of desorption with solution to resin mass ratio*

3.4 Effect of Thiourea Concentration

Figure 3 shows the effect of thiourea concentration on the desorption. When the thiourea concentration is greater than 1%, the influence of thiourea concentration on desorption ratio is weak.

3.5 Effect of the Solution to Resin Mass Ratio

Figure 4 shows the effect of the solution to resin mass ratio on desorption. When the solution to resin mass ratio is greater than 3, the influence of the solution to resin mass ratio on the desorption is small.

4 CONCLUSIONS

Strongly basic anion resin (type 717, a Chinese resin) has been used successfully to absorb gold from humic acid-gold solution, but the desorption of gold from the resin is very difficult. Several kinds of desorption solution have been used to desorb gold from the resin, and the best way has been found. The thiourea solution can desorb gold from the resin loaded with humic acid-gold effectively only when the resin is pretreated with dilute HCl solution.

The influence of a number of parameters such as time, temperature, the thiourea concentration, the solution to resin mass ratio and the pretreatment process have been studied by experiments.

The pretreatment of the resin with dilute HCl solution before the desorption process is absolutely necessary. The gold can not be desorbed from the resin with thiourea solution without the pretreatment process even with temperature as high as 80°C.

The desorption ratio increases rapidly in the first hour and there is little desorption after 2 hours when it reaches 95.4%.

The influence of temperature on the desorption process is very strong, the higher the temperature the greater the desorption. When the temperature is over 70°C, the desorption ratio increases only slightly with increasing temperature.

The desorption increases with increasing thiourea concentration. When the thiourea concentration is greater than 1%, the influence of thiourea concentration on desorption ratio is weak.

The desorption increases as the solution to resin mass ratio increases. When the solution to resin mass ratio is greater than 3, the influence of the solution to resin mass ratio on the desorption ratio is small.

Acknowledgements

The authors wish to thank the National Science Foundation of China for the financial support of this work.

References

1. J. Peng and Z. Fan, "Selected Papers of Engineering Chemistry and Metallurgy (China) ", Science Press, Beijing, 1994, p. 70.
2. J. Peng and Z. Fan, *Gold J. (China)*, 1993, **14 (9)**, 24.
3. Z. Fan and J. Peng, *Gold J. (China)*, 1995, **16 (10)**, 32.

KINETICS OF RHENIUM SORPTION FROM SULPHURIC ACID MEDIUM USING WEAK BASE ANIONITE AH-21

E. I. Ponomareva, A. N. Zagorodnyaya, Z. S. Abisheva, D. N. Abishev, A. A. Zharmenov

Institute of Metallurgy and Ore Beneficiation
National Academy of Sciences of the Republic of Kazakhstan
29/33 Shevchenko St.
Almaty 480100
Kazakhstan

1 INTRODUCTION

It is common knowledge that ion exchange involves five stages in series[1-4]. The description of the kinetics of the process has always been a problem. Some simplification is usually tolerated. In particular, the concept of the limitation stage is used. Research performed on the kinetics of ion exchange has produced definite conclusions that stages of the process are limited by diffusion in a stationary layer of liquid which encircles ionite grain (external diffusion)[1-3]. For assessment of the limitation stage of ion exchange and the calculation of the diffusion coefficients, a number of equations and criteria has been proposed[1-6].

$$F = 1 - \frac{6}{\pi^2} \sum_{n=1}^{\infty} \frac{1}{n^2} - \frac{\exp\left(\overline{D}^2 \pi^2 n^2 t\right)}{r_0^2} \tag{1}$$

$$F = \frac{2A}{W} \sqrt{\frac{\overline{D}t}{\pi}} \tag{2}$$

$$F = 1 - \sum_{n=1}^{\infty} \frac{6\alpha\,(\alpha+1)e^{-q_n^2 F_0}}{9 + 9\alpha + q_n^2\,\alpha^2} \tag{3}$$

$$F = 1 - \sum_{n=1}^{\infty} B_n \exp\left(1 - \mu_n^2 F_0\right) \tag{4}$$

where F - is the degree of exchange; D - is the diffusion coefficient in ionite; t - is the sorption time; r - is the mean radius of anionite grain; A - is the anionite surface; W - is the anionite volume; α - is the ratio of ions quantity in the solution to ions quantity in the ionite at equilibrium; F_0 - is the Fourier criterion equal to $\dfrac{\overline{D}t}{r^2}$; q_n - are the roots of

equation $\mathrm{tg}q_n = \dfrac{3q_n}{3 + \alpha\,q_n}$; $B_n = \dfrac{6B_i^2}{\mu_n^2\left(\mu_n^2 + B_i^2 - B_i\right)}$; B_i - is the Bio criterion which

determines the extent of boundary layer effect on the diffusion in the ionite; μ_n - are the roots of characteristic equation $\text{tg}\mu_n = \dfrac{\mu_n}{1 - B_i}$; B_i and μ_n values are looked up in tables by plotting $\ln(1\text{-}F)$ against t.

A survey of literature on the study of kinetics of sorption has demonstrated that numerical values of diffusion coefficients are difficult to compare due to different methods and conditions of experiments and equations. Of some interest the use equations of 1-4 to calculate diffusion coefficients under equal conditions and to assess the kinetics of rhenium sorption by using anions of various structures.

Experimental data required for the description of kinetics of ion exchange processes are obtained by various methods: "interruption" method of "kinetic memory", method of flow, and method of limited volume[1,4-7].

1.1 Effect of Physicochemical Factors on the Kinetics of Rhenium Sorption

1.1.1 Materials and Experimental Procedure. The kinetics of rhenium sorption from sulphuric acid solutions were examined using a weak base anionite AH-21 of gel-like and porous structures. This ionite contains primary and secondary amines as the ionogenic groups. During anionite synthesis, isooctane was introduced to form porosity, Ionites having different levels of divinyl benzene (DVB) were specially synthesized for the investigations. The effect of various parameters on the kinetics of sorption (with the exception of DVB) has been studied using ionites AH-21-G8 and AH-21-MP20 (one part of isooctane by weight). The above ionites have equal values of their exchange capacity with respect to rhenium and have been used to recover rhenium from sulphuric acid solutions at operations in the former Soviet Union.

The experiments were performed according to well-known methods. Ionites were allowed to stand for 24 hours, filtered and placed in stock solution, at designated temperature. Samples were withdrawn at intervals by specially designed pipettes with a filter to avoid capture of ionite. Time of sampling was 1-2 seconds. Overall volume of withdrawn solution was not more than 3-5% of the total solution. The constant stirring speed (a number of stirrer's revolutions) was 5 rev. per second.

From the experimental data the rate of sorption (V), degree of exchange (F), diffusion coefficients (D) were calculated and curves $F - \sqrt{t}$, $F\text{-}r^2$, $\ln(1\text{-}F)\text{-}t$, $D\text{-}t$ were plotted. Numerical values of D calculated by equation 1 were used in plotting D versus t. Diffusion coefficients calculated according to the equations 1-4 and all factors under investigation are listed in the Table.

1.1.2 Determination of the Limitation Stage of Rhenium Sorption. The limitation stage of the process was determined by the method for interruption of phase contacts. To accomplish this, 1g each of ionites was brought into contact with the solution containing 0.0005 kmol/m^3 rhenium and 0.25 kmol/m^3 sulphuric acid. Within 30 minutes of the experiment start, ionites were separated from the solution by filtration and within 1 hour of separation they were returned to the solution. The experiments without interruption of phase contacts were performed simultaneously. Time dependence of the degree of exchange and the rate of sorption with and without interruption was determined from the data obtained.

An increase in the rate of sorption after the recommencement of contacts between ionites of both structures and the solution proves the crucial role of interdiffusion kinetics.

Linear dependence of F on $\sqrt{t}$ and nonlinear dependence of $\ln(1\text{-}F)$ on t also confirm the predominance of interdiffusion kinetics. These criteria have allowed the approximate determination of the limiting stage. Therefore, to prove that ion exchange is going on in one or another region, a set of other parameters (such as radius of ionite grain, DVB content in ionite, rhenium concentration, process temperature) has been studied.

1.1.3 Radius of Ionite Grain. The studies were conducted with ionites which being in the swollen state have radius of (cm): 0.0375, 0.0565, 0.0815 (AH-21-G8), 0.017, 0.024, 0.034 (AH-21-MP20). Solutions containing (kmol/m^3) rhenium $5*10^{-4}$, sulphuric acid 0.25 were used for the experiments. The system: ionite-solution was brought to equilibrium.

The rate of rhenium sorption with porous anionites is much greater than the rate of gel ionite at either particle size (Figure 1). With decreasing radius of ionite grains, a rate of sorption decreases with time in all cases. At the beginning sorption basically occurs at the surface of grain with minor diffusion difficulties associated with exchanging ions. A decrease in the rate results, first, from the exchange process occurring inside a grain when steric problems are significant and, second, from decreasing rhenium concentration in the solution. The dependence of F on r^2 proves that during the first 30 minutes, sorption is limited by external diffusion (Figure 2). This is exemplified by the nonlinear dependence of the degree of exchange on the square of radius at the beginning of sorption (Figure 2, curves 1-3(a) and 5-8(b)). Subsequently, the parallelism of linear dependencies is observed: the process mainly occurs inside a grain. During first 1.5 hour, values of $\overline{D}$ increase, then they become constant and time-independent (Figure 3). However, the smaller a grain radius, the greater $\overline{D}$ value. This also proves that the process occurring inside of the ionite grain plays a crucial role.

The dependencies of B_t and $\ln(1-F)$ on t, F on $\sqrt{t}$ also prove that sorption of perrhenate ions is limited by internal diffusion.

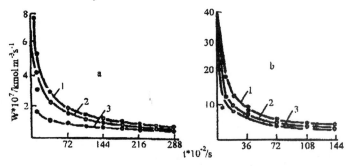

Figure 1 *Dependence of the rate of sorption on the radius of AH-21-G8 (a) and AH-21-MP20 (b) grain. (a) 1 - 0.0375, 2 - 0.0565, 3 - 0.0815 cm; (b) 1 - 0.017, 2 - 0.024, 3 - 0.034 cm*

Consequently, the studies made into the effect of a radius of ionite grain on the kinetics of rhenium sorption from sulphuric acid solutions have indicated that irrespective of anionite structure rhenium sorption in the opening stage is limited by external diffusion and later - by internal diffusion.

1.1.4 DVB Content. AH-21-G ionite contains 2, 6, 10, 12, 14, 16, 20 mass percent of DVB and AH-21-MP ionite containing 8, 16, 20 mass percent of DVB were used for the investigation. The conditions of the experiments were similar to the previous set of experiments.

At any content of DVB, the rate of perrhenate-ions sorption decreases rapidly with time. At 3 hours contact, the rate becomes constant and its value is not high (Figure 4).

As this takes place, an increase in DVB amount in gel anionite from 2 to 20 mass % results in a 4 times decrease in the rate. For ionites of porous structure, an inverse dependence of sorption rate on DVB content in the ionite as compared to gel-like structure has been found. The rate of exchange goes up as DVB increases. The mechanism found can be explained by the different volume of pores falling at definite ionite weight. Thus, according to the authors[8] data, total pore volumes of gel ionite containing 2 mass % of DVB in terms of cm^3/g are 0.0038 and 20 mass % - 0.0014 respectively. But pore volumes of porous ionite contains 8 mass % are 0.270 cm^3/g and at 20 mass % of DVB - 0.345 cm^3/g. Diffusion coefficients decrease with increasing content of DVB in gel-like ionite structure. An increase in DVB content in porous ionites results in an increasing numerical value of diffusion coefficients (Figure 5). The mechanism obtained agrees well with the rate curves (Figure 4).

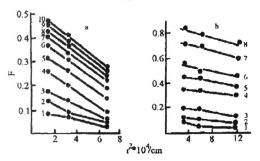

Figure 2 *Dependence of F ionites AH-21-G8 (a) and AH-21-MP20 (b) on r^2 at various times values / $s*10^2$: (a) 1 - 6, 2 - 12, 3 - 18, 4 - 36, 5 - 72, 6 - 108, 7 - 144, 8 -180, 9 - 216, 10 - 252; (b) 1 - 3, 2 - 6, 3 - 12, 4 - 18, 5 - 24, 6 - 36, 7 - 72, 8 - 144*

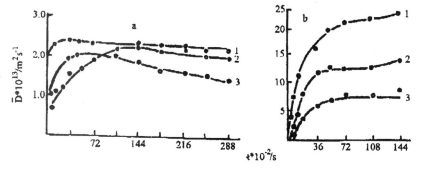

Figure 3 *Time dependence of rhenium diffusion coefficients at various radius of AH-21-G8 (a) and AH-21-MP20 (b) ionites grains / $m*10^6$: (a) 1 - 375, 2 - 565, 3 - 815: (b) 1 - 17, 2 - 24, 3 - 34*

The behaviour of B_r, and $\ln(1-F)$ versus t and that of F versus $\sqrt{t}$ with DVB content in ionites of both structures indicates that ion exchange is limited by internal diffusion. While analyzing the data obtained on the effect of DVB content in AH-21 ionites of gel-like and porous structures on the kinetic of the process, one can state the following. Rhenium sorption irrespective of the ionite structure, occurs in the mixed diffusion region.

This process is of greater intensity (2 - 2.5 times) with ionites of porous structure than gel-like structure. In kinetic properties, porous ionites having high content of DVB (16-20 mass %) are close to gel-like ionites containing 2-4 mass % of DVB.

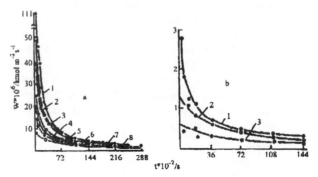

Figure 4 *Time dependence of the rate of rhenium sorption at various content of DVB in AH-21-G (a) and AH-21-MP (b) ionites; (a) 1 - 2, 2 - 6, 3 - 8, 4 - 10, 5 - 12, 6 - 14, 7 - 16, 8 - 20 mass %; (b) 1 - 8, 2 - 16, 3 - 20 mass %*

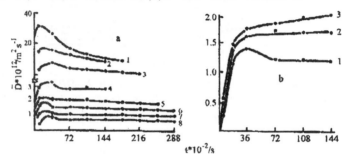

Figure 5 *Time dependence of coefficients of rhenium diffusion at various content of DVB in AH-21-G (a) and AH-21-MP (b) ionites. (a) 1 - 2, 2 - 6, 3 - 8, 4 - 10, 5 - 12, 6 - 14, 7 - 17, 8 - 20 mass %; (b) 1 - 8, 2 - 16, 3 - 20 mass %*

1.1.5 Rhenium Concentration in Stock Solution. This set of experiments has shown that the rhenium concentration varies from $5.5*10^{-4}$ to $55.6*10^{-4}$ kmol/m^3 at fixed content of sulphuric acid (0.25 kmol/m^3). Amount of perrhenate ions per 1 gram of ionite has been constant during all experiments. This has been achieved by changing ionite batch from 0.245 to 4.9 g (in term of dry weight) with similar volumes of solution. In the initial stage, the rate of perrhenate ions sorption greatly depends on rhenium concentration in the solution with respect to both structures of AH-21 ionites (Figure 6). In the course of time, the difference in values of rate is reduced due to the decreasing rhenium concentration in the solution and the approximation of the system to the equilibrium. F increases irrespective of the ionite structure with increasing rhenium concentration in the external solution (Figure 7).

The dependence of B_t and $\ln(1-F)$ on t, F on $\sqrt{t}$ point out that the process is limited by internal diffusion when the amount of rhenium in the solution reaches $3*10^{-4}$ kmol/m^3. With increasing initial concentration of metal, the process is shifted to the mixed region.

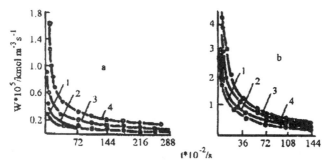

Figure 6 *Dependence of the rate of rhenium sorption with AH-21-G8 (a) and AH-21-MP20 (b) ionites on the concentration of rhenium in the solution; 1 - 5.5, 2 - 10.4, 3 - 34.2, 4 - 55.6*10⁻⁴ / kmol m⁻³*

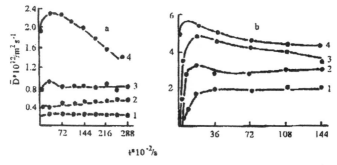

Figure 7 *Time dependence of coefficients of rhenium diffusion in AH-21-G8 (a) and AH-21-MP20 (b) ionites at various concentration of rhenium in the solution; 1 - 5.5, 2 - 10.4, 3 - 34.2, 4 - 55.6*10⁻⁴ / kmol m⁻³*

1.1.6 Temperature. The effect of this factor on the kinetics of rhenium sorption has been studied in the temperature range of 287 to 438 K. An increase in temperature is favorable to increase the rate of perrhenate ions sorption with respect to ionites of both structures (Figure 8).

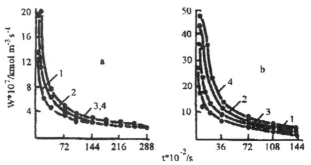

Figure 8 *Time dependence of the rate of rhenium sorption by using AH-21-G8 (a) and AH-21-MP20 (b) ionites at various temperature; 1 - 287, 2 - 298, 3 - 328, 4 - 438 K*

On the basis on temperature dependence, activation energy is calculated by the equation:

$$D = Ke^{-E/R_t} \tag{5}$$

The value of activation energy was determined by taking the logarithm of this equation and plotting it versus $1/T$

$$E = \text{tg}\,\alpha\,R \tag{6}$$

The dependence of $\ln \overline{D}$ on $1/T$ is linear. The parallelism of straight-lines ($\ln \overline{D}$ on $1/T$) for chosen time allows the use of equation 5 to calculate activation energy. The values of activation energy, 11.2 kJ/mol (AH-21-G8) and 30.7 kJ/mol (AH-21-MP20) have confirmed the diffusion nature of rhenium sorption.

1.1.7 Diffusion Coefficients. The use of equations 1 - 4 to calculate diffusion coefficients has been studied using AH-21 ionite of gel-like structure as an example. The calculation of coefficients has been accomplished by changing all parameters under investigation during rhenium sorption. The data obtained is presented in the Table.

The values of diffusion coefficients calculated according to equations 1, 2, 4 are of the same order but those calculated by equation 3 are of an order lower. Since the change in the concentration of ions subjected to sorption in external solution is taken into account, we believe that the values of diffusion coefficients calculated by this equation are the most plausible.

1.2   Petrographic Investigation of Ionites Saturated with Rhenium

Based on the study of the effect of various physicochemical factors and associated anions on the kinetics of perrhenate-ions sorption using slightly basic ionite AH-21 and on the analysis of the data obtained, we can say that this process occurs in the mixed region when initially external diffusion prevails and then internal. Sorption basically proceeds in the surface layer at partial penetration of sorbed ions deeply into the ionite. The calculations and microscopic investigations into the determination of the depth of perrhenate-ions penetration have been accomplished to confirm the above.

The depth of penetration is determined by computations[9] was 0.012 cm. The rhenium property of formation of colored complexes with ammonium rhodanite in sulphuric acid media was used for microscopic investigations. Metallographic specimens saturated with rhenium were prepared according to the method developed at our Institute. A diameter of ionite grain and its colored part was measured for quantitative evaluation. It has been found that depending on rhenium content in ionite, the depth of its penetration is variable from 0.01 cm (5 mass % of rhenium) to 0.08 cm (30 mass % of rhenium). The depth of rhenium penetration increases with decreasing radius of ionite grain. The data of microscopic investigations are in agreement with computations.

2 SUMMARY

Research on the kinetics of perrhenate-ions sorption by using AH-21 ionite of both gel-like and porous structures has resulted in the conclusion that rhenium sorption occurs in the mixed region: initially the process is limited by external diffusion and later - by internal diffusion. The dependencies found; B_t and $\ln(1-F)$ on t, F on $\sqrt{t}$, F on r^2 proves this theory. Rhenium sorption using porous ionites proceeds faster than with gel-like ionites. The values of diffusion coefficients calculated by the equations 1, 2, 4 are of the same order but those calculated by equation 3 are of an order lower. The values of apparent activation energy of perrhenate sorption with AH-21 ionite of both structures indicate that the process under investigation occurs in diffusion region.

Table   *Average Values of Diffusion Coefficients Calculated by Equations 1 - 4*

Parameters	Average values of $\overline{D}*10^{-12}/m^2s^{-1}$			
	1	*2*	*3*	*4*
Radius of grain in swollen state / cm				
0.0375	1.6500	1.4600	0.0700	1.5500
0.0565	2.2500	1.9200	0.1060	3.6100
0.0815	2.1000	2.0000	0.0820	6.7800
DVB content/mass %				
2	15.0000	15.1900	2.9600	32.1600
6	11.0000	6.1800	2.1000	11.6600
8	5.0000	4.1100	0.4370	4.6300
10	2.7000	2.3970	0.1980	2.5800
12	1.9000	1.6640	0.1370	2.1400
14	1.4100	1.0300	0.0640	1.3200
16	1.0000	0.6844	0.0410	1.5800
20	0.6000	0.2782	0.0390	1.3400
Rhenium concentration $* 10^{-4}$ / kmol m^{-3}				
5.0	0.2206	0.1876	0.0111	0.3871
10.0	0.4889	0.3564	0.0222	1.1583
34.0	0.7738	0.6297	0.0345	1.1960
55.0	1.1554	0.9445	0.0589	1.4430
112.0	1.7680	1.6314	0.0960	1.6540
Temperature / K				
287	1.1450	0.8700	0.0619	1.9531
298	1.1840	1.0300	0.0652	1.6634
328	1.7130	1.3900	0.1120	2.0265
438	1.9115	1.7600	0.1131	2.6515

References

1. F. Helfferich, "Ion Exchange", McGraw-Hill, New York, 1962.
2. J. A. Marinsky, "Ion Exchange", New York, 1966.
3. "New Problems in Current Electrochemistry", Moscow, 1962, p. 95.
4. G. E. Boyd, A. V. Adamson and A. S. Miers, "Chromatographic Method for Separation of Ions", Moscow, 1949.
5. M. M. Tunitsky, "Methods of Physicochemical Kinetics", Moscow, 1972.
6. Yu. A. Kokotov and V. A. Pasechnik, "Equilibrium and Kinetics of Ion Exchange", Leningrad, 1970.
7. Yu. M. Popkov, M. D. Kalinina, M. I. Nikolaeva and N. M. Tunitsky, *J. Physical Chemistry*, 1973, **47(1)**, 194.
8. D. M. Romanova, K. B. Lebedev, E. I. Krikova, *J. Applied Chemistry*, 1974, **47(9)**, 2003.
9. V. N. Laskorin, P. V. Pribytko and L. I. Vodolazov, "Synthesis and Properties of Ion Exchange Materials", Moscow, 1968, p. 152.

ION EXCHANGE FOR NITRIDE FUEL CYCLE

M. Nogami, M. Aida, K. Takahashi and Y. Fujii
Research Laboratory for Nuclear Reactors, Tokyo Institute of Technology
O-okayama, Meguro-ku, Tokyo 152, Japan

A. Maekawa, S. Ohe, H. Kawai and M. Yoneda
The Kansai Electric Power Co., Inc.
Nakanoshima, Kita-ku, Osaka 530-70, Japan

1 INTRODUCTION

Solvent extraction has been used for the reprocessing of nuclear fuels. The Purex process, which utilizes nitric acid (HNO_3) solutions and TBP (tributyl phosphate) as the media, has been developed for the reprocessing of metal fuels, and used for the reprocessing of oxide fuels (UO_2, PuO_2) from commercial reactors (light water reactors) and fast breeder reactors (FBR's). This process is appropriate for the recovery of highly pure uranium and plutonium. Recently attention has been placed on the partition of the high level radioactive waste (HLRW). Ion exchange may be more appropriate in principle than solvent extraction for the separation of HLRW, since it can separate different chemical species one after another.

On the other hand, nitride fuels (UN, PuN) have been considered as the promising fuels for FBR's, because of the high fissile density, high thermal conductivity, high breeding ratio, etc. However, there is a problem in actually using nitride fuels, i.e. the major stable isotope in natural nitrogen, ^{14}N, generates ^{14}C whose half-life is 5730 years, by the following reaction:

$$^{14}N + n \rightarrow {}^{14}C + p$$

where n and p are neutron and proton, respectively. To avoid this problem, ^{15}N is expected to be used as an appropriate material for nitride fuels, but isotope separation of ^{15}N becomes necessary in this case. Nitrogen is a relatively light element and chemical exchange is a suitable method for isotope separation of such a light element. ion exchange is one of the processes of chemical exchange isotope separation. Our recent study on the nitrogen isotope separation by ion exchange has been reported elsewhere.[1]

The second problem in the nitride fuel cycle is in the reprocessing. Since the highly enriched ^{15}N is costly, it should be recovered at reprocessing in addition to uranium and plutonium. For this purpose, the HNO_3 system is not appropriate, since the recovered nitrogen may be mixed with nitrogen in HNO_3 and it is too costly and impractical to use HNO_3 with highly enriched ^{15}N. Therefore another solvent which does not contain nitrogen should be selected. In this respect, hydrochloric acid (HCl) would be appropriate because of its strong solubility for many kinds of metals.

The HCl system has not yet been applied to the reprocessing of spent fuels so far. In this study the authors discuss whether it is possible or not to apply an ion exchange system with HCl to the nitride fuel reprocessing under this background.

2 EXPERIMENTAL AND RESULTS

2.1 Synthesis of Resin

The resin used in the reprocessing should have chemical properties of high resistance against radiation and fast ion exchange rates. Among various types of anion exchange resins, resins containing aromatic functional groups with conjugated bonds are more resistant to radiation than other types of synthetic organic ion exchange resins.[2] In this study, tertiary pyridine anion exchange resin, which has the most simple structure among these, was chosen.[3] The mixture of the monomers, 80 wt.% 4-vinylpyridine and 20 wt.% divinylbenzene, was polymerized. In order to produce pores which enable the high ion exchange rates, pore producing solvents were added to the monomers. In addition, the resin was embedded in porous silica beads with diameter of 44~77 μm. These silica beads enabled the resin to avoid swelling and shrinking regardless of chemical form and to bear high pressure during use.

2.2 Radiation Resistance

Irradiation experiments, using ^{60}Co gamma rays, were conducted in order to ensure the stability of this tertiary pyridine resin against radiation. Most of the commercially available synthetic organic ion exchange resins show apparent radiation damage at or above 0.1 MGy of absorbed radiation. These resins are totally unusable at the absorbed dose of 10 MGy[2] but the pyridine resin was found to usable up to 10 MGy irradiation in HCl solutions regardless of their concentration,[4] while it had considerable damage in HNO$_3$ solutions. The stability in the two solutions was opposite to that of the conventional quaternary ammonium resin. The decomposition of the resin might be due to the breakage of the principle chains. If the resin, whose principal chains have conjugation systems, is developed, it would be more resistant against radiation.

2.3 Ion Exchange Selectivities

The ion exchange selectivities of tertiary pyridine resin for major spent fuel elements were experimentally examined in HCl solutions at different concentrations of 1~9 M (mol/dm³) at 298K.[3] The adsorption selectivity was evaluated by the distribution coefficient, D, defined as

$$D = q/C$$

where q and C are the concentrations of the metal ions in resin (mmol / effective resin volume) and in aqueous phase (mM), respectively. As previously mentioned, attention was placed on uranium and FP's. The latter elements range from selenium to the light rare earths. In addition, several other elements frequently used as construction materials or contained as impurities, such as Ti, Fe, etc. have been examined in the present work. The results for different elements were shown in Figure 1.

As anticipated, it was found that the resin does not absorb ions of the majority of HLRW, an alkali metal element (Cs), alkaline earth elements (Sr, Ba), and rare earth elements (La, Nd etc.). All of the second transition elements have shown to be adsorbed in the resin, especially palladium with a large distribution coefficient at lower concentrations of HCl. Uranium, U(VI), has been confirmed to be adsorbed at high concentration of HCl, e.g. 6 M and desorbed with a dilute, e.g. 0.1 M HCl solution. The

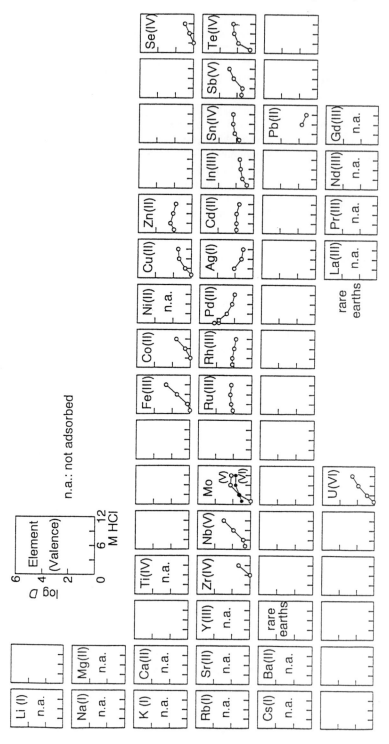

Figure 1 *Adsorption of various ions on tertiary pyridine resin in HCl solutions*

results indicate that the group of alkali metals, alkaline earths and rare earths and the other of second transition elements can be separated from uranium by changing the concentration of eluent HCl solution. The strongly adsorbed second transition elements can be eluted with solutions containing stronger ligands such as thiourea (H_2NCSNH_2).

Ion exchange can easily realize a large number of multistage separation and even in the cases of a small difference in D values of two elements, the separation can be attained. Thus even Nb(V) or Mo(V) might be separated from U(VI). In fact, our preliminary chromatographic experiment revealed that Nb(V) and Te(VI) were separated from U(VI). In the case of Mo(V), oxidation to Mo(VI) would be necessary. In HCl systems, stainless steels can not be used as construction materials because of corrosion. Ti is now promising because it has experience in the uranium isotope separation by ion exchange.

These tendencies of adsorption are basically similar to those of commercial strong base anion exchange resin. It is due to the high concentration of the acid where tertiary amine pyridine resins perform like quaternary ammonium type resin.

In the present work, transuranium elements of Np, Pu, Am and Cm have not been studied. From the reported information on the selectivities of these elements, it is anticipated that Np and Pu can be adsorbed in the ion exchange resin and these two and U can be separated from one another. Am and Cm would be difficult to separate from rare earth elements. There may be a possibility, however, that Am and Cm are separated from rare earths when the eluting solution is concentrated to ca. 10 M chloride of the rare earths since it has been reported that Am and Cm can be adsorbed in an anion exchange resin in 10 M LiCl solution. Further work would be done on the recovery of Am and Cm in the next stage of the present program on development of ion exchange reprocessing technology.

This system can be expected to be applied not only to nitride fuels but also to the present oxide fuels if an appropriate dissolution process of oxide fuels is developed. If alkali metals, alkaline earths and rare earths are separated from light platinum elements, the former three groups would be used as radiation sources and the latter would be used as catalysts.

3 CONCLUSIONS

A tertiary pyridine type anion exchange resin was synthesized for the reprocessing of nitride fuels. This resin has significant resistance against radiation and high ion selectivities for spent fuel elements in HCl solution. The results indicate that the ion exchange process is promising for the first decontamination process in nitride fuel reprocessing.

References

1. H. Ohtsuka, M. Ohwaki, M. Nomura, M. Okamoto and Y. Fujii, *J. Nucl. Sci. Tech.*, 1995, **32**, No.10, 1001.
2. K. K. S. Pillay, *J. Radioanal. Nucl. Chem.*, 1986, **102**, No.1, 247.
3. M. Nogami, M. Aida, Y. Fujii, A. Maekawa, S. Ohe, H. Kawai and M. Yoneda, *Nucl. Tech.*, in press.
4. M. Nogami, Y. Fujii and T. Sugo, *J. Radioanal. Nucl. Chem.*, 1996, **203**, No.1, 109.

NON-EQUILIBRIUM OPERATION IN A DUAL TEMPERATURE ION EXCHANGE SEPARATION SYSTEM

A. A. Zagorodni and M. Muhammed

Department of Inorganic Chemistry
Royal Institute of Technology
100 44 Stockholm
Sweden

1 INTRODUCTION

Ion exchange separation with temperature variation is a widely known method. However, there is no complete information about the dependence of the separation process on the separation cycle parameters, due to the large amount of experimental work required. It is possible, of course, to try to obtain this dependence through mathematical modelling. However, the modern state of the ion exchange dynamic modelling needs experimental confirmation of each system.

We have recently demonstrated the feasibility of separating metal ions from acidic solution using a temperature dependent ion exchange process[1,2]. The applicability of Amberlite IRC-718 iminodiacetic resin (*Rohm & Haas*) to *Cu/Zn* mixture separation has been shown. We present here experimental data on the dependence of the separation parameters on the fractions volume in the separation cycle. The degree of separation can be adjusted by controlling the fraction's volume passed through the columns at each temperature. This can decrease the separation cost without causing any product amount and/or quality detriment.

2 EXPERIMENTAL

Amberlite IRC-718 iminodiacetic ion exchange resin was received from *Rohm & Haas* (USA). The resin was washed with *HCl* then *NaOH* solution, three times to remove organic matter present.

$CuSO_4 \cdot 5H_2O$ and $ZnSO_4 \cdot 7H_2O$ (*KEBO Lab*, Sweden) and H_2SO_4 (*MERK*, Germany) of analytical grade were used as received. All solutions were prepared using deionised water. The composition of the feed solution remained constant in all series of experiments carried out. *Zn* to *Cu* millimolar ratios equalled 102:12. The total concentration of SO_4^{2-} was kept constant at 0.14 mol/L level. pH was maintained to 1.8 with H_2SO_4.

The concentrations of *Cu* and *Zn* were determined by AAS technique using *Perkin Elmer 603*. The pH was controlled using a combined glass electrode. The uncertainty of spectrochemical analysis was less than 1 % for *Cu* and 2 % for *Zn*.

The separation experiment was carried out in a thermostated glass column providing the heating/cooling of both resin and solution phases. The column was loaded with a weighted amount (around 4 g) of air dried resin Amberlite IRC-718 (H^+-form). The feed solution was passed through the column at a constant flow rate of 0.5 mL/min, corresponding to a velocity 0.37 m/h in the column. The temperature was changed in turn (at 15 and 75°C). The effluent was collected in 25 mL samples. Every sample was analyzed for *Cu*, *Zn* and pH.

Figure 1 shows the automatic experimental system which was used for the investigation of the ion exchange separation processes at variable temperatures. The automatic system included an ion

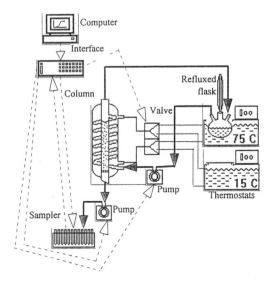

Figure 1 *The automatic experimental system*

exchange column, a computer and three computer operated parts: a thermostating system, a feed system and a sample collecting system.

An IBM-PC computer with an I/O card and a specially developed relay interface were used.

The thermostating system contained two thermostats and two three-way solenoid valves. At the desired time, the solenoid valves changed the flow of the thermostatic liquid for the column temperature switching. The valves were coupled and operated by the computer.

The feed system contained a flask for the feed solution, two peristaltic pumps (*Masterflex*, USA) and a recycling tube. The flask was placed in the thermostat which had a higher temperature to prevent solvated gas from being fed to the column. A refluxing condenser was used to prevent evaporation. The first pump helped move the feed solution through the column. It kept a constant flow rate in the column. The second pump delivered the feed solution into the column. Its pumping speed was greater than the speed of the first pump. The excess solution was returned to the flask by a recycling tube. The double-pumps system was made to maintain a constant solution level inside the column. Both pumps were operated by the computer.

The sample collecting system was a standard automatic sampler (*7000 Ultrorac Fraction Collector*, Sweden), which was operated by the computer. The back connection was used to protect solution leakage during sampler operation.

The software was written in Borland Pascal 7.0. The computer controlled the experiment, according to conditions specified in the input.

3 RESULTS AND DISCUSSION

Separation experiments with different fraction volumes were performed. A view of a typical dual temperature separation cycle is shown in Figure 2. The curve in this figure corresponds to the effluent concentration. Unfortunately, it is not possible to choose one general parameter for the effective characterisation of the separation process for two ions. Different authors used different ways for dual temperature separation process characterisation. *Bailly & Tondeur*[3] used purity of product and ratio of the flow rate recycled to the feed flow rate. *Ivanov* et al.[4] used a factor of the solution purification. Several authors[5-9] used the separation factor to characterise the process totally. This parameter is the ratio of the gross product concentrations from the hot and cold half-cycles. The following characteristics were applied for our separation system. The gross concentrating factor of copper (according to Reference 3) is given by:

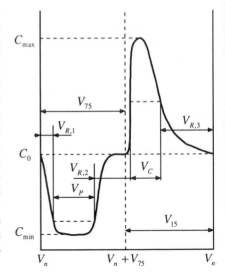

Figure 2 *The scheme of the dual temperature separation cycle. V_n is an initial point of the separation cycle number n; V_{15} and V_{75} are volumes of the feed solution, passed at 15 and 75°C. $V_{n+1}=V_n+V_{75}+V_{15}$*

$$CF_{Cu} = \frac{\overline{C}_{Cu,15}}{\overline{C}_{Cu,0}} \qquad (1)$$

where C_0 is the feed solution concentration. Average (gross) concentration of any ion in the effluent fraction ($\overline{C}_T$) was calculated by equation (2):

$$\overline{C}_T = \frac{\int\limits_{V_T} C dV}{V_T} \qquad (2)$$

where C is the effluent concentration, V_T is the volume of the feed solution passed at one temperature. The index T was 15 or 75 for two different fractions.

The purity of Zn solution results from the Cu concentration in the system. The gross zinc purification factor was calculated as follows:

$$PF_{Zn} = \frac{C_{Cu,0}}{(C_{Cu,0} + C_{Zn,0})} \cdot \frac{(\overline{C}_{Cu,75} + \overline{C}_{Zn,75})}{\overline{C}_{Cu,75}} \qquad (3)$$

We propose the parameter SF given by the equation (4) as

$$SF_{Zn}^{Cu} = \sqrt{\frac{\overline{C}_{Cu,15}}{\overline{C}_{Zn,15}} \cdot \frac{\overline{C}_{Zn,75}}{\overline{C}_{Cu,75}}} \qquad (4)$$

The square root was included in equation (4) for the scale standardization. The separation factor (4) shows the distribution of two ions between two collected fractions.

The external characteristics of the process were determined as:

$$CF_{Cu,max} = \frac{C_{Cu,max}}{C_{Cu,0}} \qquad (5)$$

$$PF_{Zn,max} = \frac{C_{Cu,0}}{(C_{Cu,0} + C_{Zn,0})} \cdot \left\{ \frac{(C_{Cu} + C_{Zn})}{C_{Cu}} \right\}_{max} \qquad (6)$$

$$SF_{Zn,max}^{Cu} = \sqrt{\left\{\frac{C_{Cu}}{C_{Zn}}\right\}_{max} \cdot \left\{\frac{C_{Zn}}{C_{Cu}}\right\}_{max}} \qquad (7)$$

The external parameters (5)-(7) correspond to gross parameters (1), (3), (4).

The first experimental set included the experiments with equivalent volumes of the feed solution passed at each temperature, i.e. $V_{15} = V_{75}$. The dependence of the separation parameters on the fraction volume is shown in Figure 3. As can be seen from the results presented in Figure 3, the gross separation parameters (*CF*, *PF* and *SF*) have a maximum at 225 mL of the fractions volume. The maximum is explained by the existence of two competing processes:

1. The difference in the resin bed loading increases with increasing the solution volume passed at each temperature. This results in separation improvement.

2. The increase of the volume passed at each temperature results in a decrease of the difference between the total concentration in effluent fraction and the concentration of the feed solution.

The separation parameters CF_{max}, PF_{max} and SF_{max} increase with the fraction volume because they depend only on the first process. However, there is a maximum reasonable volume passed at each temperature. This volume corresponds to the attainment of the ion exchange equilibrium at every step of the process. For these experiments the volume of equilibrium attainment is 525 mL.

A real separation process depends on technological needs. For example, only an effective concentration of copper or a minimal contamination of *Zn* by *Cu* may be required. A few examples of different technological needs are the following:

I. For *Zn* maximum decontamination; *Cu* concentration degree is not important. A reasonable method includes: passing a certain volume of the solution at 75°C and passing the solution at 15°C until attainment of the equilibrium. Figure 4 shows the dependence of PF_{Zn} and $PF_{Zn, max}$ on the volume passed at 75°C.

II. Only a *Cu* maximum concentration is required; *Zn* purification is not important. A reasonable method includes: passing the solution at 75°C until attainment of the equilibrium and passing a certain volume of the solution at 15°C. Figure 4 shows the dependence of the CF_{Cu} and $CF_{Cu,max}$ on the volume passed at 15°C as well.

As seen from Figure 4, both I and II processes have their own maximums of efficiency (for the

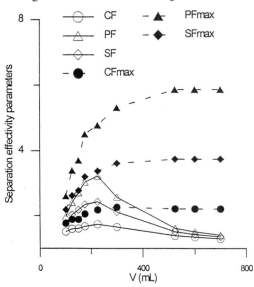

Figure 3 *The dependence of the separation efficiency parameters on the fractions volume for experiments with equivalent volumes of the feed solution passed at each temperature* $(V=V_{15}=V_{75})$

gross separation), which locate far before attainment of equilibria.

III. There are requirements of product concentration and/or purification over certain minimums. Three fractions of the effluent should be collected. The third fraction is recycled into the same column. Figure 2 shows the fraction volumes in the three fraction process: V_P is the volume of the purified fraction, V_C is the volume of the concentrated fraction and $V_R = V_{R,1} + V_{R,2} + V_{R,3}$ is the volume of the recycled solution. Recycling is possible without the separation cycle destabilization if the following equation is true for every ion:

$$\int_{V_{R,1}} CdV + \int_{V_{R,2}} CdV + \int_{V_{R,3}} CdV = C_0 V_R \quad (8)$$

For example, we have chosen the cycle including equilibrium reached at every step. Figure 5 shows the dependencies of the separation characteristics and the fractions volumes collected on the established minimal values of *CF* and *PF*.

IV. There are requirements for the gross properties of the collected fractions, i.e.

$$PF_{est} = \frac{\int_{V_P} PFdV}{V_P} \quad (9)$$

and/or

$$CF_{est} = \frac{\int_{V_C} CFdV}{V_C} \quad (10)$$

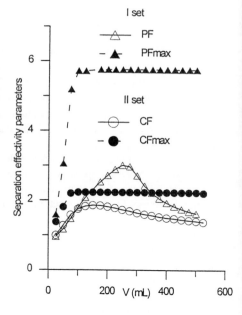

Figure 4 *The dependence of the separation efficiency parameters on the fractions volume for experiments with attainment of equilibria reaching at one temperature:*
I. Equilibria reached at 15°C (V=V$_{75}$);
II. Equilibria reached at 75°C (V=V$_{15}$)

Figure 6 shows the dependencies of the collected volumes on the established values for the same separation cycle.

4 CONCLUSIONS

The parameters for the characterization of the degree of separation of two ions are proposed. The dependence of the separation efficiency on the dual temperature cycle characteristics is presented.

Non-equilibrium dual temperature operations are preferable for the effective separation of *Cu* from *Zn* on Amberlite IRC-718 resin. Relationships between the separation cycle parameters and technological requirements are shown.

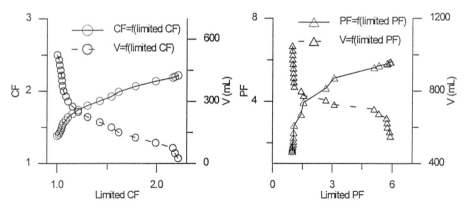

Figure 5 *The dependence of the separation characteristics and the fractions volumes on the established minimal values of CF and PF*

Acknowledgements

The work is funded by TFR (Swedish Research Council for Engineering Sciences. We would like to thank Dr. N. Nikolaev for discussions.

References

1. A. A. Zagorodni and M. Muhammed, *Proceedings of the Fourth International Conference on Ion Exchange Processes "ION-EX '95"*, in print.
2. A. A. Zagorodni, D. N. Muiraviev and M. Muhammed, *Separation Science and Technology*, submitted.

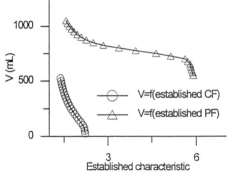

Figure 6 *The dependencies of the fractions volumes on the established gross properties*

3. M. Bailly and D. Tondeur, *Journal of Chromatography*, 1980, **201**, 343.
4. V. I. Ivanov, V. D. Timofeevskaya and V. I. Gorshkov, *Reactive Polymers*, 1992, **17**, 101.
5. T. J. Butts, N. H. Sweed and A. A. Camero, *Ind. Eng. Chem. Fundam.*, 1973, **12**, 467.
6. G. Grevillot, *A. I. Ch. E. Journal*, 1980, **26**, 120.
7. K. S. Knaebel and R. L. Pigford, *Ind. Eng. Chem. Fundam.*, 1983, **22**, 336.
8. G. Carta and R. L. Pigford, *Ind. Eng. Chem. Fundam.*, 1986, **25**, 677.
9. T. Szanya, L. Hanak, R. Mohilla and P. Szolcsanyi, *Hungarian Journal of Industrial Chemistry*, 1988, **16**, 261.

ION EXCHANGE PROCESSES FOR THE RECOVERY OF VERY DILUTE ACETIC AND RELATED ACIDS

F. L. D. Cloete and A. P. Marais

Department of Chemical Engineering
University of Stellenbosch
Private Bag X1 Matieland
7602 South Africa

1 INTRODUCTION

Acetic acid and its higher homologues occur in many industrial waste waters, often mixed with several other classes of organic compounds[1,2]. All these organics increase the COD of the effluent and incur processing charges for treating the water to acceptable limits for discharge to the environment[3].

The process proposed in this article[4], is based on the well-known reaction between a weakly basic ion exchange resin in the free-base form and an acid. Acetic acid was used to represent a mixture of carboxylic acids for a preliminary study of the process.

The distribution coefficient is extremely high, since the extraction is based on the neutralisation of a base by an acid. The ion exchanger is totally insoluble and no toxic solvents are added to the waste to cause problems further downstream.

The elution of acids from the resin is simply achieved by raising the pH using a suspension of lime in a fluidised or agitated bed[2]. The lime suspension dissolves as reaction proceeds and a solution of the highly soluble calcium acetate in the range 12-20% is formed. The resin has been converted back to the free base form for recycling to the extraction phase.

A spray drying step is the only further operation required to produce crystals of calcium acetate, if there is a market for this as a de-icing salt[6].

The solution of calcium acetate can be converted to acetic acid by mixing with sulphuric acid and filtering off the calcium sulphate precipitated. The solution of 10-20% acetic acid could then be concentrated by conventional solvent extraction and/or distillation.

2 PREVIOUS WORK

Much previous work has been aimed at developing a process for recovering acetic acid from dilute solution using solvent extraction based on long-chain amines and other functional groups. The use of a lime slurry to strip the extracted acid was attempted, but led to severe phase separation problems[6,7].

Although most extractants have extremely low solubilities, the extractant and/or diluent lost in the aqueous raffinate can contaminate the environment. Althouse and

Tavlarides[7] included the toxicity factor in their assessment of 50 possible extractants as an "unresolved problem".

The logical development of the work of Althouse and Tavlarides[7] is to use a resin for the recovery of acetic acid, which is then eluted with lime slurry. This is the basis of work by Reisinger and King[8]. There are considerable similarities between some of their results and this work. It is significant that work by King and others on solvent extraction over a decade has led them to examine ion exchange/adsorption as an alternative method of recovery.

A further possible advantage in the use of an adsorbent, such as an ion exchange resin, over a liquid extractant would be where effluent streams contain suspended solids. Solvent extraction plants normally require complete removal of solids from feed streams, using filtration followed by clarification, to reduce loss of organic phase[9].

Effluent streams containing significant amounts of suspended solids would be more easily handled using ion exchange, provided that a fluidised or agitated bed contactor were used. Such techniques have been used in the recovery of uranium from ore slurries, even up to 50% suspended solids[9,10].

3 EXPERIMENTAL PROCEDURES AND RESULTS

3.1 Preliminary Tests of Resins

The ion exchange resins considered were weak base (Duolite A-368 and Duolite A-378), and intermediate base (Duolite A-375), from the Rohm and Haas range.

Samples of resin were each recycled through eight cycles of adsorption of acetic acid and elution with lime slurry[11]. Results showed that 93% of the acetic acid taken up from solution by Duolite A-375 was recovered as calcium acetate crystals. Comparative values for Duolite A-368 were 78% and for Duolite A-378, 75%. Values for the elution of acetic acid from Amberlite IRA-35 at 60°C with dolomitic lime slurry were obtained by Reisinger and King[8], which compare well with those reported here for Duolite A-375.

These recovery values showed the superiority of intermediate base resin (Duolite A-375) over the two weak base resins, which were not considered further.

3.2 Preliminary Tests on an Actual Effluent using Duolite A-375 Resin

The effect of a petrochemical effluent on the total ion-exchange capacity of Duolite A-375 was tested using a sample which contained a mixture of acetic, propionic, butyric and valeric acids and non-acids. The total capacity of Duolite A-375 was first found to be 1.4 equiv/L, slightly lower than the nominal value of 1.6 equiv/L. The total initial strength of the acids was 0.21 M, and acetic acid comprised about 70% of the acids.

A sample of 100 mL of Duolite A-375 resin was put through eight cycles of adsorption of acids from effluent and elution with lime. The various acids were recovered in the same proportions as those occurring in the effluent.

These tests were followed by continuous tumbling in a large excess of effluent for four weeks, giving about 1000 hours' total exposure to effluent. After elution finally with a strong base, (NaOH), the total capacity of the resin was measured with a strong acid as 1.36 eq/L, which was still 97% of the value measured initially for that resin.

3.3   Conversion of a Solution of Calcium Acetate to Acetic Acid

The conversion of the solution of calcium acetate eluted from the resin with lime to a solution of acetic acid was simply achieved by mixing with 98% sulphuric acid. The overall reaction of sulphuric acid with calcium acetate gives a precipitate of gypsum and a solution of acetic acid.

The precipitate of white gypsum crystals formed was readily filtered on a Buchner filter and was washed with distilled water to recover acetic acid. The moisture content of the final filter cake was about 38%, and this accounted for a loss of about 5% of acetic acid.

The concentrations of acetic acid solutions finally produced, incorporating the cake washings, were in the range 10-25%. The calcium content of this solution was measured as 300 ppm, while sulphate was estimated at 200 ppm from the solubility product of calcium sulphate.

3.4   Equilibrium Data for Acetic Acid on Duolite A-375 Resin

Data on the equilibria between acetic acid and resin were obtained using a solution of 0.15 M analytical quality acetic acid. Various volumes of this feed solution were tumbled for 24 hours in sealed glass jars with samples of 125 mL of Duolite A-375 resin in the free-base form at 25°C. Data plotted in Figure 1 showed strong adsorption of acetic acid, until a plateau was reached at a loading of about 1.2 equivalents per liter of resin.

Figure 1   *Equilibrium distribution of acetic acid between Duolite A-375 and dilute solution*

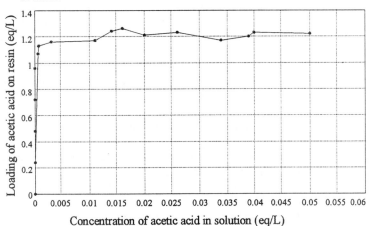

3.5   Kinetics of Adsorption of Acetic Acid on Duolite A-375 Resin

Some kinetic studies were done on the adsorption of acetic acid to help design a fluidised bed contactor.

Samples of 50 mL of resin in the free-base form were tumbled in a sealed glass jar with 500 mL of 0.15 M acetic acid, as carried out previously, but the reaction was stopped at various times and the resin filtered off. The filtrate was titrated to determine the loading of acid on the resin at the time of stoppage. The resin was then eluted and used again for a further measurement. The reactions took place at 25°C. Equilibrium loading was reached in about one hour, as shown in Figure 2.

Figure 2 *Kinetics for adsorption of acetic acid with Duolite A-375 from dilute solution*

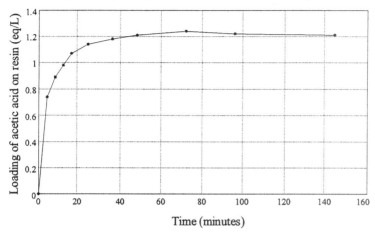

3.6 Kinetics of Elution of Acetic Acid from Duolite A-375 Resin with Lime Slurry

Experiments similar to the studies of adsorption were carried out using 35 mL samples of resin fully loaded with acetic acid. These were added to glass jars together with 500 mL of distilled water and a 10% excess of powdered lime. The jars were sealed and tumbled at 25°C for various times, after which the solution was filtered off and titrated for acetic acid, as described above.

A reaction time of 20 minutes seems adequate to elute most of the acid, but Figure 3 also shows that about 5% does not appear to come off. This did not affect the performance of the resin in any cumulative sense.

Figure 3 *Kinetics for elution of acetic acid from Duolite A-375 with calcium hydroxide slurry*

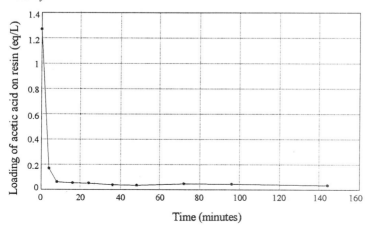

4 DESIGN OF PROPOSED PROCESS

Data obtained on solutions of pure acetic acid were used to suggest a notional process for recovery from a dilute waste. The block diagram of this process is given in Figure 4, which also includes a mass balance for 1000 kg/h of acetic acid contained in a dilute feed solution. The recovery of acid from a 1% solution was taken as 85% and the loss of acid in the final gypsum filter cake as 5%. An excess of 10% of the stoichiometric equivalent of lime was used to ensure complete elution of acid from the resin.

Figure 4 *Process to recover acetic acid from a dilute effluent*

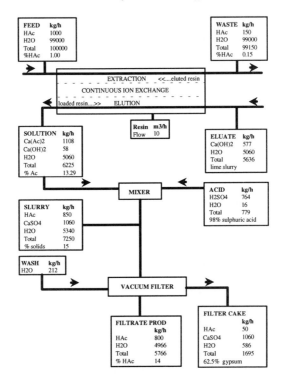

Marais[11], suggests a resin inventory of 45 m^3 for a feed flow of 180 m^3/h, or a resin duty of 4 m^3/h per m^3 of resin inventory. This value is comparable to the resin duty of 4.6 m^3/h per m^3 of resin used by the very large Rössing Uranium plant based on fluidised beds[12], and is considered realistic.

5 COSTING OF PROPOSED PROCESS

An approximate cost analysis of the process for South African conditions has been done by Marais[11]. The main points are summarised in Table 1 for a production rate of 800 kg of acetic acid per hour, or 5760 tons per year.

Table 1 *Cash Flow Estimates for Process in Figure 4*

| | | Capital cost of plant : | R5.33million (1993 values) | |
| | | R1=0.182 Pound sterling | 1 Pound sterling = R5.5 | |

Description	Price (R/ton)	Turnover (ton/a)	Amount (R/a)	AMOUNT (Rmil/a)
Acetic acid(glacial)	2000	5760		11.52
Gypsum	150	9649		1.45
TOTAL SALES				12.97
Fixed costs				
Maintenance				2.67
Labour			1200000	1.20
Local tax, insur.				1.07
Depreciation				0.53
Variable costs				
Lime	200	4151	830280	0.83
Sulphuric acid	200	5498	1099560	1.10
Resin make-up	25000	1	32397	0.03
Misc. op.costs contingency				1.00
TOTAL COSTS				8.43
EARNINGS				4.54
TAX(45%)				2.04
PROFIT				2.49
note 1. costs of capital included in 'profit'				
note 2. cost of concentrating acid not included				

The most significant item is the value of acetic acid sold, which depends on its market price. Reducing the market price from R2000 to R1500 per ton reduces the earnings from R4.54million to R1.66million. The presence of higher acids such as propionic, butyric and valeric, which command a better price than acetic acid, could be a useful supplement to the sales.

The main operating costs are for lime and sulphuric acid, both of which are required in approximately stoichiometric quantities. They are both readily available at low cost in the industrial areas of South Africa.

Labour costs are expected to be low, since the plant items are simple and the process would be operated on a continuous basis. The process is a by-product recovery and effluent reduction operation, and as such would be attached to a large factory with an established management, laboratory, personnel services and existing physical infrastructure.

Gypsum is produced by the process and must be either sold or disposed of. However the gypsum would not contain heavy metals and should be useful as a raw material. There is a market in South Africa for gypsum as a cheap soil additive to provide sulphur, where this is deficient and normal solid fertilisers are not used.

An example of another market for gypsum for the manufacture of plaster board is the contract between British Gypsum and National Power for the sale of one million tons of gypsum per year, produced by scrubbing sulphur dioxide from the flue gas from Drax power station with lime slurry[13].

The costing data incorporates contingency amounts of R1milllion as operating cost and as part of the generous R2.67milllion allowance for maintenance. A comparatively low sales value of R150/ton was used for gypsum, which makes a small contribution to the income of the plant.

6 DISCUSSION

It is clear from Figure 1 that not all ion exchange sites on A-375 resin were loaded with acetic acid. This resin, containing mostly tertiary amine active sites, had only 80% of its total capacity utilised. The reactions with the ion exchange resin here are essentially protonation and de-protonation reactions rather than ion exchange reactions.

Preliminary measurements of the adsorption of acids from a petrochemical effluent and their elution with lime, followed by a month's continual exposure to the effluent revealed no serious effects on the capacity of the resin. This shows that the process can also work with a real effluent, within the limits of the test. A longer trial operation would be useful.

All the homologues of acetic acid were eluted together using lime. Subsequent dehydration and separation of specific acids using distillation should prove straightforward.

The kinetics of both the adsorption and elution reactions are relatively fast. Continuous processes to handle both these parts of the recovery process can readily be designed for large-scale operation.

The handling and dewatering of gypsum slurry has been demonstrated on a very large scale at the Drax power station[13], where centrifuges are used to produce a gypsum cake with only 8% moisture. Use of this separation technique would improve the recovery values of acetic acid assumed in the flowsheet of Figure 4.

The estimates of cash flow give an indication of the importance of the various parts of the process based on South African costs. They also show that the operation could be profitable, even making conservative assumptions. Operation of such a process on a pilot plant scale would provide much more accurate data for the design and economic assessment of a full-scale plant.

7 CONCLUSIONS

Medium base ion exchange resins are the basis of a potentially economic technique for selectively recovering acetic acid and its homologues from dilute aqueous solutions below 1%.

Further studies on a pilot-plant scale are needed to verify and refine the flowsheet and the assumptions on which the cost estimates were based.

The ability to handle feeds containing suspended solids and the absence of toxic solvents in effluent streams offer advantages over processes based on solvent extraction.

Acknowledgements

The authors wish to thank Sastech Ltd., a member of the Sasol group of companies, for financial support of this project.

Figures 1, 2 and 3 are reprinted with permission from Reference 4. Copyright 1995 American Chemical Society.

References

1. P. M. Kohn, "Process Technology and Flowsheets", McGraw-Hill, New York, 1979, 275.
2. N. L. Ricker, E. F. Pittman, and C. J. King, *J. Sep. Process Technol.*, 1980, **1(2)**, 23.
3. R. W. Helsel, *Chem. Eng. Prog.*, 1977, **73**, 55.
4. F. L. D. Cloete and A. P. Marais, *Ind. Eng. Chem. Res.*, 1995, **34**, 2464.
5. B. A. Hendry, *Water Sci. and Technol.*, 1982, **14**, no. 6/7 Part 2, 535.
6. D. L. Wise and D. Augenstein, *Solar Energy*, 1988, **41(5)**, 453.
7. J. W. Althouse and L. L. Tavlarides, *Ind. Eng. Chem. Res.*, 1992, **31**, 1971.
8. H. Reisinger and C. J. King, *Ind. Eng. Chem. Res.*, 1995, **34**, 845.
9. R. C. Merritt, "The Extractive Metallurgy of Uranium", Colorado School of Mines, Golden, 1971, Chap. 6.
10. F. L. D. Cloete, *J. S. Afr. Inst. Min. Met.*, 1981, **81**, 66.
11. A. P. Marais, M. Ing. Thesis, Department of Chemical Engineering, University of Stellenbosch, Stellenbosch, South Africa, 1994.
12. F. L. D. Cloete, "Ion Exchange Technology", D. Naden, M. Streat, eds., Ellis Horwood Ltd., Chichester, Society of Chemical Industry, London, p. 661.
13. M. Bloxham, *The Chemical Engineer*, 1995, **587**, 32.

CONTINUOUS ION EXCHANGE - SELECTIVE SORPTION OF METAL IONS

C. H. Byers and D. F. Williams

Chemical Technology Division
Oak Ridge National Laboratory
Oak Ridge, TN 37831-6268
USA

1 INTRODUCTION

Continuous and semi-continuous ion exchange have been proposed and implemented over the past 40 years in an effort to overcome the shortcomings of fixed-bed operation for constant-duty sorption. These include need for multiple columns, complicated piping and operation schedules, and tankage for blending, analyzing, and equalizing batch-to-batch variations. Fixed beds are often grossly oversized for the separation requirement in order to meet cycle times of the remainder of the process. Fixed beds remain a mainstay of constant-duty sorption because of their perceived simplicity, reliability, and ability to resolve complex multicomponent mixtures. The various continuous concepts have achieved improvements in throughput, resin utilization, and ability to handle suspended solids[1,2], but this progress was gained at the expense of separative resolution. Until recently, high-purity separation of multicomponent streams on a continuous basis could be achieved only by using a network of fixed beds. In this study we show that a recently developed continuous ion exchanger (CIEx) based on same concept as the continuous annular chromatograph (CAC) can be applied to the high-purity separation of multicomponent streams in a more efficient manner.

The use of a slowly rotating annular bed, with fixed feed and eluent positions, to continuously separate components as a function of effluent angular position was first proposed by Martin[3], later advocated by Giddings[4], but was not realized as a practical separation technique until demonstration of the CAC by Scott[5]. The CAC concept which has been described many times[6], depends on stationary feed and eluent stations in a slowly rotating annular bed. In a chromatographic multicomponent separation each component forms a distinct, unchanging helical sorption band on the annular surface. Rotation transforms the unsteady-state fixed-bed process, in which a sorption wave travels down the column, into a steady-state process characterized by a standing sorption wave. As is shown in Figure 1a, one observes that only the annular bed moves – the feed, eluent, effluent distribution, and sorption wave remain stationary. Since 1976 the CAC has been applied to a variety wide variety of separations, including fractionation of metal[7,8], sugar[9], protein[10], and amino acid mixtures[11]. These studies and others have established that almost any chromatographic separation can be run truly continuously on the CAC.

The leap to continuous annular sorption studies requires the reversal of the roles of the feed and eluent: i.e., flood the column with feed and use a minimum flow of eluent to separate the sorbed components (Figure 1b). This mode of operation, which has broad

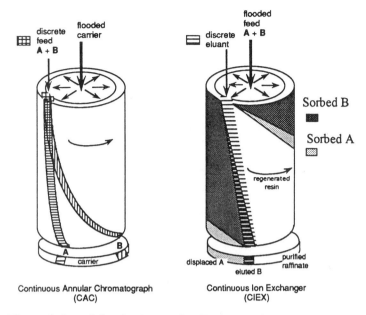

Continuous Annular Chromatograph
(CAC)

Continuous Ion Exchanger
(CIEX)

Figure 1 *The evolution of the Continuous Ion Exchanger from the continuous annular chromatograph idea*

applicability to sorption and ion exchange processes is the focus of this investigation and we distinguish it from chromatographic applications by referring to continuous ion exchange, or *CIEx*.

In establishing the concept of the CIEx two preliminary studies were undertaken whose objective was to translate column technology to continuous annular application. The two investigations are, a) the recovery of lanthanides from a dilute nitric acid solution, and b) the simultaneous isolation of lead from an aqueous metal mixture and the recovery and concentration of the lead-free metal product.

2 THE LANTHANIDE EXPERIMENTAL PROGRAM

In the first case, perhaps more significant than the obvious application of rare-earth recovery for the specialty materials industry is that the separation of lanthanides by ion exchange as a model for a host of radiochemical clean-up separations. The lanthanides and associated rare earths comprise a considerable fraction of the fission-product activity and also serve as excellent predictors of the behavior of the tripositive actinides[12]. The focus of this study is the recovery of semidilute levels of lanthanides as a group from dilute nitric acid streams, without the use of complexing agents. Here the chemical basis for separation is straightforward – the Ln^{3+} species have a relatively high ionic charge and are most strongly held by the negatively charged resin sites. However a gap in the ion-exchange literature still exists, because most of the published studies are restricted to dilute levels of lanthanide[13]. Although the few investigations of more concentrated systems did establish a Langmuirian sorption pattern for the rare earths[14], they are not directly applicable to this study because of differences in acid type and concentration.

Therefore batch shake tests were conducted to fill this gap in the sorption data and fixed-bed tests were conducted to establish a benchmark for comparison with the CIEx test results.

2.1 Experimental

Except for the Nd/Pr lanthanide mixture, all chemicals were derived from reagent grade sources of purity greater than 99%. The lanthanide nitrate (25% Pr, 75% Nd), part of a commercial sample was analyzed and shown to be substantially free of metallic contaminants (>99% pure). In all experiments, a single lot of narrow size fraction (37-45mm wet diameter) strong-acid cation resin (Dowex 50W-X8) was used. Absorption spectrophotometry was used to measure Nd (574, 740nm), Pr (444nm), Cu (710nm), and nitrate (302nm) concentrations. The lanthanide absorption peaks are distinct and very narrow. The broader nitrate and copper peaks do not overlap significantly; so spectral deconvolution is not necessary. Both batch and flow-cell absorption measurements were automated using a PC linked to a HP8452 UV- visible diode array spectrophotometer. For both fixed-bed and CIEx column tests elution of sorbate was achieved with the optimal nitric acid concentration of $5\underline{M}$[15], and at a superficial velocity between 2 and 3 cm/min.

Experimentation began with batch equilibration tests. Various weights of the Dowex in acid form were mixed with 50 mL of 10 mM $Nd(NO_3)_3$ in 0.1 M HNO_3. Equilibrium was typically achieved after overnight agitation in a shaker bath at ambient conditions (~25°C). The distribution coefficient at the equilibrium neodymium concentration was calculated in the usual manner[6]. The dimensionless coefficient is the product of the weight-based distribution coefficient and the measured bulk density of the resin (ρ_B = 0.53 g dry resin per milliliter of wet resin bed).

The fixed-bed test consisted of measurement of the solute breakthrough curve during loading of the column to saturation and measurement of the concentration profile in the subsequent elution step. A Pharmacia XK16140 chromatography column (1.6-cm ID), and an Alltech 325 positive displacement metering pump provided an accurate and stable 3 cm/min superficial throughout the experiments. The throughput (50% breakthrough in bed volumes) served as the measure of the dimensionless distribution coefficient. The slope of the curve at 50% breakthrough is related to the controlling mass transfer coefficient, k. To a linear first approximation, Glueckauf,[16] predicts that

$$k \sim N \text{ (ve/L)} \quad \text{where} \quad N = \frac{V_{0.5} \, V_{0.159}}{\left(V_{0.5} \, V_{0.159}\right)^2} \quad \text{and} \quad V_X = \text{throughput at } \{C/C_0 = x\} \tag{1}$$

and v is the superficial velocity, e is the void fraction, and L is the column length.).

Finally CIEx tests were undertaken. The elution profile (i.e., the angular distribution of lanthanide exiting the annular column) is the critical CIEx performance parameter and the primary experimental measurement. This profile is obtained using a spectrophotometric flow cell by continuously withdrawing a side stream from one of the effluent nozzles as it rotates through the high-acid elution arc. The detailed construction of the annular bed used for these experiments has been described previously[17]. Resin is confined by a 3.25 x 3.5 in. annulus and rotated at a slow uniform rate during a particular experiment. The effluent is collected from a set of 90 tubes that are equally distributed below the annular resin support. In addition to the parameters of feed condition, bed length, and rotation rate, one experiment was conducted with a buffer stream to prevent any mixing between the feed and eluent. Steady operation of the CIEx column is maintained by the constant feed (30 mL/min) and eluent (3 mL/min) flows provided by two Alltech 325 pumps.

2.2 Modeling

To achieve the goal of predicting the steady-state performance of CIEx separations, the defining transport equations must be supported by the attendant equilibrium and kinetic models and a practical solution scheme devised. The diffusive transport equations for this case have recently been considered in detail by Byers and Williams[6]. As with a number of studies they showed that one may directly relate time and angular position by applying the transformation[18] $\theta = \omega \cdot t$. This is the mathematical representation of the analogy developed in Figure 1 (i.e., the time evolution of fixed-bed operation is mapped onto the steady-state angular distribution of continuous annular sorption).

Since the elution profile governs both the separative purity and the eluent consumption, it is the primary focus of our predictions. The boundary conditions for determining the elution profile from a fully loaded bed, with eluent arc commencing at $\theta = 0$, are:

$$Z = 0, Q > 0: \qquad C - Pe \frac{\partial C}{\partial Z} = \left\langle \begin{array}{l} 0 \text{ for rare earths / Cu} \\ \text{ratio of eluent / feed acid for } H^+ \end{array} \right\rangle \tag{2}$$

$$Z = 1, \text{ all } Q: \qquad \frac{\partial C}{\partial Z} = 0 \tag{3}$$

$$Q = 0, \ 0 < Z < 1 \qquad C = 1, Q = 1 \tag{4}$$

Except for the equilibrium relationship, all the parameters implicit in the diffusive transport are readily estimated. An estimate of 8.7×10^{-4} cm²/s for the dispersion coefficient is obtained from standard correlations[9], and a value of 12.3 s^{-1} for the mass transfer coefficient follows from application of equation 2 to the fixed-bed breakthrough curve (Figure 2).

At trace levels of rare earth in the presence of monovalent acid the weight-based distribution coefficient is approximated by

$$D\,[\text{mL / g}] = 223.6\,[\text{Acid}]^{-2.6} \quad \text{for } [\text{Acid}] < 3\,\underline{M} \tag{5}$$

$$D\,[\text{mL / g}] = 23.07 - 3.499\,[\text{Acid}] - 17.95\,[\text{Acid}]^2 + 0.049\,[\text{Acid}]^3 \text{ for } [\text{Acid}] > 3\,\underline{M} \tag{6}$$

For nontrace rare earth levels, a Langmuir fit describes the rare-earth sorption

$$Q^* = \left\{ \frac{a\,c}{1 + b\,c} \right\} \frac{\rho_b}{q_0} \tag{7}$$

where $a = D$ {trace}, $b = D$ {trace}$/q_\infty = D/4.7$ meq/g, $c = $ rare earth conc. [meq/mL]
$Q^* = $ dimensionless equilibrium sorption referenced to feed condition $= q^*/q_0$
The finite-difference code GEARB [19] was used to solve the above system of partial differential equations by the standard "method of lines" [20].

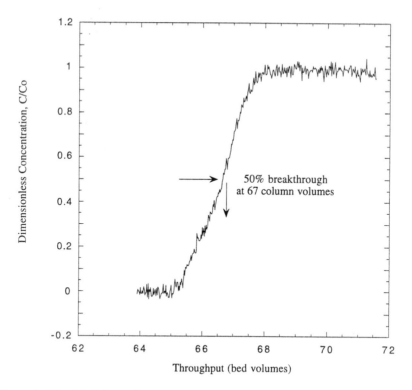

Figure 2 *Fixed-bed throughput curve*

2.3 Lanthanide Results

Distribution coefficients determined from shake tests and the fixed-bed trial are summarized in Equations 5 and 6. The sorption associated with these measurements is also in close agreement with the values predicted by equation 7. The location of the 50% breakthrough point for the fixed-bed trial (Figure 2) corresponds to a distribution coefficient of 126 mL/g, and the slope of the breakthrough curve is consistent with a value of 12.3 s^{-1} for the rate parameter k (Eq. 1).

The elution profile from the fixed-bed trial is an important benchmark for evaluating both CIEx performance and the modeling results. Even for a relatively concentrated (5 $\underline{M}$) acid eluent, the effluent profile (Figure 3) exhibits "tailing" and required four bed volumes to fully strip the sorbed lanthanide. The computed profile displayed in Figure 6 was derived from the finite-difference solution of equations 5 and 6 at conditions analogous to the fixed-bed test. Except for bed length all of the modeling parameters were set to match the actual experimental conditions. It was necessary to select a shorter column (L = 2 cm) because of the need to maintain a fine spatial grid for the purposes of computational accuracy. Computed results converged after a reduction in the grid spacing to 0.02 cm. The good agreement between the two profiles for different bed lengths is due to the nearly constant pattern elution profile of the favorably sorbed lanthanide species.

The effluent-concentration vs. collection-angle CIEx measurements correspond to the effluent history from a fixed-bed trial. Formal correspondence between the results

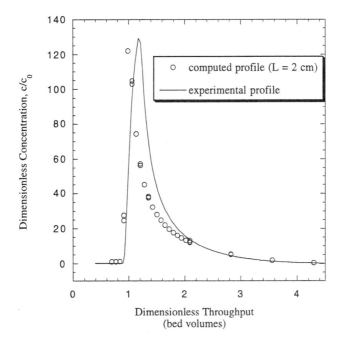

Figure 3 *Breakthrough curve for lanthanide bed experiment*

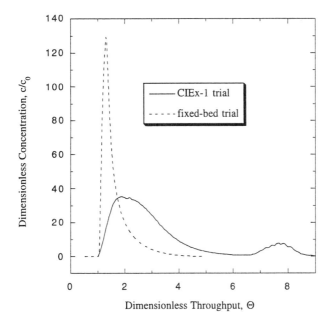

Figure 4 *Comparison of CIEx with fixed bed elution*

from these two types of experiments is achieved if plots are made in terms of the appropriate dimensionless variables. Figure 4 compares the dimensionless elution profile for the fixed bed with that of the CIEx-1 trial. Both curves have the same shape, but the fixed-bed profile is considerably sharper and the CIEx profile contains a small secondary peak at extended throughput. Repeated trials under the CIEx-1 conditions showed precisely the same features. Initially it was postulated that mixing at the interface between feed and elution streams might cause transport of lanthanide down the trailing edge of the elution arc and into the region where the secondary peak was observed.

The final CIEx experiment, CIEx-3, demonstrates the continuous separation and recovery of copper and lanthanides from a dilute nitric acid stream. The mode of operation in this case was slightly different because the strong sorption of lanthanides displaced copper from the column, as schematically depicted in Figure 1. The effluent profile of the copper is essentially a breakthrough curve followed by a saturation plateau, while the lanthanide profile is basically the same elution profile observed previously (Figure 5). It is also evident that this trial should have been conducted at a slightly slower rotation rate, since it is desirable to allow the lanthanide band to break through and completely displace the copper from the column, thereby freeing the elution arc from the presence of any copper and achieving complete separation of copper from lanthanide.

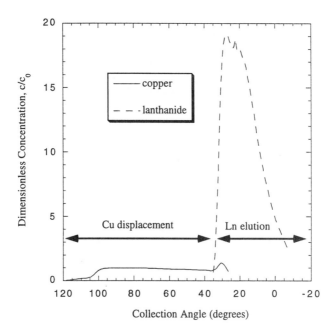

Figure 5 *Separation of copper and lanthanides*

2.4 Discussion

The preceding results establish the concept of the CIEx but leave several issues for further study. Let us focus upon the resolution (sharpness) of the elution profile as a measure of performance, since elution or regeneration costs dominate the economics of most sorption-based separations. Figure 4 shows the fixed-bed profile to be much sharper than the CIEx profile, but this must be viewed in context. First, the fixed bed used has roughly one-tenth the cross-section of the CIEx annular device with carefully fashioned low-volume inlet and outlet piping, thus minimizing external band-broadening influences. Secondly, the effect of discrete collection of the product from the CIEx; each of the 90 collection ports below the resin support acts as a mixing chamber that broadens the effluent profile. It can be shown that about half of the CIEx broadening can be attributed to this latter factor. For larger units, this is less of a problem because of the better angular resolution (<1°) achieved by placing more collection ports in the extended circumference at the bottom of the column. Previous CAC studies have not identified a basic parameter or mechanism related to transport in the resin bed that would argue for superior performance in either device. However, there are instances in which an annular bed may offer significant advantages. For cases where separative performance is seen to degrade rapidly as the fixed-bed diameter increases, the use of an annular bed provides an equivalent flow area at an effectively smaller bed diameter (Figure 1). Taylor dispersion for the annular geometry is reduced because the velocity profiles develop predominantly in the radial coordinate and the resulting concentration gradients are flattened by transverse diffusion across a shallow radial gap. For those few separations that are sensitive to dispersion (e.g., those involving sharp displacement fronts) this factor may be a significant benefit.

Regardless of how separative performance factors align themselves, sometimes a particular contacting scheme is favored because of the nature of the sorption application. Truly continuous operation eliminates the need for multiple columns and blend tanks, provides for automated operation, and keeps material inventories to a minimum. However stable continuous operation requires a relatively constant flow rate and feed composition. CIEx is distinguished from the other continuous contacting schemes by its capability of performing high-resolution multicomponent separations. A major benefit is that the mechanical attrition of resin associated with the other continuous moving-bed contactors is eliminated by CIEx.

Previous work with CAC units have extended the scope of operation from laboratory-scale to large preparative operations[11]. The barriers to CIEx scale-up are practical rather than conceptual. The model developed in this paper demonstrates that prediction of CIEx performance can be accomplished within the established fixed-bed framework. CAC units as large as 2 ft in diameter that use as much as 75% of the available area as annular bed have been used successfully for chromatographic separations. It should also be possible to configure a nested annular bed in order to increase the bed cross section and yet retain a thin annular gap. There is no barrier to constructing extremely large rotating annular beds. Probably the most important need is for an automated effluent collection system. In the ideal system product would not be collected in discrete angular ports; instead cuts could be made at any position automatically with the aid of on-line instrumentation. Previous attempts at improving product collection disturbed the symmetry of flow profiles in the bed and caused mixing of the separated bands. The design of a more flexible automated collection system will require some ingenuity and experimentation.

3 LEAD-NICKEL CIEx EXPERIMENTS

Wastewater generated by metal plating and circuit manufacturing facilities contain high levels of toxic heavy metals such as lead, chromium, nickel, as well as other metals, whose recovery is a vital aspect of environmental remediation, as well as providing opportunities for recycle feed materials. Precipitation and adsorption are the most commonly used methods to achieve these goals[21]. Recently Lee and Hong[22] have offered an interesting technique for approaching this problem in which a combined ion exchange adsorption and precipitation process is used in a single ion exchange column. Our goal is to upgrade the proposed process to the CIEx and make it continuous.

The ion exchange-precipitation combined process, the separation and recovery of lead and a mixture of other metals is achieved. For analytical ease of operation, nickel represents the mixture of metals which might be present in a normal process. A solution containing Pb^{2+} and Ni^{2+} ions is fed to a cation exchange resin of sodium or acid form. The Na^+ ions in the resin are exchanged with Pb^{2+} and Ni^{2+} in the solution because the separation factors for Pb^{2+} and Ni^{2+} are much greater than that for Na^+. The solution depleted of Pb^{2+} and Ni^{2+} ions passes through the column whose first product is a dilute NaCl solution. Since Pb^{2+} will displace Ni^{2+} and $NiCl_2$ will pass out of the column for some time period before we ultimately get breakthrough of both metals. Optimally one would cease feeding the column just before Ni^{2+} begins to break through. At that point a solution containing NaCl at high concentration is added to the resin bed causing the displacement of the adsorbed Pb^{2+} and Ni^{2+} at high concentration. The released ions will interact with high concentration Cl^- ions in the solution. Because the solubility product of $PbCl_2$ is much smaller than that of $NiCl_2$, the Pb^{2+} ions precipitate out as $PbCl_2$ and are caught in the resin matrix, while Ni^{2+} ions stay in the solution and passes from the bed as a high-concentration product solution. Next, a sodium acetate (NaAc) solution at high concentration is fed to the column to dissolve $PbCl_2$. The concentrated lead acetate is recovered in the effluent stream. Simultaneous separation and concentration are achieved. This concept has been proposed previously in connection with desalination by Popper et al[23], Page et al[24] and a number of other workers, before the work of Lee and Hong [22] on toxic metal wastes.

Figure 6 shows a conceptual drawing of the operation of a CIEx with the annulus shown in two dimensions, thus allowing a descriptive view of each portion of the operation. Here we have shown our experimental situation in which chromium and lead are separated. The lead displaces chromium, so the operation can cover a feed segment that allows the lead to load the column. In the adjacent segment NaCl solution is used as an eluent, precipitating the lead as a chloride and, by mass action, replacing the remaining chromium with sodium. A sodium acetate eluent which enters in the sector adjacent to the NaCl will dissolve the precipitate and cause its removal from the column as an acetate. The resin is regenerated to the sodium form or perhaps the acid form.

4 SUMMARY

This study has established CIEx as a viable option for larger-scale separations that require more efficient use of eluent. Two experimental studies showed that the CIEx can be used for a variety of ion exchange applications. This study has also demonstrated that CIEx performance can be predicted within the same basic framework of equations that describe fixed-bed operation. More widespread use of CIEx will require both continued experimental trials to refine equipment design to take full advantage of the automation potential of CIEx.

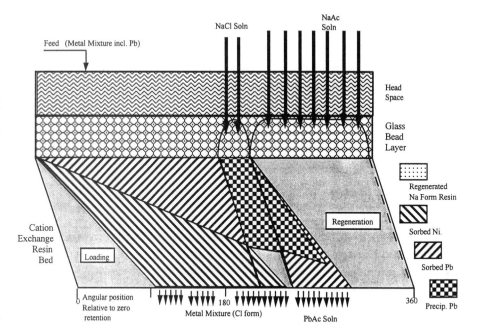

Figure 6 *CIEx operation with metal mixture including lead*

Acknowledgement

The authors would like to thank W. G. Sisson for his advice on set-up and operation of the continuous ion exchanger. The support of the Efficient Separations and Processing Integrated Program (ESPIP) of the Department of Energy for this research is appreciated.

References

1. M. Streat, and D. Naden, "Ion Exchange and Sorption Processes in Hydrometallurgy", Wiley, New York, 1987.

2. M. Streat, "Ion Exchange for Industry", Halsted, New York, 1988.

3. A. J. P. Martin, *Discuss. Faraday Soc.* 1949, **7**, 332.

4. J. C. Giddings, *Anal. Chem.*, 1962, **34**, 37.

5. C. D. Scott, R. D. Spence, W. G. Sisson, *J. Chromatogr.*, 1976, **125**, 381.

6. C. H. Byers and D. F. Williams *I. E. C. Research,* 1995, In Press.

7. J. M. Begovich, and W. G. Sisson, *Hydrometallurgy*, 1983, **10**, 11.

8. J. Wolfgang, C. H. Byers, G. Schhmuckler, and G. Bart, *Sep. Sci. Technol.*, 1995, in press.

9. A. J. Howard, G. Carta, and C. H. Byers, *Ind. Eng. Chem. Res.*, 1988, **27**, 1873.

10. G. F. Bloomingburg and G. Carta, *Chem. Eng. J.,* 1994, **55**, B19.

11. J. P. DeCarli, G. Carta, and C. H. Byers, *Advanced Techniques for Energy-Efficient Industrial-Scale Continuous Chromatography*, Oak Ridge National Laboratory: ORNL/TM-11282, 1989.

12. F. A. Cotton, and G. Wilkinson, "Advanced Inorganic Chemistry", Interscience: New York, 1972, p. 594.
13. F. W. E. Strelow, R. Rethemeyer and C. J. C. Bothma, *Anal. Chem.*, 1965, **37**, 106.
14. E. R. Tompkins, and S. W. Mayer, *J. Am. Chem. Soc.*, 1947, **69**, 2859.
15. F. Nelson, T. Murase, and K. A. Kraus, *J. Chromatogr.*, 1964, **13**, 503.
16. E. Glueckauf, *Trans. Faraday Soc.*, 1955, **51**, 34.
17. J. M. Begovich, C. H. Byers, and W. G. Sisson, *Sep. Sci. Technol.*, 1983, **18**, 1167.
18. P. C. Wankat, *A. I. Ch. E. J.*, 1977, **23**, 860.
19. A. C. Hindmarsh, *GEARB: Solution of Ordinary Differential Equations Having Bounded Jacobeans*, Lawrence Livermore National Laboratory: UCID-30059, 1975.
20. C. A. Silebi and W. E. Schiesser, *Dynamic Modeling of Transport Process Systems*, Academic, New York, 1992, p. 310.
21. J. W. Patterson, *Wastewater Treatment Technology,* Ann Arbor Science, MI, 1975, p. 23.
22. H. Y. Lee and S. K. Hong, *A. I. Ch. E. Journal,* 1995, In Press.
23. K. Popper, R. J. Bouthilet, and V. Slamecka, *Science,* 1963, **141**, 1038.
24. B. W. Page, G. Klein, F. Golden, and T. Vermeulen, *A. I. Ch. E. Symp. Ser., No. 152*, 1975, **71**, 121.

CHROMATOGRAPHIC SEPARATION OF RARE EARTH ELEMENTS BY THE CHELATING RESIN FUNCTIONALIZED WITH NITRILOTRIACETATE OR THE RELATED AMINOCARBOXYLATE LIGAND

T. M. Suzuki, M. Kanesato and T. Yokoyama
Tohoku National Industrial Research Institute
4-2-1, Nigatake, Miyagino-ku, Sendai, Japan

Y. Inoue and H. Kumagai
Yokogawa Analytical Systems Inc.
2-11-19, Nakacho, Musashino-shi, Tokyo, Japan

1 INTRODUCTION

Ion exchange chromatography is one of the important procedures to obtain highly purified rare earth elements (REE) in industry as well as in the analysis of individual REE.[1,2] Typically, a cation exchanger is used as the column stationary phase with the combination of a complexing agent as a mobile phase. The separation efficiency in this system is mainly attributed to the selectivity of the complexing agent in the mobile phase. On the contrary, when we use a selective ion exchanger as the column stationary phase, simple acid may be used as eluent in place of a complexing agent. Chelating polymer resin is one of the attractive candidates for the selective stationary phase in the chromatographic separation of REE.

Structure of the NTA and CMA resins

We have prepared the chelating resins having the functional group structurally similar to nitrilotriacetic acid (NTA)[3,4] and diethylenetriaminepentaacetic acid (DTPA).[5,6] The resin matrix is macroreticular type styrene-10%-divinylbenzene copolymer beads. The selectivity of the chelating resin has been estimated by the distribution coefficient (Kd). The present chelating resins were applied to the chromatographic separation of adjacent REE pairs. In addition, NTA-resin based on fine particle polymer matrix (10 μm) was prepared and applied to a rapid ion chromatographic separation of a series of REE.

2 EXPERIMENTAL

2.1 Preparation of the Chelating Resins

Lysine-N^α,N^α-diacetic acid was immobilized onto cross-linked polystyrene beads (32-60 mesh) by a one-step reaction as given in Scheme 1.[3] Nitrogen analysis indicated that the NTA resin contains 1.1 mmol of the ligand per gram resin. In a similar manner, lysine-N^α,N^α-diacetic acid was introduced into the cross-linked glycidylmethacrylate gel beads with a diameter of 10 μm.

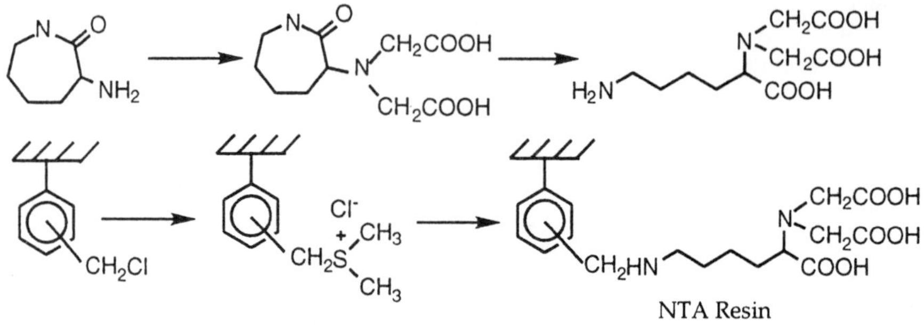

Scheme 1 *Preparation of the NTA resin*

The CMA resin was prepared by N-carboxymethylation of the dien-resin in which diethylenetriamine (dien) is linked to the polystyrene resin uniquely through the imino nitrogen (Scheme 2). Nitrogen analysis indicated that the CMA resin contains 1.65 mmol of the functional group per gram resin.

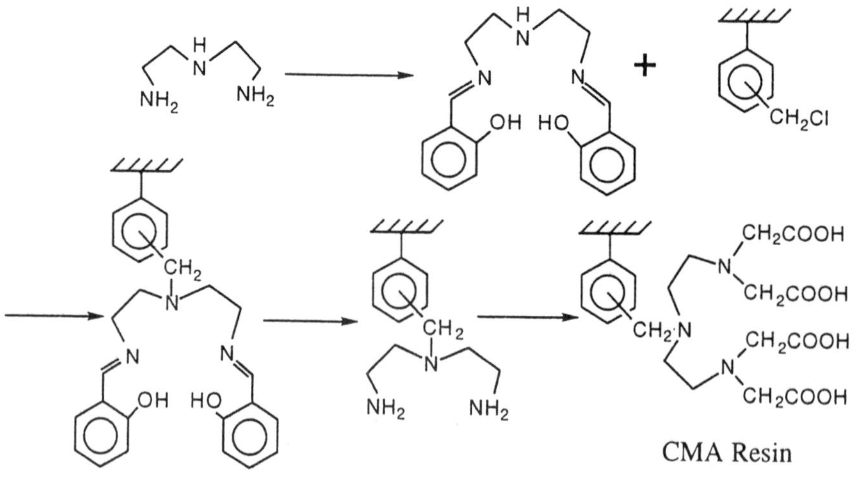

Scheme 2 *Preparation of the CMA resin*

2.2 Distribution Coefficient

The distribution coefficient is defined as, Kd = amount of metal ion retained in one gram of resin / amount of metal ion remaining in 1 mL of solution. A 1 g portion of dry resin was added to a 100 mL of buffered metal ion solution (1 mM). After shaking 7 days at 25 °C, the equilibrium pH and the amount of metal ion remaining in the solution were determined. The Kd value was then calculated.

2.3 Column Separation of REE

A typical procedure is as follows; the chelating resin beads were swollen with water and poured into a jacketed glass column (ϕ 10 X 464 mm). An aqueous solution of mixed REE of equimolar concentration (pH=3) was mounted on the column and the sorbed metal ions were then eluted with an acid solution. The temperature of the column was kept constant by circulating thermostated water inside the jacket. The effluent was fractionated into small portions and the amount of metal ion was determined. The metal ion concentration was determined by a SEIKO ICP-atomic emission spectrometer SPS-1200A.

3 RESULTS AND DISCUSSION

3.1 Distribution of REE between the Chelating Resin and Aqueous Solution

The present chelating resins effectively adsorb a series of REE and the metal ions retained can be released quantitatively from the resin by elution with 0.2 M HCl.
The reaction of the chelating resin with REE can be expressed as follows

$$M^{n+} + RH_m \overset{K}{\rightleftharpoons} RH_{m-n}M + nH^+ \tag{1}$$

where RH_m denotes the resin group.
The equilibrium constant (K) of above reaction can be given by

$$K = \frac{[RH_{m-n}M][H^+]^n}{[M^{n+}][RH_m]} \tag{2}$$

The distribution coefficient (Kd) of the metal ion between the resin phase and the aqueous phase is defined by

$$Kd = \frac{[RH_{m-n}M]}{[M^{n+}]} \tag{3}$$

From equations (2) and (3), the following relationship can be obtained.

$$\log Kd = \log K + \log [RH_m] + n\,pH \tag{4}$$

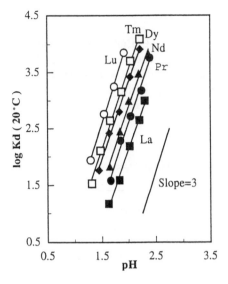

Figure 1 *The log Kd of the NTA resin for several REE as a function of pH*

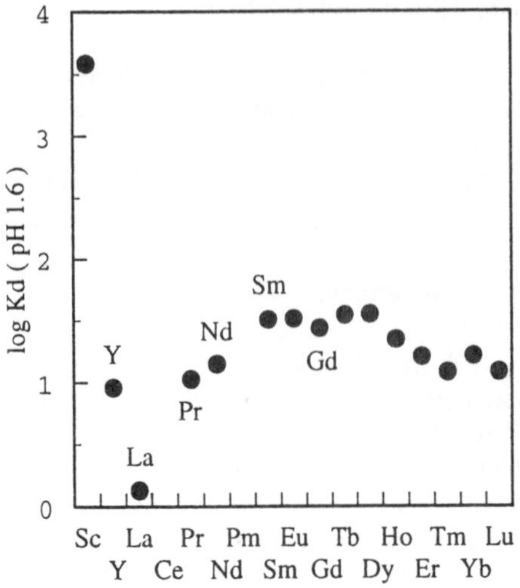

Figure 2 *The plots of Kd of the CMA resing for fifteen REE at pH 1.6*

Equation (4) predicts that the plot of log Kd against pH gives a straight line of slope n under the condition where the amount of functional group is much larger than that of the metal ion. The Kd value was measured as a function of pH under the condition where the resin being excess over metal ion. A linear plot of log Kd vs. pH was obtained with a slope of approximately 3 for the CMA and NTA resins. These observations indicate that the adsorption of a REE ion is accompanied by the release of three protons. Figures 1 and 2 show the log Kd values at the given pH. The Kd value increases with the increase in atomic number for the NTA resin whereas a plateau was observed for the CMA resin. These trends are in good accordance with the stability constant sequence of the NTA and DTPA complexes in solution.[7] For example a linear relationship holds for log Kd at the given pH (pH=1.8) and log K_{ML} of the corresponding homogeneous NTA complexes in solution (Figure 3). In the NTA resin, the ligand is remote from the polymer matrix so that it can behave in a similar manner to that of the ligand in a monomeric system.

3.2 Chromatographic Separation of REE

A chromatographic separation of Pr(III) and Nd(III) was carried out by using the column of the NTA resin with 0.1 M nitric acid as the mobile phase. During elution, the column was kept at 65 °C by circulating thermostated water inside the jacket. The chromatogram is given in Figure 4. The elution order is consistent with the order of Kd value at the given pH i.e., Pr(III) which forms less stable complex with NTA resin eluted much faster than Nd(III). In a similar manner, mutual separation of La/Pr, Y/Gd and Nd/Sm was achieved by the column of the CMA resin with a buffer solution as the mobile phase. The chromatographic conditions were empirically evaluated in terms of resolution

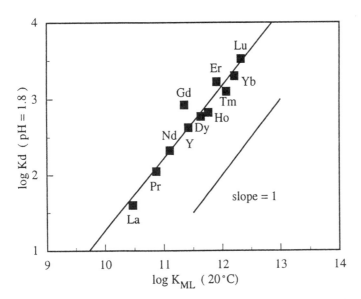

Figure 3 *Plots of log Kd values of the NTA resin (at pH 1.8) against the formation constants (K_{ML}) of rare earth metal complexes of nitrilotriacetic acid*

(Rs), with varying column length, pH of the eluent, the flow rate and the column temperature. The pH of eluent is a most important factor which affects the chromatographic separation. When the pH is too low, the separation was unfavorable due to a small difference in the retention time, whereas the elution at higher pH leads to a significant band broadening. Better resolution was attained with an increased column length, and a reduced flow rate. An apparent improvement has also attained by raising the column temperature from 25 to 45 °C. A typical chromatogram for Y/Gd is shown in Figure 5. A favorable separation has been realized under appreciably high loading using the CMA resin as the stationary phase. Traditional ion exchange systems, where a chelating reagent is contained in the mobile phase, essentially involve a restriction of solubility of the eluent.[8] However, a chelating reagent is not always necessary in the present system, and hence the trouble due to precipitation during column operation can be avoided.

3.3 Ion Chromatography in an Analytical Separation of REE

We have introduced a lysine-N^α,N^α-diacetate into cross-linked glycidylmethacrylate gel of small diameter (ϕ 10 µm). The NTA-gel was packed to a column (ϕ 4.6 X 150 mm) and examined as the stationary phase for the ion chromatographic separation of fourteen kinds of REE. The concentration of nitric acid as the mobile phase was linearly increased from 0.02 M to 0.08 M for 25 min. The REE eluted was detected by the post column derivatization method with Chlorophosphonazo III as a coloring reagent. The chromatogram is shown in Figure 6. A series of REE was favorably separated within 20 min. The order of elution follows the order of the atomic number. Clear separation of Eu

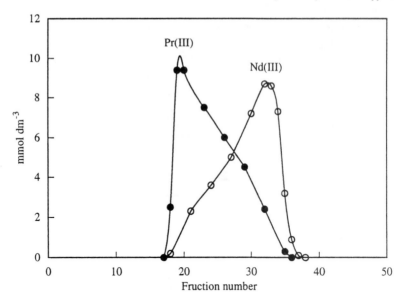

Figure 4 *Chromatogram for the separation of Pr(III) and Nd(III) with NTA resin column. Metal ion : 1 mmol each, Eluent : 0.1 mol dm^{-3} HNO$_3$*

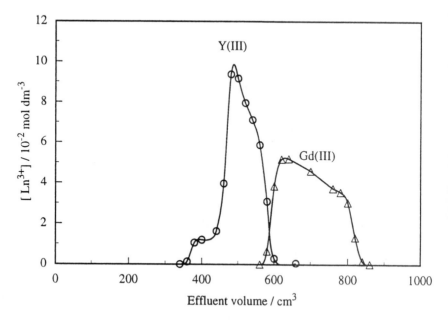

Figure 5 *Separation of Y/Gd with the CMA resin. The pH of the eluent : pH 1.2. Flow rate : 0.25 cm^3 min^{-1}. Column size : ϕ 1.0 X 95 cm. Column temperature : 45 °C. Amount of the each metal ion : 10.0 mmol*

and Gd was not achieved despite modifications of gradient elution condition. The unique feature of the present system is that the mutual separation of REE can be achieved merely by elution with mineral acid which offers a great advantage for the following detection of individual metals. For example, ICP-AES and ICP-MS interfaced with ion chromatography have been applied to simultaneous detection of REE.[9,10] However large amount of organic reagents or salts in the mobile phase may cause spectral interferences and clogging of the ICP torch and/or sampling orifice. In addition, concentrated salt solution tends to influence the nebulization efficiency owing to their high viscosity. In the present system, in contrast, the sample solution does not contain any salts or chelating reagents which can minimize such drawbacks in the ICP-AES or ICP-MS detection.

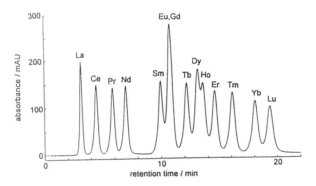

Figure 6 *Ion chromatographic separation of REE on the NTA-resin column. Gradient elution with nitric acid followed by post column colorimetric detection. Column temp. : 40 °C*

References

1. J. E. Powell, "Handbook on the Physics and Chemistry of Rare Earths", K. A. Gschneider, Jr. and L. Eyring, eds., North Holland, 1979, Vol. 3, p. 81.
2. D. T. Gjerde and J. S. Fritz, "Ion Chromatography", 2nd Edition, Hutig, 1986.
3. T. Yokoyama, M. Kanesato, T. Kimura and T. M. Suzuki, *Chem. Lett.*, 1990, 693.
4. T. Yokoyama, S. Asami, K. Kanesato and T. M. Suzuki, *Chem. Lett.*, 1993, 383.
5. T. M. Suzuki, T. Yokoyama. H. Matsunaga and T. Kimura, *Bull. Chem. Soc. Jpn.*, 1986, **59**, 865.
6. M. Kanesato, T. Yokoyama and T. M. Suzuki, *Bull. Chem. Soc. Jpn.*, 1989, **62**, 3451.
7. A. E. Martell and R. M. Smith, "Critical Stability Constants", Prenum Press, New York and London, 1979, Vol. 3, p. 33.
8. J. K. Marsh, *J. Chem. Soc.*, 1957, 978.
9. K. Yoshida and H. Haraguchi, *Anal. Chem.*, 1984, **56**, 2580.
10. K. Kawabata, Y. Kishi, O. Kawaguchi, Y. Watanabe and Y. Inoue, *Anal. Chem.*, 1991, **63**, 2137.

STUDIES ON THE COMPARATIVE PERFORMANCE OF LOW PRESSURE ION-EXCHANGE CHROMATOGRAPHY MEDIA FOR PROTEIN SEPARATION

P. R. Levison, C. Mumford and M. Streater

Whatman International Ltd.
Springfield Mill
James Whatman Way
Maidstone
Kent ME14 2LE

1 INTRODUCTION

Ion-exchange chromatography is routinely carried out in the downstream processing of commercially important bioproducts. Proteins are based on copolymers of amino acids[1] and thus may be regarded as polyions. At a given pH they will bear either a positive or negative charge dependent on their isoelectric point, pI, which is influenced by their primary, secondary, tertiary and quaternary structures. Protein purification by ion-exchange utilises anion or cation-exchangers, functionalised with amines or acids respectively[2] and traditionally attached to polysaccharide matrices including cellulose, agarose and dextran[3]. More recently ion-exchangers based on composite polymers have been introduced[4]. Protein separations can be carried out in either a positive or negative ion-exchange step, where either the target or contaminants are retained respectively[5]. Biochemical applications of process-scale ion-exchange liquid chromatography include the isolation of uridine phosphorylase from *Escherichia coli*[6], prochymosin from *Escherichia coli*[7], L-asparaginase from *Erwinia spp.*[8], monoclonal antibodies[5,9], albumin from human plasma[10], proteins from hen egg-white[11,12], immunoglobulin G from goat serum[13], and DNA modifying enzymes from microbial sources[14].

When establishing an ion-exchange protocol, the protein chromatographer is faced with a myriad of related adsorbents from different vendors bearing similar functional groups, but attached to different matrices using proprietary chemical techniques. In the field of affinity chromatography, there have been several small studies comparing aspects of the base matrix and how it affects the performance in the affinity process[15-18]. However for ion-exchange the influence of the base matrix has not been widely reported. While there are a plethora of publications on ion-exchange protein separations, there has been little attention to screening various ion-exchangers for the same separation. We reported a limited comparison of some cellulose and agarose ion-exchangers[2,19] and more recently have carried out a detailed evaluation into the process-scale purification of hen egg-white proteins on the anion-exchange cellulose Express-Ion Exchanger™ Q and the agarose Q-Sepharose Fast Flow[20].

In the present study we report a comparative screen of various commercially available ion-exchangers obtained from various vendors. The data is descriptive rather than prescriptive but highlights differences in the biochemical application of the products.

2 MATERIALS AND METHODS

2.1 Ion-Exchange Media

Ion-Exchange media were obtained as follows:- DE52, QA52, CM52, SE52, Express-Ion D, Express-Ion Q, Express-Ion C and Express-Ion S were obtained from Whatman International Ltd., Maidstone, UK. DEAE-Sephacel, DEAE-Sepharose CL6B, DEAE-Sepharose Fast Flow, Q-Sepharose Fast Flow, Q-Sepharose HP, CM-Sepharose Fast Flow, S-Sepharose Fast Flow, SP-Sepharose HP, DEAE-Sephadex A-25, DEAE-Sephadex A-50, QAE-Sephadex A-25, QAE-Sephadex A-50, CM-Sephadex C-25, CM-Sephadex C-50, SP-Sephadex C-25 and SP-Sephadex C-50 were obtained from Pharmacia Biotech, St Albans, UK. Matrex DEAE A-200 Cellufine, Matrex DEAE A-500 Cellufine, Matrex DEAE A-800 Cellufine, Matrex CM C-200 Cellufine and Matrex CM C-500 Cellufine were obtained from Amicon, Stonehouse, UK. DEAE Thruput, CM Thruput and Q Thruput were obtained from Sterogene, Carlsbad, USA. Fractogel EMD TMAE-650, Fractogel EMD DEAE-650, Fractogel EMD DMAE-650 and Fractogel EMD SO_3-650 were obtained from Merck, Poole, UK. Poros 50 HQ and Poros 50 HS were obtained from Perseptive Biosystems, Freiburg, Germany.

2.2 Methods

Each ion-exchange medium was handled according to the media manufacturers instructions. The media were cycled in either 0.5M-HCl then 0.5M-NaOH for anion-exchangers, or 0.5M-NaOH then 0.5M-HCl for cation-exchangers. In the case of Fractogel 0.2M-NaOH and 0.2M-HCl was used. Each precycled medium was collected by filtration and used for subsequent testing. Each medium was tested for regains, small-ion capacity, protein binding capacity (bovine serum albumin for anion-exchangers and lysozyme for cation-exchangers) and desorption efficiency and column flow rate. The testing was carried out according to standard QC test methods used by Whatman International Ltd in accord with their ISO9001 accreditation.

3 RESULTS AND DISCUSSION

3.1 Celluloses

Test data on the cellulose-based ion-exchangers are summarised in Table 1. The regains expressed as grams of wet exchanger/gram dry exchanger give a measure of the swelling of the particle and the differences in regain values between ionised and deionised forms indicates dimensional stability as a function of pH. The small-ion capacity expressed as mequivalents/dry gram is a measure of the number of ionizable groups on the exchanger and for a monovalent ion, mequivalents equal mmoles. The protein capacity expressed both in terms of mg protein/dry gram exchanger and mg/mL column volume indicates the practicality of the ion-exchanger but as we have previously reported, protein capacity is variable, dependent on molecular mass and pH[21]. A partial desorption value indicates the binding strength of the protein and this influences chromatographic resolution and selectivity of the medium. Linear velocity was determined at two standard pressures in a laboratory column, but as we have reported[20], linear flow rates reduce ~ 5-fold when you scale-up to a process column (i.e. 45 cm i.d.) due to wall-effects.

The data presented in Table 1 demonstrates significant differences between each of the media tested although they are all based on a cellulosic backbone.

3.2 Agaroses

Test data on the agarose-based ion-exchangers are summarised in Table 2. As was seen for celluloses, there is significant variability between the products.

3.3 Dextrans

Test data on the dextran-based ion-exchangers are summarised in Table 3. It is evident that the large pore size A-50 and C-50 exchangers will give rise to higher protein binding than their A-25 and C-25 equivalents on a dry weight basis. However the significantly higher swelling of these materials (regains) results in a lower working protein capacity compared to their A-25 and C-25 equivalents.

3.4 Composite Polymers

Test data on the composite polymer-based ion-exchangers are summarised in Table 4.

4 CONCLUSIONS

The purpose of this work was not to be prescriptive in terms of recommending particular ion-exchange media. Rather the data presented here, clearly demonstrates that when comparing some 38 different anion and cation-exchange media there are significant performance differences between them. These differences are manifest both when different manufacturers prepare similar chemistries on similar matrices i.e. cellulose or agarose and also when similar chemistries are prepared on differing matrices i.e. dextran versus composite polymers.

Due to the variability of ion-exchangers for a common set of performance tests both physical and functional it is recommended that protein chromatographers rigorously screen media prior to scale-up in order to optimise process throughput and commercial viability of the purification.

References

1. L. Stryer, "Biochemistry" 2nd edition, W. H. Freeman and Company, San Francisco, 1981, p.13.
2. P. R. Levison, "Cellulosics: Materials for Selective Separation and Other Technologies", J. F. Kennedy, G. O. Phillips and P. A. Williams, eds., Ellis-Horwood Ltd., Chichester, 1993, p. 25.
3. E. F. Rossomando, *Methods Enzymol.*, 1990, **182**, 309.
4. L. L. Lloyd and J. F. Kennedy, "Process Scale Liquid Chromatography", G. Subramanian, ed., VCH, Weinheim, 1995, p. 99.

5. P. R. Levison, "Process Scale Liquid Chromatography", G. Subramanian, ed., VCH, Weinheim, 1995, p. 131.
6. K. Weaver, D. Chen, L. Walton, L. Elwell and P. Ray, *Biopharm.*, 1990, July/August, 25.
7. F. A. O. Marston, S. Angal, P. A. Lowe, M. Chan and C. R. Hill, *Biochem. Soc. Trans.*, 1988, **16**, 112.
8. C. R. Goward, G. B. Stevens, I. J. Collins, I. R. Wilkinson and M. D. Scawen, *Enzyme Microb. Technol.*, 1989, **12**, 229.
9. A. Jungbauer, F. Unterluggauer, K. Uhl, A. Buchacher, F. Steindl, D. Pettauer and E. Wenisch, *Biotechnol. Bioeng.*, 1988, **32**, 326.
10. G. K. Sofer and L. E. Nystrom, "Process Chromatography A Practical Guide", Academic Press, San Diego, 1989, p. 61.
11. P. R. Levison, S. E. Badger, D. W. Toome, M. L. Koscielny, L. Lane and E. T. Butts, *J. Chromatogr.*, 1992, **590**, 49.
12. P. R. Levison, S. E. Badger, R. M. H. Jones, D. W. Toome, M. Streater, N. D. Pathirana and S. Wheeler, *J. Chromatogr.*, 1995, **702**, 59.
13. P. R. Levison, M. L. Koscielny and E. T. Butts, *Bioseparation*, 1990, **1**, 59.
14. J. M. Ward, L. J. Wallace, D. Cowan, P. Shadbolt and P. R. Levison, *Anal. Chim. Acta*, 1991, **249**, 195.
15. M. Baeseler, H. -F. Boeden, R. Koelsch and J. Lasch, *J. Chromatogr.*, 1992, **589**, 93.
16. P. Fuglistaller, *J. Immunol. Methods*, 1989, **124**, 171.
17. Y. Kamiya, T. Majima, Y. Sohma, S. Katoh and E. Sada, *J. Ferment. Bioeng.*, 1990, **69**, 298.
18. J. Tharakan, F. Highsmith, D. Clark and W. Drohan, *J. Chromatogr.*, 1992, **595**, 103.
19. P. R. Levison and F. M. Clark, "Pittsburgh Conference", 1989, Atlanta, Abstracts, p. 754.
20. P. R. Levison, R. M. H. Jones, D. W. Toome, S. E. Badger, M. Streater and N. D. Pathirana, *J. Chromatogr.*, 1996, in the press.
21. P. R. Levison, "Preparative and Process-Scale Liquid Chromatography", G. Subramanian, ed., Ellis-Horwood, Chichester, 1991, p. 146.

Table 1 Test Data on Cellulose-based Ion-exchangers

Media Grade	Regains (g/dry g)		Small-ion Capacity (mequiv/dry g)	Protein Capacity* (mg/dry g)		Column Packing Density (dry g/mL)	Protein Capacity (mg/mL)	Flow Rate (cm/h)	
	Ionised	Deionised		absorp.	% desorp.			50 cm H$_2$O/cm	75 cm H$_2$O/cm
DE52	3.05	2.55	0.98	700	37.6	0.18	126	155	221
Express-Ion D	1.92	1.88	0.98	290	28.4	0.23	67	425	614
DEAE-Sephacel	5.40	5.13	1.52	1110	24.9	0.10	111	166	205
DEAE A-200 Cellufine	3.80	2.75	1.03	930	37.9	0.16	149	387	556
DEAE A-500 Cellufine	4.24	4.55	1.34	630	36.4	0.12	76	519	759
DEAE A-800 Cellufine	5.25	4.97	0.97	740	42.0	0.10	74	238	316
QA52	2.72	2.64	1.20	730	32.9	0.22	161	77	259
Express-Ion Q	1.78	1.82	0.89	350	31.9	0.25	87	477	696
CM52	3.50	2.42	1.02	1150	-	0.15	173	147	210
Express-Ion C	1.97	1.78	0.85	660	-	0.24	158	455	643
CM C-200 Cellufine	4.95	4.71	1.19	700	-	0.13	91	152	187
CM-C-500 Cellufine	3.48	2.48	0.62	890	-	0.17	151	705	984
SE52	3.31	-	0.89	1270	65.9	0.18	229	98	127
Express-Ion S	2.28	-	0.92	640	54.3	0.22	141	378	530

*Bovine serum albumin - anion exchangers
Lysozyme - cation exchangers

Table 2 *Test Data on Agarose-based Ion-exchangers*

Media Grade	Regains (g/dry g)		Small-ion Capacity (mequiv/dry g)	Protein Capacity* (mg/dry g)		Column Packing Density (dry g/mL)	Protein Capacity (mg/mL)	Flow Rate (cm/h)	
	Ionised	Deion-ised		absorp.	% desorp.			50 cm H₂O/cm	75 cm H₂O/cm
DEAE-Sepharose CL-6B	7.33	7.40	2.22	1120	6.3	0.07	78	90	112
DEAE-Sepharose Fast Flow	7.15	8.10	1.89	602	35.6	0.08	48	470	677
DEAE Thruput	7.78	6.90	2.02	1100	5.2	0.08	88	513	726
Q-Sepharose Fast Flow	4.67	4.86	1.45	648	33.9	0.12	78	579	835
Q-Sepharose HP	5.89	6.20	1.65	770	26.9	0.09	69	102	141
Q Thruput	8.79	8.15	0.39	690	70.9	0.06	41	668	837
CM Sepharose Fast Flow	8.63	8.57	1.74	1520	-	0.07	106	408	585
CM Thruput	10.50	8.38	1.91	990	-	0.06	59	439	596
S Sepharose Fast Flow	4.45	-	1.40	1095	64.7	0.12	131	499	718
SP Sepharose HP	5.04	-	1.47	710	66.6	0.09	64	121	166

*Bovine serum albumin - anion exchangers
Lysozyme - cation exchangers

Table 3 *Test Data on Dextran-based Ion-exchangers*

Media Grade	Regains (g/dry g)		Small-ion Capacity (mequiv/ dry g)	Protein Capacity* (mg/dry g)		Column Packing Density (dry g/mL)	Protein Capacity (mg/mL)	Flow Rate (cm/h)	
	Ionised	Deion-ised		absorp.	% desorp.			50 cm H₂O/cm	75 cm H₂O/cm
DEAE Sepharose A-25	3.78	3.34	3.36	570	16.3	0.17	97	544	775
DEAE Sephadex A-50	> 10	> 10	3.48	1080	3.5	0.03	32	43	46
QAE Sephadex A-25	3.26	4.58	2.44	280	4.8	0.17	48	179	282
QAE Sephadex A-50	> 10	> 10	2.80	1040	3.4	0.04	42	61	62
CM Sephadex C-25	5.25	1.59	4.35	900	-	0.12	108	462	642
CM Sephadex C-50	> 10	8.71	4.76	860	-	0.02	17	234	237
SP Sephadex C-25	4.26	-	1.87	1722	27.6	0.15	259	571	788
SP Sephadex C-50	> 10	-	1.39	1870	43.9	0.02	37	236	238

*Bovine serum albumin - anion exchangers
Lysozyme - cation exchangers

Table 4 *Test Data on Polymer-based Ion-exchangers*

Media Grade	Regains (g/dry g)		Small-ion Capacity (mequiv/dry g)	Protein Capacity* (mg/dry g)		Column Packing Density (dry g/mL)	Protein Capacity (mg/mL)	Flow Rate (cm/h)	
	Ionised	Deionised		absorp.	% desorp.			50 cm H₂O/cm	75 cm H₂O/cm
Fractogel EMD TMAE-650	2.31	2.37	0.28	410	38.6	0.19	78	509	703
Fractogel EMD DEAE-650	2.32	2.26	0.50	390	47.6	0.21	82	445	621
Fractogel DMAE-650	2.36	2.26	0.56	200	86.4	0.21	42	460	653
Poros 50 HQ	1.66	1.63	0.64	80	94.9	0.24	19	340	505
Fractogel EMD SO₃⁻-650	2.47	-	0.66	560	39.4	0.21	118	347	487
Poros 50 HS	1.18	-	0.41	220	62.6	0.32	70	147	209

*Bovine serum albumin - anion exchangers
Lysozyme - cation exchangers

ADSORPTION KINETICS OF L-PHENYLALANINE ON A GRAFT COPOLYMER FIBRE ION EXCHANGER

A. Kärki and E. Paatero
Laboratory of Industrial Chemistry, Lappeenranta University of Technology
FIN-53851, Lappeenranta, Finland

M. Sundell
Department of Polymer Technology, Åbo Akademi University
FIN-20500, Åbo, Finland

1 INTRODUCTION

Interest to use fibres in ion exchange applications is based on their faster kinetic performance compared to spherical resins. This has been reported among others by Petruzzelli et. al[1,2] who have found the kinetics of the fibrous ion exchanger to be one order of magnitude faster compared to the equivalent spherical material. Furthermore, the fibrous geometry gives a packing configuration that enables higher liquid flow rates and the potential for significantly higher mass transfer rates[3].

In the present work, dynamic adsorption characteristics of grafted polyolefin ion exchange fibres prepared by electron beam pre-irradiation technique are studied for chromatographic applications using a packed bed of differential length for breakthrough measurements. Spherical crosslinked polystyrene-divinylbenzene resins are used as reference materials. The differential bed technique was introduced by Tien and Thodos[4] who studied adsorption kinetics of oxalic acid on a strong basic ion exchanger. The method has several advantages compared to other techniques. In comparison with the batch method, the differential bed simulates the column conditions better in particular in the presence of extra-particle mass transfer resistance. In comparison with a long bed, the experiments in the differential reactor take shorter time and the mathematical treatment is more simple due to the differential reactor hypothesis[4,5].

The use of high energy radiation induced grafting offers a convenient method for the preparation of polymer supported reagents. Grafted ion exchange fibres can be prepared either by pre-irradiation or by mutual grafting techniques. The pre-irradiation method consists of irradiation of the polymer in an inert atmosphere and subsequent contact with the monomer to be grafted. Compared to the mutual grafting technique, this method produces much less homopolymer, which in some cases can be hard to remove from the product. Nevertheless, the majority of studies in this field concern the mutual grafting technique using γ-irradiation[6-9].

The kinetic characterization of the fibres in this work is carried out using L-phenyl-alanine (Phe) as a model substance. This neutral amino acid has a dipolar nature and exists as cationic (Phe+), neutral (Phe±) or anionic (Phe-) form depending on the pH of the solution (isoelectric point at pH 5.48). It is therefore an appropriate substance for observing the penetration differences of neutral and cationic species in the different cation exchangers.

2 EXPERIMENTAL

2.1 Preparation of Graft Copolymers

The ion exchange fibres were prepared from two different polyolefin fibres (Steen 21 μm PP-fibre, Wetekam 100 μm high density PE-fibre) using electron beam pre-irradiation technique. For the grafting, the matrix polymer was exposed to a predetermined radiation dose by passing the polymer through an electron accelerator under nitrogen (<220 ppm O_2). Immediately after this, the polymer was grafted by immersing it in nitrogen purged styrene. The styrene grafted polyolefin fibre was cut in 5 mm long pieces and sulphonated with chlorosulfonic acid. The ion exchange fibres are commercially manufactured under the trade name SMOPEX 101 (SmopTech Oy Ltd., Finland).

Conventional strong acidic PS-DVB resins Dowex 50 WX8 of two size fractions 100 - 200 and 200 - 400 dry mesh (Sigma Chemical Co.) were used as reference.

2.2 Material Properties

The ion exchangers were converted to Na^+-form with 1 M NaOH and washed with distilled water until the pH was below 8. The water content of Na^+-form ion exchangers were determined as follows. First the interstitial water was removed from swollen material by centrifugation at 2500 rpm for 10 minutes. The exchanger was dried at 68 °C and 100 mbar for 16 h. The water content was calculated on the basis of the mass difference. The average diameters of water swollen Na^+-form ion exchangers were determined from microscope photographs. Compared to air dried material the PP-fibre and the PE-fibre were found to swell in water in average by factors of 1.85 and 1.40, respectively. The ion exchange capacities were determined in dry H^+-form according to the customary method described in literature[10]. The properties of the ion exchangers are presented in Table 1.

2.3 Kinetic Measurements

2.3.1 Equipment. The measurement system consisted, in the following order, of a peristaltic pump (Masterflex 7550-62), a feed selection valve, a 5 cm glass column (Pharmacia Biotech XK 50/20), a spectrophotometer (HP 8452A) with a 30 μL flow through cell, and a balance at the outlet of the system. The pump and the balance were interfaced to a PC for flow control and flow rate measurements, respectively.

Table 1 *The Properties of the Ion Exchangers*

Ion exchanger	Ion exchange capacity meq/g (dry H^+-form)	Water content % (Na^+-form)	Average diameter μm (Na^+-form)	
			air dry	H_2O swollen
grafted PP	3.85	78	61.6	114
grafted PE	2.89	43	237	331
Dowex 50 WX8	5.11	44	-	207
	5.10	45	-	111

2.3.2 Data Treatment. The differential bed method has previously been applied by Leaver et. al.[5] for the investigation of adsorption kinetics of albumin on a crosslinked cellulose ion exchanger. They obtained the mean concentration of protein in stationary phase from the breakthrough curve by subtracting the sum of the mass of protein eluted from the column and the mass of protein in the liquid holdup from the mass fed to column as shown in equation (1).

$$q = \frac{L}{m}\left[\int_0^t (C_0 - C)\, dt - C_0\, t_M\right] \tag{1}$$

where q is the concentration in the ion exchanger, L is the volumetric flow rate, m is the dry mass of the ion exchanger, C_0 is the feed concentration, C is the outlet concentration and t_M is the liquid holdup time in the bed.

In the present study, the adsorption kinetics of Phe is studied using the nonretaining form (Phe⁻) as reference to obtain the amount of its retaining forms (Phe⁺, Phe±) in the ion exchanger. The reference breakthrough curves of Phe were measured in 0.2 M NaOH where the Phe is in anionic form and will be rejected by the cation exchanger. The breakthrough curves for kinetic measurements were carried out in citrate buffers, where the total concentration of Phe was the same as in the reference solution. The swelling-shrinking changes between 0.2 M NaOH and Na-citrate buffers are here regarded insignificant since the breakthrough points occur identically in the reference as well as in the pH 5.0 and in the pH 6.0 citrate breakthrough curves at high flow-rates (see Figure 1). The amount of Phe adsorbed at breakthrough volume V is obtained from the mass balance over the column by equation (2).

$$n(V) = \int_0^V (C_{ref}(V) - C(V))\, dV \tag{2}$$

where $n(V)$ is the amount of adsorbed species at breakthrough volume V, $C_{ref}(V)$ is the concentration at column outlet for the reference (nonadsorbing solute) breakthrough curve, and $C(V)$ the concentration of adsorbing species at breakthrough volume V.

2.3.3 Bed Characteristics. Measurements were carried out using bed heights of 3 mm and 10 mm. The liquid volume of the bed for the calculation of linear velocity ($U_L = L \times Z/V_L$) was determined from the mass balance of nonadsorbing species using 2 mM Phe in 0.2 M NaOH. The 10 mm bed gives an indication about the packing properties of the material in a column. For 114 µm PP-fibre, the volume capacity of tightly packed bed was 0.74 meq/mL and the bed porosity was 0.42. For Dowex 50 WX8 100 - 200 dry mesh, the bed porosity was 0.43, which is closely the same as for the PP-fiber. The bed volume capacity of the resin was, however, considerably higher being 2.26 meq/mL; the 10 mm bed contained 4.11 g (14.6 meq) of PP-fibre and 9.70 g (44.5 meq) of 207 µm Dowex 50 WX8 in Na⁺-form. Correspondingly, the 3 mm bed contained 1.40 g (5.00 meq) of PP-fibre, 1.95 g (5.31 meq) of PE-fibre and 3.656 g (16.8 meq) of 111 µm Dowex 50 WX8.

2.3.4 Procedure. The L-phenylalanine concentration in the citrate buffers was 2 mM and the Na⁺-concentration was 0.098 M. The pH of the Na-citrate buffers was adjusted with 1 M HCl between 2.0 - 6.0. The measurement cycle consisted of 1) equilibration of the bed with the Na-citrate buffer, 2) loading of the ion exchanger with the L-phenylalanine containing Na-citrate buffer, 3) regeneration of the ion exchanger with 0.2

M NaOH, and 4) measurement of the reference breakthrough curve. The measurements were carried out at room temperature (21 ± 1 °C) with bed heights of 10 mm and 3 mm and the linear velocity varied in the range of 0.14 - 1.01 mm/s. The concentration of Phe was determined at 258 nm on the UV-spectrophotometer. The flow rate was continuously measured with the balance. The dry weights of the bed materials were determined in Na^+-form after the kinetic measurements.

3 RESULTS AND DISCUSSION

A set of breakthrough curves of Phe at different pH values against the reference breakthrough curve for 3 mm bed of PP-fibre are shown as examples in Figure 1. The adsorbed amounts of Phe calculated using equation 2 and based on breakthrough curves are presented in Figures 2, 3 and 4 at pH 5.0 , pH 3.5 and pH 2.0, respectively.

At pH 5 the adsorption of Phe in the bed of same height is clearly faster with the PP-fibre than with the 8 % crosslinked resin (Figure 2), despite the considerably smaller mass and capacity of the fibre bed. This can be understood according to the results in Table 1. The water swollen PP-fibre contains 78 % H_2O and its dimensions enlarge in average 1.85-fold. This suggests very open structure of the PP-fibre. Phe is mainly in neutral form at pH 5.0 and diffuses faster into the fibre than into the resin. At pH 3.5 and pH 2.0 (Figures 3 and 4, respectively) where the cationic binding of Phe^+ is more dominant, the absolute adsorption rates at the beginning of the breakthrough with PP-fibre and resin are equal in the bed of same height. Due to the considerably smaller number of ion exchange sites in the fibre bed the adsorption rate decreases faster because of closeness to equilibrium.

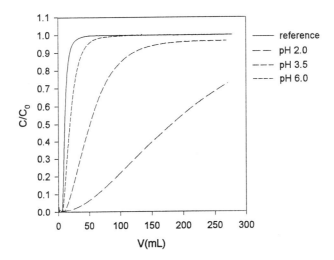

Figure 1 *Breakthrough curves of 2 mM L-phenylalanine with the linear velocity of 0.80 mm/s for 0.2 M NaOH(reference) and citrate buffers with PP-fiber based ion exchanger in 3 mm bed*

The adsorption kinetics with the 114 μm PP-fibre compared with the Dowex 50 WX8 of both size fractions is less controlled by particle diffusion. This can be seen for example with the 10 mm bed at pH 3.5 (Figure 3). With the PP-fibre, the amount of adsorbed Phe increases with increasing linear velocity at a given breakthrough volume, which is a good sign that film diffusion control is more dominant. With the 207 μm resin, the order of the curves is reversed. Also with the 111 μm resin at pH 2.0, the trend of the loading curve (Figure 4) clearly shows that as the equilibrium state approaches particle diffusion control becomes stronger than with the 114 μm PP-fibre. With the 331 μm PE-fibre at pH 2.0, the adsorbed amount of Phe at a given breakthrough volume strongly decreases with increasing linear velocity. The larger diameter explains the stronger particle diffusion control with the PE-fibre. In addition diffusion is slower in the PE-fibre obviously due to the less open sructure indicated by the swelling values in Table 1. The water swollen PE-fibre ion exchanger in Na^+-form contains 43 % H_2O compared to 78 % for the PP-fibre and it swells by factor of 1.40 compared to 1.85 for the PP-fibre. The less open structure of PE-fibre is partly a consequence of crosslinks formed between polymer chains during irradiation and partly a consequence of crystallinity of the structure. According to literature, the dominating effect in polyethylene[11] is crosslinking and compared to polypropylene the crosslink yields are 4 - 5 times greater[12]. It has been observed that the crystallinity of the fibre decreases during grafting[13]. In the present case, PE is grafted more than PP and consequently the crystallinity of PE has decreased more than that of PP during irradiation.

In Figure 5, the adsorption rates on volume basis are compared for the studied materials as a function of loading. Although the driving force in the liquid side is not the same for the three different materials during breakthrough at a given degree of loading, the results show how the absolute adsorption rate develops in the bed of same height with comparable linear velocities. Figure 5 indicates that the maximum value for volume based adsorption rate on 3 mm bed at pH 2.0 with the PP-fibre and the 111 μm resin are

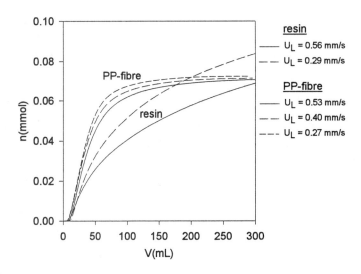

Figure 2 *Adsorption of L-phenylalanine on 10 mm bed of 114 μm PP-fibre ion exchanger and 207 μm Dowex 50 WX8 in pH 5.0 citrate buffer*

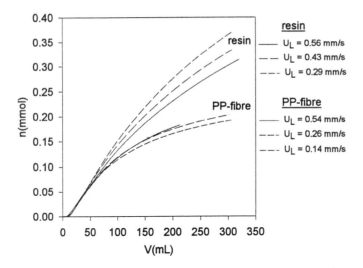

Figure 3 *Adsorption of L-phenylalanine on 10 mm bed of 114 μm PP-fibre ion exchanger and 207 μm Dowex 50 WX8 in pH 3.5 citrate buffer*

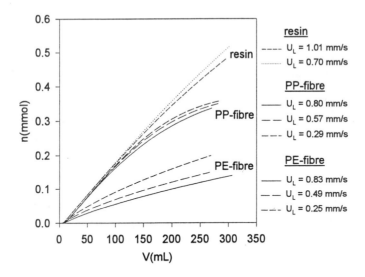

Figure 4 *Adsorption of L-phenylalanine on 3 mm bed of 111 μm Dowex 50 WX8, 114 μm PP-fibre and 331 μm PE-fibre ion exchanger in pH 2.0 citrate buffer*

equal. The adsorption rate develops faster with resin compared to the PP-fibre reflecting larger total outer surface area and consequently larger fraction of ion exchange sites on the outer surface. As the degree of loading increases the accessibility of ion exchange sites of PP-fibre becomes better, which is due to the more open structure of the fibre.

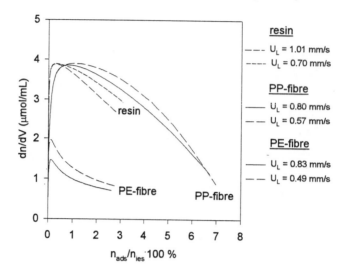

Figure 5 *Adsorption rate of L-phenylalanine along breakthrough curve on 3 mm bed of 111 μm Dowex 50 WX8, 114 μm PP-fibre and 331 μm PE-fibre ion exchanger in pH 2.0 citrate buffer*

4 CONCLUSIONS

The comparison of polyolefin based fibrous ion exchangers with traditional resins on the basis of swelling and dynamic adsorption characteristics show that the PP-fibre ion exchanger has a more open structure than the PE-fibre and conventional 8% crosslinked resins. This allows the adsorbing species to diffuse in the matrix more freely and therefore the adsorption with PP-fibre is more film diffusion controlled compared to the other materials studied. The open structure of the PP-fibre arises from the relatively small extent of crosslinking reactions and from the crystallinity decreasing effect of grafting reactions in the matrix during irradiation. It was found with L-phenylalanine that especially at pH values close to the isoelectric point, the adsorption kinetics of the PP-fibre is faster than the more dense structure possessing PE-fibre and Dowex 50 WX8 resin. In a more open matrix such as PP-fibre, absorption and penetration phenomena become especially important in relation to ion exchange and adsorption phenomena.

The adsorption kinetics of the PP-fibre ion exchanger in relation to the capacity of the bed is competitive with conventional 8 % crosslinked polystyrene resins. Although the volume capacity of the fibre bed is approximately one third of the resin bed, it has potential for fast flow separations due to the more dominating role of film diffusion control. In addition, the flexible nature of the randomly packed fibre bed allows adjustable bed tightness and porosity.

Acknowledgements

The experimental assistance by P. Puomi and T. Sahlberg is gratefully acknowledged.

Nomenclature

C	liquid outlet concentration	mol/L
C_0	liquid feed concentration	mol/L
C_{ref}	liquid outlet concentration of reference substance	mol/L
L	liquid flow rate	L/s
m	mass of dry resin	g
n	amount of substance	mol
n_{ads}	amount of adsorbed substance	mol
n_{ies}	amount of ion exchange sites	mol
q	resin concentration	mol/g
t	time	s
t_M	liquid holdup time in the bed	s
U_L	linear velocity	m/s
V	volume	L
V_L	liquid volume of a bed	L
Z	bed height	m

References

1. D. Petruzzelli, G. Tiravanti, L. Liberti, G. Sergeev, and V. S. Soldatov, in "In Ion Exchange Processes, Advances and Applications", A. Dyer, M. J. Hudson, P. A. Williams, eds., Cambridge, 1993.
2. D. Petruzzelli, A. Kalinitchev, V. S. Soldatov, and G. Tiravanti, *Ind. Eng. Chem. Res.*, 1995, **34**, 2618.
3. A. Singh and N. G. Pinto, *Reactive Polym.*, 1995, **24**, 229.
4. C. Tien and G. Thodos, *Chem. Eng. Sci.*, 1960, **13**, 120.
5. G. Leaver, J. A. Howell, and J. R. Conder, *J. Chromatogr.*, 1992, **590**(1), 101.
6. M. J. Sundell, Doctoral Thesis, Åbo Akademi University, Åbo, 1994.
7. A. Chapiro, "Radiation Chemistry of Polymeric Systems", Interscience Publishers, New York, 1962, Chap.12.
8. V. T. Stannett, *Radiat. Phys. Chem.*, 1990, **35**(1-3), 82.
9. V. S. Soldatov, A. A. Shunkevich, and G. I. Sergeev, *Reactive Polym.*, 1988, **7**, 159.
10. K. Dorfner, ed., "Ion Exchangers", de Gruyter, New York, 1991.
11. L. Mandelkern, in "The Radiation Chemistry of Macromolecules", M. Dole, ed., Academic Press, New York, 1972, Vol. 1, Chap. 12.
12. V. D. McGinniss, in "Encyclopedia of Polymer Science and Engineering", H. F. Mark et. al., eds., Wiley and Sons, New York, 1986.
13. M. Lindsjö, K. Ekman, and J. Näsman, *J. Polym. Sci., Part B: Polym. Phys.* (in press).

THE USE OF POWDERED ION EXCHANGE RESINS IN PHARMACEUTICAL FORMULATIONS

S. A. Bellamy

Rohm and Haas
Croydon
Surrey
UK

1 INTRODUCTION

Ion exchange resins have been used in Pharmaceutical formulations for a number of years either as the active ingredient (e.g. Sodium Polystyrene Sulfonate USP or Cholestyramine Resin USP) or excipients used for controlled/sustained release, taste masking or as a tablet disintegrant.

Whilst there has been little change in the type of products offered to pharmaceutical companies, their uses and, more importantly, the regulatory aspects have developed significantly and this has been more marked in recent years.

2 PRODUCTS

In general, the resins that are typically used in pharmaceutical formulations are powdered with a particle size of approximately 10-150 μm (100-500 Mesh) although there are some applications where ion exchange resins beads are used; for example in gelatin capsules. From the viewpoint of their chemistry, the resins are basically ground versions of the more common water treatment resins. However, these resins are not simply ground water treatment resins as this would not give the required purity of final product. Instead, the whole bead intermediates are specially produced using well defined starting materials and without the use of recycled solvent or acid. This latter point is particularly important in today's regulatory environment. The best known products are shown in Table 1 together with their most common applications whilst Table 2 shows how these products compare with some of the better known water treatment resins.

The history of using ion exchange resins in pharmaceutical applications can be traced back to the 1950's and 1960's with perhaps the greatest interest, in terms of papers and patents published, in the period 1960 to 1975. Since this period, there has been a steady stream of papers on the use of ion exchange resins for a variety of different applications including some of the more recent types of formulations such as ocular and "patch" type delivery systems.

Table 1

USP 23 / NF 18 monograph	Resin type	Application type
Sodium Polystyrene Sulfonate USP	SAC Styrenic/DVB Na⁺ form	Active ingredient. Sustained or Controlled release. Taste masking.
Cholestyramine Resin USP	SBA Styrenic/DVB Cl⁻ form	Active ingredient. Sustained or Controlled release. Taste masking.
Policrilin Potassium NF	WAC Methacrylic K⁺ form	Tablet disintegrant. Taste masking.
Policrilin	WAC Methacrylic H⁺ form	Taste masking Controlled release

Table 2

USP 23 / NF 18 monograph	Resin type	Product name	Water treatment resin
Sodium Polystyrene Sulfonate USP	SAC Styrenic/DVB Na⁺ form	Amberlite® IRP-69	Amberlite® IR 120 Na
Cholestyramine Resin USP	SBA Styrenic/DVB Cl⁻ form	Duolite® AP-143	Amberlite® IRA 404 Cl
Policrilin Potassium NF	WAC Methacrylic K⁺ form	Amberlite® IRP-88	No equivalent
Policrilin	WAC Methacrylic H⁺ form	Amberlite® IRP-64	Amberlite® IRC-50

In terms of the present formulation excipient market, the use of ion exchange resins occupies a very small part of what is a rapidly growing and very diverse market. Compared with 20 years ago, the formulation chemist now has a wide range of different carriers and excipients for a particular active ingredient which are designed to release the product at a controlled rate in a known area of the body. There are also a number of excipient suppliers who offer contract development services to pharmaceutical companies for their speciality products.

3 WHAT FUTURE FOR ION EXCHANGE RESINS.....????

The future of the use of ion exchange resins in the pharmaceutical formulations industry is linked to the advantages and disadvantages, either real or perceived, as compared to other excipients available today. These can be summarised as follows :-

Advantages :
- Long history of use in a variety of applications.
- Existing monographs - USP/NF & JP.
- Drug Master Files in most major markets.
- No adsorption into the body.
- Easily explained chemistry of adsorption and desorption.

Disadvantages :
- Limited product range available.
- Generally regarded as an "old" technology.
- Limited toxicological data on extractables.
- Mouthfeel an issue with liquid formulations.

In broad terms, the above points will have a different significance depending on the type of formulation under consideration. If a pharmaceutical company is looking to develop a New Chemical Entity (NCE) then it is likely that they will look to the most effective means of delivery irrespective of whether this is a known technology or not although the obvious preference would be for a known system. The New Drug Application (NDA) submission would then include both the NCE and delivery system as a complete package. If, however, a pharmaceutical company is looking to develop a new formulation of any existing drug (line extensions) then there will be a preference to use established excipients and drug delivery systems in order to gain approval for the Amended New Drug Application (ANDA) as quickly as possible. In this case, reference to an existing excipient Drug Master Files (DMF) is very useful.

With the ever increasing cost in developing NCE's and the number of important products coming off patent, then the development of line extensions of existing active ingredients, either as new formulations or as an over-the-counter (OTC) products, there is an increasing interest in considering the use of ion exchange resins as drug supports.

4 TECHNOLOGY ASPECTS

The way in which the exchange resins are typically used in pharmaceutical formulations is significantly different to that traditionally used in other areas. The majority of ion exchange resins in use today are in the whole bead form and employed in large fixed bed installations, primarily in an aqueous media with the obviously exception of those resins used in catalysis reactions. Other powdered resins, such as those used in the nuclear industry, are also used in an aqueous solution. In the formulations area, the formulation chemist may be interested in simply mixing dry powders in order to make a tablet or sachet (which may include a resin as a tablet disintegrant or active ingredient). In those cases where the ion exchange resin is used as a drug support, the loading can be performed in either an aqueous or solvent phase but invariably in a batch wise operation. The resulting resinate will then be dried and may be re-ground in order to obtain the desired final product. On paper, the use of a packed column would appear to offer some advantages particularly in optimising the low loading capacities (5-50% of the theoretical capacity) typically seen although the physical handling of fine powders in at times fraught with difficulties, particularly if washing of the resin is required.

5 REGULATORY ISSUES

Perhaps the major change that has occurred in recent times is the regulatory requirements now imposed by both customer and the various national authorities but in particular the Food and Drug Administration (FDA). Despite the fact the many of the resins offered today have been used in large doses (up to 15g/person/day) for many years, the recent advances in analytical chemistry have meant that impurities of 10's ppb can now be detected routinely and can be relatively easily identified by systems such as GC-MS.

In the past, the powdered ion exchange resins were classified as Bulk Pharmaceutical Chemicals (BPC's) which were considered separately to the active drug substances such as the penicillins, cephalosporins etc. This no longer applies today although a clear concenus as to how the production of BPC's is viewed is difficult to determine as seen by recent articles from the FDA. There are certain to be additional constraints placed upon a manufacturer of the resins with less of a difference being made between BPC's and other active drug substances. The auditing of the facilities alone is now a lengthy procedure taking up to several days. Adherence to current Good Manufacture Practices (cGMP) is strongly advised for those manufacturers remaining in this area. Whilst this is almost second nature to the large pharmaceutical companies, it is a lesson worth taking the time to learn carefully for the ion exchange manufacturers who are often considered as the pharmaceutical industry's poorer cousins in the chemical industry.

In addition to those changes listed above, there has also been significant changes in terms of the way in which DMF's are deposited in the various countries.

The "older" system involved the ion exchange resin manufacturer preparing a DMF for their product and then depositing it with the appropriate authorities. This dossier of information was never seen by the formulator and was used when the formulator submitted their dossier for marketing authorisation (either as a NDA or ANDA). The ion exchange manufacturer then wrote a "letter of authorisation" to the medical authorities reviewing the formulators dossier which effectively allowed the reviewer to use the resin manufacturer's information. This is the system which is presently used in the US.

In the absence of any European monographs, the USP/NF monographs were often cited as conformity although the registration of the formulation and submission of the ion exchange manufacturer's DMF was still required in each and every European country. Not every country had the same requirements which required a great deal of work as did the maintenance of the files which had been deposited. This has improved with the arrival of the European equivalent of the drug part of the FDA; the European Medicines Agency (EMA) although a coherent Europe wide system for DMF and market authorisations is still some way in the future. The structure of the European DMF for ion exchange resins is significantly different to that used in the US in that the file is divided into two parts; a "Applicant's" part which must be supplied with the demand for market authorisation of the formulator and an "Active Ingredient Manufacturer's" (AIM) part which is still supplied to the authorities without being inspected by the formulator. The typical information contained in the two parts is shown below in Table 3.

Table 3 *European DMF information*

Information	AIM Restricted part	Applicant Open part
Name and site of manufacture	✔	✔
Specifications and routine tests		✔
Nomenclature		✔
Description		✔
Manufacturing method :		
Brief outline (flow chart)		✔
Detailed description	✔	
In-process QC	✔	
Process validation	✔	
Development chemistry		
Evidence of structure		✔
Potential isomerism		✔
Physico-chemical characterisation		✔
Analytical validation		✔
Impurities		✔
Batch analysis		✔
Stability (where necessary)		✔

The differences between the two systems does give rise to a contradiction in that information which would normally be supplied to a pharmaceutical company in Europe would not be available to the US subsidiary of the same company. This will hopefully be resolved some time in the future........

One important point that has to be taken into consideration by the resin manufacturer is that the use of ion exchange resin as BPC's is being reviewed particularly in terms of the allowable levels of impurities and are being treated in a similar way to the more classical drugs. As mentioned earlier, the improvements in analytical techniques are being applied to ion exchange resins and there are additional criteria being imposed for monomers and extractables, both water and organic volatiles.

6 A CHALLENGING FUTURE....????

The challenge for ion exchange resin manufacturers in the next decade or so will be three fold as outlined below.
- To keep abreast and to anticipate changes in the rules concerning the production of BPC's and analytical techniques.
- To be innovative in an area where it is becoming increasing more difficult to introduce new chemical species.
- To be cost effective in an industry which is increasing becoming aware of the cost of the products it supplies both on a national and international level.

Perhaps the most interesting challenge in the next decade or so will be trying to educate those companies who are not yet aware of the possibilities of using ion exchange resins in formulations and to do this in an environment which is constantly changing in terms of acquisitions, mergers and marketing ventures.

ANION RESINS AND THE LAW OF MASS ACTION

S. A. Fisher

Puricons, Inc.
Malvern, PA 19355
U.S.A.

1 INTRODUCTION

Lin Yutang, a delightful Chinese essayist that I read, in translation, in my teens wrote, "Each day rose like an unchalked blackboard and we did not know what it was going to bring forth..."[1]. That was the way it was every day when I first started working with ion exchange resins. Fortunately, there are still days like that after fifty years.

2 IN THE BEGINNING

No, I was not there at the very beginning as Roger Kressman was. This tale only goes back to the summer of 1946. Dr. Guy Alexander, who later wrote an ACS monograph[2] about his experiences that summer, worked in a little laboratory that opened off the big student laboratory where I was teaching. He was trying to separate perrhenate from molybdate using Amberlite IR-4B. Rhenium is a rare bird in the earth's crust and one of the places it is concentrated to a luxuriant one part per thousand is in the flue dust from molydenite roasters.

Zeolitic water softeners were widely used in my native state, Wisconsin. Indeed, I was born on a limestone ridge. But watching Guy was my introduction to the existence of polymers that could function as ionic substrates in chromatographic analysis. I was so fascinated by what Guy was doing that when he left to become a professor at the end of the summer I petitioned to take up his project as my Ph. D. project.

3 AN UNTRIED STRUCTURE

But this story really begins sometime about a year later at the mailboxes in the front hall of the old chemistry building at the University of Wisconsin. My major professor, Dr. V. W. Meloche, handed me the contents of a little package he had just opened. "Why don't you see what this will do", he said. The box contained a bottle marked Amberlite XE-75. It was a pint of the first semi-works batch of Type I styrene-based quaternary resin made under the McBurney patent[3].

Only I did not know that at the time. All the accompanying letter told me was that this was an experimental anion resin and could be regenerated completely with 4%

sodium hydroxide. That was a daunting bit of information for my source material, molybdenum trioxide, was only soluble in alkaline solution. Fortunately, my investigative spirit was not further inhibited by reading the first Technical Notes[4] which said that the maximum operating pH was 10. They were not printed until 1948.

It apparently never occurred to me to try to characterize the polymer for exchange capacity by the routine that every sample that comes in the door of my laboratory now endures. I did determine the loss on drying at 110 °C (45.07% in chloride form).

My interest obviously lay not in the characteristics of the substrate but in whether it could be used to separate rhenium and molybdenum. Indeed, in view of the manufacturer's comment, I needed to determine if it would have any interest in them under the alkaline conditions necessary to dissolve the flue dust. Experiments with individual solutions of the two players were immediately encouraging. I calculate now that either one could be loaded on a column of the resin to a level of about 2 meq per dry gram chloride form without detectable leakage from a solution of 0.05N sodium hydroxide. Or, to put it in modern terms, about 0.5 equivalents per liter.

Did they come off in 4% sodium hydroxide? Molybdenum eluted very slowly in 100 bed volumes, rhenium not at all. The failure to detect any rhenium in the 4% sodium hydroxide caused a consideration of a theory that these little yellow beads had a functionality capable of reducing the rhenium to a valence not detected by my method of analysis. It was only after cooking a bit of the resin with concentrated sulfuric acid that I was assured that rhenium as perrhenate (ReO_4^-) was still present on it.

Even with 10% sodium hydroxide it took 30 bed volumes to get all the molybdenum off. A similar volume of sodium hydroxide removed about 15% of the rhenium from a column loaded to saturation. Obviously, I had stumbled onto a material that would perform my separation. All I had to do was persuade it to give me back the rhenium, which was a material urgently wanted by the government for use in radar tubes, in a reasonable volume.

3.1 Elution of Perrhenate with Chloride

The fumbling search for a reagent to strip off the rhenium is told in surprising detail in Analytical Chemistry[5]. Suffice to say I tried a number of anions and finally settled on hydrochloric acid in the 7-8 N range. While the use of 50 bed volumes of eluting agent may sound preposterously inefficient, the resulting solution represented not only a molybdenum-free product but a 100-fold concentration of the rhenium. Dr. Albert Preuss, who I had gotten hooked on ion exchange in almost the same way as Dr. Alexander had hooked me, subsequently approached the problem in a more scientific manner and cut my 50+ bed volume process down to 25 by using perchloric acid[6].

What I really came here to talk about are some of the aspects of the chloride elution studies that are discussed in the thesis[7] but have never been published. Once I had discovered that hydrochloric acid would at least strip detectable quantities of perrhenate from the column, I set out on a systematic investigation of the effect of hydrochloric acid concentration on the elution efficiency.

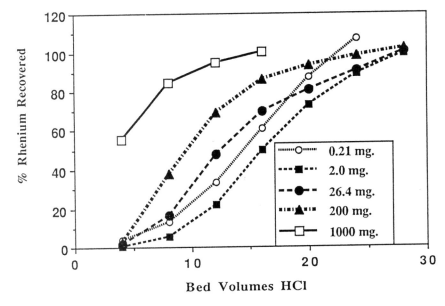

Figure 1 *Elution of perrhenate with hydrochloric acid*

It must be appreciated that this was done on very small columns. Twenty five milliliters (2.5×10^{-5} m³) of resin in something approximating the hydroxide form in 12 mm I.D. columns. In this testing, columns loaded with 22.7 mg of rhenium applied as a solution of potassium perrhenate were eluted. Why such a low loading? It had previously been determined that at moderate loadings the volume of acid required to strip the column of the perrhenate was essentially independent of the amount of rhenium present. That point is substantiated by the curves in Figure 1. All these columns had been loaded with solutions containing 2 g of molybdenum and the specified amount of rhenium as perrhenate. All had been eluted with the same amount of sodium hydroxide to remove the molybdenum prior to the hydrochloric acid elution. Except for the very low loading and the column loaded to saturation, all required the same volume of acid to recover the perrhenate.

But back to my hydrochloric acid experiments. Why was 22.7 mg chosen for loading? Like so many odd numbers in experimental work the reason for it is lost. I suspect it reflects primarily my frugality. Not only was potassium perrhenate worth $10 per gram but my total supply for the project was 10 grams, equivalent in value to a month's salary before taxes for a Research Assistant in the late 1940's. Elution conditions for each column were carefully controlled at 2 mL/min by a complicated feed bottle recommended by Rohm and Haas and a capillary outflow backed up by a screw clamp. Fractions were cut by hand, often during all night bridge games on Saturday nights. Analysis of them, spectrophotometrically, for rhenium produced the interesting set of curves in Figure 2.

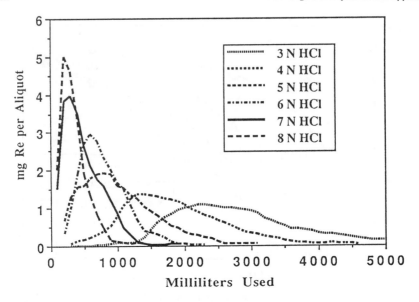

Figure 2 *Effect of hydrochloric acid concentration on the elution of perrhenate*

3.2   Chloride-Perrhenate Equilibrium

While I was accumulating these data, the American Chemical Society printed the section on "The Separation of Rare Earth, Fission Product and other Metal Ions and Anions by Adsorption on Ion-Exchange Resins" in the Journal of the American Chemical Society in late 1947.   This probably increased the available literature on the use of synthetic ion exchange materials by at least a factor of ten.   I read every word, certainly more than once.   An article by Mayer and Tompkins on "A Theoretical Analysis of the Column Separation Process"[8] seemed particularly applicable to what I was doing.

I am not going to go through the mathematical development in the original paper as it takes more than my allotted space.   Further, I believe that anyone who works in ion exchange who has not read the 1947 papers should read them now and marvel at the exacting work that was done with the then available equipment.   Suffice to say that Mayer and Tompkins constructed their theory on the assumption that the ion exchange column operated on the same model as a fractionating column.   Like the fractionating column they assumed that equilibrium was attained at a finite number of stages or theoretical plates.   Using equilibrium measurements, they were then able to predict the separability of difficult to separate ions.

Now, it is well known that I am an experimentalist rather than a mathematician.   If I had read Mayer and Tompkins before I started work I too would probably have started with batch equilibrium studies.   But I had worked only with columns as Alexander had. Having the relative volumes of the elution peaks shown in Figure 2 as a function of hydrochloric acid concentration (Table 1) could this be related to the chloride activity?

Table 1 *Effect of Hydrochloric Acid Concentration on Rate of Elution of Perrhenate*

N of HCl	Peak Volume		Total Volume	Peak Conc.
	mL	*Bed Volumes*	*mL*	*mg Re/L*
3	2100	84	5200	11
4	1300	52	3600	14
5	700	28	2000	19
6	500	20	1600	30
7	250	10	800	40
8	150	6	800	50

Let us assume a very simple mechanism for the elution of perrhenate with chloride of the form:

$$R\text{-}ReO_4 + Cl^- \rightleftharpoons R\text{-}Cl + ReO_4^-$$

where $R\text{-}ReO_4$ and $R\text{-}Cl$ represent the concentrations on the resin and Cl^- and ReO_4^- the ion concentrations in solution. Under equilibrium conditions, we would have an old fashioned equilibrium constant of the form:

$$K = \frac{(a_{R\text{-}Cl})\,(a_{ReO_4^-})}{(a_{R ReO_4})\,(a_{Cl^-})}$$

What Mayer and Tompkins contributed in their development of column theory was the hypothesis that the ratio of the activity of the ion being eluted in the resin phase and in solution phase is directly proportional to the volume at which the maximum concentration in the elution curve occurs:

$$a_{R ReO_4} / a_{ReO_4^-} = Vmax$$

The difficulty in testing this lies in the fact that we seldom have the requisite information concerning the activities of the players in either the resin or the solution phase. A search of the literature of the era unearthed a treasure trove of activities in Lewis and Randall[9] for all of the hydrochloric acid solutions that had been used to elute perrhenate. Further, as may be seen in Table 2, the chloride activity shows an interesting 10 fold increase over the less than 3-fold range of ion concentration.

Table 2   *Chloride Activity$^{(a)}$ in Hydrochloric Acid Solutions*

N of HCl	a_{Cl^-}	$1/a_{Cl^-}$	V_{max}
3	4.05	0.247	2100
4	7.36	0.136	1300
5	12.56	0.0797	700
6	20.4	0.0490	500
7	32.6	0.0307	250
8	50.4	0.0196	150

(*a*)   Chloride activities from Lew and Randall[9].

If Mayer and Tompkins hypothesis is correct, a plot of V_{max} against $1/a_{Cl^-}$ should produce a straight line. As Figure 3 shows, this is precisely the case here. Further, if we simply multiply V_{max} by the activity coefficients ignoring the question of what the magnitude of the units are, all the values fall in the neighborhood of 8500 which suggests, correctly, that under these conditions the activity of the chloride ion in the resin phase is a constant.

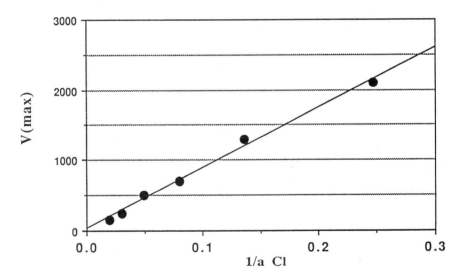

Figure 3 *Variation of elution peak with chloride activity*

4 A WAY TO COMPARE RESINS?

Even if we limit ourselves to Type I quaternary resins there may be more than 100 different commercial polymers, very few of which are absolutely identical in polymeric structure. Selection among them is frequently made based on equilibrium studies. But equilibrium measurements do not take into account the kinetic effect of operation at high flow rates. So we devise kinetics tests which involve large volumes of water and sophisticated measuring capabilities.

It is possible that the old Mayer-Thompkins work could be used to develop small scale bench tests that could be used to predict differences between resins? If we took a fixed volume of each resin in the form of our projected regenerant and loaded it with a fixed amount of the ion of interest and then eluted it at a fixed rate the relative positions of the maximum in the elution curves should only be a function of the differences in resin affinities for the ions. Shouldn't it? No, it is not necessary to work with perrhenate. The 20 g per liter sodium nitrate solution routinely used in anion analysis might be an appropriate regenerant, and, perhaps, a one equivalent percent chloride loading using the same number of milliequivalents of resin. Obviously, there are still blank blackboards to fill.

Acknowledgement

This work was sponsored in part by the Wisconsin Alumni Research Foundation. The author dedicates it to the memory of the late Professor Norris F. Hall without whose guidance and support she would not be here.

References

1. L. Yutang, "Vigil of a Nation", Circa 1939.
2. G. B. Alexander, "Chromatography", American Chemical Society, Washington, D.C., 1977.
3. C. B. McBurney, U.S. Patent 2,291,573, April 1952.
4. Rohm & Haas Technical Notes, "Amberlite IRA-400", Philadelphia, PA, 1948.
5. S. A. Fisher and V. W. Meloche, *Anal. Chem.*, 1952, 24, 1100.
6. V. W. Meloche and A. F. Preuss, *Anal. Chem.*, 1954, 26, 1911.
7. S. A. Fisher, Ph. D. Thesis, University of Wisconsin, 1949.
8. S. W. Mayer and E. R. Tompkins. *J. Am. Chem. Soc.*, 1947, 69, 2866.
9. Lewis and Randall, "Thermodynamics and the Free Energy of Chemical Substances", McGraw Hill Book Co., New York, 1923.

CONCENTRATION PROFILES AND KINETICS FOR FASTER BIOMOLECULE SEPARATIONS

J. R. Conder, B. O. Hayek and Q. Li

Chemical Engineering Department
University of Wales Swansea
Swansea SA2 8PP

1 INTRODUCTION

The separation of proteins and large molecules is inherently a slow process. During the last ten years, however, advances in selective separation media have made the process potentially much faster than before. This has been achieved by developing more rigid hydrophilic adsorption matrices coupled with improvements in the kinetics of mass transfer between the solution and adsorbed phases. These improvements have been accompanied by better mathematical models of the adsorption process to provide a basis for prediction of process performance.

Improvements in the inherent speed and performance of the process are now putting a premium on the performance of the associated equipment. The separation process is started by creating a concentration change in the solution of protein or biomolecules in contact with the separation medium. It is usually assumed that the change is instantaneous and step-shaped. In practice input concentration profiles are never perfectly sharp steps. Mixing and diffusion chambers and non-plug velocity profiles occur in both the flow system which introduces the concentration step and in sampling lines used to monitor the concentration profile. The effect of non-sharp steps is masked when mass transfer processes are slow but has a large effect on performance when mass transfer is fast.

Several models of the kinetics of the adsorption process have been published in which the concentration step is assumed to be perfectly sharp[1-5]. In this paper we describe a model of the adsorption kinetics that includes the effect of non-sharp concentration steps on the observed concentration profiles. The part of the model that describes the adsorption process alone is based on models presented by Horstmann and Chase[3] and Leaver et al[5] for sharp concentration steps. We also describe some practical applications of the present analysis. Both stirred cells and packed beds are considered. The model provides a basis for process measurement and process prediction of fast biomolecule separation processes.

2 THEORY

It is assumed that
1. The particles are spherical and of uniform radius R and porosity ε with ion-exchange groups uniformly distributed over the whole surface, internal and external.

2. The kinetics are controlled (i) by mass transfer through an extra-particle film, with a concentration-independent film coefficient k_f, and (ii) by diffusion in the intra-particle pores. The effective pore diffusivity D_e, based on the whole of the pore volume associated with the porosity ε, is uniform and independent of protein concentration.
3. The surface reaction is fast enough not to affect the kinetics. The adsorption equilibrium is described by the Langmuir equation, $q = q_m \, c/(K_d + c)$.
4. Axial dispersion is negligible in a packed column.

For both stirred cells and packed beds, the model starts with a set of two differential equations with their associated boundary and initial conditions. One equation allows computation of the temporal change in concentration at a point along a pore from a knowledge of the initial concentrations in the vicinity. The other equation allows computation of the temporal change in concentration in a stirred cell or at a position along a packed bed from a knowledge of the existing concentrations in the bulk liquid and at the exit of the pores. The first of these equations is obtained from conservation of mass of protein diffusing through the pore-occupied fraction of a spherical shell of a particle, combined with the Langmuir isotherm, giving[5]

$$\frac{\partial c_i}{\partial t} = \frac{D_e \left(\dfrac{\partial^2 c_i}{\partial r^2} + \dfrac{2}{r} \dfrac{\partial c_i}{\partial r} \right)}{1 + \dfrac{1-\varepsilon}{\varepsilon} \dfrac{q_m K_d}{\left(K_d + c_i \right)^2}} \tag{1}$$

There are two boundary conditions associated with this equation. At the centre of the particle, $\partial c_i/\partial r = 0$, $r = 0$; and, at the surface of the particle, $\varepsilon \, D_e \, \partial c_i / \partial r = k_f (c - c_s)$, $r = R$. The rest of the analysis depends on whether a stirred cell or packed bed is under consideration.

2.1 Stirred Cell

In the case of a stirred cell, the second differential equation to accompany equation (1) is obtained from conservation of mass of protein in the bulk liquid, assuming perfect mixing:

$$\frac{\partial c}{\partial t} = -\frac{3vk_f}{VR_{sv}} \left(c - c_s \right) \tag{2}$$

The surface volume mean radius R_{sv} is the appropriate mean value appearing in this equation if the model, which assumes that the particles are monodisperse, is applied to an experimental system of polydisperse particles.[6]

2.2 Packed Bed

In the case of a packed bed, the second differential equation arises from conservation of mass of protein in an element of bed length:

$$\frac{\partial c}{\partial t} = -u \frac{\partial c}{\partial x} - \frac{3}{R_{SV}} \frac{(1-\varepsilon_b)}{\varepsilon_b} k_f (c - c_s) \qquad (4)$$

2.3 Apparatus Effects on c(t) Profile in Stirred Cell

In a stirred cell the main source of apparatus effects on the measured concentration-time profile c(t) is the time delay and mixing that occurs as liquid is carried from the stirred cell through a filter and sampling line to the detector which measures the protein concentration. Spreading of protein solution in a narrow, tubular line occurs due to, first, laminar flow and, second, sudden changes in diameter of the line at fittings and connections which result in the formation of mixing and diffusion chambers. The effect of laminar flow may be minimised by having a small tube diameter. Laminar flow mixing, if significant, results in a symmetrical spreading[7] of a concentration step profile. The response profile to a step change with no adsorbent can be used to identify the mechanism of spreading. The response profile in Figure 1 shows that the spreading is exponential rather than sigmoid. This confirms expectation[7] that mixing chambers rather than laminar flow are the normal cause of spreading with narrow tubing.

To develop a model for mixing in the sampling line, consider a package of liquid reaching the detector at time t. This is made up of liquid elements which left the stirred cell at various previous times λ and spent various residence times t_R in a single exponential mixing chamber representing the combined contributions of all mixing chambers in the line. The response of the mixing chamber to unit concentration change at the inlet is $[1 - \exp(-gt)]$ where g is the reciprocal time constant. A liquid element ∂t_R of concentration $c(\lambda)$ in the stirred cell thus makes a contribution to the profile at the detector given by

$$c(\lambda) \frac{\partial}{\partial t_R} \left(1 - e^{-gt}\right) \partial t_R = c(\lambda) g e^{-gt} \partial t_R$$

If the delay time in the sampling line is t_o, then $\lambda = (t - t_o - t_R)$ and the concentration profile observed at the detector is

$$\bar{c}(t) = \int_{t_R=0}^{t_R=\infty} c\left(t - t_o - t_R\right) g e^{-gt} \partial t_R \qquad (4)$$

To predict a concentration profile $\bar{c}(t)$ for comparison with experimental profiles from a stirred cell, equations (1) and (2) are solved numerically and the solution, regarded as $c(\lambda)$, is substituted in equation (4) to generate $\bar{c}(t)$. A numerical and computational implementation of this procedure will be described elsewhere[6].

2.4 Apparatus Effects on c(t) Profile in Packed Bed

In a packed bed the main source of apparatus effects on the concentration-time profile is lack of a perfectly sharp step profile in the concentration step at the column inlet. The concentration change should be instantaneous, but this is prevented in practice by mixing in tube fittings and in pumps and/or valves used to generate the concentration

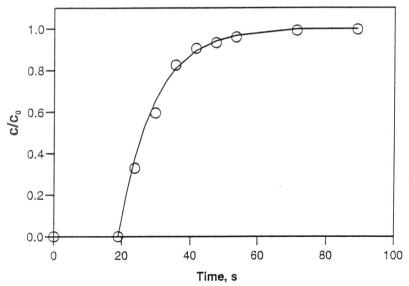

Figure 1 *Response of a stirred cell and filtered sampling loop and UV detector to a step $c_o = 0.44$ kg/m³ of bovine serum albumin, pH 7.5, 5 kg/m³ NaCl. The experimental points are well fitted by the continuous curve which represents the mixing chamber equation $\bar{c}(t) = c_o \{1 - exp[-g(t - t_o)]\}$ with $g = 0.10$ s⁻¹ and $t_o = 19$ s*

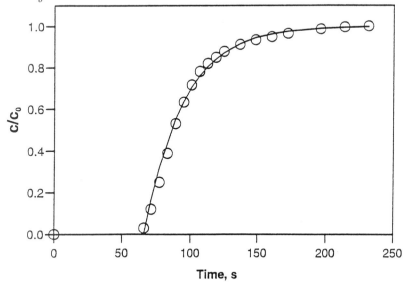

Figure 2 *Response of a Waters LC 3000 pump unit and 440 UV detector, with the column shorted out, to a step $c_o = 1.5$ kg/m³ myoglobin at $u = 7.5$ mm/s, pH 5.7, 0.01 M phosphate buffer. The experimental points are not quite perfectly fitted by the continuous curve which represents the mixing chamber equation (5) with $g = 0.023$ s⁻¹ and $t_o = 77.4$ s*

step. This is illustrated by Figure 2, where a small contribution to the profile from laminar flow spreading is seen in the slightly sigmoid modification of the basically exponential shape of the curve.

If the concentration step at the column inlet was perfectly sharp, the boundary condition associated with equation (3) would be $c(0, t \le 0) = 0$, $c(0, t > 0) = c_0$. To modify the model for non-sharp steps this boundary condition is modified to,

$$c(0, t > 0) = c_0 \{1 - \exp[-g(t - t_0)]\} \tag{5}$$

where the mixing chamber and delay constants g and t_0 are determined for a given flow rate with the column shorted out or replaced by a short length of capillary tube, or by carrying out a run with the column present but under nonsorbing conditions of pH and salt concentration.

3 ILLUSTRATIVE APPLICATIONS

3.1 Model Concentration Profiles with a Stirred Cell

The general effects of slow mixing in the sampling loop attached to a stirred cell may be illustrated by replacing the numerical solution to equations (1) and (2) by a simplified analytical expression which describes the adsorption curve approximately and substituting this into equation (4). Consider a stirred cell containing suspended adsorbents. If a sharp concentration step c_0 of protein is introduced the output is a c(t) curve in which c falls at a decreasing rate from c_0 to some lower asymptotic value c_f. Assume as a typical example that the adsorption curve is represented approximately by

$$c(t) = (c_0 - c_f) e^{-\gamma t} + c_f \tag{6}$$

where γ is the reciprocal time constant for the adsorption process in the stirred cell. Substituting this equation into equation (4) and expressing the result in terms of the dimensionless variables

$$C = \bar{c}/c_0, \quad T = \gamma(t - t_0) \text{ and } R = \gamma/g$$

we obtain the concentration profile observed by a detector in a sampling line attached to the stirred cell:

$$C = \frac{c_f}{c_0} \left[1 - e^{-T/R} + \frac{1 - \dfrac{c_f}{c_0}}{1 - R} \left(e^{-T} - e^{-T/R} \right) \right] \tag{7}$$

This profile is plotted in Figure 3 for $c_f/c_0 = 0.5$ with the ratio R of the time constants for the mixing and adsorption processes as parameter. When the mixing is fast, the response is a diffuse step followed by a diminishing concentration region whose shape and time

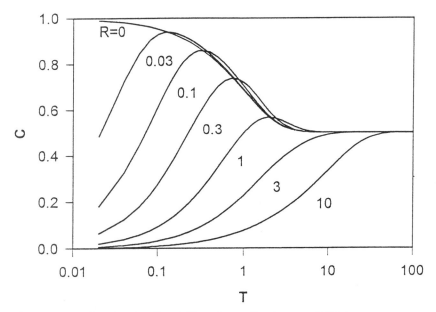

Figure 3 *Response of a stirred cell and filtered sampling loop and UV detector to a step of protein when the adsorption is described by equation (6) with $c_f/c_o = 0.5$*

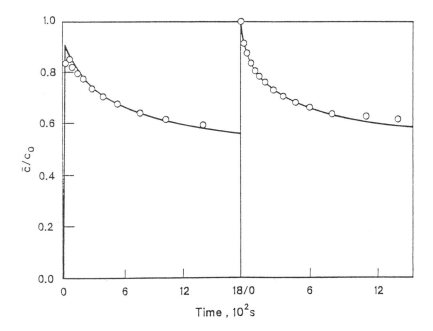

Figure 4 *Adsorption of BSA on 15 μm Bakerbond WP PEI particles in a stirred cell, with (a) BSA ($c_o = 0.77$ kg/m³) added after the adsorbent is dispersed and (b) particles added after developing the protein step ($c_o = 0.93$ kg/m³), pH 7.5, 5 kg/m³ NaCl. The points are experimental. The curves are theoretical simulations by numerical solution of equations (1), (2) and (4)[8]*

constant reflect the adsorption process. When mixing is slow relative to adsorption, the curve reflects the mixing process alone. The adsorption process can be studied with reasonable accuracy provided mixing is fast enough to give $R < 0.1$.

3.2 Effect of Particle Size on Adsorption in a Stirred Cell

Experiments have been conducted on the adsorption of bovine serum albumin (BSA) on a wide-pore hydrophilic polymer-coated silica gel in its polyethylene-imine anion-exchange form (J. T. Baker, Ltd.). The shape of the plots in Figure 4 shows that when the adsorbent has a particle size of 15 μm, the adsorption is slow enough for the adsorption to be observed with little interference from mixing effects. The Figures also show, however, that the degree of interference depends on the starting protocol used. When protein is added to the adsorbent already dispersed in the solution in the cell, the initial part of the curve is affected by the protein addition process, which lasts several seconds. It is preferable to reverse the order of events by developing the protein step first; subsequent addition and dispersion of the adsorbent (added as a slurry) takes about a second. Mixing in both the stirred cell and sampling line are then sufficiently fast in comparison with the rate of adsorption in this system for mixing to be neglected in modelling the process; the theoretical simulation curves in Figure 4b, obtained by numerical solution of equations (1) and (2), are virtually unchanged by using equation (4). (The residual lack of fit in this Figure is due to other model assumptions listed at the start of Section 2 above.) This second protocol also has the advantage of minimising problems caused by build-up of adsorbent on the filter at the entrance to the sampling line.

When smaller, 5μm, adsorbent particles are used the adsorption process is an order of magnitude faster. The resulting concentration profiles depend on both adsorption and mixing processes and are thus insensitive to the adsorption model used. Figure 5 shows that equally good fits are obtained with the approximate adsorption curve expression, equation (6), and with numerical solution of equations (1) and (2), both used in conjunction with equation (4) for the mixing process. The value of R indicates that the time constant for the adsorption process is only slightly greater than that of the mixing process. This is too small to allow useful information on the adsorption process to be obtained. With small particles, it is essential to design the equipment for minimal, i.e. fast, mixing.

These are two extreme cases. In intermediate cases the adsorption process can be studied provided mixing is taken into account in the theoretical simulation, as already described.

3.3 Validation of Non-sharp Inlet Profile Model for a Packed Bed

Experiments have been carried out with a Waters LC 3000 pump unit and 440 UV detector. Measurements with the column shorted out allowed both the delay time t_o and the reciprocal time constant g for mixing in the line to be determined (Figure 2). The products (Ft_o) and (F/g) were independent of interstitial velocity u (range 2.5-25 mm/s) and of concentration step size c_o, as expected. Concentration profiles (breakthrough curves) for myoglobin (Mb) on a wide-pore hydrophilic polymer-coated silica in its carboxymethyl cation-exchange form are shown in Figure 6. It is seen that the theoretical simulation is much improved by incorporating the non-sharp inlet concentration profile as a boundary condition in the model, according to the procedure described in Section 2.4.

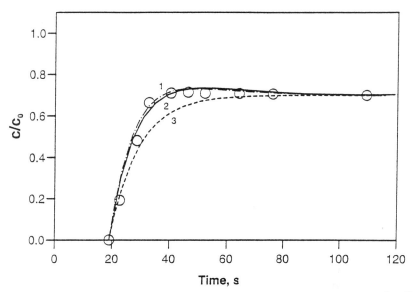

Figure 5 *Adsorption of BSA on 5 μm Bakerbond WP PEI particles in a stirred cell for c_o = 1.5 kg/m³, pH 7.5, 5 kg/m³ NaCl The points are experimental. Curve 1 is predicted from the approximate model of Section 2.1 (equations (4) and (6)) with c_f/c_o = 0.70 and R = 0.8. Curve 2 is the theoretical simulation obtained by numerical solution of equations (1), (2) and (7). Curve 3 is reproduced from Figure 1 where no adsorbent is present*

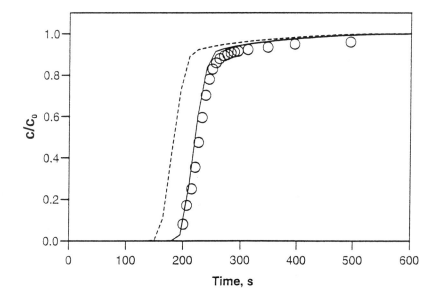

Figure 6 *Adsorption of Mb on 40 μm Bakerbond WP CBX in a packed bed for c_o = 4.0 kg/m³ at pH 5.7, 0.01 M phosphate buffer, u = 7.5 mm/s. The points are experimental. Theoretical simulations: --- assuming sharp concentration step at column inlet; — allowing for non-sharp concentration step*

Curve 2 does not have a common starting point with Curve 1 in this case but is shifted almost parallel to it because the direction of curvature of the equilibrium isotherm makes the boundary self-sharpening as it passes through the column.[9]

The procedure we have described for non-sharp inlet profiles is valid provided the mixing volume of the line between the column exit and the detector is small compared with that upstream of the column, as is likely in most practical cases. Failure of this provision could be accommodated by allowing for post-column mixing in the same way as described for a stirred cell (equation (4)). It is also necessary that mixing chamber volumes due to poor flow distribution at the ends of the column should be small. The condition for this is that the pressure drop over frits at the ends of the column should exceed the pressure drop over the packing. If this conclusion does not hold the constants g and t_o required for the simulation will be better determined in a run carried out under nonsorbing conditions of pH and salt concentration than by shorting out the column.

Nomenclature

c	concentration of protein in bulk/liquid, kg/m^3
c_f	equilibrium (final) liquid phase concentration, kg/m^3
c_i	liquid phase concentration at particle radius r in a pore, kg/m^3
c_o	amplitude of a liquid concentration step, kg/m^3
c_s	liquid phase concentration at surface of particle (pore entrance), kg/m^3
$\bar{c}$	liquid phase concentration at detector, kg/m^3
D_e	effective diffusivity of protein in pores, m^2/s
g	reciprocal of time constant for mixing in flow line, s^{-1}
k_f	extra-particle liquid film mass transfer coefficient, m/s
K_d	dissociation constant, kg/m^3
q	concentration of adsorbed protein in adsorbate, kg/m^3 solid
q_m	adsorption capacity in Langmuir equation, kg/m^3 solid
r	radial co-ordinate in adsorbent particle, m
R	ratio of mixing to adsorption time constants $= \gamma/g$
R_{sv}	surface-volume mean radius of particle, m
t	time, s
t_o	delay time in flow line, s
t_R	residence time in mixing chamber, s
T	dimensionless time ratio $= \gamma(t - t_o)$
u	interstitial velocity of liquid through bed, m/s
v	volume of adsorbent in stirred cell, m^3
V	volume of liquid in stirred cell, m^3
x	axial co-ordinate in packed bed, m
γ	reciprocal of effective time constant for adsorption, s^{-1}
ε	particle porosity (void fraction)
ε_b	bed voidage
λ	time when liquid enters sampling line from stirred cell, s.

Acknowledgements

The authors wish to thank the Science and Engineering Research Council for financial support for this work. We are also grateful to Dr D. J. Gunn for helpful discussion of part of the modelling.

References

1.   B. H. Arve and A. I. Liapis, *Biotech. Bioeng.*, 1988, **31**, 240; *ibid.*, 1988, **32**, 616.
2.   J. H. Petroupoulos, A. I. Liapis, N. P. Kolliopoulos, J. K. Petrou and N. K. Kanellopoulos, *Bioseparation*, 1990, **1**, 69.
3.   B. J. Horstmann and H. A. Chase, *Chem. Eng. Res. Des.*, 1989, **67**, 244.
4.   G. L. Skidmore, B. J. Horstmann and H. A. Chase, *J. Chromatog.*, 1990, **498**, 113.
5.   G. Leaver, J. A. Howell and J. R. Conder, *J. Chromatog.*, 1992, **590**, 101.
6.   B. O. Hayek and J. R. Conder, in preparation.
7.   J. C. Sternberg, "Advances in Chromatography, Vol. 2," J. C. Giddings and R. A. Keller, eds., Edward Arnold, 1966, Chap. 6, p. 205.
8.   B. O. Hayek, Ph. D. thesis, University of Wales Swansea, 1994.
9.   J. R. Conder and C. L. Young, "Physicochemical Measurement by Gas Chromatography", John Wiley and Sons, Chichester, 1979, p. 372.

TEMPERATURE-SENSITIVE ION-EXCHANGE SYSTEMS : EQUILIBRIUM OF ION-EXCHANGE ACCOMPANIED BY COMPLEXATION EITHER IN SOLUTION OR IN RESIN PHASE

D. Muraviev, A. Gonzalo, M. J. González and M. Valiente

Unitat de Química Analítica, Departament de Química
Universitat Autònoma de Barcelona
E-08193 Bellaterra (Barcelona)
Spain

1 INTRODUCTION

The influence of temperature on the affinity of ion exchangers towards target ions can be used for designing practically reagentless and ecologically clean separation processes based upon parametric pumping[1-3] and related dual-temperature ion-exchange (IX) fractionation techniques.[4-10] A high heat of IX reaction is the necessary condition for identifying temperature-sensitive IX systems. IX reactions on conventional type of resins (e.g. sulfonic) are characterized by small enthalpy changes, when no complex formation or association is involved.[11,12] Much higher enthalpy values are observed in those IX systems where complex formation occurs either in solution or on the resin phase.[13] However, the information available about these characteristics in IX systems is very limited; it is of particular relevance to solve the following problems: (1) the selection of commercially available chelating ion exchangers suitable for separation of certain ion mixtures by applying temperature-responsive IX fractionation techniques, and (2) the preselection of the ligands with promising characteristics for use in the synthesis of chelating resins with temperature responsive selectivity. Solution to these problems will stimulate further development and wider application of reagentless IX separation techniques, which can allow the design of ecologically clean IX technologies. The present study was undertaken (1) to obtain information on Zn^{2+}-Cu^{2+} IX equilibrium at different temperatures in systems involving (a) iminodiacetic (IDA) resin and metal sulfate solutions, and (b) sulfonate cation exchanger and metal ions solution containing stoichiometric amount of IDA acid, and (2) to develop a novel approach for predicting temperature dependencies of equilibrium of IX accompanied by complexation either in solution or on resin phases.

2 EXPERIMENTAL

The source and the quality of all chemicals used in this work as well as the analytical methods applied were identical to those described in our previous papers.[7-13] All solutions were prepared with doubly distilled water and were degassed by using an ultrasonic bath and a vacuum pump. The pH of stock solutions was adjusted to 1.9 ± 0.05 with H_2SO_4 and kept constant by controlling it with a Crison pH meter (model 507) supplied with a combined glass electrode. The concentrations of metal ions were determined by the ICP

N°	Resin		Solution				temp. range studied, K
	type	initial ionic form	C_0, M	Zn:Cu*	pH	media	
S_1	iminodiacetic Lewatit TP-207	H+	0.165	8.5	1.9	SO_4^{2-}	293-353
S_2	sulfonic Lewatit SP- 112	H+ or Na+	0.165	8.5	1.9	0.165 M IDA (SO_4^{2-})**	293-353
S_3	sulfonic Lewatit SP-112	H+ or Na+	0.165	8.5	1.9	SO_4^{2-}	293-353

*) Molar ratio
**) Equimolar mixture of (Cu, Zn)SO$_4$ and IDA.

Table 1 *System Characteristics. C_0 is the Total Metal Concentration in Solution*

technique using an ARL Model 3410 spectrometer with minitorch. The emission lines used for the spectrochemical analysis were 224.7 nm for Cu^{2+} and 206.191 nm for Zn^{2+}. The uncertainty of metal ions determination was <1.5%. The main characteristics of the systems (S_1 to S_3) under investigation are presented in Table 1. S_2 may be considered as a spectral symmetrical reflection of S_1. As observed, S_2 and S_3 differ by the absence of the complexing agent in the solution phase in S_3.

The IX equilibrium at different temperatures was studied under dynamic conditions in thermostatic glass columns[7-13] (of 1.1 cm i.d. for Lewatit TP-207 and 1.4 of i.d. for Lewatit R 252 k and Lewatit SP-112) which permitted the thermostatic conditioning of both resin and solution. The columns were charged with 2 g of each ion-exchanger. The stock solutions were passed at constant flow rate (1.6 or 3.0 mL/min) through the resins in H+- or Na+-form pre-equilibrated with H$_2$SO$_4$ solution at pH=1.9. The achievement of IX equilibrium in the systems under study was determined by comparison of the metal concentration in the column outlet with that of the initial feed solution. After the eluate sample was collected with concentrations of Cu^{2+} and Zn^{2+} close to those of the initial solution, the flow of solution was stopped and then resumed after a certain period of time. The coincidence of the initial concentration with that of the sample collected after the break was considered as the criterion indicating the equilibrium in the system. After equilibration the resin was rinsed with twice distilled water, and the metal stripping was carried out with 1 M H$_2$SO$_4$, followed by the analysis of Cu^{2+} and Zn^{2+} in the resulting eluate. The results of the stripping solution analysis were used to determine both capacity of the resin towards copper and zinc and the equilibrium separation factor, α, expressed as follows:

$$\alpha_{Zn}^{Cu} = \frac{q^*_{Cu}}{q^*_{Zn}} \cdot \frac{C^*_{Zn}}{C^*_{Cu}} \tag{1}$$

where q^* and C^* are the equilibrium concentrations of metal ions in resin and solution phases respectively. The relative uncertainty on α determination did not exceed 7%. The

	System		
T, K	S_1	S_2	S_3
293	85.0	0.185	1.12
313	73.3	0.144	1.08
333	56.2	0.112	1.01
353	30.0	0.097	0.94

Table 2   α *vs Temperature*

separation factor is a parameter used to describe the IX equilibrium of the system under study.

Thermostripping experiment was carried out as follows. After equilibration of the resin at high temperature (60°C) with the initial solution, the excess of equilibrium solution was removed from the column so that its level coincided with that of the resin bed. The temperature was the decreased to the preselected value of 20°C. After equilibration of the system at the lower temperature, the same initial solution was passed through the column at constant flow rate (0.4 mL/min) and collected in portions where concentrations of Cu and Zn were determined.

3 RESULTS AND DISCUSSION

Table 2 presents α values determined at different temperatures in S_1, S_2 and S_3.

As follows from the results shown in Table 2, S_1 and S_2 are characterized by strong dependencies of α on temperature, which is observed to be far less in S_3. IDA resin is highly selective towards Cu^{2+} (in S_1 α >>1) while sulfonic resin in the presence of IDA acid in solution phase preferentially sorbs Zn^{2+} (in S_2 α <<1). The selectivities of the resins in S_1 and S_2 correlate with the sequence of the corresponding stability constants of Cu^{2+} and Zn^{2+} complexes with IDA acid (CuIDA and ZnIDA).[14] Formation of a far stronger CuIDA either in the resin (S_1) or in the solution phases (S_2) shifts the IX equilibrium in the system under consideration so that less stable ZnIDA is accumulated in the solution (S_1) or in the resin (S_2) phases. In the absence of complexation, sulfonic resin demonstrates practically zero selectivity towards the ionic couple under study (α≈1 in S_3).

Prior to discussions of temperature dependencies of α in S_1 and S_2 we will consider the main features of these two systems and demonstrate their "symmetry" to each other. Qualitatively this symmetry follows from the composition of the respective phases given in Table 1. Indeed, in S_1 metal ions form IDA complexes on the resin phase while the solution phase contains Cu^{2+} and Zn^{2+} sulfates. In S_2 CuIDA and ZnIDA complexes exist in the solution while Cu^{2+} and Zn^{2+} polysulfates are located in the resin. To quantify such approach in S_1, the IX reaction between Cu^{2+} and Zn^{2+} on IDA resin (R-IDA) can be written as follows:

$$R\text{-}IDA^{2-}\,Zn^{2+}\;+\;Cu^{2+}\;\Leftrightarrow\;R\text{-}IDA^{2-}\,Cu^{2+}\;+\;Zn^{2+} \tag{2}$$

Reaction (2) can be represented as two simultaneous independent equilibria which describe complexation of a single ionic species by IDA resin:

$$R\text{-}IDA^{2-} + Zn^{2+} \Leftrightarrow R\text{-}IDA^{2-} Zn^{2+} ; \quad \beta_{R\text{-}IDAZn} \tag{3}$$

and

$$R\text{-}IDA^{2-} + Cu^{2+} \Leftrightarrow R\text{-}IDA^{2-} Cu^{2+} ; \quad \beta_{R\text{-}IDACu} \tag{4}$$

where stability constants are defined as follows:

$$\beta_{R\text{-}IDAZn} = \frac{q^{*}_{Zn}}{q^{*}_{R\text{-}IDA}\, C^{*}_{Zn}} ; \quad \beta_{R\text{-}IDACu} = \frac{q^{*}_{Cu}}{q^{*}_{R\text{-}IDA}\, C^{*}_{Cu}} \tag{5}$$

with $q^{*}_{R\text{-}IDA} = q^{*}_{Zn} + q^{*}_{Cu}$ being the total IX capacity of IDA resin.

Since IX reaction (2) can be obtained by subtracting eq.(3) from eq.(4) and taking into account the expression in (5) one can rearrange eq.(1) as:

$$\alpha^{Cu}_{Zn}(S_1) = \frac{\beta_{R\text{-}IDACu}}{\beta_{R\text{-}IDAZn}} \tag{6}$$

obtaining a direct relationship between the separation factor an equilibrium constants.

On the other hand, IX reactions in S_2 may be considered as simultaneous combination of reactions:

$$ZnIDA + Cu^{2+} \Leftrightarrow CuIDA + Zn^{2+} ; \quad K = \frac{\beta_{CuIDA}}{\beta_{ZnIDA}} \tag{7}$$

and

$$(R\text{-}SO_3^-)_2\, Zn^{2+} + Cu^{2+} \Leftrightarrow (R\text{-}SO_3^-)_2\, Cu^{2+} + Zn^{2+} \tag{8}$$

with,

$$\beta_{CuIDA} = \frac{C^{*}_{CuIDA}}{C^{*}_{IDA}\, C^{*}_{Cu^{2+}}} ; \quad \beta_{ZnIDA} = \frac{C^{*}_{ZnIDA}}{C^{*}_{IDA}\, C^{*}_{Zn^{2+}}} \tag{9}$$

In this case, the C^{*}_{Cu} and C^{*}_{Zn} of eq.(1) have the respective values:

$$C^{*}_{Cu} = C^{*}_{Cu^{2+}} + C^{*}_{CuIDA} ; \quad C^{*}_{Zn} = C^{*}_{Zn^{2+}} + C^{*}_{ZnIDA} \tag{10}$$

After substituting (10) in (1) and expressing C_{CuIDA} and C_{ZnIDA} from (9) one can obtain:

$$\alpha_{Zn}^{Cu} = \left[\frac{q^*_{Cu}}{q^*_{Zn}} \cdot \frac{C^*_{Zn^{2+}}}{C^*_{Cu^{2+}}} \right] \cdot \left[\frac{\left(1 + \beta_{ZnIDA} \, C^*_{IDA}\right)}{\left(1 + \beta_{CuIDA} \, C^*_{IDA}\right)} \right] \qquad (11)$$

Equation (11) is similar to that reported by Gorshkov et al.[15,16] As follows from eq.(11) α in S_2 represents the product of two terms. The first one is in fact α in S_3 (absence of IDA in the solution phase). The second term depends on pH of the equilibrium solution and on the total concentrations of metal ions and complexing agents. In S_2 (see Table 1) $C^*_{CuIDA} \gg C^*_{Cu^{2+}}$ and $C^*_{ZnIDA} \gg C^*_{Zn^{2+}}$, therefore:

$$\frac{C^*_{CuIDA}}{C^*_{Cu^{2+}}} = \beta_{CuIDA} \, C^*_{IDA} \gg 1 \; ; \quad \frac{C^*_{ZnIDA}}{C^*_{Zn^{2+}}} = \beta_{ZnIDA} \, C^*_{IDA} \gg 1 \qquad (12)$$

Also, as observed in Table 2, the first term in eq.(11) (α in S_3) practically equals 1, hence after considering equations of eq.(12) one obtains:

$$\alpha_{Zn}^{Cu}(S_2) = \frac{\beta_{ZnIDA}}{\beta_{CuIDA}} \qquad (13)$$

Comparison of eq.(13) and eq.(6) testifies to the validity of the above assumption about the symmetry of S_1 to S_2.

From data collected in Table 2 one can estimate the differential enthalpies of IX reactions which proceed in S_1 and S_2 assuming the conventional Arrhenius dependencies of α on T. This estimation gives a value of -8.6 kJ/mol for S_1 and -9.35 kJ/mol for S_2.

The enthalpies, ΔH, of CuIDA and ZnIDA formation reactions were determined by Anderegg[17] and Bonomo et al.[18] to be: -18.6 (CuIDA),[17] -9.2 (ZnIDA)[17] and -16.6 kJ/mol (CuIDA).[18] Hence, the corresponding differential enthalpy of reaction (7) appears to be $\Delta(\Delta H)_{Cu-Zn}$= -9.4 kJ/mol[17] or -7.4 kJ/mol.[17,18] Comparison of these $\Delta(\Delta H)_{Cu-Zn}$ values with those determined experimentally in S_1 and S_2 (see above) shows that $\Delta H(S_1)$ agrees well with the average of the $\Delta(\Delta H)_{Cu-Zn}$ given above ($\Delta(\Delta H)_{av}$ = -8.5 kJ/mol). $\Delta H(S_2)$ perfectly agrees with $\Delta(\Delta H)_{Cu-Zn}$= -9.4 kJ/mol.[17] Because of the agreement of $\Delta H(S_1)$ and $\Delta H(S_2)$ with the value of the differential standard enthalpy of reaction (7), the following approach may be suggested to predict temperature dependencies of IX equilibrium accompanied by complexation in systems of S_1 or S_2 characteristics.

The prediction is based on applying the data of heats of complex formation for different metal ions with ligands of the same type as the chelating resin under interest (IDA in our case). The influence of the temperature on the thermodynamic equilibrium constant, K, can be described by the following equation:

$$\ln \frac{K(T_2)}{K(T_1)} = \frac{\Delta H}{R} \left(\frac{1}{T_1} - \frac{1}{T_2} \right) \qquad (14)$$

which is valid when ΔH is independent of the temperature in the given temperature interval.

As it has been previously reported,[13] when ΔH does not depend on the composition of the resin phase, then $\mathbf{K}$ in eq.(14) can be substituted by α, and the logarithms of $\alpha(T_2)/\alpha(T_1)$ can be calculated from the ΔH value for different temperature intervals. A comparison of calculated $\alpha(T_2)/\alpha(T_1)$ values using the corresponding experimental data of Table 2 is shown in Table 3.

As seen from the results given in Table 3 the calculated $\ln(\alpha(T_2)/\alpha(T_1))$ values agree well with those determined experimentally both in S_1 and in S_2, this testifies to the validity of the approach proposed.

The same approach can be used for the preselection of chelating ion exchangers suitable for dual-temperature IX separation of certain ion mixtures (see Introduction). This preselection can be based on $\Delta(\Delta H)$ values for complex formation of metal ions under interest with the respective resin analogues (monomer ligands). The estimation made in our previous paper[13] has shown that $\Delta(\Delta H)$ values should not be less than ~ 10 kJ/mol to provide remarkable temperature dependence of the resin selectivity (e. g. $\alpha(T_2)/\alpha(T_1) > 2$) towards a given ion couple. As an example, the applicability of IDA resin for separation by dual-temperature IX fractionation techniques can be predicted from the heats of IDA complex formation with Ni^{2+} (-21 kJ/mol),[17] Co^{2+} (-8.95 kJ/mol),[17] Cu^{2+} and Zn^{2+} (see above). Ion couples such as: Co^{2+}-Cu^{2+} ($\Delta(\Delta H) = -9.8$ kJ/mol), Zn^{2+}-Ni^{2+} ($\Delta(\Delta H) = -11.9$ kJ/mol) and Co^{2+}-Ni^{2+} ($\Delta(\Delta H) = -12.2$ kJ/mol) must be characterized by a quite strong temperature dependence of α on IDA resin and therefore can be effectively separated by the above methods. However, for Ni^{2+}-Cu^{2+} ($\Delta(\Delta H) = -2.3$ kJ/mol), a far weaker α vs T dependence can be expected and the effectiveness of metal ions separation has to be much lower. Finally, for Co^{2+}-Zn^{2+} ($\Delta(\Delta H) = -0.25$ kJ/mol), no remarkable influence of temperature on α can be observed, hence no separation on IDA resin can be expected in this case.

| | $\ln(\alpha(T_2)/\alpha(T_1))$ | | | |
$T_2 - T_1$	*exp. (S_1)*	*exp. (S_2)*	*calc.[1]*	*calc.[2]*
353-293	---	-0.65	-0.66	-0.52
353-313	---	-0.40	-0.41	-0.32
353-333	---	-0.15	-0.19	-0.15
333-293	-0.41	-0.50	-0.47	-0.36
333-313	-0.26	-0.25	-0.22	-0.17
313-293	-0.15	-0.25	-0.25	-0.19

[1] $\Delta(\Delta H) = -9.6$ kJ/mol.   [2] $\Delta(\Delta H) = -7.4$ kJ/mol

Table 3 *Calculated and Experimentally Determined $\ln(\alpha(T_2)/\alpha(T_1))$ Values for Zn-Cu Exchange in S_1 and S_2 (see Table 1) at Different Temperatures*

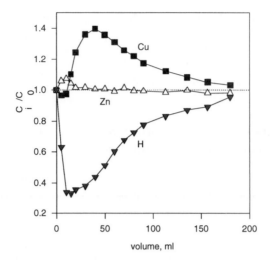

Figure 1 *Thermostripping elution curve obtained for S_1. Stripping with stock solution at 20°C from resin pre-equilibrated with the same solution at 60°C; C_i and C_0 are concentrations of ions in "i" solution sample and in stock solution, respectively*

The validity of the proposed approach has been demonstrated above (see the data obtained in S_1) and the results are illustrated in Figure 1 where the thermostripping elution curve obtained in S_1 is presented.[13]

As seen in Figure 1, selective thermostripping of Cu^{2+} against Zn^{2+} from IDA resin indicates that this ion exchanger can be used for dual-temperature fractionation of the ion couple under consideration.

Furthermore, another possible application of the proposed approach follows from the results obtained in S_2 and deals with its predictive ability in the preselection of ligands having greater promise for use in the synthesis of chelating resins with temperature-dependent selectivity. This preselection must be based on the fact that $\Delta(\Delta H)$ values for the reaction between the chosen ligand and certain metal ion couples should not be less than 10 kJ/mol to provide a sufficient variation of the respective α values in a suitably narrow temperature range. The experimental testing of water soluble ligands preselected can be carried out in systems of S_2 characteristics.

Acknowledgement

This work was supported by Research Grant N°. EV5V-CT94-556 from the Commission of the European Communities, Programme Environment 1990-1994. D. M. is visiting Professor by the Programa de Sabáticos of the Spanish Ministry of Science and Education (MEC). A. G. is a recipient of a fellowship from CIRIT (Commission for Science and Technology of Catalunya). Bayer Hispania Industrial, S.A., is gratefully acknowledged for kindly supplying with samples of Lewatit resins.

Nomenclature

α separation factor
β formation constant
C^* equilibrium concentration in solution phase
K   equilibrium constant
q^* equilibrium concentration in resin phase
T   temperature

References

1. H. T. Chen, "Handbook of Separation Techniques for Chemical Engineers", P. A. Schweitzer, ed., McGrawHill, New York, 1979, p. 467.
2. D. Tondeur and G. Grevillot, "Ion Exchange: Science and Technology", A. E. Rodrigues, ed., NATO ASI Series 107, Martinus Nijhoff, Dordrecht, 1986, p. 369.
3. G. Grevillot, "Handbook for Heat and Mass Transfer", N. P. Cheremisinoff, ed., Gulf Publishing, West Orange (NJ), 1985, Chapter 36.
4. B. M. Andreev, G. K. Boreskov and S. G. Katalnikov, *Khim. Prom-st.*, 1961, **6**, 369.
5. V. I. Gorshkov, A. M. Kurbanov and N. V. Apolonnik, *Zh. Fiz. Khim.*, 1971, **45**, 2969.
6. M. Bailly and D. Tondeur, *J. Chromatogr.*, 1980, **201**, 343.
7. D. Muraviev, J. Noguerol and M. Valiente, *React. Polym.*, in press.
8. D. Muraviev, J. Noguerol and M. Valiente, *Proceedings of the ION-EX'95 Conference*, Wrexham, Wales, September 10-14, 1995 (in press).
9. D. Muraviev, J. Noguerol and M. Valiente, *Hydrometall.*, in press.
10. R. Khamizov, D. Muraviev and A. Warshawsky, "Ion Exchange and Solvent Extraction", J. Marinsky and Y. Marcus, eds., Marcel Dekker, New York, 1995, p. 93.
11. G. E. Boyd, J. Schubert and A. W. Adamson, *J. Am. Chem. Soc.*, 1947, **69**, 2818.
12. A. J. Groszeck, "Ion Exchange for Industry", M. Streat, ed., Ellis Horwood Ltd., Chichester, 1988, p. 286.
13. D. Muraviev, A. Gonzalo and M. Valiente, *Anal. Chem.*, 1995, **67**, 3028.
14. J. J. Christensen and R. M. Izatt, "Handbook of Metal Ligand Heats", Marcel Dekker, New York, 3rd. ed., 1983.
15. V. I. Gorshkov, M. V. Ivanova and A. M. Kurbanov, *Zh. Fiz. Khim.*, 1974, **48**, 1237.
16. V. I. Gorshkov, D. N. Muraviev and G. A. Medvedev, *Zh. Fiz. Khim.*, 1977, **51**, 2680.
17. G. Anderegg, *Helv. Chim. Acta*, 1964, **47**, 1801.
18. R. P. Bonomo, R. Cali, F. Riggi, E, Rizzarelli, S. Sammartano and G. Siracusa, *Inorg. Chem.*, 1979, **18**, 3417.

REMOVAL OF HYDROXYACETONE FROM PHENOL CATALYZED BY A STRONG ACID ION-EXCHANGE RESIN

V. Milanesi, A. Franzoni, P. Nota, R. Penzo

ENICHEM
Research Centre
Via Taliercio 14
46100 Mantova
Italy

1 INTRODUCTION

The production of phenol from cumene is mainly carried out by the oxidation of cumene to cumene hydroperoxide, followed by the acid cleavage of the hydroperoxide to phenol and acetone.

In addition to the principal products, several side products are formed, such as α-methylstyrene, α-methylstyrene dimers, cumyl phenols, dicumyl peroxide, and so on. Moreover, along with these by-products, a number of minor impurities are formed, either due to impurities in the feed cumene or as a result of further side reactions occurring at various stages of the process. Hence, the cumene hydroperoxide cleavage reaction product is subjected to a series of purification steps in which phenol and acetone are recovered.

While most of the impurities can be removed by conventional operations, some of them are extremely difficult to separate from phenol. One such impurity is hydroxyacetone (HYAC). Theoretically it would appear to be a simple matter to separate hydroxyacetone (b.p. 145°C/760 mmHg) from phenol (b.p. 182°C/760 mmHg) by distillation. In practice, complete separation has been found to be almost impossible because of polar interaction between the compounds, and a chemical treatment is usually necessary to reduce hydroxyacetone to less than 5 ppm, as requested in most phenol specifications.

It is reported in the literature that hydroxyacetone can be removed from phenol by treatment with strong acid ion-exchange resin[1-3], but no systematic kinetic study has been reported. Therefore a programme of investigation on the kinetics of disappearance of hydroxyacetone in phenol medium, catalyzed by the strong acid ion-exchange resin Dowex M 15, was undertaken.

2 EXPERIMENTAL

The phenol used was produced in the Enichem plant of Mantova, and contained typically less than 100 ppm of total impurities.

In all experiments, the starting concentration of hydroxyacetone was increased to about 200 ppm by adding the requisite amount of commercial hydroxyacetone (Fluka Chem., assay: 95 %).

The strong acid ion-exchange resin used was the commercial catalyst Dowex M 15 (Dow Co.). The properties of the catalyst are listed in Table 1.

Table 1 *Properties of Ion-Exchange Resin Dowex M 15*

Properties	
Matrix	Polystyrene-DVB, macroporous
Functional group	Sulphonic acid
Minimum total capacity (wet), eq/L	1.8
Water retention, % w/w	60-64
Particle size, < 16 mesh, % w/w	5 max
Particle size, > 40 mesh, % w/w	3 max

The resin was initially washed with deionized water at 40°C to remove any impurities present, and then dehydratated under vacuum (5-10 mmHg) at a temperature of 100°C for 24 hours. The resulting resin had a content of water lower than 1 %.

The experiments were carried out in a 500 mL, fully baffled, mechanically stirred reactor.

In all experiments, phenol and the ion-exchange resin were first added in the reactor and heated to the reaction temperature while stirring. The hydroxyacetone was then added to the reactor, taking this moment as the starting time of the reaction. Samples were withdrawn at regular intervals and analyzed to follow the disappearance of hydroxyacetone. Each run was carried out to reach at least 90 % conversion of hydroxyacetone.

An experiment was carried out at 90°C, adopting a starting concentration of hydroxyacetone of about 2000 ppm. The samples withdrawn during this run were analyzed to determine the reaction products.

3 ANALYSES OF PRODUCTS

The determination of 2- and 3-methylbenzofuran was carried out by GC analysis.

GC conditions: GC Carlo Erba mod. HRGC 5300, equipped with split/splitless injector and FID detector. Column: fused silica capillary column Quadrex Series Bonded Phase, 25 m length, 0.53 mm internal diameter, 3 μm film thickness, 007 Carbovax 20 M stationary phase. Carrier gas: Helium, 2 mL/min, split/splitless 1/5. Temperatures: oven: 90-220°C, 1 min at 90°C, 5°C/min; injector: 240°C; detector: 240°C. Internal standard: n-hexadecane. Sample preparation: add 1 % w/w of water and 0.1 % w/w of internal standard to the phenol sample. Injection volume: 1 μL. Response factor of methylbenzofuran: 1.25.

The determination of the heavy compound $C_{15}H_{14}O_2$ was carried out by GC and GPC analyses.

GC conditions: GC Carlo Erba mod. HT-HRGC 5300, equipped with cold on-column injector and FID detector. Column: fused silica capillary column Mega, 15 m length, 0.32 mm internal diameter, 0.1-0.15 μm film thickness, SE54 stationary phase. Carrier gas: Helium, 2 mL/min. Temperatures: oven: 60-340°C, 1 min at 60°C, 20°C/min, 20 min at 340°C; detector: 360°C. Internal standard: n-hexadecane. Sample preparation: chloroform solution containing 30 % w/w of phenol sample and 0.1 % w/w of internal standard. Injection volume: 1 μL. Theoretical response factor of $C_{15}H_{14}O_2$: 1.3.

GPC conditions: LC Hewlett Packard 1090 M, equipped with UV/Visible Diode Array Detector and ACS Evaporative Mass Detector. Columns: three different porosity columns system, PL-Gel styrene-divinylbenzene stationary phase, each column of 300 mm length and 4.6 mm internal diameter. Filling porosity: 50 A, 100 A, and 500 A. Solvent: tetrahydrofuran. Injection volume: 100 μL. Measurement wavelength: 280 ± 2 nm, with reference to 450 ± 50 nm. External standard: bisphenol. EMD response factor: 1.

The identification of products was carried out by GCMS analysis.

GCMS conditions: mass spectrometer Hewlett Packard mod. 5988 A coupled with a GC Hewlett Packard mod. 5890, equipped with split/splitless injector. Measurements: electronic impact ionization, 70 eV, 25-350 m/z range, 200°C source temperature; chemical ionization, with methane as the reactant gas.

4 DETERMINATION OF THE KINETICS

The kinetics of disappearance of hydroxyacetone in phenol was studied in the temperature range 50-92°C.

The catalyst loading was varied in the range 0.3-2 % (w/w), and the starting concentration was about 200 ppm for all the experiments.

The speed of agitation was varied from 200 to 800 rpm, without any effects on the kinetics. This proved that there was no external mass-transfer resistance associated with the disappearance of hydroxyacetone in phenol.

The influence of the catalyst particle size was also tested to find out the presence of intraparticle gradient, carrying out kinetic runs with resin beads respectively ≥ 45 and ≤ 20 mesh, but no difference was observed. The disappearance of hydroxyacetone in the presence of the strong acid ion-exchange resin Dowex M 15 thus appeared to be kinetically controlled.

Under the experimental condition examined, the reaction was found to follow first-order kinetics with respect to hydroxyacetone. Experimental data were fitted using the equation system:

$$C = C^{\circ} \cdot e^{-kt}$$

$$k = k_* \cdot e^{-\frac{Ea}{R}\left(\frac{1}{T} - \frac{1}{T_*}\right)}$$

where C is the dependent variable, C°, T and t are the independent variables, k_* and Ea are the parameters of the model. The parameters were computed by a weighted non linear least squares analysis on the basis of a direct search method. The objective function to be minimized was the summation of the squares of the differences between calculated and experimental concentration values. A good agreement between calculated and experimental data was generally found, as shown in Figure 1.

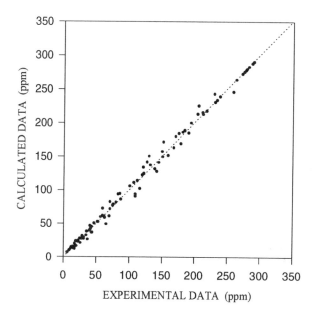

Figure 1 *Experimental vs predicted concentrations of hydroxyacetone*

5 DETERMINATION OF THE REACTION PRODUCTS

The disappearance of hydroxyacetone in phenol, catalyzed by the strong acid ion-exchange resin Dowex M 15, gave predominantly a heavy product having molecular formula $C_{15}H_{14}O_2$ and m.w. 226.27. The additional products formed in the reaction were the two isomers of methylbenzofuran (MBF). The overall reaction scheme can be represented as follows:

At 90°C, the complete conversion of hydroxyacetone gave a selectivity with respect to $C_{15}H_{14}O_2$ of about 90 %. The concentration profiles of reactant and products are reported in Figure 2.

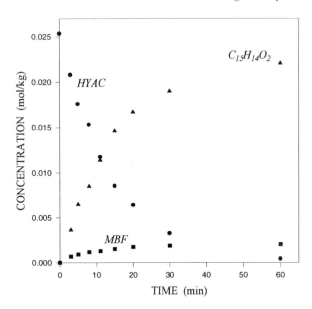

Figure 2 *Concentration profile of the reactant (HYAC) and products ($C_{15}H_{14}O_2$ and MBF) at 90°C*

The reaction of condensation of hydroxyacetone with phenol to give methylbenzofuran is well known[4-9]. On the contrary, the reaction leading to $C_{15}H_{14}O_2$ is very scantly described in the literature. The only available information[10] concern the acid catalyzed reaction of phenol with chloroacetone to give the $C_{15}H_{14}O_2$ compound with structure corresponding to 1,2-bis-[4-hydroxy-phenyl]-propene. We have recently started work to verify if this structure corresponds to our compound.

6 TREATMENT SECTION

Based on the kinetic model, an industrial reaction unit was sized to obtain almost the complete removal of hydroxyacetone at 50°C.

The plant data were in good agreement with the expected values. Feeding the reactor with a phenol stream containing about 50 ppm of hydroxyacetone as a maximum, typically 10-20 ppm, under the adopted reaction conditions, the complete conversion of hydroxyacetone was obtained.

The resin, charged in 1990, has been satisfactorily working until now, and has treated about 100000 kg of phenol / 1 kg of dry resin. The gradual loss of activity shown by the catalyst during these five years, has been successfully overcome by increasing the reactor temperature.

The adoption of this catalytic treatment has also allowed a substantial improvement of other phenol quality characteristics.

7 CONCLUSIONS

The proposed kinetic model is satisfactory in order to interpret all the laboratory experimental data and also the industrial plant data. This model permits the simulation of the plant unit with good results and represents a useful tool to optimize the operating conditions.

Nomenclature

$C°$: initial concentration of hydroxyacetone, ppm
C : concentration of hydroxyacetone, ppm
k : kinetic constant, min^{-1}
k_* : kinetic constant at the reference temperature, min^{-1}
t : time, min [(catalyst mass (g) / solution mass (g)) · sampling time (min)]
Ea: activation energy, cal/mol
R : gas constant, 1.98717 cal/K/mol
T : temperature, K
T_* : reference temperature, K

References

1. Universal Oil Products Company, British Patent 1,108,584, 1968.
2. Union Carbide Company, British Patent 1,381,398, 1975.
3. Phenolchemie, French Patent 726,817, 1969.
4. BP Chemicals Limited, European Patent 0,004,168, 1979.
5. Distillers Company Limited, British Patent 916,536, 1963.
6. BP Chemicals Limited, British Patent 1,394,452, 1975.
7. Universal Oil Products Company, British Patent 1,108,327, 1968.
8. G. Messina et al, *La chimica e l'industria*, 1983, **65**, N.1, 10.
9. R. C. Elderfield, V. B. Meyer, "Benzofuran and its Derivatives", Eterocyclic Compounds, Wiley, New York, 1951, Vol. 2, p. 1.
10. Zaheer et al., *J. Chem. Soc.*, 1954, 3360.

RECOVERY OF $^{60}Co(CN)_6{}^{3-}$ AND $S^{14}CN^-$ FROM SALINE WATER

G. Yan*, D. Ø. Eriksen and T. Bjørnstad

Institute for Energy Technology
P.O. Box 40, N-2007 Kjeller, Norway

*Department of Chemistry
University of Oslo
Norway

1 INTRODUCTION

To enhance oil-recovery, injection of water to maintain field pressure is usually the primary choice. Each production well may thus be surrounded by several injection wells. However, fluids from each injection well may reach more than one production well, depending on the topography. To achieve the optimum field performance, there is a need for means to determine the flow pattern of water in the reservoir. The best method of obtaining this information is to add compounds to trace the injected water and measure them when being produced together with the oil. In addition to knowledge of flow-paths in the reservoir, by simulations of the tracer-pulse response one may deduce other important information of the reservoir from the produced water. Thus, use of tracer methods is part of the oil companies' basis for taking economical decisions. Formation-water, present originally in the rock, will usually differ from injected water in the salt-content and -composition, as injected water is usually filtered and de-oxygenated sea-water.

The definition of an ideal tracer is a compound that behaves exactly as the substance studied under the conditions of interest. This implies that a compound to be used for tracing water in a water-flooded oil-field must fulfil the following requirements:
-    No absorption to rock.
-    No partition (or distribution) to the oil present in the reservoir.
-    Long time thermal stability. North-Sea reservoirs have temperatures in the range 70 - 150°C and injected water may use from one to three years to flow from the injector- to the producer-well.
-    Due to the high dilution, factors in the order of 10^9 - 10^{12} are typical, the tracers must be detectable in minute quantities.

These requirements implies that no reaction with species present in the injection-water must take place. Also, the species must be water soluble at pH 6 - 8, pH of formation-water. In addition, health and safety precautions as well as price-level are considerations to be taken into account when choosing tracers. Among those commonly used as water tracers in flooded reservoirs are $Co(CN)_6{}^{3-}$ and SCN^-.[1] The former may be radioactively labelled with the following nuclides: ^{56}Co, ^{57}Co, ^{58}Co, ^{60}Co or ^{14}C whereas the latter compound may be labelled with ^{14}C or ^{35}S. All the cobalt nuclides are γ-emitters. The nuclei ^{14}C and ^{35}S are both pure β-emitters with quite similar β-energies.

To reduce detection limits and tracer-costs pre-concentration methods are valuable tools and often necessary in order to measure mixtures of tracers. In concentrating radioactive tracers we have the possibility of using a carrier.

2 OBJECTIVES

In this study the aim was to develop a method for quantitative analysis of one litre sea-water containing a mixture of SCN^- and $Co(CN)_6^{3-}$ by pre-concentration into samples small enough to be measured by liquid scintillation counting (LSC). The separation of the two tracers is crucial since presence of radioactive Co will disturb the β-measurements of SCN^-. The sample for LSC should not exceed 8 mL as typically 12 mL scintillation cocktail is added to dissolve the tracer containing, eluant brine. A requirement for the separation of $Co(CN)_6^{3-}$ is that no chemicals should be used that will lead to high quenching of ^{14}C-spectrum. Pre-concentration procedures of $S^{14}CN^-$ from sea-water are published by Bjørnstad et al.[2] Their best method is based on absorption on a Bio-Rad AG1-x8 column and elution with 2.8M $NaClO_4$. Preferentially, this procedure should be included in these separations. To measure γ-emitting Co-nuclei, measurements directly on resins or feed-solution is possible, but if labelled with ^{14}C it is imperative to have the hexacyanocobaltate complex in a solution of small volume.

3 EXPERIMENTAL

3.1 Design of Experiment

The main hypothesis was to take advantage of the difference in valence charge of the anions to obtain a separation. Since a separation method for thiocyanate was available and in use in our laboratory, the primary step would be to perform screening tests to find a selective separation of $Co(CN)_6^{3-}$ by using ^{60}Co-labelled tracer. Secondly, test for stripping properties to avoid strong quenching solutions, for allowing use of $Co(^{14}CN)_6^{3-}$. Thirdly, checking the absorption characteristics of SCN^- in the $Co(CN)_6^{3-}$ -method, and finally, use the two procedures together to separate a solution containing both tracers, labelled with ^{60}Co and ^{14}C, respectively.

Thus, several probable pre-concentration procedures by solvent extraction and ion-exchange of $^{60}Co(CN)_6^{3-}$ were performed. Table 1 lists the different solvent extraction systems tested. The extractants tested were primary-, secondary and ternary amines, i.e. anion exchangers. Per cent extracted as a function of pH was measured. Table 2 lists the anion-exchanger resins tested. To fulfil the objectives, high distribution ratios and high separation factors from SCN^- are imperative. Thus, those extractants having the best extraction properties of $^{60}Co(CN)_6^{3-}$ were tested towards stripping properties and extraction properties of SCN^-.

3.2 Radioactivity Measurements

^{60}Co was measured with a lead-shielded Ge(Li)-detector with corresponding electronics and coupled to a PC acting as a multi-channel analyser through Ortec-interface and MAESTRO-software from Ortec. ^{14}C was measured with a Beckman LS 3000 liquid scintillation counter.

Table 1 *Screening Tests of Some Probable Solvent Extraction Pre-concentration Procedures*

Extractant	Concentration	Stripping agents tested		
Primene JM-T	10% in kerosene			
Amberlite LA-2	10% in kerosene	3.6M H_2SO_4	6M NH_3	1M $K_2(COO)_2$
Amberlite LA-2, without pre-treatment with 0.1M H_2SO_4	10% in kerosene			
Alamine 336	10% in kerosene	3.6M H_2SO_4	6M NH_3	1M $K_2(COO)_2$

Table 2 *Some Probable Anion Exchangers for Pre-concentration. Stripping Agents Tested are Shown*

Extractant	Particle size (mesh)	Stripping agents tested	
Amberlite IR45*	50 - 100		
Lewatit MP60*	50 - 100		
Dowex-1 x2	50 - 100	4M NH_4Cl+2M HCl	12M HCl
		65% HNO_3	
Dowex-1 x2	200 - 400	NH_4Cl (saturated)	65% HNO_3
Dowex-2 x8	100 - 200	70% $HClO_4$	HI (conc.)
		96% H_2SO_4	12M HCl
		32.5% HNO_3	
		2.5M NH_4NO_3	5M NH_4NO_3
		7.5M NH_4NO_3	10M NH_4NO_3

* Cross-linking is unknown.

3.3 Solvent Extraction Procedures

The organic phases were pre-treated with 0.05M H_2SO_4. 5 mL of sea water pH-adjusted by addition of HCl and containing trace amounts of ^{60}Co-labelled $Co(CN)_6^{3-}$ was contacted with 5 mL organic phase in a separatory funnel. Phase ratios were held constant at 1. The funnel was shaken mechanically for a predetermined time, 3 minutes. After phase separation, 1 mL samples of both phases were measured for radioactivity. Distribution coefficients, D, and per cent extraction, $\%E$, were determined as follows:

$$D = \frac{\dfrac{R_{org}}{v_{org}} - B}{\dfrac{R_{aq}}{v_{aq}} - B} \qquad (1)$$

$$\%E = \frac{100\% \, D}{\dfrac{V_{aq}}{V_{org}} + D} \qquad (2)$$

where R is sample count rate, B background count rate, v sample- and V phase-volume and subscripts *org* and *aq* denote organic and aqueous phases, respectively.

Loaded organic phases were tested for stripping ability with the same experimental procedure as above.

3.4 Anion-exchange Procedures

Resins to be tested for absorption of $Co(CN)_6^{3-}$ were immersed in water for 24 hours. The same column, $\emptyset = 5$ mm, $l \approx 25$ mm, was used for all tests. After transfer to the column, the resin was washed with 10M HCl followed by pure water until pH ≈ 7 on exit. 100 - 1000 mL of sea water with trace amounts of ^{60}Co-labelled $Co(CN)_6^{3-}$ was passed through the column. We define per cent absorbed, $\%A$, as

$$\%A = \frac{\left(R_{feed} - B\right)\dfrac{V_{feed}}{v_{feed}} - \left(R_{raf} - B\right)\dfrac{V_{raf}}{v_{raf}}}{\left(R_{feed} - B\right)\dfrac{V_{feed}}{v_{feed}}} 100\% \tag{3}$$

where R denotes sample count rate, B background count rate, v sample volume, V phase volume and subscripts *feed* and *raf* denote feed (or input) solution and raffinate, respectively. Stripping yield, y_s, was determined as the fraction of tracer in eluate relative to the amount in the feed solution. This method requires a 100% absorption, but only those systems were tested for stripping. Care must be taken in assuring similar counting efficiencies for the feed and eluate.

$$y_s = 100\% \frac{\left(R_s - B\right)\dfrac{V_s}{v_s}}{\left(R_{feed} - B\right)\dfrac{V_{feed}}{v_{feed}}} \tag{4}$$

index s refers to sample.

The resin used for SCN$^-$-separation, Bio-Rad AG1x8, was pre-treated the same way, but a column $\emptyset = 6$ mm, $l = 300$ mm was used. When ^{14}C-labelled SCN$^-$ was used, amounts in raffinate and portions of eluate were measured relative to the feed solution. The separation factor in stripping between $^{60}Co(CN)_6^{3-}$ and S^{14}CN$^-$ was accordingly defined as

$$\alpha \, _{Co}^{SCN} = \frac{y_{SCN}}{y_{Co}} \tag{5}$$

Flow rates were not measured properly in all tests, but was approximately 0.5 mL/min.

4 RESULTS

4.1 Solvent Extraction

Of the four extraction systems tested, only the tertiary amine extractant, Alamine 336 from Henkel Corp., i.e. trioctylamine[3], extracted the cyanide-complex strongly in the pH-range 1 to 7. The primary amine, Primene JM-T from Rohm and Haas, extracted only minor amounts at pH = 1, while the secondary, Amberlite LA-2 from Rohm and Haas, showed decreasing extraction property with increasing pH. These results are shown in Figure 1.

Loaded phases of Amberlite LA-2 and Alamine 336, both as 10% by volume in kerosene, were tested for stripping. The following stripping agents were tested: 1.8M H_2SO_4, 3.6M H_2SO_4, 6M NH_3 and 1M $K_2(COO)_2$. The latter stripped both loaded phases, whereas 6M NH_3 also could be used for stripping Amberlite LA-2.

4.2 Anion Exchange

4.2.1. Hexacyanocobaltate. Table 3 summarises the absorption of $^{60}Co(CN)_6^{3-}$ for the resins listed in Table 2. Per cent absorption is determined from equation 3. The strong base resins were superior to the weak base ones. In Table 2 the stripping agents tested are listed. Strong acids and oxidising agents are usually strong quenchers and should be avoided. In Table 4, a comparison of stripping characteristics between Dowex-1 and 2 is made. The difference in cross-linking between the resins is considered to have minor importance.

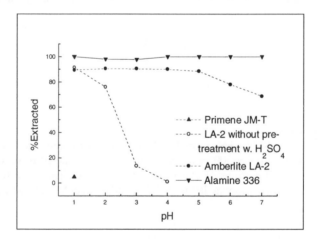

Figure 1 *%E for extraction of $Co(CN)_6^{3-}$ with four solvent extracting agents*

In Figure 2 the remaining fraction of tracer on the Dowex-2 column after elution of 10 mL of either NH_4NO_3 (2.5, 5.0, 7.5 and 10M) or HNO_3 (7.25 and 14.5M) is shown.

4.2.2 Thiocyanate. As Bio-Rad AG1 is equivalent to Dowex-1, its performance towards $Co(CN)_6^{3-}$ was known. It was therefore necessary to test the performance of Dowex-2 x8 towards absorption of SCN^- and the separation factor between the two tracers obtained for various stripping (or washing) solutions. Absorption break-through curves for Dowex-2, 0.5 mL column, and Bio-Rad AG1, 8.5 mL column, is shown in Figure 3. Further, Table 5 lists the stripping agents and the separation factors obtained.

Table 3 *%A for Five Different Anion-Exchange Systems*

Extractant	Producer	Resin type	%A
Amberlite IR45	Rohm and Haas	Weak base	40.7
Lewatit MP60	Bayer AG	Weak base	2.6
Dowex-1 x2 (50-100 mesh)	Dow Chemical	Strong base	95.9
Dowex-1 x2 (200-400 mesh)	Dow Chemical	Strong base	100
Dowex-2 x8 (100-200 mesh)	Dow Chemical	Strong base	100

Table 4 *Comparison of Absorption Strength of Dowex-1 and 2*

Stripping agent	Stripping yield (%)	
	Dowex-1	*Dowex-2*
12M HCl	24.7	63.4
14.5M HNO$_3$	52.3	80.3

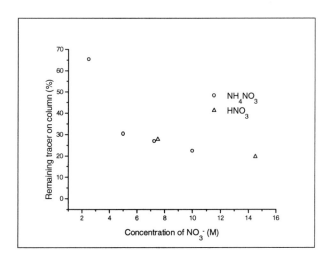

Figure 2 *Stripping of Co(CN)$_6^{3-}$ from Dowex-2 with NH$_4$NO$_3$ (2.5, 5.0, 7.5 and 10M) or HNO$_3$ (7.25 and 14.5M)*

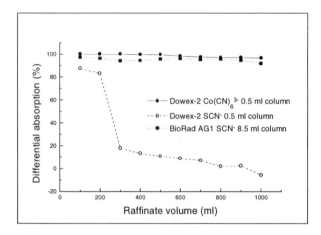

Figure 3 *Absorption break-through curves for $Co(CN)_6^{3-}$ and SCN^- on Dowex-2 x8 (100 - 200 mesh) and for SCN^- on Bio-Rad AG1 x8 (50 - 100 mesh)*

Table 5 *Stripping Agents Tested for 0.5 mL Dowex-2 x8 Column and the Separation Factors Obtained between $^{60}Co(CN)_6^{3-}$ and $S^{14}CN^-$*

Resin	Stripping agent	Yield (%) $^{60}Co(CN)_6^{3-}$	$S^{14}CN^-$	Separation strip factor, $\alpha\ {}^{SCN}_{Co}$
Dowex-2 x8	Na_3PO_4 (sat.)	≈0	3.4	-
	1M NH_4NO_3	5	85	17.0
	1.5M NH_4NO_3	7	95	13.6
	0.1M $NaClO_4$	2	60	30.0
	0.2M KI	5	88	17.6

5 CONCLUSIONS

The results show a higher extraction of the hexacyanocobaltate complex with strong basic extractants than with weak ones. This is valid both for the liquid extraction agents and for the solid anion exchangers tested.

Dowex-1 x2 extracts the hexacyanocobaltate complex more strongly than Dowex-2 x8 as shown in Table 4. It may be stripped from Dowex-2 x8 by employing 5M NH_4NO_3. By using a modern liquid scintillation cocktail, e.g. 4 mL sample to 10 mL Optiphase Supermix from Wallac Oy , this concentration gives acceptable quenching.

The difference in cross-linking between the Dowex-1 and -2 resins used, would favour the kinetics of the x2-resin.[4] However, the x8-resin show the highest stripping yield. We therefore conclude that the difference in absorption strength is due to the difference in functional groups.

Thiocyanate will be washed out selectively from hexacyanocobaltate on Dowex-2 x8 with 10 mL 0.1M $NaClO_4$. This strip-solution added to the raffinate from the cobaltate absorption may be used as feed for the thiocyanate absorption on Bio-Rad AG1.

Recovery of hexacyanocobaltate from 1000 mL sea-water on a 0.5 mL Dowex-2 x8 column is >99%, stripped with 10 mL 5M NH_4NO_3 it is >75%. Recovery of thiocyanate is > 65% by absorption of 1000 mL sea-water on a 8.5 mL Bio-Rad AG1 column and stripping with 10 mL 4M $NaClO_4$. This latter yield may be improved by decreasing flow rate of eluant.

Hexacyanocobaltate may thus be concentrated 75 times, whereas thiocyanate may be concentrated 65 times by the method here reported.

The method for concentrating hexacyanocobaltate may also be employed in analysis of non-radioactive tracers. By neutron activation analysis of the resin with absorbed cobaltate detection limits in the order of 20 nM of Co is easily obtained. Miller et al.[5] uses combustion of Bio-Rad AG1 x8-resin and inductively coupled plasma connected to mass spectrometry to obtain detection limits in the range of ppt (< 20 nM) of Co.

Acknowledgements

The authors are indebted to the IFE Tracer Research Co-operation, ITRC, for allowing us to publish procedures and results obtained during the course of this project. The work is supported by the following oil-companies: Amerada Hess Norge, BP Exploration, Conoco, Mobil Exploration Norway, Petrobras Norge, Phillips Petroleum Company Norway, Saga Petroleum and Statoil.

References

1. T. Bjørnstad, IFE Research Report, 1991, IFE/KR/E-91/009.
2. T. Bjørnstad, E. Brendsdal, O. B. Michelsen, S. A. Rogde, *Nucl. Instr. Meth.*, 1990, **A299**, 629.
3. D. Flett, J. Melling, M. Cox, "Handbook of Solvent Extraction", T.C. Lo, M.H.I. Baird and C. Hanson, eds., John Wiley & Sons, New York, 1983, Chap. 24, p. 629.
4. F. Helfferich: "Ion Exchange", McGraw-Hill Book Company, Inc., New York, 1962, Chap. 6, p. 250.
5. J. F. Miller, C. O. Sheely, J. W. Wimberley, R. A. Howard, US Patent 5,246,861 Sep. 21, 1993.

EXTRACTION OF CEPHALOSPORIN C FROM AQUEOUS SOLUTIONS WITH HYPERSOL MACRONET RESINS

S. Belfer and N. Daltrophe

The Institutes for Applied Research
Ben-Gurion University of the Negev
P.O. Box 653, Beer-Sheva 84105
Israel

1 INTRODUCTION

Large-scale chromatographic separation is of increasing interest both in research and in applied fields. Purification of biologically produced materials is currently gaining the attention of many research teams, and there are a vast number of reports on the development of large-scale purification of downstream feed stocks.

Antibiotics are probably the most important pharmaceutical products processed by means of ion-exchange technologies. This presentation deals with cephalosporin antibiotics, which occupy a unique place in the historical development of β-lactam drugs. Among the variety of systems described for the isolation of this class of antibiotic from fermentation broths, ion-exchange is the simplest and most economical, with total recovery of the antibiotic activity[1-3].

Typical product concentration in fermentation broths for the production of cephalosporin C ranges from 5 to 15 g/L. This low product concentration combined with the fact that the components of the medium may remain in solution after cell separation usually dictate a multi-stage product recovery operation[4]. After an initial isolation step, the product undergoes a gross purification step. Unit operations, such as adsorption extraction and precipitation, are used in primary recovery. Following this step, secondary purification or a polishing step may be necessary to eliminate contaminants not amenable to removal by the preceding operations. In the large-scale production of cephalosporin C, this polishing step includes selective adsorption on polystyrene-divinylbenzene co-polymers. The advantages of polystyrene non-functional adsorbents lie in their chemical and physical stability throughout the pH range 1-14, their large capacity, and their response to a variety of eluent conditions, including pH, type of organic modifier, water-organic modifier ratio and presence of counterions[5].

Macroporous Amberlite XAD-2[1], XAD-4[4], XAD-16[6] and specially produced XAD-180[7] have been used commercially for the purification of cephalosporin C. Parallel to the development of large-scale processes, publications concerning the synthesis and application of new polystyrene adsorbents for the extraction of cephalosporin C are constantly appearing in the literature[8-10]. These papers are devoted to elucidating the structure-performance correlation. It is recognized that hydrophobic interactions play a significant role in the organic matter extraction by polymer systems, particularly in the extraction of antibiotics by ion exchangers[11-15]. The constant interest of chemists in improving the synthesis—and hence the performance—of sorbents for the purification of

Table 1 *Characteristics of Hyper-crosslinked Macronet Resins (as given by manufacturer)*

Resin	Functionality	Volume capacity (eq)	Surface area (m²/g)	Pore volume (mL/g)	Pore diameter (Å)
MN-100	Weak base anion exchanger	0.1 - 0.2	800 - 1000	1 - 1.1	850 - 950
MN-150	Weak base anion exchanger	0.1 - 0.3	800 - 1000	0.6 - 0.8	300 - 400
MN-200	None		800 - 1000	1 - 1.1	850 - 950
MN-400	Strong base anion exchanger	0.2 - 0.4	800 - 1000	1 - 1.1	850 - 950
MN-500	Strong acid cation exchanger	0.8 - 1.0	800 - 1000	1 - 1.1	850 - 950
XAD-4	None		872	1.14	62

cephalosporin C undoubtedly serves as an indicator of the need for better sorbents.

Recently, Purolite International made available a new class of resins, known as Hypersol-Macronets[16]. These resins are based on cross-linked polystyrene prepared according to a procedure initially developed and patented by Russian investigators[17,18]. Although the ability of the new adsorbents to extract different chemical species (metals ions, organic matter, vapors) has been studied[19,20], there are no data concerning the extraction of cephalosporin C. Thus, the data required should include, first of all, the equilibrium isotherm data, in particular the type and shape of the isotherm, and the numerical values of the isotherm parameters and their dependence upon the conditions of adsorption. The conventional technique for the determination of isotherm data involves batch experiments, in which the ultimate adsorbate concentration in a solution of known initial concentration is determined after contact with a given amount of adsorbent.

The goal of the present research was to study adsorption and desorption of cephalosporin C in batch technique and to perform some column experiments. In addition, some experiments were performed to evaluate the ability of the resins to extract proteins.

2 EXPERIMENTAL

2.1 Equilibrium Adsorption Data

Cephalosporin C solutions of different concentrations were prepared by dissolving the Zn salt of cephalosporin C. Buffered solutions at pH 2 and pH 3.5 were prepared using citric acid and NaOH solution, respectively. For pH 7, Zn salt of cephalosporin C was first dissolved in acetic acid, and then NaOH was added to pH 7. About 0.5 g of dried resin was placed in an Erlenmeyer flask. Then, 10 mL of cephalosporin C solution of known concentration were added to the flask, which was placed on a shaking machine at constant temperature until equilibrium was attained. Usually, the suspension was gently mixed by inversion over a period of 24 h. The equilibrium amount of sorption was calculated from the differences in cephalosporin C concentrations before contact with the resin and after a given time. The concentration was determined with a UV spectro-photometer; absorbance at 260 nm was read.

2.2 Column Performance

Column runs were done using a 10 mm x 200 mm glass column in a down-flow fashion at room temperature (22-25°C). The resin was placed in the column, and the bed was washed several times with buffer solution. Cephalosporin C solution, 1,000 mL at pH 2.5, was passed through the column at a rate of 1 mL/min. The effluent samples were collected by a fraction collector and were examined for adsorbate concentration (optical density at 260 nm). For the determination of breakthrough curves, the beds were loaded until the cephalosporin C concentration at the outlet approached 300 ppm.

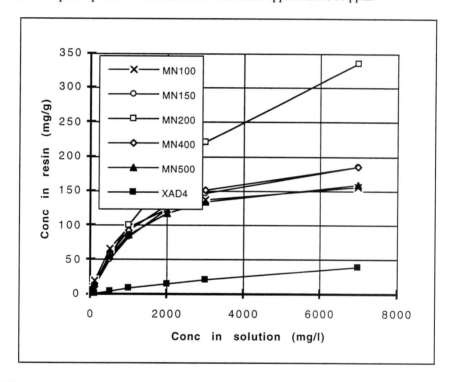

Figure 1 *Langmuir isotherms of cephalosporin C absorption at 25°C*

3 RESULTS

3.1 Equilibrium Adsorption Experiments

The characteristics of the five resins used in these experiments are given in Table 1. These five resins comprised three anion-exchange resins (two weak and one strong), a cation exchange resin and an inert resin. Commercial Amberlite XAD-4 was also used for purposes of comparison. All resins were first extracted with methanol in a Soxhlet apparatus and dried under vacuum at 40°C. The equilibrium adsorption data was obtained from batch experiments. Capacities for each resin were calculated from the experimental data for concentrations from 250 ppm to 10,000 ppm at pH 2.5 and 24 h of contact time (Figure 1). Resin capacities (mg/g) were plotted as a function of solution concentrations (mg/L). The equilibrium adsorption data were fitted to a Langmuir isotherm by least-squares regression of the linear transformation.

Figure 1 shows that for all the resins studied the amount of cephalosporin C adsorbed on the resin increased as its concentration in the liquid phase increased. The highest capacity (about 160 mg/g) was shown by resin MN-200, which has no functional groups, although differences in the capacities between the other resins were not significant. The capacity of XAD-4 (42 mg/g) was about four times lower than that of MN-200.

3.2 Effect of pH

The effect of the pH of the external solution on the capacity of the different resins for cephalosporin C was studied for an initial cephalosporin C concentration of about 1000 ppm. Data were collected after 4 h of contact of the solution with the resin (Figure 2)

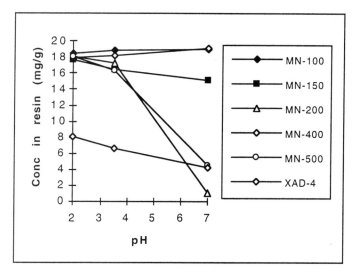

Figure 2 *Effect of pH on sorption onto different resins*

The Figure shows that at acidic pH the capacities of all the resins are similar, but the picture changes dramatically at pH 7. According to their behavior at pH 7, three types of resin were distinguished: a) those for which the capacities did not change (MN-100 and MN-400); those for which the capacities of the resins were reduced dramatically (MN-200 and MN-500); and those whose capacities decreased only insignificantly (MW-150 and XAD-4).

The effect of pH on the cephalosporin C adsorption by resins MN-150 and MN-200 was also tested in another way. The two resins were brought into contact with a 1,000 ppm solution of cephalosporin C at pH 2 and pH 7, and the capacities of the resins

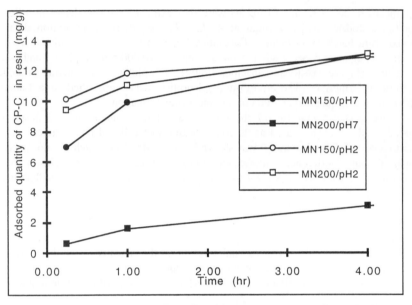

Figure 3 *Sorption as a function of time onto the resins MN-150 and MN-200 at pH 2 and pH 7*

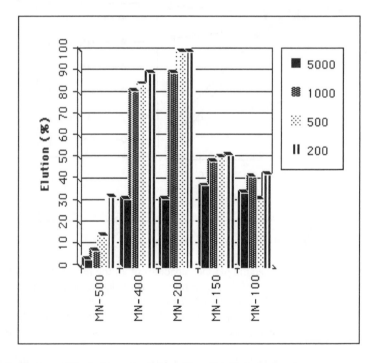

Figure 4 *Elution of Cephalosporin C from Purolite Macronet resins*

were plotted as a function of time (Figure 3). It should be noted that, unlike the above-mentioned experiments, this set of experiments were carried out over a short period (only 4 h) that was not sufficient to facilitate equilibrium. Therefore, it is not surprising that the results obtained for resin MN-150 were slightly higher than those for MN-200. At pH 7 the amount of cephalosporin C taken up by resin MN-200 decreased significantly, while the capacity of resin MN-150 was only slightly reduced.

3.3 Desorption

Experiments on desorption from loaded resins were performed by a batch technique. Loading was performed from solutions of the same cephalosporin C concentration as in the sorption experiments. In the desorption tests, 0.5 g loaded resin were treated with 10 mL of eluent—in this case methanol. The results of one-step desorption are shown in Figure 4. The histogram shows that the resins MN-200 and MN-400 could be almost completely eluted when they were loaded from a relatively low cephalosporin C concentration (200-1,000 ppm). The degree of desorption decreased sharply when loading was performed from a 5,000-ppm solution of cephalosporin C: only 30-40% of cephalosporin C was recovered. Under the same conditions, however, the other two resins, MN-100 and MN-150, which had shown much poorer elution, gave about 40% recovery even at low loading. For the resin MN-500, only 5% recovery of the antibiotic was obtained when the resin was loaded from solution of 500 ppm of cephalosporin C.

3.4 Column Performance

The column operation carried out with the inert resin, MN-200. Wetted resin, 5g, was placed in a column (1 cm ID and 25 cm high), the resin bed being about 18 mL. The break-through of loading was observed after 58 bed volumes (BV), when the rate of

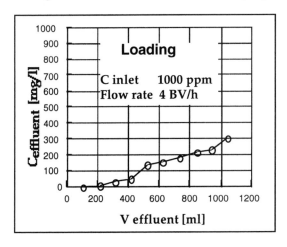

Figure 5 *Column performance with resin MN-200: Loading*

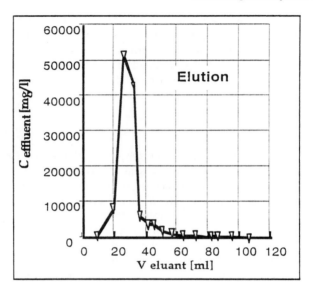

Figure 6 *Column performance with resin MN-200: Elution*

loading was about 4 BV/h (Figure 5). The concentration of cephalosporin C in the inlet was 1,000 ppm, and at the end of the sorption step the concentration of effluent was about 200 ppm. After washing the resin bed with water (about 500 mL), elution with methanol was performed. The elution curve exhibited a very sharp peak of average concentration of about 36,000 ppm. In the following run, the same loading conditions were used, but washing was carried out with only 100 mL of water before elution was performed. In this experiment the concentration of eluted cephalosporin C reached 52,000 ppm (Figure 6).

4 DISCUSSION

Cephalosporin C is an amphoteric derivative of 7-aminocephalosporanic acid; its formula is:

$$NH_2\text{-}CH\text{-}(CH_2)_3\text{-}CO\text{-}NH \qquad COOH \qquad CH_2OCOCH_3 \qquad COOH$$

CEPHALOSPORIN C

Generally, the adsorption of cephalosporin C on a polymer matrix is a typical case of sorption of a polar organic compound from an aqueous solution. The theory of adsorption of such compounds on charged or inert polymers has been previously described[12-14, 21,22]. According to the theory, the hydrophobic interaction of the nonpolar skeleton and the aromatic part of the sorbate are operative in the adsorption of organic compounds. However, interactions such as acid-base or donor-acceptor cannot be ruled out. Biochemicals containing the soft donor thiol or thioether groups, as for instance cephalosporin, may be preferentially recovered from solutions by using soft acceptors[23].

Depending on the pH of the medium, cephalosporin—having both amino and carboxyl groups—may be found in solution in cationic ($^+H_3NPCOOH$), anionic (H_2NPCOO^-), or dipolar ($^+H_3NPCOO^-$) form. At acidic pH, the cationic form is dominant, and the whole molecule may be regarded as inert. In this case, hydrophobic interactions rather than ion exchange are responsible for adsorption (non-specific adsorption in conventional terminology). Therefore, it was not surprising that the highest adsorption of cephalosporin C was demonstrated by the inert resin.

Our results show that the sorption of cephalosporin C from acidic solutions on different Purolite Macronet resins is much higher than on the XAD-4 resin. The most logical explanation for the better performance of Purolite resins could be simply the larger pore size of these resins (Table 1). It is likely that the origin of the Purolite polystyrene matrix (determined by its synthesis) is the major factor determining the mechanism of van der Waals' interaction between the molecule of antibiotic and the polymer under our particular sorption conditions.

The desorption data show significant differences that require explanation. Complete elution was achieved from the inert resin (MN-200) and from the resin with the strong amine functionality (MN-400). Two possible situations might arise in the case of these two resins:

1. Hydrophobic interaction, which is the driving force of cephalosporin C adsorption on the inert matrix at acidic pH, becomes less pronounced when the pH is shifted in the elution step; and

2. The presence of positively charged amino groups in resin MN-400 should alter the penetration of adsorbate molecules deeply inside the polymer due to the electrostatic repulsion between the charged groups of the solute and polymer.

In both cases, the adsorbate molecules are retained on the surface and can therefore be easily eluted. In contrast, the cation-exchange resin, with its high negative charge, favors the formation of an acid base complex between the amino group of cephalosporin C and the sulfo group on the surface and inside the polymer. In this case, the cephalosporin C molecules appear to be entrapped, and recovery is hampered.

With regard to the resin with the weak amine functionality, its ability to retard the penetration of molecules inside the polymer is lower than in the strong anion exchanger; therefore the cephalosporin C molecules may be taken up rather strongly by polymer and not easily eluted.

The presence of functional groups on the surface of the studied Macronet resins influences the sorption of cephalosporin C even more strongly at pH 7. The two resins having the same porous characteristics, i.e., the strong anion exchange resin (MN-400) and the weak anion exchanger (MN-100), showed the highest sorption. The weak base anion exchange resin with smaller pores (MN-150) had a lower capacity. At pH 7, the net charge of cephalosporin C becomes negative as a result of the ionization of the carboxyl groups. The electrostatic attraction between the amino groups of the resin and acidic groups of the solute is thus the driving force for adsorption. Consequently, the ability of the inert resin to take up the cephalosporin C declines. A similar decrease in the capacity of polystyrene adsorbents with a change from acid to neutral solutions has been reported for cephalosporins by Salto and Prieto[15] and for amino acids by Pietrzyk[5]. Since access to the functional groups depends on the pore size, it is understandable that resin MN-150 (having a smaller pore size) shows a lower capacity than MN-100.

In the column operation, resin MN-200 showed excellent applicability for the purification of cephalosporin C at acidic pH. The results of the loading are given in Figure 5. Operating at rate of loading equal to 4 BV/h, we were able to process 1 liter of a solution of 1 g/L of cephalosporin C over 12 h using 18 mL of resin. In comparison with equilibrium data, the capacity (240 mg/g) increased dramatically. Complete elution was performed with methanol, and a very high concentration of cephalosporin C (36,000 ppm) was obtained in a small volume (10 mL). In the next run, superior results were obtained: the concentration of cephalosporin C in one fraction reached a value of 52,000 ppm (Figure 6).

In addition to the work described above, some experiments were undertaken to evaluate the ability of the resins for protein extraction. Solutions of two proteins, bovine serum albumin (BSA) and γ-globulin, were prepared in Tris/HCl buffer. The extraction of a single protein was studied. The resins showed a reasonable capacity for BSA, about 40-60 mg/g from a 5,000 ppm solution of protein at pH 4.8 and 6.9. When the extraction of γ-globulin from a solution of the same concentration was performed, only the two resins MN-500 and MN-200 were able to extract this protein. Further work on proteins separation is now in progress.

Acknowledgement

We thank Dr. J. Dale of Purolite International Ltd. for his help in providing samples of the resins.

References

1. S. C. O'Connor, in "Methods in Enzymology", J. H. Hash, ed., Academic Press, New York, 1975, p. 299.
2. W. Voser, *J. Chem. Technol. Biotechnol.*, 1982, **32**, 109.
3. P. A. Belter, *A. I. Ch. E., Symp. Ser.*, 1984, **80**(233), 110.
4. P. Grammot, W. Rothchild, C. Sauer and J. Katsahian, in "Ion Exchange, Science and Technology", A. E. Rodrigues, ed., NATO ASI Ser, Martin Nijhoff Publishers, Dordrecht, 1986, p. 441.
5. Z. Iskandarani and D. J. Pietrzyk, *Anal. Chem.*, 1981, **53**, 489.
6. B. Rowatt and D. C. Sherrington, in "Ion Exchange Advances", M. J. Slater, ed., 1992, p. 198.
7. M. Pirotta., *Angew. Makromol. Chem.*, 1982, **197**, 109.
8. K. Ando, T. Ho, H. Teshima, and H. Kusano, in "Ion Exchange for Industry", M. Streat, ed, Ellis Horwood, Chichester, 1988, p. 232.
9. T. Hagaki, T. Morita, J. Watanabe and H. Teshime, *Proceeding of the Ion Exchange Conference*, Tokyo, 1991, p. 241.
10. J. L. Casillas, M. Martinez, F. Addo-Yobo and J. Aracil, *Chem. Eng. J.*, 1993, **52**, B71.
11. G. F. Payne, N. N. Payne, Y. Ninomiya and M. L. Shuler, *Sep. Sci. Technol.*, 1989, **24**, 457.
12. G. F. Payne and Y. Ninomiya, *Sep. Sci. Technol.*, 1990, **25**, 1117.

13. N. F. Kirkby, N. K. H. Slater, K. H. Weisenberger, F. Addo-Yobo and D. Doulia, *Chem. Eng. Sci.*, 1986, **41(8)**, 2005.
14. F. Addo-Yobo, N. H. Slater and C. N. Kenney, *Chem. Eng. J.*, 1988, **39**, B9.
15. I. Salto and I. G. Prieto, *J. Pharm. Sci.*, 1981, **70**, 994.
16. J. Dale (personal communication).
17. V. A. Davankov, S. V. Rogozhin and M. P. Tsyurupa, Patent USSR 299165, US 37294957, C.A. 75,6841B, 1971.
18. V. A. Davankov and M. P. Tsyurupa, *React. Polym.*, 1990, **13**, 27.
19. V. A. Davankov and M. P. Tsyurupa, *Angew. Makromol. Chem.*, 1980, **91**, 127.
20. M. P. Tsyurupa, L. A. Maslova, A. I. Andreeva, T. A. Mrachkovskaya and V. A. Davankov, *React. Polym.,* 1995, **25(1)**, 69.
21. N. Maity, G. F. Payne and J. L. Chipchoskyh, *Ind. Eng. Chem.*, 1991, **30**, 2456.
22. N. Maity, G. F. Payne, M. V. Ernest and R.L. Albright, *React. Polym.*, 1992, **17**, 273.
23. A. A. Garcia, *Biotechnol. Prog.,* 1991, **7(1)**, 33.

DUSTING OFF THE PHENOLIC RESINS

S. A. Bellamy
Rohm and Haas, Lennig House
Croydon, Surrey, UK

E. Zaganiaris
Rohm and Haas France S.A.
B.P. 48, 02301 Chauny Cedex, France

1 HISTORICAL DEVELOPMENT

The ion exchange properties of the phenolic resins (formaldehyde-phenol polycondensates) were first described in 1935[1,2] by Adams and Holmes. Further studies resulted in the synthesis of cation exchange and anion exchange resins by modifying the phenolic polycondensates to include sulfonic acid groups or amine groups, making them some of the first ion exchange materials to be produced. The early Phenolic resins were commercialised in the UK, Germany and the United States and were used extensively up until the development of the styrenic resins at the end of the 1940's. The styrenic cation resins did possess several advantages over the Phenolic ion exchangers such as a higher capacity and a greater resistance to oxidation and the Phenolic ion exchangers were quickly replaced in most common water softening and deionisation applications. The Phenolic ion exchangers were then relegated to a limited number of applications where their unique properties still gave them a technical edge over their more common styrenic and acrylic counterparts. During this period, many of the resin manufacturers discontinued this type of product line and the situation today is that Rohm and Haas, through the acquisition of Duolite International in 1984, is the only major resin manufacturer which still offers these products, albeit, in a limited product range.

2 RESIN CHARACTERISTICS

In order to understand why the Phenolic ion exchangers were so quickly replaced by the styrenic resins then an understanding of their chemical and physical characteristics is important as it is these differences which set the Phenolic ion exchangers apart in certain applications which will be reviewed later.

The chemical structure of the styrenic and Phenolic matrices are shown below in Figure 1. Whilst there are certain similarities to the styrenic matrix, the Phenolic matrix is relatively hydrophilic in nature and is inherently more porous than standard macroporous styrenic matrices. The presence of the -OH groups gives a slight cationic ion exchange capacity in addition to rendering the surface less hydrophobic. These groups are only ionised at a pH > 9.5 to give the phenolate ion.

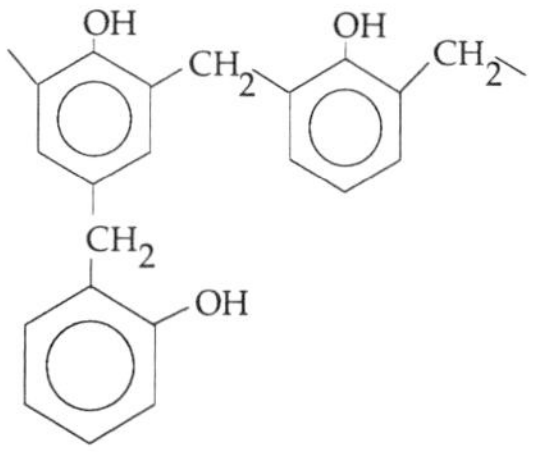

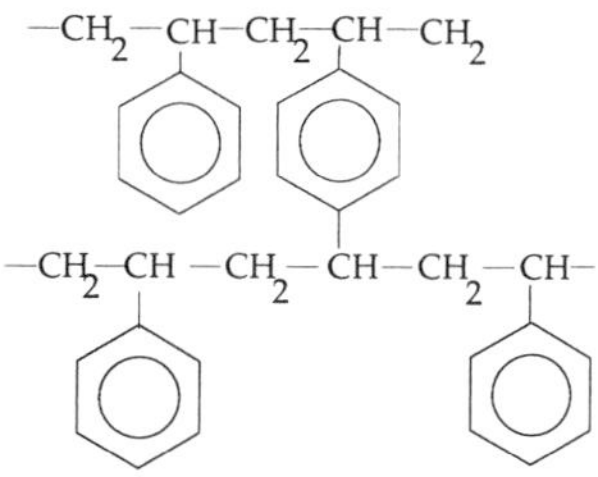

Figure 1 *Phenolic matrix*   *crosslinked polystyrene matrix*

Unlike most common ion exchange resins, the Phenolic ion exchangers are granular due to the fact that they are derived by grinding large pieces of polycondensates made by bulk condensation. Whilst less appealing to the eye than the more typical bead form resins, the granules have certain advantages such a lower pressure drop compared to a bead product of a similar particle size. A non-functionalized Phenolic resin also has a higher density (1.11 as compared to the more usual 1.04 for the non-functionalised aromatic adsorbents such as AMBERLITE XAD-16).

Over time, the Phenolic ion exchangers were replaced in many applications by the styrenic resins for a number of reasons. The Phenolic ion exchangers were generally more expensive than the corresponding styrenic resins and the range of Phenolic ion exchangers were limited to weak base anion exchange and cation exchange resins. Other advantages of the styrenic resins included a higher volumetric capacity for the cation exchangers, and a better resistance to oxidation and temperature.

The range of Phenolic ion exchangers and adsorbents which are commercially available at present are :-

- DUOLITE XAD-761   Adsorbent resin
- DUOLITE XAD-765 *   Partially aminated adsorbent resin
- DUOLITE A-568   Tertiary amine
- DUOLITE A-561   Tertiary amine with higher total exchange capacity
- DUOLITE A-7*   Secondary amine
- DUOLITE C-3*   Methylene sulphonic acid

(* - conforms compositionally with the FDA 21 CFR 173.25 monograph)

In order to understand why Phenolic resins are still used today in certain applications, then an understanding of the technical advantages over the styrenic resins is required. These can be summarised as follows :-

- Chemical specificity,
- Good Osmotic shock and attrition resistance,
- Generally lower swelling than styrenic resins,
- High adsorption capacity particularly for large molecules,
- Highly porous structure,
- Low pressure drop and good filtration efficiency,
- High density

The ideal application of Phenolic resins is in those where the advantages significantly outweigh the disadvantages as listed above. In general, this can be divided into two broad categories; purification of biological or natural process streams and the highly selective recovery of species in aqueous solutions. Applications of Phenolic resins include:

- Glucose syrups decolorisation[5]
- Citric acid purification
- Metals removal or recovery
- Kraft pulp effluent treatment[5]
- Monosodium Glutamate decolorisation
- Enzyme immobilisation[6,7]
- Plant extracts

Two of these applications are discussed below in some detail.

2.1 Citric Acid Purification

Citric acid is produced by fermentation of sugar, with molasses being the raw material. The citric acid is first precipitated as the Ca salt by adding lime and then it is recovered by adding H_2SO_4 whereby Ca is precipitated as $CaSO_4$. The citric acid thus obtained contains as impurities Ca^{2+}, K^+, Na^+, Mg^{2+}, and SO_4^{2-}. It is purified with a strong acid cation (SAC) and a weak base anion (WBA) exchanger.

Among the criteria for choosing the appropriate ion exchange resins are: low concentration of the ionic impurities in the product, low citric acid losses and high decolorisation of the citric acid. In order to choose the best performing WBA resin, first, a comparative test was performed to evaluate the SO_4^{2-} removal, using a synthetic citric acid solution containing only H_2SO_4 as impurity. DUOLITE A 561, a Phenolic type resin, was compared with AMBERLITE IRA 92, a styrene-DVB type resin. Following these experiments, real citric acid solution was used to determine the overall performance- deionisation and decolorisation- of a Phenolic versus a styrenic resin system.

2.1.1. Experimental. Table 1 describes the characteristics of the resins used in this work while Tables 2 & 3 describe the operating conditions.

Table 1 *Ion Exchange Resins*

	Duolite A-561	*Amberlite IRA-92*	*Amberlite 252 Na*	*Amberlite XAD 761*	*Amberlite XAD 1180*
Total exchange capacity (eq/L R)	1.8	1.7	1.85		
Moisture (%)	57	46	50	64	64.1
Effective size (mm)	0.45	0.45	0.5		
Mean diameter (mm)	0.65	0.65	0.7	0.65	0.55
Uniformity coeff.		1.5	1.4		
Specific surface area (m²/g)				100	530
Porosity (mL/g)				0.5	1.1
Porosity (mL/mL)				0.33	0.64
Average pore diameter (A)(*)				171	126

(*) Calculated from porosity, surface area and polymer density.

Influent Composition

Citric acid	239	g/L
H_2SO_4	4.9	g/L

Loading Step

Flow Rate	4	BV/h
Temperature	40	°C

Regeneration Step

Regenerant	2% NaOH	
Level	100	g/L R
Flow Rate	8	BV/h
Temperature	20	°C

Influent Composition

Citric acid	3950	meq/L
SO_4^{2-}	50	meq/L
Cl^-	3	meq/L
Na^+	53	meq/L
Ca^{2+}	40	meq/L
Mg^{2+}	6	meq/L
K^+	118	meq/L
Fe^{3+}	35	ppm

Loading Step

Flow Rate	4	BV/h
Temperature	40	°C

Regeneration Step

Acid Regenerant	5% HCl	
Level	100	g/L R
Flow Rate	4	BV/h
Mode	Counter-flow	
Temperature	20	°C
Regenerant	2% NaOH	
Level	100	g/L R
Flow Rate	8	BV/h
Temperature	20	°C

Table 2 *Operating Conditions, Synthetic Solution*

Table 3 *Operating Conditions, Real Solution*

2.1.2 Results and Discussion. The sulphate removal by the WBA resins was evaluated using a synthetic solution containing only citric acid and H_2SO_4 in the concentrations indicated in Table 2. Two WBA resins were evaluated; DUOLITE A 561, a Phenolic resin having tertiary amine functional groups and AMBERLITE IRA 92, a styrenic resin also with tertiary amine functional groups. The chemical structure of these two resins is shown in Figure 2, whereby it is seen that in the Phenolic resin, the functional groups are highly branched polyamine groups rather than individual tertiary amine groups in the case of the styrenic resin.

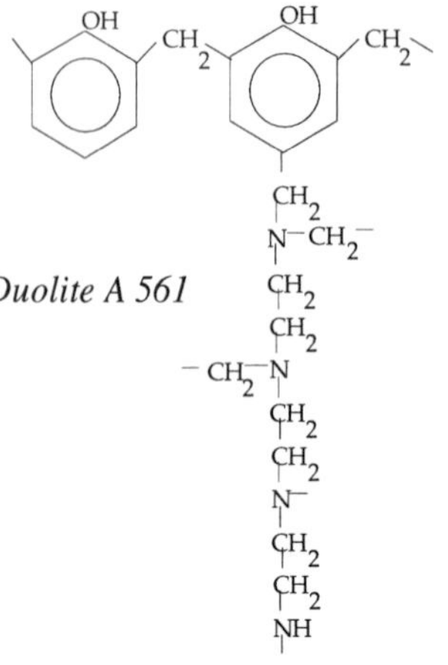

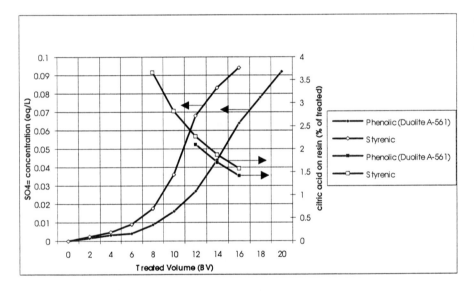

Duolite A 561 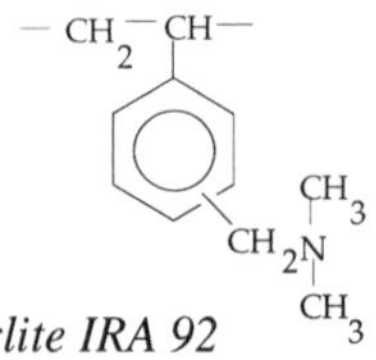   *Amberlite IRA 92*

Figure 2

Figure 3 *Citric acid demineralization*

Figure 3 illustrates the loading cycle for both resins. The key difference between these resins is that the Phenolic resin DUOLITE A 561 loads proportionally more sulphate ions that citrate ions compared to the styrenic resin AMBERLITE IRA 92.

The practical advantage of this higher selectivity of sulphates over citrate ions of Duolite A 561 is that it results into lower citric acid losses. In fact, from Figure 3, if we take as the end of the cycle 10 BV for AMBERLITE IRA 92 and 12 BV for DUOLITE A 561, we have 2.8% citric acid losses for the former and 2% citric acid losses for the latter resin. This higher selectivity of sulphates over citrate ions of DUOLITE A 561 is due probably to the fact that the polyamine functional groups have a more favourable spacing of the amine groups[3]. This resin shows in fact a SO_4^{2-}/NO_3^- selectivity of the order of 100 compared to 3 of a styrenic resin of the AMBERLITE IRA 92 type[3].

Following the above results, it was decided to test these resins using a real citric acid solution (from a citric acid plant) and to determine the decolorisation performance as well. In order to enhance decolorisation, the system SAC/Adsorbent/WBA was employed. As adsorbent, AMBERLITE XAD 761, of Phenolic type, and AMBERLITE XAD 1180 of Styrene/DVB type were employed. The description of these resins is given in Table 1 while the operating conditions are given in Table 3.

Figure 4 gives the decolorisation after the Adsorbent resins. These results suggest the use of the Phenolic adsorbent in the final resin system. In fact, an overall decolorisation of 75% was achieved by using the system AMBERLITE 252 Na/AMBERLITE XAD 761/DUOLITE A 561 (Figure 5).

Figure 4 *Citric acid decolorization*

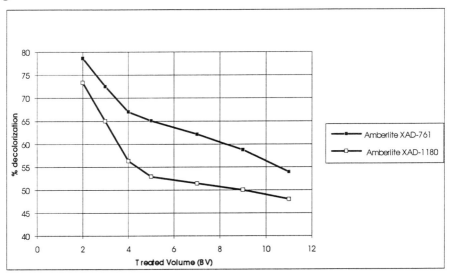

Figure 5 *Citric acid decolorization*

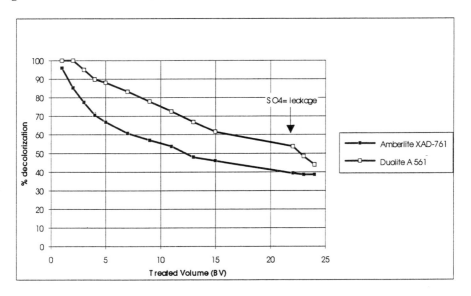

2.2 Metals Removal

Certain transition metals, such as Cu^{2+}, Co^{2+}, Ni^{2+}, Zn^{2+}, Hg^{2+}, Pb^{2+} form stable complexes with ammonia and amines like ethylenediamine, diethylenetriamine or triethylenetetramine. As seen from the chemical structure of DUOLITE A 561 given above, or of DUOLITE A 7 given below (Figure 6), these resins can, in principle, form stable complexes with transition metals and therefore they can be used to selectively remove these metals from solutions without altering the rest of the ionic background.

In the following experiments, it was studied the removal of Cu^{2+} and Ni^{2+} from solutions at pH = 4.

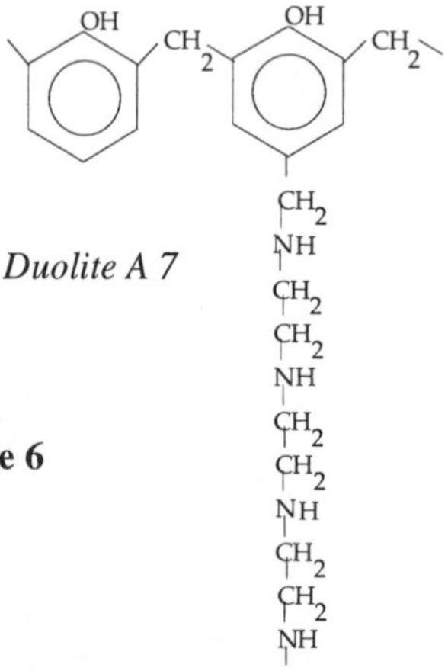

Duolite A 7

Figure 6

2.2.1 Experimental. Influent composition: Cu^{2+}: 150 ppm, Ni^{2+}: 60 ppm, pH = 4. The solution was allowed to pass through the resin at 4 BV/h. Regeneration of the resin was done with 2 BV of 10% HCl followed by 5 BV of water rinse followed by 1.5% NH_4OH regeneration to put the resin into the free base form (unless otherwise indicated in the text) and then finally with 5 BV water rinse.

2.2.2 Results and Discussion. The leakage curves obtained are illustrated in Figure 7. As seen, Cu^{2+} removal was very efficient while Ni^{2+} leaked through well before. This correlates with the stability constants of these metals with ethylenediamine[4] which are of the order $Cu^{2+} > Ni^{2+}$.

When the resin is found in the protonated form, the operating capacity decreases, as shown in the same figure.

Figure 7 *Cu²⁺ and Ni²⁺ removal from water with Duolite A7*

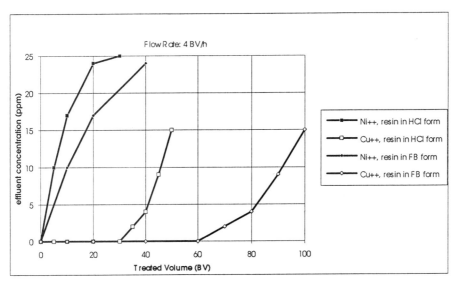

References

1.   B. A. Adams, and E. L. Holmes, *J. Soc. Chem. Ind. (London)*, 1935, **54**, 1T.
2.   B. A. Adams, and E. L. Holmes, UK Patent 450,308, 1936.
3.   D. Clifford, *Reactive Polymers,* 1983, **1**, 77.
4.   L. E. Orgel, "An Introduction to Transition-Metal Chemistry Ligand-Field Theory",
     Methuen/John Wiley & Sons, p. 85.
5.   "Macroporous Condensate Resins as Adsorbents" *Ind. Eng. Chem., Prod. Res. Dev.*,
     1975, **14**, 108.
6.   A. P. Ison, *Biocatalysis*, 1990, **3**, 329.
7.   R. H. M. Stouffs, CPC International Inc., EPA 0 014 866, 1980.

Author Index

Subject Index